A Bird-Finding Guide to Panama

A Bird-Finding Guide to
PANAMA

George R. Angehr,
Dodge Engleman, and
Lorna Engleman

Published in association with the
Panama Audubon Society,
a BirdLife International Partner

Comstock Publishing Associates
a division of
Cornell University Press
Ithaca, New York

This work was previously published in Panama as *Where to Find Birds in Panama: A Site Guide for Birders* copyright © 2006 by Panama Audubon Society/Sociedad Audubon de Panamá, Panamá.

United States edition first published 2008 by Cornell University Press
First printing, Cornell Paperbacks, 2008

Printed in the United States of America

Library of Congress Cataloging-in-Publication Data

Angehr, George Richard.
[Where to find birds in Panama]
A bird-finding guide to Panama / George R. Angehr, Dodge Engleman, and Lorna Engleman. – U.S. ed.
p. cm.
"Previously published in Panama as Where to find birds in Panama : a site guide for birders, copyright 2006 by Panama Audubon Society/Sociedad Audubon de Panamá"–T.p. verso.
Includes bibliographical references and indexes.
ISBN 978-0-8014-4650-4 (cloth : alk. paper) – ISBN 978-0-8014-7423-1 (pbk. : alk. paper)
1. Birding sites–Panama–Guidebooks. 2. Bird watching–Panama–Guidebooks. 3. Panama–Guidebooks. I. Engleman, Dodge. II. Engleman, Lorna. III. Sociedad Audubon de Panamá. IV. Title.

QL687.P3A54 2008
598.072'347287–dc22

2007044420

Cornell University Press strives to use environmentally responsible suppliers and materials to the fullest extent possible in the publishing of its books. For further information, visit our website at www.cornellpress.cornell.edu.

Cloth printing 10 9 8 7 6 5 4 3 2 1

Paperback printing 10 9 8 7 6 5 4 3 2 1

Design and production: Editora Novo Art, S.A., Panamá; style edition: Montserrat de Adames; Graphic designer: Pedro Antonio Argudo F.; www.editoranovoart.com

Acknowledgments

We would especially like to thank the US Agency for International Development (USAID) in Panama, and its director, Kermit Moh, for sponsoring this guide and related products through the Panama Canal Integrated Watershed Management Project. The late Leopoldo Garza of US-AID deserves particular thanks for his support from the earliest phases of the project until his untimely death in 2005. The project was managed by the Academy for Educational Development (AED), and we would also like to thank Brian Rudert, Chief of Party, and Roberto Ibañez and Gina Castro of AED for their help. Rosabel Miró, Karl Kaufmann, Ofelia Rodríguez, George Angehr, Glenda Bonamico, and Rosa Montañez all contributed substantially to developing the ideas behind the proposal. Glenda Bonamico, Yenifer Díaz, Rosabel Miró, and Carmen Contreras managed or assisted in administrating the project for the Panama Audubon Society. We would particularly like to thank William Adsett, Karl Kaufmann, and Rosabel Miró for carefully reviewing the text and making many valuable suggestions. Darién Montañez designed the cover, and prepared the maps in collaboration with George Angehr. Oscar Vallarino from the Comisión Interinstitucional de la Cuenca Hidrográfica del Canal assisted with the maps of the Panama Canal area, and Enrique Samudio from COPEG-MIDA with the maps for Changuinola, Chiriquí Grande, and El Real. This guide was developed through the experience of many members of the Panama Audubon Society and other birders in Panama over the course of many years, and we would like to thank them for sharing their knowledge with us, especially William Adsett, Maria Allen, Guido Berguido, Bob Brown, Alberto Castillo, Dan Christian, Paul Coopmans, Francisco Delgado, Daniel George, John Guarnaccia, Michael Harvey, Karl Kaufmann, Howard Laidlaw, George Ledec, Mark Letzer, Horace Loftin, Rosabel Miró, Delicia, Pedro, Darién, and Camilo Montañez, Jacobo Ortega, Clemente Quirós, Robert Ridgely, Ghislain Rompré, Gilles Seutin, José Tejada, Gary Vaucher, and Venicio Wilson. In addition to George Angehr, field research for some sites was carried out by the following people: Altos del María (William Adsett), Boquete (Dan and Kay Wade and Cora Herrera), Cerro Azul (William Adsett, Karl Kaufmann, and Rosabel Miró), Cerro Campana, Fortuna, the Oleoducto Road, and Chiriquí Grande (Karl Kaufmann and Rosabel Miró), Changuinola (Karl Kaufmann, Rosabel Miró, and Darién Montañez), Cerro Santiago (William Adsett, Loyda Sánchez, and Dan Wade), Cerro Hoya National Park (William Adsett and Loyda Sánchez), El Real-Pirre Station (Euclides Campos, with support from Guido Berguido), Isla Coiba and Rancho Carolina (José Carlos García), Puerto Armuelles (William Adsett and Dan Wade), and Volcán, Cerro Punta, and Santa Clara (Loyda Sánchez). Students Margelys

Barría, Justo Camargo, Ovidio Jaramillo, Dinora López, Angel Sosa, and Maribel Tejada assisted on some of these field trips. Other individuals who assisted us with information or in other ways include Nariño Aizpurúa, Carlos Alfaro, Javier Araúz, Raúl Arias, Isidro Barría, Carla Black, Michael Braun, Robb Brumfield, Marcial Caisamo, John Collins, Jim Cone, Berta de Castrellón, Charlotte Elton, Oscar Gaslin, Gabriel Jácome, Andrea Kayser, Andrew Kratter, James Kushlan, Harmodio Menbache, Adalberto Montezuma, Antoní Morale, Ricaurte Moreno, Luís Naar, Santos Navas, Norma Ponce, Dídimo and Armando Quiróz, Tom and Ina Reichelt, Angel Rodríguez, Danilo Rodríguez, Iñaki Ruíz, Genober Santamaría, Beatriz Schmitt, and Norita Scott. Since 2003, the Panama Records Committee, consisting of George Angehr, Dodge and Lorna Engleman, Darién Montañez, and Robert Ridgely, has reviewed reports of new species for Panama. We would like to thank them, as well as everyone who has provided reports, trip lists, and other information to the Panama Audubon Society, for their contributions to this guide. George Angehr is particularly indebted to the Smithsonian Tropical Research Institute for providing office space and logistical support, and to the staff of STRI's Tropical Sciences Library for assistance in obtaining many obscure references on bird distribution in Panama.

About the Panama Audubon Society

The Panama Audubon Society (PAS), known in Spanish as the Sociedad Audubon de Panamá, was founded in 1967 as a branch of the Florida Audubon Society. Now independent, PAS has since developed into the leading organization in Panama devoted to the conservation of birds and their habitats, and is the country partner of BirdLife International. In 1995, under the auspices of BirdLife, PAS initiated one of the first Important Bird Areas (IBA) programs in the Americas, which led to the identification of 49 globally and 39 nationally important sites in Panama. In the two highest priority IBAs, PAS has succeeded in having the Upper Bay of Panama, a critical area for more than 1 million migratory shorebirds, designated as a Wetland of International Importance under the Ramsar Convention, and as a hemispheric site within the Western Hemisphere Shorebird Reserve Network; and in establishing private reserves now totaling 283 hectares (700 acres) at El Chorogo, a critical site for endemic species of the western Pacific lowlands. PAS conducts four Christmas Bird Counts each year, and has collaborated with the Hawk Mountain Sanctuary in monitoring the enormous migrations of soaring raptors that pass through Panama each year. The Society also conducts programs on environmental education, bird observation, and conservation throughout Panama. Our office is located in Llanos de Curundu (Curundu Flats), just west of Metropolitan Nature Park. To find it, follow the directions for getting to Metropolitan Nature Park from Av. Omar Torrijos on p. 67 Instead of turning off on Av. Juan Pablo II for the park, take a left at the traffic signal just past this turn off. Take a left at the first intersection, where there are tennis courts on the right. Turn left at the third intersection, where our office is the first building on the right, number 2006-B. PAS holds meetings on the second Thursday of each month at 6:30 PM. at Metropolitan Nature Park headquarters on Av. Juan Pablo II. Call for details on speakers and other planned events, including birding trips.

In addition to this guide, the Panama Audubon's recent publications include:

- *Annotated Checklist of the Birds of Panama* (2006), by George R. Angehr
- *Panama Canal Birding Trail Map* (2006), by Darién Montañez
- *Guía de Las Aves de Panamá* (second edition in Spanish, 2005), by Robert S. Ridgely and John A. Gwynne (published in collaboration with ANCON)
- *Common Birds of Panama City/¿Qué vuela ahí?* (2004), by Jorge Ventocilla, with illustrations by Dana Gardner (published in collaboration with the Smithsonian Tropical Research Institute)
- *Directorio de Áreas Importantes para Aves en Panamá/Directory of Important Bird Areas in Panama* (2003), by George R. Angehr

For more information, or to obtain copies of our publications, please contact the Panama Audubon Society at the addresses below:

mailing address: Sociedad Audubon de Panamá
Apartado 0843-03076
Panamá, República de Panamá

street address: Casa 2006-B, calle 1ª este
Llanos de Curundu
Corregimiento de Ancón
Ciudad de Panamá

phone: (507) 232-5977
fax: (507) 232-5977
e-mail: audupan@cwpanama.net
info@panamaaudubon.org
website: www.panamaaudubon.org

To report bird records: Records of rare birds and unusual sightings, as well as trip lists, would be greatly appreciated. To submit them, please contact George Angehr, Chair, Panama Records Committee, at angehrg@si.edu, or send them to the Panama Audubon Society at the addresses above.

Updates: The Panama Audubon Society provides updates to information in this guide on its website (www.panamaaudubon.org). If you are planning a trip, check this site to obtain the most recent information.

Table of Contents

List of Maps

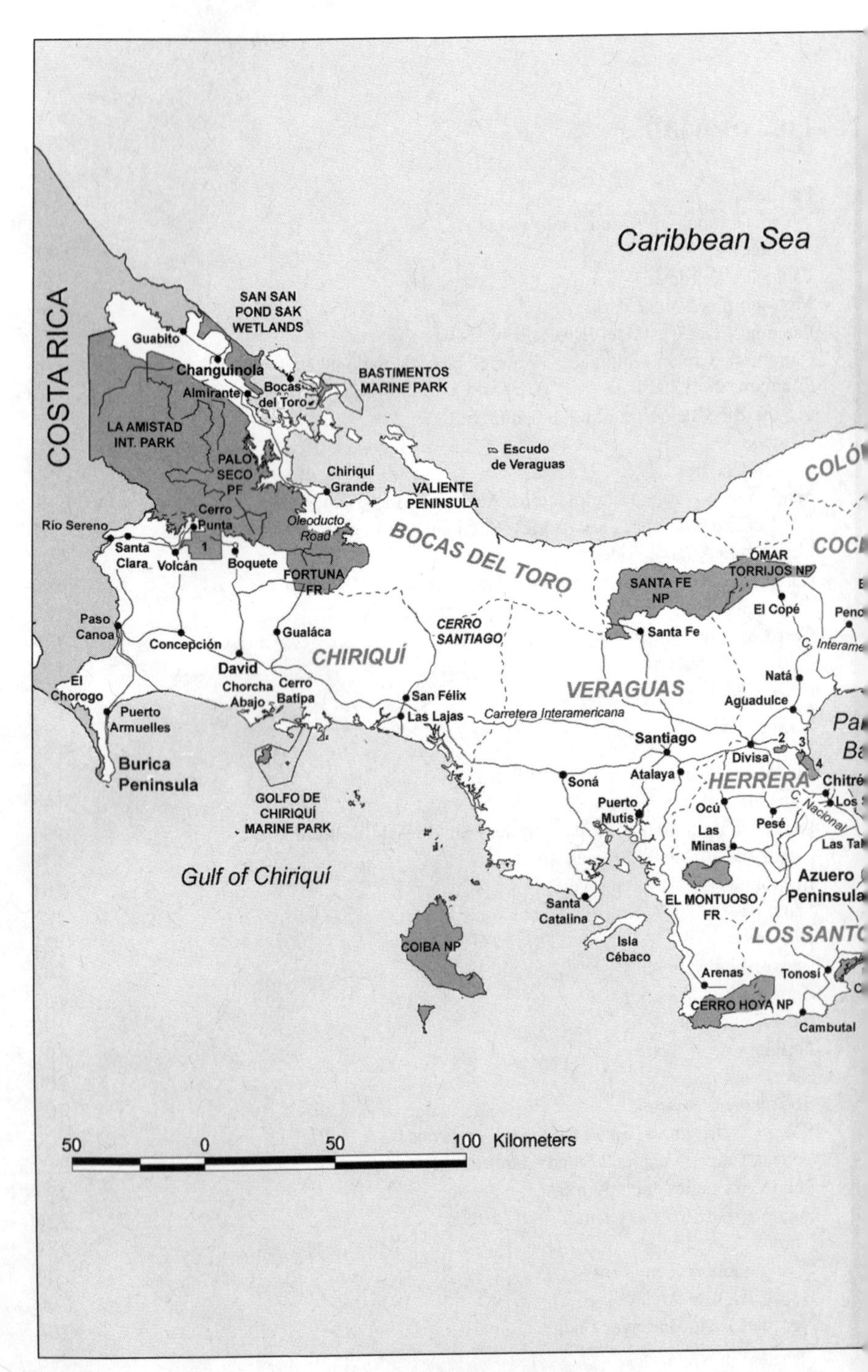

Panama. Abbreviations: FR = Forest Reserve, Int. Park = International Park, NP = National Park, PF = Protection Forest, WR = Wildlife Refuge.

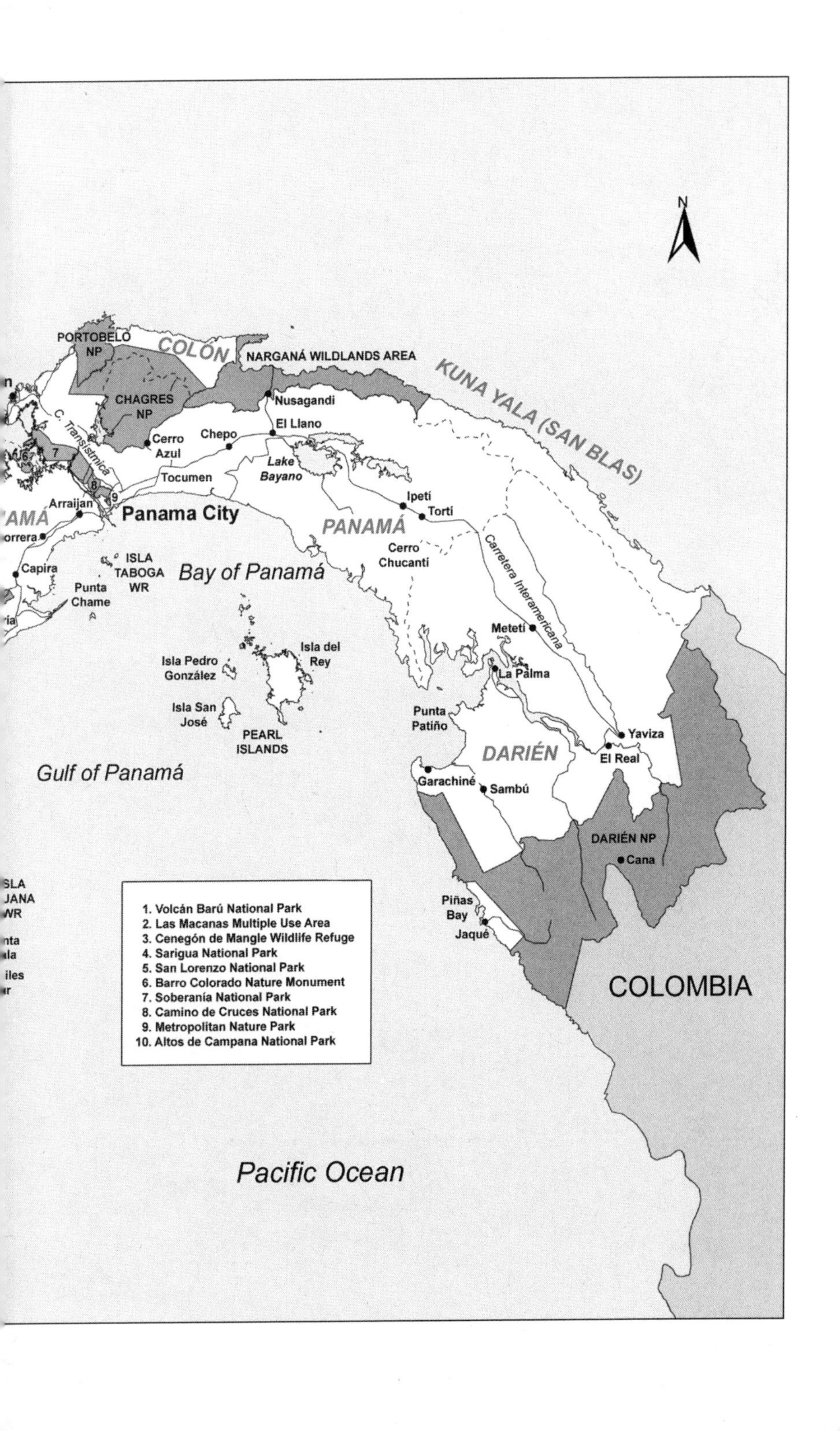
PORTOBELO NP
COLÓN
NARGANÁ WILDLANDS AREA
KUNA YALA (SAN BLAS)
CHAGRES NP
Nusagandi
El Llano
C. Transístmica
Cerro Azul
Chepo
Lake Bayano
Tocumen
Ipetí
Tortí
Arraiján
Panama City
PANAMÁ
Cerro Chucantí
Carretera Interamericana
Capira
ISLA TABOGA WR
Bay of Panamá
Punta Chame
Metetí
Isla del Rey
Isla Pedro González
La Palma
Isla San José
PEARL ISLANDS
Punta Patiño
Yaviza
DARIÉN
El Real
Gulf of Panamá
Garachiné
Sambú
DARIÉN NP
Cana
Piñas Bay
Jaqué
COLOMBIA
1. Volcán Barú National Park
2. Las Macanas Multiple Use Area
3. Cenegón de Mangle Wildlife Refuge
4. Sarigua National Park
5. San Lorenzo National Park
6. Barro Colorado Nature Monument
7. Soberanía National Park
8. Camino de Cruces National Park
9. Metropolitan Nature Park
10. Altos de Campana National Park
Pacific Ocean

Panama for the birder

Panama's major attraction for the birder is its astonishing diversity of birdlife for a country of its size. As of 2006, a total of 972 species had been recorded within its boundaries, in an area a little smaller than the state of South Carolina (74,949 sq km/28,950 sq mi). This is more than the entire continental United States and Canada combined, and more than any other country in North America except Mexico, which is 26 times Panama's size. These include 12 national endemics and 107 regional endemics, as well as many mainly South American species that range no further north. Most of the best birding areas are reachable from Panama City within a one-day drive on good roads, or are within a one-hour's flight, making it possible to attain a very large bird list within a short period of time.

As part of the Neotropical biogeographic realm, Panama has many kinds of birds that may be unfamiliar to visitors from temperate areas. These include tinamous, tropicbirds, boobies, frigatebirds, guans, curassows, chachalacas, sungrebes, sunbitterns, jacanas, parrots, potoos, trogons, motmots, puffbirds, jacamars, barbets, toucans, furnariids, woodcreepers, antbirds, tapaculos, cotingas, manakins, sharpbills, caciques, and oropendolas. Other families are far more diverse in the Neotropics than they are farther north, including cuckoos, hummingbirds, tyrant flycatchers, wrens, and tanagers. Families that occur in the temperate zone often have their own distinctive representatives in Panama, and its diversity is further augmented by the presence of many northern migrants during the northern winter. Seabirds of both the Atlantic and the Pacific reach its shores, including some that breed in South Temperate and Antarctic waters.

Panama is easily reached by international flights originating in North America and Europe. In many areas the infrastructure for tourism is good, with comfortable accommodations and excellent restaurants close to good birding areas. Most of the country is safe for travel (with the exception of a few urban areas and remote areas near the Colombian frontier), and Panamanians are friendly to visitors, especially in rural areas. Panama retains more of its forest cover than most other countries in Central America, and much of it is protected by a network of national parks and other reserves. For those wishing a wilderness experience, the Darién features some of the wildest country remaining in the Americas. Despite all this, Panama is still relatively unknown to international tourists, and visitors can avoid the hordes now common in better-known destinations such as Costa Rica.

This new guide is intended to provide the most up-to-date information to the visiting birder, enabling you to find the best birding spots easily and quickly. Although it is mainly designed for the independent traveler, it should be very useful to those on organized tours as well, and also to the growing number of birders resident in the country. This guide provides information not only on the birds present at each site, but also detailed directions for how to get there, as well as information on nearby accommodations and places to eat.

Plan of the Book

This guide is divided into four parts: the introduction; a guide to birding sites; a section on where to find individual species; and a species list for Panama.

Introduction

The introduction includes four main sections: a discussion of Panama's environment and the distribution of its birdlife; conservation issues; information for the birder, with specific information about the birding experience in Panama, equipment, and other basics; and more general information on tourism, including practicalities of travel, transportation, accommodations and meals, and health issues.

Birding sites

This part of the guide is divided into three sections: Panama City and the Canal Area[1]; Western Panama; and Eastern Panama. The first section includes sites within about a 1 1/2 hour drive from Panama City, including some outside the Canal corridor itself, including Campo Chagres at Lake Alajuela, Tocumen Marsh and Cerro Azul-Cerro Jefe to the east of Panama City, and Cerro Campana to its west. The second section includes all of western Panama west of Cerro Campana, and the third all of eastern Panama east of Tocumen and Cerro Azul-Cerro Jefe. Within each section, sites are generally discussed in order of their distance from Panama City: in the first section they are covered approximately from south to north (but with Tocumen, Cerro Azul-Cerro Jefe, and Cerro Campana discussed separately at the end); in the second from east to west; and in the third from west to east.

Each site description contains an initial section which includes a general description of the site; birds of interest found there; and a description of trails and other localities within the site and how to find them. This is followed by a section on how to get to the main site, which includes directions from a major nearby city or highway. The next section includes information on accommodations and places to eat near the site, with an emphasis on hotels and restaurants most likely to appeal to or be suitable for

[1] In this guide, we use the term "Canal Area" to refer to the area of the former US-administered Canal Zone, which extended 8 km (5 mi) on either side of the Panama Canal, and to areas immediately adjacent to it. This is also often referred to as the *Región Interoceánica,* or Interoceanic Region.

birders. Finally, a complete bird list is provided for each major site, based on species that have actually been recorded there. (Lists are sometimes not provided for smaller sites or ones that are of interest for only a few species.)

The following codes are used in the species lists. * = a species that is believed to be a vagrant at the site and which is not to be expected. R = rare; a species that regularly occurs but is very unlikely to be recorded on a short visit. X = formerly present, but thought to no longer occur. ? = present status uncertain. For a few sites additional codes are used for species that may occur in only certain areas of the site.

Knowing how frustrating it can be to lose an hour's birding time after taking a wrong turn and getting lost, we have given directions to sites in some detail, including road or trail distances and major landmarks. We have often given the names of specific stores or other buildings at intersections. Of course, some of these will change over time, but in most cases the directions should be sufficiently complete so that they can still be followed if some of the details have changed. Road distances are given in kilometers (or meters) only, since that is how they will be indicated on road signs and the odometer of most vehicles available in Panama. We do, however, provide English equivalents for elevations, areas, and temperatures. Odometer readings may differ significantly between vehicles, so road distances should be considered to be approximate.

Species accounts

The third section contains information on where to find selected species of interest, including national and regional endemics and rarities. It also includes information on species newly recorded in Panama since the publication of the most recent edition of the *Guide to the Birds of Panama,* by Robert Ridgely and John Gwynne, in 1989.

Species list

For reference, we also include a list of the 972 species recorded to date in Panama. More detailed information is provided in the *Annotated Checklist of the Birds of Panama* (2006), by George Angehr, and published by the Panama Audubon Society. The Panama Audubon Society also publishes field cards suitable for keeping your daily lists.

Panama's Environment and Birdlife

Geography

The country of Panama is long, thin, and sinuous. Over 600 km long, it is only about 50 km wide at its narrowest point. The isthmus has the shape of a shallow horizontal S, with the large blocky Azuero Peninsula jutting from its southern side. Many visitors find the geography of Panama disorienting. Contrary to the assumptions of many, the isthmus runs east and west instead of north and south. And while the

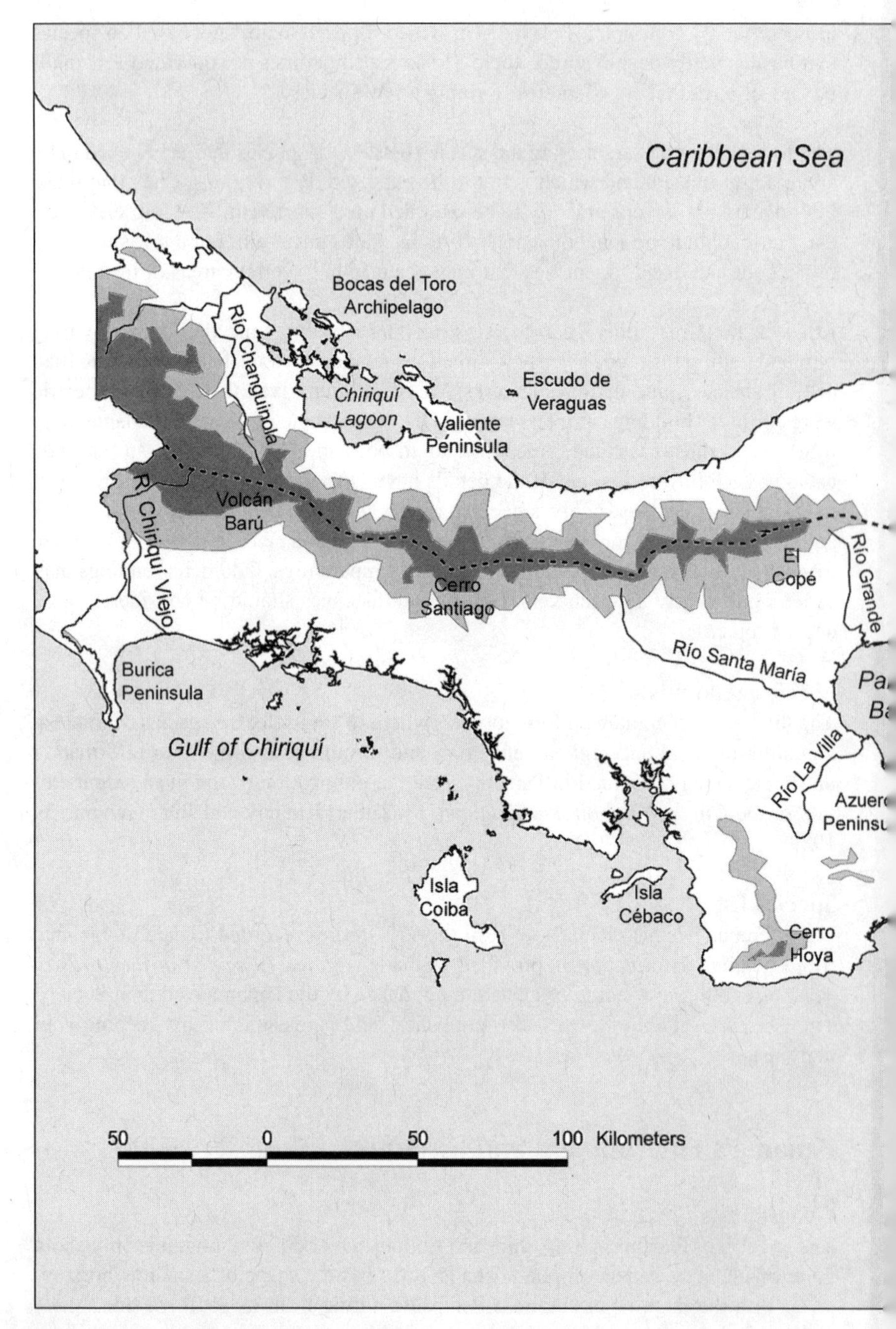

Panama's topography and major rivers. Light shading indicates land over 600 m (2,000 ft), and dark shading land over 1,200 m (4,000 ft).

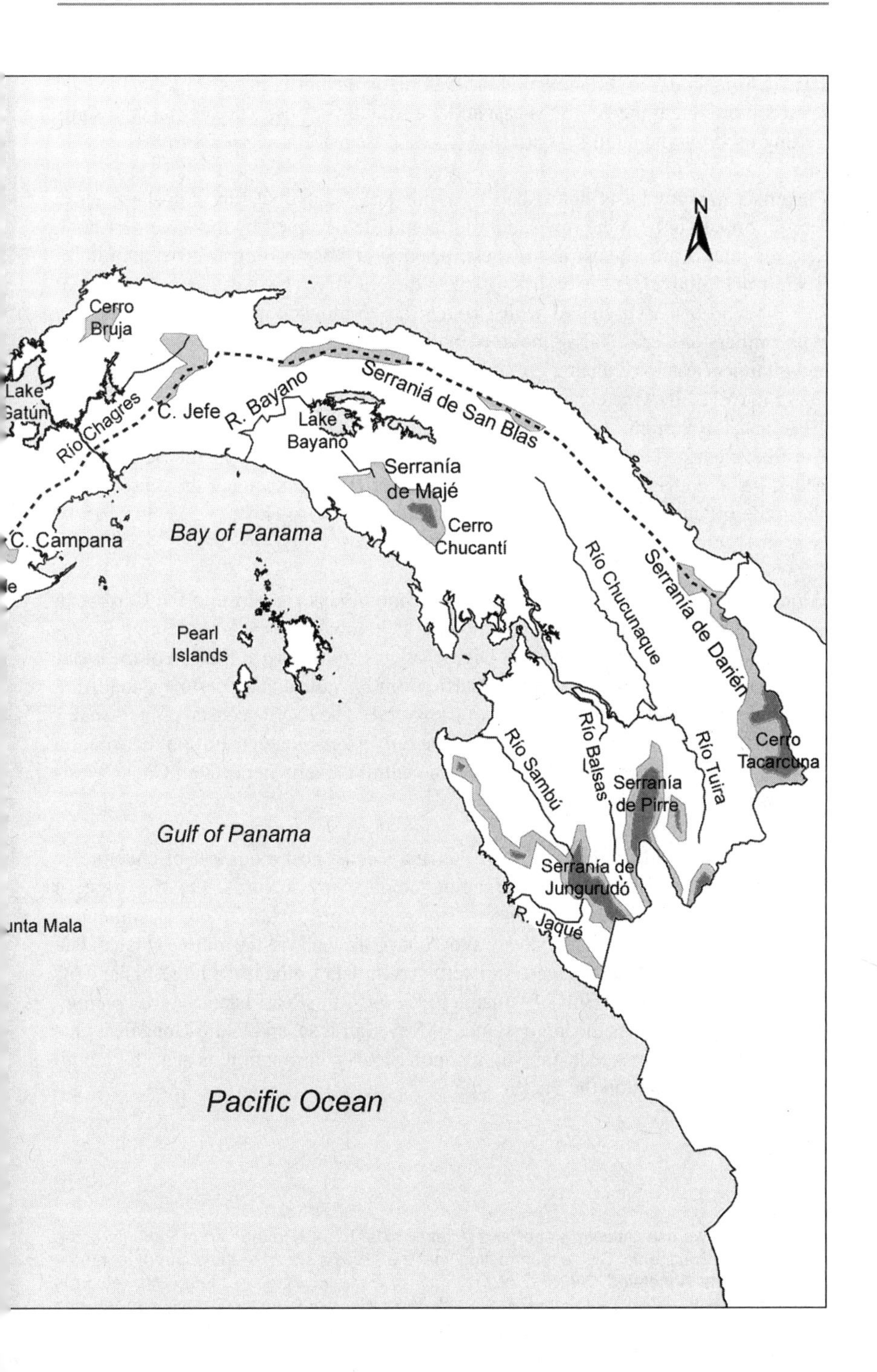
N
Cerro Bruja
Lake Gatún
Río Chagres
C. Jefe
R. Bayano
Lake Bayano
Serraniá de San Blas
Serranía de Majé
Cerro Chucantí
C. Campana
Bay of Panama
Pearl Islands
Río Chucunaque
Serranía de Darién
Cerro Tacarcuna
Río Sambú
Río Balsas
Serranía de Pirre
Río Tuira
Gulf of Panama
Serranía de Jungurudó
R. Jaqué
Pacific Ocean

Pacific is generally to the south and the Atlantic to the north, in some places the land is so contorted that the Pacific is east and the Atlantic is west! All this can take some getting used to when you first arrive.

Panama's topography is dominated by mountain chains that run almost its entire length, broken only at the low saddle in its middle occupied by the Panama Canal. The mountains are highest in the west, peaking at Volcán Barú, whose summit, at 3,475 m (11,400 ft), is the highest point in the country. This western range, part of the Talamanca massif shared with Costa Rica, terminates just east of Penonomé, with outliers at Cerro Gaital above El Valle de Antón and at Cerro Campana. The mountains of eastern Panama are lower, the highest peak being Cerro Tacarcuna at 1,875 m (6,152 ft). One chain, including Cerro Jefe, Cerro Bruja, the Serranía de San Blas, and the Serranía de Darién, which includes Cerro Tacarcuna, runs along the Caribbean coast. There are three other ranges on the Pacific slope, the Serranías de Majé, Pirre, and Jungurudó. In the west, another range runs along the west side of the Azuero Peninsula, reaching its highest point on Cerro Hoya (1,559 m/5,115 ft) near its southern end.

Along most of Panama's length, the highest mountains run close to the Caribbean coast, so that the Pacific slope is much wider than the Atlantic[2]. An exception is central Panama, where the Continental Divide swings southward at the end of the western mountains, running close to the Pacific from El Valle and across the Canal Area before heading back north to the east of Cerro Jefe. The Pacific coastal plain is mainly between 40-100 km (25-60 mi) wide. The only extensive lowlands on the Atlantic slope are in western Bocas del Toro, and in central Panama in northern Coclé, western Colón, and the Canal Area.

Various islands and island groups off Panama's coast host a number of endemic or relict species, and also harbor breeding colonies of seabirds. On the western Caribbean coast is the Archipelago of Bocas del Toro (Islas Colón, Bastimentos, Cristóbal, Popa, and Solarte, and Cayo Agua), as well as the more isolated Isla Escudo de Veraguas. On the western Pacific coast, Isla Coiba is the largest island off Central America. In the Gulf of Panama to the east, the Pearl Islands (*Archipiélago de Las Perlas*) includes the large islands of Rey, San José, and Pedro González, plus dozens of smaller ones. Isla Taboga, a popular tourist destination, is near the mouth of the Panama Canal on the Pacific side.

[2] In this guide we use "Atlantic slope" and "Atlantic side" to refer to the part of Panama to the north of the Continental Divide, contrasting with the "Pacific slope" to the south of it, in conformity to the *Annotated Checklist of the Birds of Panama* (Angehr 2006). This region is referred to as the "Caribbean slope" in *A Guide to the Birds of Panama* (Ridgely and Gwynne 1989). We use "Caribbean Sea" and "Caribbean coast," however, to refer to the body of water to the north of Panama.

Climate and seasons

Lying between 7° and 10° north of the Equator, Panama is entirely in the Tropical Zone. The lowlands are warm all year round, while the highlands are cooler. Average daytime temperatures in the lowlands range between about 82-90°F (28-32°C), although it can get hotter in the driest parts of the Pacific slope. As in most of the tropics, there is more variation between daytime and nighttime temperatures than there is seasonally. Nighttime temperatures, ranging from 71-76°F (22-24°C), are usually quite comfortable. In the highlands, temperatures vary with altitude. There is already a noticeable cooling in the foothills above 600 m (2,000 ft). In the highlands of western Panama daytime temperatures are usually in the low 70s°F (20s°C), or even cooler during rains. At night they can fall below 50ºF (10°C), and frost sometimes occurs on the highest slopes.

In the tropics seasons are defined on the basis of rainfall, rather than temperature. In most of Panama the dry season (known in Spanish as the *verano,* or "summer," because it is sunnier and somewhat warmer) runs from mid-December until late April. Strong trade winds blow from the northeast, and rain may fall briefly once or twice a week. Sometimes, however, an entire month may go by with no appreciable rainfall. During the wet season, known in Spanish as the *invierno* ("winter"), rain may fall almost daily. In either season, most rain falls as relatively brief showers, usually less than an hour, although these can be extremely intense. Mornings are usually clear, rainstorms being most frequent in the mid to late afternoon. All-day rains are infrequent, and are most common in the rainiest months of October and November. There is often a one-to-four-week lull in the rains in July and August, known as the *veranillo,* or "little summer." Panama lies to the south of the main track of Caribbean hurricanes, and has not been hit by one in more than 100 years, although it may sometimes be affected by peripheral rains.

The general rainfall pattern described above varies markedly according to locality. On the Pacific slope the dry season is generally longer and more intense than it is on the Atlantic. The *Arco Seco* ("Dry Arc"), in the coastal parts Coclé, Herrera, and Los Santos, is the driest part of Panama. Most of Darién and the western side of the Azuero Peninsula are somewhat wetter than the rest of the Pacific slope. Rainfall is less seasonal on the Atlantic slope, and some areas in the foothills may receive more than 4,000 mm (13 ft) a year. In Bocas del Toro there are two distinct wet seasons, from May through August and in December. February and September are drier with about half as much rain as the wettest months.

The highlands also have a somewhat different rainfall pattern from the lowlands. In the dry season the trade winds may bring all-day mist and blowing drizzle to areas near the Continental Divide such as El Copé, Santa Fe, and Fortuna. In the western highlands around Boquete and Volcán afternoons often bring a light drizzle known as *bajareque*. In contrast to the lowlands, the odds of having a bright sunny day all day long are greatest during the wet season, though all-day rains are possible too.

During the dry season the trade winds produce upwelling in the Gulf of Panama by pushing warm surface waters away from the coast around the Gulf of Panama and allowing nutrient-rich cold water to rise. This creates an extremely productive marine ecosystem in the Gulf, so that most seabird and wader nesting colonies, as well as the major staging and wintering areas for migratory shorebirds, are found here. The trade winds blow towards shore on the Caribbean, and are blocked by high mountains on the western Pacific coast. Upwelling does not occur in these areas, so that seabirds, waterbirds, and shorebirds are much less numerous there than they are around the Gulf of Panama.

Major habitats

Panama's rugged topography and diverse climates have produced a variety of different habitats. In general, because the Pacific slope is drier, it is more open than the Atlantic side, in part because it has undergone more deforestation.

Lowland deciduous and semideciduous forests. In these forest types many of the trees lose their leaves during the dry season to conserve water. The trees are generally lower in stature than in wetter forest types. These forests once covered much of the Pacific lowlands, but have now been mostly cleared for agriculture and cattle. Deciduous forest has been reduced to a few small patches, while the most extensive areas of surviving semideciduous forest are in the Canal Area and in Darién.

Lowland evergreen forests. These are the forests most commonly referred to as "rainforests." Here most trees retain their leaves through the dry season. In older forests of this type, the canopy is nearly continuous. The canopy is usually high, reaching about 30 m (100 ft), with some emergent trees reaching 45 m (150 ft). Little light reaches the forest floor, so that undergrowth is often sparse. There is often a heavy growth of lianas (woody vines), and branches are clothed with many epiphytes including bromeliads and orchids. In Panama, these evergreen forests are found mainly on the Atlantic slope.

Submontane and montane forests. Higher elevations are cooler and usually wetter than the lowlands, and the forests are different in composition. Forests here are usually evergreen, with extensive growth of epiphytes, and the canopy is often lower than in the lowlands. In the western highlands some trees typical of the North Temperate Zone such as oaks occur. *Cloud forest* may be found at higher elevations on ridges where clouds form almost every day. In this very humid environment most surfaces may be coated with a luxuriant growth of mosses, ferns, and other small plants, and the soil is soggy and sponge-like. *Elfin forest* is a kind of cloud forest that occurs on windswept ridges, where the forest canopy may be only a few meters high and tree trunks are twisted and contorted by the wind.

Paramo. Paramo is a wet alpine grassland that occurs in tropical mountains. Although common at high elevations in Costa Rica, Panama has very little true paramo. Near the summit of Volcán Barú there are grassy and open areas, but this is due largely to the rocky terrain, and is not true paramo. Panama's largest area of paramo is on its second highest peak, Cerro Fábrega, but this is remote and not easily accessible. Occasionally very windswept slopes at lower elevations may have only herbaceous vegetation or knee-high scrub.

Second-growth forest. In much of Panama true primary forest that has never been cut (or at least not been cut for hundreds of years) can only be found in more remote areas. Tall secondary forest, however, may have most of the bird species present in primary forest, with a few exceptions. Forest that is younger and hence shorter, and forest that is highly fragmented, however, may lack most primary forest species and instead be inhabited by species that prefer second growth.

Young second growth and scrub. In traditional tropical agriculture, fields are usually cleared and burned in the dry season when vegetation dries out most easily. Often fields are allowed to regenerate for several years before being cleared and burned again, resulting in areas of young second-growth forest up to a few meters tall. Areas that are repeatedly cleared and burned, and some coastal areas, may be maintained in scrub indefinitely. Many species of birds are characteristic of such young second growth and scrub.

Savanna. Savannas are characterized by scattered trees with the intervening areas occupied by grass. Although Panama probably once had some natural savanna in the driest parts of the country, most savanna today has been produced by human activity, particularly repeated burning. Savanna is most common on the Pacific slope.

Mangroves and swamp forests. Mangroves are trees specially adapted to live in salt water. There are extensive areas of mangrove forest on the Pacific coast of Panama around the Gulf of Panama and in Chiriquí, and smaller areas on the Caribbean coast. Red Mangrove (*Rhizophora*), which has arching stilt roots, dominates the seaward fringes of these forests, with Black Mangrove (*Avicennia*) and White Mangrove (*Laguncularia*) growing in more inland areas. In Bocas del Toro there are extensive freshwater swamp forests dominated by *Orey* (*Campnosperma*), and in Darién others dominated by *Cativo* (*Prioria*).

Lakes, rivers, and freshwater wetlands. Panama's rugged topography is not conducive to the formation of long rivers or extensive wetlands. The largest lakes in the country are the artificial reservoirs of Lake Gatún on the Río Chagres and Lake Bayano on the Río Bayano. In the highlands the largest lake is the Fortuna Reservoir at the Fortuna hydroelectric project in Chiriquí. The Volcán Lakes and Lake La Yeguada, which are natural lakes, are much smaller. Other large rivers besides those previously mentioned include the Río Changuinola in Bocas del Toro, Chiriquí Viejo in Chiriquí, La Villa in Herrera, Grande in Coclé, and the Chucunaque, Tuira, Sambú, Balsas, and Jaqué in Darién. The largest natural freshwater wetlands surviving are those of the Río San San and Pond Sak near Changuinola in Bocas del Toro. At one time there were extensive coastal wetlands in Chiriquí, Herrera, and eastern

Panamá Province, but these have now mostly been drained for rice and sugar plantations and pasture or converted to shrimp farms. Some of the best surviving marshes on the Pacific slope are Las Macanas in Herrera and Tocumen in eastern Panamá Province, although these have also been much reduced.

Coastal habitats. Besides mangroves, several other coastal habitats are of importance to birds. In Panama, the most significant of these are the enormous coastal mudflats of the Pacific coast, mainly in the upper Bay of Panama and in the Bay of Parita. Because of the very large tidal range here, sometimes over 5 m (16 ft), and the very gentle slope of the terrain, these intertidal flats may be exposed for distances of up to 3 km (2 mi) from the shore at low tide. These mudflats are a very rich feeding area for over a million migratory shorebirds that visit Panama every year, plus other birds such as herons and ibises. Sandy and rocky beaches can be found in many places along both coasts of Panama.

Human-made habitats. Although human-made habitats generally have much less bird diversity than similar natural habitats, some kinds of birds favor them. Gardens and parks in urban areas can be good for some hummingbirds, open-country flycatchers, and tanagers. Pastures and other grasslands have raptors, seedeaters, finches, meadowlarks, blackbirds, and other species. Rice fields, especially when flooded, may be good for some ducks, egrets, and ibises, and shrimp ponds for these species and shorebirds. Exotic plantations, such as teak, Caribbean pine, and eucalyptus, however, are usually very poor for birds compared to native forests.

Bird distribution in Panama

Bird distribution in Panama is strongly influenced by differences in climate, elevation, and biogeographical history. The isthmus was once part of an archipelago stretching between North and South America. As a continuous land bridge gradually formed, one of the last gaps was in the area of the present Panama Canal. Even today there is a distinct break in bird distribution in this area, which is one of the lowest parts of the Continental Divide between the Atlantic and Pacific slopes anywhere in the Americas. Many species with northerly ranges get no farther south than the Canal Area, and southerly ones no farther north.

Rainfall has a strong influence on bird species distributions, with some mostly restricted to the wetter Atlantic slope, and others to the drier Pacific. One common distributional pattern is for birds with a mainly Atlantic slope distribution to range onto the Pacific slope in eastern Panama and Darién, which are somewhat wetter. Such species may also occur a short distance onto the Pacific slope where there are lower points in the mountain chain, such as Fortuna, Santa Fe, and El Valle. Some of these also have isolated populations in wetter areas on the Burica and Azuero Peninsulas.

Birds typical of open country are found principally on the western and central parts of the Pacific slope, which is not only drier but has also been subjected to more deforestation than the Atlantic side. However, increased deforestation in recent years has been allowing many of these species to spread onto the Atlantic slope in Bocas del Toro, northern Coclé, western Colón, and the Canal Area, and to spread eastward into Darién and to higher elevations in foothills and highlands. A number of species of open country and disturbed areas have colonized Panama in recent decades, including White-tailed and Pearl Kites, American Kestrel, Glossy Ibis, Southern Lapwing, Yellow-breasted Flycatcher, Cattle Tyrant, and Grayish Saltator.

The cooler and wetter foothills and highlands have their own distinctive avifaunas. Some birds typical of the foothills zone start to appear above about 400-500 m (1,300-1,650 ft), sometimes lower, especially on the wetter Atlantic side. (Some foothills birds may even appear in the lowlands here.) The foothills zone extends up to about 1,200 m (4,000 ft). While some species whose elevational ranges begin in the foothills are found all the way to the top of the higher peaks, other species have distinct upper limits and are essentially confined to the foothills zone. Highland species begin to appear above 900 m (3,000 ft). Some species that are found only at higher elevations in the western part of the western highlands in Chiriquí and Bocas del Toro may be found at significantly lower elevations toward the east in Veraguas and Coclé. In eastern Panama, although a few basically highland birds are found in the western part of the ranges at Cerro Azul and Cerro Jefe, many are confined to the higher eastern ranges of Cerro Tacarcuna, Cerro Pirre, and the Serranía de Jungurudó.

Panama's offshore islands also have distinctive avifaunas. Although islands in general have a lower number of species than do equivalent areas on the mainland, several of Panama's islands or island groups have species not found elsewhere in Panama. The Stub-tailed Spadebill has an isolated population in the Bocas del Toro Archipelago, being otherwise found from northwestern Costa Rica northwards to southern Mexico. White-fringed Antwren is found only on the larger islands (and islands immediately adjacent to them) of the Pearl Archipelago, next appearing in northern Colombia. Coiba Spinetail is endemic to its namesake island, and Escudo Hummingbird to tiny Isla Escudo de Veraguas in Bocas del Toro. Coiba and the Pearl Islands each have large numbers of endemic subspecies as well.

Endemics

One of Panama's major attractions for the birder is its large number of regional and national endemics. We refer here to "regional endemics" as those species which qualify as "restricted range species" under the definition of BirdLife International, that is, those with a total world range of less than 50,000 sq km (19,300 sq mi; approximately the size of Costa Rica). Panama has a total of 107[3] of such regional endemics. Of these, 12 are national endemics, species that have so far been found only within Panama's borders. (Some of these may occur in Colombia as well, but

[3] The *Annotated Checklist of the Birds of Panama* gives this figure as 105 since Costa Rican Swift, Costa Rican Pygmy-Owl, Streak-breasted Treehunter, and Canebrake Wren were inadvertently left out, and Buffy Tuftedcheek and Plain Wren were included in error.

have not yet been recorded there. In any case, the areas in which they might occur in that country are not accessible to birders.) The best localities to find each of the regional and national endemics are discussed in Section 5 of this guide (p. 294). Regional endemics are indicated in the text and lists in **bold**.

Birdlife International has identified more than 200 "Endemic Bird Areas" throughout the world, in which two or more restricted-range species occur together, as well as "Secondary Areas," in which only one restricted-range species occurs. Panama includes parts of five such Endemic Bird Areas, plus one Secondary Area. These are:

- Central American Caribbean Slope (Atlantic slope lowlands from Honduras to Panama)
- Costa Rica and Panama Highlands (Highlands of Costa Rica and western Panama)
- South Central American Pacific Slope (Pacific slope lowlands of Costa Rica and Panama)
- Darién Lowlands (Lowlands of eastern Panama and western Colombia)
- Darién Highlands (Highlands of eastern Panama and extreme western Colombia)
- Isla Escudo de Veraguas Secondary Area

Panama national endemics

Brown-backed Dove*	*Leptotila battyi*
Azuero Parakeet*	*Pyrrhura eisenmanni*
Veraguan Mango	*Anthracothorax veraguensis*
Escudo Hummingbird*	*Amazilia handleyi*
Glow-throated Hummingbird	*Selasphorus ardens*
Stripe-cheeked Woodpecker	*Piculus callopterus*
Coiba Spinetail*	*Cranioleuca dissita*
Beautiful Treerunner	*Margarornis bellulus*
Yellow-green Tyrannulet	*Phylloscartes flavovirens*
Green-naped Tanager	*Tangara fucosa*
Pirre Bush-Tanager	*Chlorospingus inornatus*
Yellow-green Finch	*Pselliophorus luteoviridis*

* species not presently recognized by the AOU, although recognized by other references.

Migration

Some 149 species (15% of Panama's avifauna) are regular migrants, and another 70 species or so are casual migrants or vagrant individuals of migratory species. The overwhelming majority of these come from the North Temperate Zone, although some come from the South Temperate Zone or are intratropical migrants. Some species are transients, passing through Panama en route to wintering areas in South America, and are absent during most of the northern winter. Others are winter residents, present in Panama throughout the northern winter. In some migrant species, a few individuals may stay through the summer. These are often juveniles who wait until their second breeding season after fledging to attempt to breed.

Although some northern migrants arrive as early as August, the peak of migration is in September and October, with the return flight mostly from March to May. The most spectacular migratory movements are shown by some species of small land birds, raptors, and shorebirds.

Small land birds. The passage of some small land birds through Panama can be quite impressive. Some migrants come through in waves, and may be very abundant for a few days and much less common through the rest of the migratory period. Swallows, especially Barn and Cliff Swallows, may pass through coastal areas in thousands, with a continuous stream of birds passing by certain points along the coast. Eastern Kingbirds migrate in flocks, sometimes descending en masse on fruiting trees (fruit rather than insects being their favored food on migration). Dickcissels may sometimes be seen in flocks of thousands in grassy areas. During migration and while wintering, most migrants are more common in secondary habitats, including scrub and early second growth, rather than in deep forest. However, there are a few migrants that can be found in the forest interior.

Migratory raptors. The passage of migratory raptors through Panama is one of the great wildlife spectacles of the Americas. In some species that winter in South America, virtually the entire North American population funnels through the narrow isthmus in a relatively brief period, making this one of the best places to observe them. The greatest numbers belong to three species of soaring raptors, Turkey Vulture and Broad-winged and Swainson's Hawks. These birds rely on thermals, rising columns of warm air caused by sunlight heating the ground, during migration. They usually pass through in large flocks numbering in the thousands. During the southward passage, most enter Panama on the Atlantic slope in Bocas del Toro, then cross over the Continental Divide in central Panama to continue their passage on the Pacific slope in the Canal Area and eastern Panama. The return passage in March and April is less concentrated along particular routes.

The migration of these species is heaviest from early October to mid-November, with the peak being in the last two weeks of October. Broadwings pass through earliest, and Swainson's later, with Turkey Vultures present throughout this period. The numbers of birds that move through are enormous: In 2004, on the Panama Audubon Society's "Raptors Ocean to Ocean" raptor watch, conducted with support from Hawk Mountain Sanctuary, over 3.1 million raptors were counted in a six-week period at nine watch sites extending across the isthmus. The greatest single one-day movement was recorded on 14 November the following year, when over 640,000 raptors, more than 90% of them Turkey Vultures, were counted from Ancón Hill in Panama City in the course of a single day.

Several other raptors migrate through Panama in flocks, but not in the huge numbers reached by the three previous species, congregating mainly in flocks of a few hundred. The earliest migrant is Swallow-tailed Kite, which passes south in late July to early September, and returns in late January to February, and Plumbeous Kite, which moves south in early August to late September, and returns in early February to mid-March. Some members of both of these species breed in Panama, but migrate to South America during the non-breeding season. Mississippi Kite migrates south mainly in October, and north from mid-March to late April.

The most convenient places to observe these raptor migrations are at Ancón Hill, in Panama City, and at the Canopy Tower Hotel, in Soberanía National Park. The Mirador at the Gamboa Rainforest Resort, and the Escobal area on the Atlantic side of the Canal Area, also sometimes get good numbers. Raptor migrations can also be very impressive in Bocas del Toro and some other places.

Shorebirds. Panama's coastal areas are a critical migratory staging and wintering area for shorebirds that breed in North America. The most important habitats are the extensive intertidal mudflats around the Gulf of Panama, including the upper Bay of Panama, the Bay of Parita in Coclé and Herrera, the Gulf of San Miguel in Darién, and several other areas. The largest numbers of shorebirds of all are found in the upper Bay of Panama, stretching from Panama City eastward for 70 kilometers (about 40 miles). It has been estimated that over 1.3 million shorebirds pass through or winter in this area every year, and more than 320,000 have been counted here in a single day. The most numerous species is Western Sandpiper; it has been estimated that 30% of its world population uses this site. Other species that are abundant here include Semipalmated Sandpiper, Semipalmated Plover, Black-bellied Plover, Willet, and Whimbrel. A total of 36 species of shorebirds has been found in the area.

The easiest places to see large numbers of migratory shorebirds are Panama Viejo and especially Costa del Este in Panama City. Other good places include Tocumen east of Panama City, Aguadulce in Coclé, El Agallito Beach near Chitré in Herrera, and near Punta Patiño in Darién.

Seabird and waterbird nesting colonies

Panama is host to several large and many smaller breeding colonies of seabirds and of herons and other waders that can offer a fascinating spectacle during the nesting season. Most of these colonies are on islands and in coastal areas around the Gulf of Panama, but there are others scattered elsewhere.

There is a large and easily accessible colony of Brown Pelican, sometimes numbering over 1,000 pairs, on Isla Taboga near Panama City. Pelicans also nest on Isla Pacheca in the Pearl Islands, and in other localities in the Pearls and around the Gulf of Panama. Isla Iguana, off the southeastern tip of the Azuero Peninsula near Pedasí, has a colony of over 1,000 pairs of Magnificent Frigatebird. This species also nests on Pacheca and several other localities. There are large colonies of Neotropic Cormorant on Pacheca, smaller islands in the Pearls, and in Lake Bayano. Brown and Blue-footed Boobies nest on many small islands in the Gulf of Panama, and the former also nests in the Gulf of Chiriquí and off the Caribbean coast.

Other species are less widespread. The only nesting site of Red-billed Tropicbird in the southwest Caribbean is on Swan Cay near Isla Colón in Bocas del Toro. Audubon's Shearwater nests in small numbers on the Tiger Cays off the Valiente Peninsula in Bocas del Toro, but these small islets are not easily visited. The tiny rock stacks of the Islas Frailes del Sur off the southern coast of the Azuero have a large colony of Sooty Tern, numbering in the thousands, and smaller numbers of Bridled Tern and Brown Noddy, with a few dozen pairs each.

Pelagic birding

Pelagic birding in Panama is relatively poorly known, mainly because the best areas are remote from major population centers. Pelagic birds are much more diverse and numerous off the Pacific coast of Panama than off the Caribbean. A few pelagic species, particularly some storm petrels and shearwaters, can sometimes be seen well within the Gulf of Panama between Panama City and the Pearl Islands, but the better areas are near the edge of the continental shelf. On the Pacific, the shelf approaches the mainland along the southern edge of the Azuero Peninsula and in southern Darién. For the Azuero, the best option would be to drive down to Pedasí and hire a boat there to go out south of Punta Mala. A pelagic trip could also be arranged by those staying at the Tropic Star Lodge at Piñas Bay in the Darién, but this is quite expensive. Those with their own boat could also try cruising to the edge of the continental shelf south of the Pearl Islands. Among the more common pelagics are Sooty and Audubon's Shearwaters, Wedge-rumped, Black, and Least Storm-Petrels, Red-necked Phalarope, Parasitic and Pomarine Jaegers, Sabine's Gull, Bridled, Sooty, and Black Terns, and Brown Noddy. Many rarer species also occur.

Conservation

As in many other areas in the tropics, Panama's birds are under threat from many directions. The most serious problem is habitat loss, particularly deforestation. One of the major factors causing loss of forest is cutting for subsistence agriculture. Small farmers, mostly very poor, clear and burn forest to plant crops such as corn, rice, beans, or root crops. However, many tropical soils lose fertility very quickly, so that after a few years the fields are no longer productive and new forest must be cleared to replace them. Formerly older unproductive fields were often allowed to regenerate to forest, but today they are usually converted to cattle pasture, which persists indefinitely due to repeated burning. Other causes of deforestation include logging (both legal and illegal), cutting for firewood, poles, and other products, and construction of housing developments, resorts, and other infrastructure. Construction or improvement of roads can often promote deforestation by providing better access and allowing farmers to bring their crops to market more easily, and thus motivating them to clear more land.

Besides forests, other habitats are also threatened, especially wetlands. Most of the former freshwater wetlands along the Pacific coast have now been drained and converted to rice fields, other kinds of plantations, or cattle pasture. Other marshes, swamps, and wetlands are affected by agricultural chemicals and pesticides used on adjacent croplands, especially near the large banana plantations in Bocas del Toro. Mangroves are cut for timber, tannin, or for agricultural or urban development.

Although grasslands in the form of cattle pasture have expanded greatly at the expense of forest, even some grassland birds have declined. Grasshopper Sparrow, including an endemic subspecies, has not been seen in many decades in Panama, and Grassland Yellow-Finch and Yellowish Pipit are now quite scarce. The factors involved in these apparent declines are obscure, but may include too-frequent burning.

Hunting and capture for the pet trade are also threats to some birds. Although hunting is technically illegal in Panama, guans, curassows, tinamous, ducks, and quail are favored as game and are heavily hunted in some areas. Ducks and some other species are sometimes shot or poisoned for feeding on rice or other crops, and herons and other waterbirds are shot for depredations at shrimp ponds. Crested Guan and Great Curassow in particular have become very scarce anywhere near populated areas. Raptors, including Panama's national bird, the Harpy Eagle, may be shot as a supposed threat to domestic fowl, to take their talons as trophies or ornaments, or merely for "sport." Macaws and other parrots are frequently captured for the pet trade. These species are hole-nesters, and nest holes, which are usually in short supply, may be destroyed in the course of stealing nestlings to raise in captivity. Because the reproductive rate of these species is low, this can have a major effect on their populations. Macaws are also shot for their feathers, which are used in traditional costumes and dances. Today macaws in Panama are mainly confined to more remote areas, and the Scarlet Macaw is almost gone from the mainland, surviving mostly on Isla Coiba.

Pollution and contamination may also be a threat to some birds, although how severe this threat may be is not well known. Panama City and other urban areas discharge their untreated waste water, including not only raw sewage but also commercial and industrial waste, directly into the ocean or into rivers. In many agricultural areas there is heavy use of agrochemicals and pesticides, especially in rice fields and banana plantations. These may drain into adjacent wetlands and rivers and thence into coastal areas. Such pollution could be a threat to shorebirds and seabirds that feed in coastal waters, as well as to freshwater birds.

Panama has, however, made some strides in protecting its natural heritage. Over 29% of its land area has been legally protected as national parks, forest reserves, wildlife refuges, and other kinds of protected areas. And unlike some other countries, a significant amount of forest still remains outside national parks. In 2003 some 44% of the country was still covered in forest. In 1998 the National Environmental Authority (Autoridad Nacional del Ambiente, or ANAM) was established, with a broader mandate than its predecessor agency, the Institute for Renewable Natural Resources (Instituto de Recursos Naturales Renovables, or INRENARE), as part of a new Law of the Environment. ANAM's responsibilities include management of national parks and other protected areas, and also enforcement of environmental regulations on forest cutting, hunting, pollution, and contamination. Laws are in place regulating forest cutting and waste water discharge, and prohibiting hunting. New development projects are required to file Environmental Impact Assessments (EIAs).

The downside, however, is that enforcement of existing regulations is often weak. Some protected areas are little more than "paper parks," without sufficient staff, vehicles, or funds to effectively patrol them. Even in one of Panama's premier parks, Soberanía in the Canal Area, poaching is a severe problem, and deforestation threatens Achiote Road in San Lorenzo National Park, one of Panama's best-known birding areas. Political and economic considerations can also threaten protected areas. In 1995, part of the Corredor Norte highway was constructed through Panama City's Metropolitan Nature Park, prompting protests that helped raise environmental awareness. When the government proposed building another road through Volcán Barú National Park in 2002 that would have destroyed one of its best birding areas, the Los Quetzales Trail, including prime Resplendent Quetzal nesting habitat, local and international environmental organizations were able to mount a successful campaign to halt its construction.

Protection of the environment is often seen as being in conflict with economic development. But the preservation of natural areas, besides providing benefits such as watershed protection, prevention of erosion, and the maintenance of biodiversity, can be a direct and indirect source of revenue and income through ecotourism, including birding. It has been estimated that birders and other wildlife watchers spent $32 billion, producing $85 billion in total economic benefit, in the United States in 2001. In Costa Rica, tourism generated an income of $1.5 billion in 2005, and is now the leading sector of the national economy. This has contributed greatly to the high priority Costa Rica has placed on the protection of its national parks and other reserves. Although Panama is a growing destination for birders and other ecotourists, most tourism in the country is still focused on attractions such as beaches, golf courses, and casinos. As birding and other ecotourism increases in Panama, enhanced income from this source should encourage an understanding that investment in the protection of natural areas is simply good business. We hope that this guide, by publicizing the many natural attractions of Panama and making it easier for birders to have a high-quality experience here, will help promote protection of the habitats of birds and of other wildlife as well.

Information for the Birder

This guide is mainly directed to birders visiting Panama from outside the country. We have generally assumed that a visitor will have a rental car. Many birding areas can also be reached by public transportation, or by hiring a taxi, and we have not necessarily mentioned all possible options for doing so. If you do not have a car, local inquiry in towns close to birding spots may provide some options we have not discussed; ask at your hotel or at the bus station, if there is one. We have also primarily focused on hotels that are particularly well suited to serve as bases for birding, and have not attempted to be complete in our coverage of accommodations in all areas. In general, we have emphasized mid-range accommodations in terms of cost and comfort, including expensive or luxury accommodations mainly when they are especially dedicated to birding or nature tourism, and very basic accommodations when there are few other options available in the area.

This guide should be used in conjunction with one of the standard tourist guides to Panama, which have a more complete listing of accommodations and restaurants for many areas, and which may suggest more options for the budget traveler than we do here. Two of the best guides are *Lonely Planet: Panama* (third edition, 2004, by Regis St. Louis and Scott Doggett), and *Moon Handbooks: Panama* (first edition, 2005, by William Friar). In addition to general tourist information, both also provide interesting cultural and historical background. Another useful book is *Adventures in Nature: Panama* (first edition, 2001, by William Friar), which provides information on many national parks and other natural areas.

For those interested in nature or adventure tourism in general, Panama offers a host of activities in addition to birding. These include hiking, horseback riding, white-water rafting and kayaking, scuba diving and snorkeling, surfing, and sport fishing. Consult one of the guides listed above for information on these activities.

When to come

Birding is good in Panama any time of year. Northern migrants are present mostly from September to April, so this is when it is possible to see the greatest number of species. Some birders may prefer to come during the dry season, from mid-December to late April, when rainless days are most frequent in the lowlands of the Pacific slope. However, even during the rainy season rains are usually brief and limited to the afternoons, so that the prime birding hours in the morning are frequently rain-free. In the highlands, since the dry-season trade winds often bring mist and drizzle, the rainy season has a better chance of producing a rain-free day. In rainy Bocas del Toro, the best times of year may be September-October and February-March. If you are planning to travel by sea, the dry season has rougher conditions due to wind, especially in the afternoon, and September and October are probably the calmest months. Breeding activity is at its peak in the late dry season and early rainy season in April and May, since this is when many deciduous trees leaf out, producing a flush of insects that provide food for nestlings. Because calling and song is most frequent at this time, many species are then easiest to find. By June and July many birds are incubating or feeding young. Calling activity falls significantly, and the forest at times can seem almost empty of birds.

Panama's high season for tourism is during the dry season (and northern winter), that is, from December to April. Airfares may be more expensive then, and some hotels, especially in the main tourist areas, raise their rates. In some areas, particularly the

Azuero Peninsula, hotel rooms may be difficult to obtain during Carnaval (Mardi Gras), Semana Santa (Holy Week), or during local festivals and fairs, so it's best to call in advance when heading for smaller towns in this region.

Birding in Panama

In tropical lowland forest, bird activity is much greater during the first few hours of the day, so getting out early is even more important than it is in temperate areas. It is best to get to birding areas as close to first light as possible. You can bird along the forest edge or along roads until it becomes light enough to see on trails within the forest. Bird activity usually slows considerably after about 9:30 or 10:00 AM. There is another, lower peak of activity in the late afternoon after about 4:00 PM, so it is often worth going out again then after the heat of mid-day.

Birds of aquatic habitats and open areas may be in evidence all day. Once activity has slowed in forest, it can be a good strategy to check out nearby fields for raptors, open-country flycatchers, and seedeaters; marshes or swamps for herons, ducks, kingfishers, and other aquatic species; and beaches for shorebirds and coastal species. When birding areas along the Pacific coast with extensive mudflats, it is best to go within an hour or two of high tide, since otherwise most birds will be too far out on the flats to see. Tide tables can be purchased at sporting goods and hardware stores in Panama City, and can also be found at the Panama Canal Authority's website (www.pancanal.com/eng/eie/radar/tide2) or at a number of other places on the internet.

The daily pattern of bird activity may be somewhat different in montane and cloud forests. Activity may be quite low in the early morning, especially if there is thick fog or mist. Birding may improve later in the day, after the sun has burned off the fog and the temperature has risen. Later in the afternoon mist or light rain may roll in again. If conditions are bad at higher elevations, it may be worthwhile to descend to lower altitudes where the weather may be better.

Within tropical forest, many birds tend to travel together in mixed-species flocks, which can include dozens of species. You can sometimes walk for an hour or more through a forest that is seemingly devoid of birds, and then be surrounded by a swarm too numerous to keep track of individually. These flocks occur among both understory and canopy birds. When you find a flock, spend as much time with it as possible – some peripheral species may not be immediately apparent. It can also be productive to wait near fruiting trees, and, for hummingbirds and honeycreepers, near patches of flowers or trees in bloom.

Some birds persistently follow the large swarms of army ants that forage through the leaf litter in lowland and foothill forests. They feed on insects fleeing from the ants, rather than the ants themselves. These species include Rufous-vented Ground-Cuckoo, Ocellated, Bicolored, and Spotted Antbirds, Black-crowned Antpitta, Plain-brown, Ruddy, and Northern Barred-Woodcreepers, Gray-headed Tanager, and others. Learning the calls of these species, especially the more common ones such as Bicolored Antbird and Plain-brown Woodcreeper, can help you locate these swarms.

Knowledge of vocalizations is very important in diverse tropical habitats. Many species, such as tinamous, wood-quail, rails, quail-doves, cuckoos, some antbirds, tapaculos, and wrens, as well as many canopy species, are heard far more often than they are seen. Even with many less secretive species, individuals are often detected by call well before they are seen. You will increase your chances of seeing birds greatly by learning as many calls as possible. A good tape recorder and microphone can be very useful for luring a bird out of hiding by playing back its recorded call. Recording what you hear and then comparing it with reference tapes can also be a valuable aid to learning calls, as well as for identifying unknowns. Playback should not be overused, however, especially in areas frequented by many birders, since it can cause disruption to a bird's activities. Used in moderation, however, it probably will not produce serious disturbance.

Night birding can be productive for owls, nightjars, potoos, and some other species. Where there are roads you can drive along slowly scanning the roadside vegetation for eyeshine with a spotlight or other bright light. Potoos produce particularly bright eyeshine. They and owls will typically be found perched in vegetation, while nightjars will usually be seen resting on the road or the ground. Sometimes nightjars and owls can be found near streetlights where they feed on insects attracted by the light. Stop to listen for calls and play recordings of them. A calling owl can often be lured in by playback (but bear in mind the caveats mentioned in the previous paragraph).

It can be quite easy to get lost in tropical forest. If you are using unmarked trails, take note of or mark all forks so you do not take the wrong one on the way back. If you go off the trail in pursuit of a calling bird, make sure to break twigs as you go along so you can find your way back. Take note of which way streams run across a road, so if you get lost you can follow one back and find the road again.

Ticks, chiggers, and biting insects are common hazards of birding throughout Panama. See the section on "animal bites" (p. 54) for advice on keeping them at bay. Black Palms are covered with long sharp spines and often grow next to trails where you may inadvertently try to grab one for balance. Always look first before grabbing onto a tree in the forest, and watch where you put your hands and feet to avoid being bitten or stung by ants or other nasties.

Clothing and equipment

Field clothing is largely a matter of personal taste. Lightweight clothing of fast-drying material is preferable in the lowlands due both to the chance of sudden showers and the certainty of sweat. Some birders prefer long pants of tough material as protection against insects, spiny vegetation, and barbed wire, while others feel that shorts are adequate. A hat with a brim is especially important if you will be spending time in the open, and also provides protection against rain. On level trails sneakers may be adequate, but steeper trails on clayey tropical soils can be very slippery and footwear with a good tread is desirable. Heavy boots are not really necessary in Panama. In the highlands, where trails are often very soggy, and in the lowlands during the rainy season, rubber Wellington-style boots are valuable to have. They can be purchased cheaply at most hardware and many general stores in Panama. A light poncho or a small folding umbrella is handy in case of rain; a full raincoat can be very hot in the lowlands, but may be useful if you will be spending a lot of time in the cooler highlands. It is a good idea to at least bring along a plastic bag for your camera and binoculars in case you are caught in a heavy downpour.

If you will be spending a lot of time in the tropics, sealed, nitrogen-filled binoculars are recommended. Moving from an air-conditioned hotel or vehicle into humid tropical air can hopelessly fog unsealed binoculars for some time, and moist conditions promote the growth of fungus on lenses and prisms. As elsewhere, a scope is useful for birds of open areas and those on distant perches. A good tape recorder and microphone can also improve your success in the field. If you plan on doing night birding, a headlamp will enable you to use your binoculars with hands otherwise unencumbered, and will also be useful if you will be staying in areas away from major cities. A spotlight is helpful for night birding from a car.

Make sure you carry enough water for the amount of time you will be out, keeping in mind that you are likely to be perspiring quite a bit. Taking along some small snacks is a good idea in case you are out longer than you planned. Bring along insect repellent, and perhaps also some topical anesthetic ointment in case you do get bitten.

Field guides and recordings

The indispensable guide for birding in Panama is *A Guide to the Birds of Panama, with Costa Rica, Nicaragua, and Honduras,* by Robert S. Ridgely and John A. Gwynne. This guide provides detailed descriptions of all species recorded for Panama (as of 1989), as well as information on distribution, seasonal occurrence, habitat, calls, and behavior. Particularly useful is its discussion of similar species and of key field marks for distinguishing them. Gwynne's illustrations are generally excellent, the chief drawback being that not all species are depicted, and some are shown only in black-and-white drawings in the text. For brevity, we refer to this guide henceforward simply as Ridgely.

A useful supplement to Ridgely is the *Annotated Checklist of the Birds of Panama,* by George Angehr, published in 2006 by the Panama Audubon Society. The checklist gives information on the distribution and relative abundance of all species recorded in Panama in 10 regions of the country, and also summarizes information on habitat and status. A section on new records for Panama, by George Angehr and Dodge Engleman, provides information on the 33 species newly recorded for the country since the publication of Ridgely, as well as new breeding records. In addition, nine taxonomic splits are recognized, bringing the total number of species for Panama to 972, compared to the 930 included in Ridgely.

Because Ridgely does not illustrate the rarer migrants, if you are visiting Panama anytime between September and May it is essential to have one of the North American field guides that show these species. Because Panama draws migrants from both eastern and western North America, it is best to have a guide that covers the entire area. Which of these you bring is partly a matter of taste, but among the more popular guides are *The Sibley Guide to Birds,* by David Allen Sibley, the *National Geographic Field Guide to the Birds of North America* (fourth edition), and *Birds of North America,* by Kenn Kaufman. These guides can be useful even during the northern summer, because some migrants are present year round, and because they include other North American species not illustrated by Ridgely. More specialized references that provide illustrations of many migrant species include *Seabirds: An Identification Guide,* by Peter Harrison; *Shorebirds: An Identification Guide,* by Peter Hayman, John Marchant, and Tony Prater; and the Peterson Field Guide to *Warblers of North America,* by Jon Dunn and Kimball Garrett.

Other guides can be valuable supplements for birding in particular areas of the country or for particular groups. In western Panama, *A Guide to the Birds of Costa Rica,* by F. Gary Stiles and Alexander Skutch, with illustrations by Dana Gardner, is very useful, particularly since it illustrates a number of species and plumages not shown by Ridgely. In eastern Panama, and especially in Darién, *A Guide to the Birds of Colombia,* by Steven Hilty and William L. Brown, can be helpful, especially since species new for Panama regularly turn up in this relatively little-known region. *A Guide to the Birds of Mexico and Northern Central America,* by Steve N. G. Howell and Sophie Webb, has some of the best illustrations of Neotropical raptors available.

Birders who will be spending time in Panama City may wish to pick up a copy of *A Guide to the Common Birds of Panama City,* by Jorge Ventocilla with illustrations by Dana Gardner, published by the Panama Audubon Society and the Smithsonian Tropical Research Institute. The Panama Audubon Society's *Panama Canal Birding Trail,* by Darién Montañez, is a handy reference to sites for those birding in this region.

Although there are no CDs or cassettes specifically dedicated to Panama birds, *Voices of Costa Rican Birds: Caribbean Slope,* by David L. Ross, Jr. and Bret M. Whitney, and the *Costa Rican Bird Song Sampler* by David L. Ross, Jr., both published by the Macaulay Library of Natural Sounds of Cornell University, contains many species found in Panama.

Birding tour companies and guides in Panama

A growing number of international tour companies offer birding trips to Panama. In addition, there are several companies based in Panama that offer tours designed specifically for birders. While this guide is directed primarily to the independent birder, organized tours may be the most efficient way of getting to some sites, particularly in remote areas such as the Darién. Of course, even at sites that are easy to get to, having an experienced guide with local knowledge can vastly increase your success in finding birds.

The following Panama-based tour companies and individuals offer birding tours or guiding services. In addition to standard itineraries, most will be happy to custom-design a tour to suit specific interests. Most guides mentioned are bilingual in English and Spanish, but if you do not speak Spanish (or would like a guide who speaks a language other than English) you may wish to inquire about your guide's language capabilities in advance.

Tour Companies

- *Advantage Tours* (ph 232-6944; cel 6676-2466; fax 232-7517; info@advantagepanama.com; www.advantagepanama.com) offers one-day tours in the Canal Area as well as longer itineraries that can include the Canal Area, the western highlands, Bocas del Toro, Darién (including Cerro Chucantí in the Serranía de Majé), and Isla Coiba. Custom trips as well as educational programs can be designed for special interests. Advantage's highly experienced birding guides include Guido Berguido, Euclides "Kilo" Campos, and Venicio "Beny" Wilson. They also have available basic accommodations at the Soberanía Research Station and Lodge in Gamboa.

- *Ancon Expeditions of Panama* (ph 269-9415; fax 264-3713; info@anconexpeditions.com; www.anconexpeditions.com) is Panama's largest ecotourism operator. They offer day trips in the Canal Area, and short or extended itineraries that include the Canal Area, the western highlands, Bocas del Toro, eastern Panamá Province, and Darién. They can also arrange custom itineraries. The birding specialists among their excellent guide corps include Hernán Araúz, Rick Morales, Ivan Hoyos, and María "Marisín" Granados. Ancon Expeditions operates lodges and field stations at Cana and Punta Patiño in Darién, in Bocas del Toro, and near Soberanía National Park in the Canal Area.

- *EcoCircuitos Panamá* (ph/fax 314-1586; cel 6617-6566; annie@ecocircuitos.com; www.ecocircuitos.com) specializes in adventure tourism but also offers birding tours, including day trips in the Canal Area and extended itineraries that can include the Canal Area, Cerro Azul, the western highlands, eastern Panamá Province (Burbayar Lodge), Darién, and Isla Coiba. They also provide trips to Bocas del Toro and other areas of interest to birders, and offer tours featuring indigenous cultures and educational programs. Their guides include Sonia Tejada, Ariel Rodríguez, Yesenia Yepez, and Fidelino Jiménez.

- *Nattur Panamá* (ph 442-1340; fax 442-8485; Panabird@natturpanama.com; www. natturpanama.com) is operated by Wilberto "Willy" Martínez, one of Panama's most experienced bird guides. They offer day trips in the Canal Area and longer itineraries including the Canal Area, the western highlands, and Darién, and can set up customized trips for special interests. Wilberto also operates the Rancho Ecológico Willy Mazú on the Oleoducto Road in Bocas del Toro.

Guides

All of the following guides speak Spanish and English, with other languages as noted. All except Aizpurúa and Fonseca are based in the Panama City area, but can cover other parts of the country as well.

- Ariel Aguirre (cel 6637-3786; fax 558-0523; aguirreguia@yahoo.com; www.ariel birding.com) has led birding tours for some of Panama's major tour operators for many years, and speaks French and Italian in addition to Spanish and English.

- Nariño Aizpurúa (ph 771-5049; cel 6704-4251) lives in Volcán, Chiriquí, and has an excellent knowledge of the birds of western Panama.

- Deibys Fonseca (ph 771-4952; cel 6532-5350) also lives in Volcán, Chiriquí, and knows the birds of western Panama well.

- José Carlos García (ph 230-1728; cel 6647-8648; joseca_10@hotmail.com) is a professional ecotourism guide who knows birds.

- Jacobo "Jacob" Ortega (ph 235-0636; cel 6676-4464; jacobo@birdinginpanama.com, info@birdinginpanama.com; www.birdinginpanama.com) has an excellent eye for spotting birds, and also knows calls well. From near El Copé, Coclé, he is especially knowledgeable about that part of the country.

- José Tejada (cel 6504-8434; joseetejada@hotmail.com) has been one of Panama's top birding guides for many years. He is exceptionally knowledgeable about birds, and has several first records of species for the country and many other notable records.

- Venicio "Beny" Wilson (ph 261-1976; cel 6617-6461; beny@advantagepanama.com; info@advantagepanama.com; www.advantagepanama.com) is from Bocas del Toro, and is an expert on that region and on the Atlantic side of the Canal Area. Beny also works with Advantage Tours.

General Information for the Tourist

Panama's people

Panama has a total population of about 3 million, the smallest of any country of Latin America. Most of its people are *mestizo,* that is, of mixed European and American Indian ancestry. Because of Panama's role as a crossroads for many centuries, European immigrants came not only from Spain but also from France, Great Britain, Germany, Italy, eastern Europe, and many other countries. There are two groups of African descent: *afrocoloniales*, descended from slaves who arrived during the colonial era, and inhabiting mainly Panama City, parts of the

Atlantic coast, Darién, and the Pearl Islands; and *afroantillanos,* English-speaking descendents of workers from the West Indies who came to work on the Panama Railroad, Panama Canal, or banana plantations, and living mainly in Colón and Bocas del Toro. There is also a sizeable Chinese minority, as well as groups from the Middle East and South Asia, many of whose ancestors also came to work on the Panama Railroad or the Canal. The long US presence in the Canal Zone also resulted in many Americans or their descendents settling in the country, and today increasing numbers of US retirees are also coming to live in Panama on a full-time or part-time basis.

Panama has seven indigenous groups. The largest is the Ngöbe, who together with the related Buglé are collectively known as Guaymí. They live mostly in the western highlands in Chiriquí and Veraguas and in Bocas del Toro. The next most numerous group, the Kuna, live along the eastern Caribbean coast in the Comarca de Kuna Yala, and in eastern Panamá Province and Darién. The Emberá and the related Wounaan live mainly in Darién, but there are some villages in eastern Panamá Province and in the Canal Area. The Naso (also known as Teribe) and Bribrí are small groups living in Bocas del Toro. Several of these groups, including the Ngöbe-Buglé, Kuna, and Emberá-Wounaan, have been granted *comarcas,* semi-autonomous homelands in which they largely manage their own internal affairs.

Language

The official language of Panama is Spanish. English is fairly widely spoken in Panama City, and high-end hotels, most mid-range ones, and most other businesses that regularly deal with tourists will have staff that speak it. Cheaper hotels and restaurants often do not, and few taxi drivers speak English either. Away from the capital, relatively few people speak English, and in remoter areas almost none will. The exception is Bocas del Toro, where much of the population speaks a variant of West Indian English (which can itself be a bit difficult to understand!). Having at least a basic knowledge of Spanish will facilitate your travel in the country, especially if you plan on going off the beaten path. For arranging independent travel in truly remote areas, such as Darién, it is essential that at least one member of your party have good Spanish language skills.

Time

Panama is on Eastern Standard Time year round (Greenwich -5:00 hr), and is one hour later than Costa Rica. Because Panama is so close to the equator, there is little change in day length between seasons. Days are a little longer during the northern summer and shorter during the winter, but the variation is less than an hour. Sunrise is generally within a half hour of 6:00 AM, and sunset within a half hour of 6:00 PM, throughout the year. These times are a bit later in the western part of the time zone near the Costa Rican border.

Money and banks

Panama uses the US dollar for all transactions. Although Panama's currency is officially called the *balboa,* this is equal to the dollar and balboa banknotes are not used. Panama's coins, in denominations of 1, 5, 10, 25, and 50 cents, are the same size and shape as their US equivalents but differ in design. US coins are also in circulation and accepted interchangeably. Prices may sometimes be stated in dollars and sometimes in balboas, but these are exactly the same. Because of counterfeiting, many businesses will not accept $50 or $100 bills, so it is best to bring cash in denominations no higher than $20. ATMs are common in Panama City and are also available in many smaller towns, indicated by a red *Sistema Clave* sign. Cards from most major networks are accepted, including Plus, Cirrus, Master Card, and Visa.

Major credit cards are accepted at most mid-range or better hotels and restaurants, but not at the cheapest ones. However, American Express cards are not as universally accepted as Visa or MasterCard. Credit cards will also usually be accepted by travel agencies and tour companies. They may be difficult to use elsewhere, so it is wise to bring along an adequate supply of cash, especially when traveling away from Panama City or other major centers.

Taxi drivers and many small stores will often be unable, or at least reluctant, to change bills larger than $5, and $20 bills can be quite difficult to break. It is a good idea to keep a supply of $1 bills and small change on hand if you can.

Phones and internet

The country code for Panama is 507. (From the US, you must also dial the international prefix 011 before the country code.) Phone numbers within the country are seven-digit (there are no local provincial or city codes), with the exception of cellular phones, which have an additional "6" prefix. Calls to cell phones are more expensive than to regular phones.

Public phones are common in cities and are available in even some surprisingly remote areas (although they may not always work). The main provider of phone service is Cable & Wireless. Pay phones may accept either coins or pre-paid phone cards; some phones accept cards only. Phone cards in various denominations may be purchased at Cable & Wireless offices and in many stores. There are two types of phone cards: one ("Telechip") stores the value on the card electronically, and is inserted in a slot on the phone; the other ("Telechip Total") uses a coded number you must dial in. The latter is more flexible to use. Phone calls within the country are generally quite inexpensive.

Many mid-range and most high-end hotels have internet service available on premises. Internet cafes are increasingly available, and can be found in most major towns and tourist destinations. The cost is usually $0.50-$1.00 per hour.

Safety

Most of Panama is quite safe for the traveler. Panama City, like any major urban area, has some poorer areas where crime is a problem. However, the main hotel and banking district along Vía España and the restaurant areas in El Cangrejo and Bella Vista are generally safe. The older parts of town in Chorillo, Calidonia, and the Casco Viejo and Santa Ana are best avoided after dark. The city of Colón has chronic high unemployment and a correspondingly high crime rate, and should be avoided by the tourist. Muggings or thefts from cars have been reported from a few birding areas near Panama City, including Chivo Chivo Road, the Camino de Cruces Trail off Madden Road, Juan Díaz, and at Madden Dam and the road that runs past Campo Chagres. Use caution if visiting these areas.

Away from the major cities Panama is generally very safe. An unattended car is usually OK, but of course it is best not to leave temptation in the form of expensive equipment in plain view; lock it in the trunk or take it with you. If you will be leaving your car for an extended period while hiking, you can ask a nearby resident to watch it for you for the payment of a dollar or two. (It's best to pay when you return rather than in advance.)

The exception to this is some of the more remote areas of Darién near the Colombian border, which are frequented by Colombian guerillas, drug smugglers, bandits, and other lawless types. Dangerous areas include the upper Tuira Valley above El Real, and Jaqué and the Jaqué Valley. As of this writing, the main tourism destinations of Cana, Punta Patiño, Tropic Star Lodge, and El Real were considered safe. It is always a good idea to check with the local police if planning any independent travel in the Darién.

Major cities

Panama City, with a population of 860,000 (including its major suburbs of San Miguelito, Tocumen, and Pedregal) in 2005, is by far the largest city in the country. On the Pacific coast near the entrance to the Panama Canal, it is the country's transportation hub, and has the broadest range of accommodations, restaurants, banking facilities, and other amenities for tourists. Colón, near the Atlantic entrance to the Canal, is Panama's second largest city, with a population of 100,000 (including nearby Cristóbal). Unfortunately, unemployment in Colón is very high and the crime rate is as well, so it is not recommended for the international visitor. Regional centers with a good range of accommodations, restaurants, and amenities include Penonomé (population 12,000) in Coclé, Chitré (28,000) in Herrera, Santiago (47,000) in Veraguas, David (85,000) in Chiriquí, and Changuinola (26,000) and Bocas del Toro Town (3,000) in Bocas del Toro Province. Facilities for tourists can be sparse elsewhere in the country.

Airlines and airports

There are currently non-stop flights from the United States to Panama from Miami (American, COPA), Atlanta (Delta), Newark and Houston (Continental), New York-JFK, Los Angeles, and Orlando (all COPA). The only direct flights from Europe are from Madrid (Iberia and Air Madrid). Other international airlines flying to Panama, with stopovers between the US or Europe, include Avianca, Lacsa, Lloyd Aéreo Boliviano, and TACA. Some other airlines operate in Panama via codeshares with some of the listed ones.

There are presently two domestic airlines: Aeroperlas (ph 315-7500; fax 315-7580; info@aeroperlas.com; www.aeroperlas.com) and the newer AirPanama (ph 316-9000; reservaciones@flyairpanama.com; www.flyairpanama.com), formed by the merger of Mapiex and Turismo Aéreo. Both fly from Panama City to David in Chiriquí, and to Changuinola and Bocas del Toro Town in Bocas del Toro, as well as to various destinations in Kuna Yala, Darién, and the Pearl Islands. AirPanama also has flights from San José, Costa Rica to Panama City, Bocas del Toro Town, Changuinola, and David. Most domestic flights are to and from Panama City, but it is possible to fly between David and Bocas del Toro Town and Changuinola, and flights to Kuna Yala, Darién, and the Pearl Islands generally make stops at several destinations en route.

Panama's main international airport is at Tocumen, about a half hour from downtown Panama City via the Corredor Sur highway. Panama City's airport for domestic flights is officially called Marcos A. Gelabart, but is much more often referred to as Albrook, from the name of the area in which it is located, a former US air base in the Canal Area. Albrook Airport is about 20 minutes from downtown Panama City, but allow more time during times of day when traffic is heavy.

The major rental-car companies have desks at Tocumen. The taxi fare from Tocumen to Panama City is $25.00. Shared taxis called *colectivos* are also available, for $10.00 per person for three people, or $15.00 each for two. These may require a wait to find other people to share, particularly on the later flights during the day. Ask the driver to take the Corredor Sur toll road, which is considerably faster; you will have to pay $2.40 in tolls (which is well worth it) in addition to the fare. The official fare from Panama City to Tocumen Airport is $12.00, but some drivers may want to charge more. Agree on a price before getting in the cab. Again, asking to go via the Corredor Sur is worth the additional cost in tolls.

Albrook Airport is easily reached from downtown Panama City by taxi. The fare should be $1.50-$2.00 or so for a single person, but the driver may try to charge you more, so ask the price before leaving. Taxis waiting at the airport will charge you about $3.00 or so for the drive to Panama City.

Road system

The national road system in Panama is quite simple. The country's major artery is the Carretera Interamericana (Interamerican Highway, also often called the Carretera Panamericana and abbreviated on road signs as CPA), which runs along most of the length of the Pacific slope. There are only a few major side branches, including the Carretera Transístmica (Transisthmian Highway) which runs between

Panama City and Colón; the Camino Oleoducto (Oleoducto Road, or Pipeline Road, not to be confused with the road in Soberanía National Park), which runs from Chiriquí across the Continental Divide into Bocas del Toro to Chiriquí Grande, and the road that branches from it to go to Changuinola and the Costa Rican Border; and the Carretera Nacional (National Highway), which branches from the Interamericana at Divisa to run down the eastern side of the Azuero Peninsula. In the text, we refer to the main highways by their Spanish names since this is the form you will see on road signs.

At present, the Carretera Interamericana is four-lane between Panama City and Santiago, and from David to the Costa Rican border. These sections are generally in quite good condition, although some of the older parts are beginning to deteriorate. The stretch between Santiago and David is two-lane asphalt, and can be slow going, although the addition of passing lanes on some of the hills in western Veraguas has improved it considerably. East of Panama City, the road is four-lane only to the town of 24 de Diciembre, 7 km from the end of the Corredor Sur Highway. From there it is two-lane asphalt as far as Meteti in western Darién, and then gravel, often in bad condition, to Yaviza on the Río Chucunaque. Yaviza is presently the end of the road, the roadless "Darién Gap" of eastern Panama and western Colombia extending 120 km (70 mi) to the next road in South America.

Note that just west of Panama City there is a toll road called the Autopista, starting at Arraiján, that bypasses the suburb of La Chorrera, while the Carretera Interamericana itself passes directly through it. Avoid taking the Interamericana into La Chorrera, and take the Autopista instead. Tolls are $0.50 for small cars and $1.00 for light trucks.

The Transístmica remains one of the worst roads in Panama. It is two-lane asphalt, often with narrow shoulders, and in bad condition in some places. It is often clogged with smoke-spewing buses and heavy trucks, and traffic jams are frequent during rush hours. Fortunately, a branch of the Corredor Norte Highway now extends about halfway across the isthmus, and some improvements have been made to the Transístmica closer to Colón, so the trip is not as bad as it once was (but is still not pleasant).

The Oleoducto Road, which was built to service an oil pipeline across the isthmus, is two-lane asphalt and generally in good condition (except for some slumping near the Continental Divide), as is the road to Changuinola and Costa Rica. The Carretera Nacional is also two-lane asphalt, and is in generally good condition.

There are asphalt roads to many of the smaller towns in Panama, although these are sometimes badly potholed. Some birding sites are reached by gravel or dirt road, and the condition of these can vary dramatically depending on season and when the road was last repaired. Many roads that are passable to 2WD vehicles in the dry season require 4WD or at least high clearance once the rains start. Roads can sometimes deteriorate very quickly after they have been repaired, especially in high-rainfall areas. Sometimes a road that was passable in 2WD after repair may require a strong 4WD vehicle with a winch within only a year or two. When planning to use unpaved roads, it is always wise to inquire about their condition before setting out.

Most of the birding sites included in this book can be reached with an ordinary 2WD vehicle. For the others, having high clearance is generally more important than having 4WD itself. However, 4WD will provide greater security where roads are muddy, sandy, or slippery. See the individual site accounts for details on local roads.

Panama City. Panama City's main traffic arteries are fairly straightforward, mostly well-signposted, and not too difficult to navigate. However, once you get off these main routes, you can find yourself in a maze of streets, often one-way, and can easily get lost. It's best to stay on the main streets unless you have a good map and a good sense of direction. If you do get lost, keep in mind the major east-west and north-south routes described below and you will eventually find your way.

Several of Panama's major streets have two or more names. The name you will find on a map is often different from the one that everybody actually calls it. This latter is usually the name you will have to use when telling a taxi driver your destination, and the name you will be told when asking directions. Also, because Panama City does not have local mail delivery (you must pick up your mail at a post office box), many people do not know street names, but simply refer to them by local landmarks. In addition, many street and road names in the former Canal Zone have been changed since these areas reverted to Panama in 1999, so these will differ from those used in older guidebooks and maps. In general, in giving directions in this guide we try to use the name used on road signs to avoid confusion. Bear in mind that you may encounter other names for the same street. These are mentioned in the following descriptions.

The main east-west routes through the city are the following, given from south to north:

- Avenida Balboa/Corredor Sur. Av. Balboa runs along the Panama City waterfront from Punta Paitilla to the older part of town in Santa Ana. At its eastern end at Paitilla it leads to the Corredor Sur toll road, which goes to Tocumen Airport and eastern Panama, and to Vía Israel. Vía Israel becomes Vía Cincuentenario, which goes to Panama Viejo and Costa del Este. (After passing Costa del Este, Vía Cincuentenario turns north and eventually intersects Vía España and Av. Ricardo J. Alfaro.) At the western end of Av. Balboa, turning north will lead you to the Bridge of the Americas and western Panama, and to Av. Omar Torrijos and the southern Canal Area. Av. Balboa has two-way traffic. However, during rush hours it switches from three lanes in each direction to four westward (in the morning) or eastward (in the afternoon) and two the other way.

- Avenida Nicanor de Obarrio (universally called Calle Cincuenta, that is, 50th Street). This street branches off Av. Central a little west of the intersection with Av. Federico Boyd, and then runs parallel to it and Vía España a few blocks to the south. It is one-way east.

- Vía España/Avenida Central. Vía España runs through Panama City's main business and hotel district. West of the intersection with Avs. Federico Boyd and Manuel Espinosa Batista, it becomes Av. Central, and runs toward the older part of the city. It is one-way west.

- Avenida de los Mártires. This begins near Ancón in the former Canal Zone and runs westward to Amador and the Bridge of the Americas. It has two-way traffic.

- Vía Simón Bolívar (Carretera Transístmica). At its eastern end this road becomes the transisthmian highway that extends to Colón. It has two-way traffic.

- Avenida Ricado J. Alfaro (universally called Tumba Muerto). This is an extension of Av. Manuel Espinosa Batista (see below), which changes its name as it crosses Vía Simón Bolívar. The road then swings eastward and skirts the north side of the city, running past the El Dorado shopping district. It eventually crosses Vía Bolívar farther out and heads east toward Tocumen Airport. It has two-way traffic.

- Corredor Norte. This toll road runs along the north side of the city from Balboa to the Carretera Transístmica. A branch goes part way across the isthmus, paralleling the Transístmica, and meeting it in the town of Chilibre. This is the preferred route to the Atlantic side. It has two-way traffic.

The main routes that run north-south through the center part of the city are the following:

- Avenida Federico Boyd/Avenida Manuel Espinosa Batista. Av. Federico Boyd begins at Av. Balboa opposite the Miramar Hotel, and runs past Av. Nicanor de Obarrio to the point where Vía España becomes Av. Central. At this point it changes its name to Av. Manuel Espinosa Batista, running up to a traffic circle at the intersection with Vía Bolívar, where it changes its name once again to become Av. Ricardo J. Alfaro.

- Avenida Brasil. This starts at a traffic circle on Vía Israel, running north and intersecting with Av. Nicanor de Obarrio, Vía España, Vía Simón Bolívar, Av. Ricardo J. Alfaro, and the Corredor Norte.

In addition to the Carretera Transístmica and the Corredor Norte that run north from the eastern end of the city, another road runs north from its western end, Avenida Omar Torrijos H. (This was formerly called the Gaillard Highway. The Gaillard Highway, as Calle Gaillard, still retains its old name north of Summit.) It begins near the western end of the Corredor Norte and extends to Gamboa.

Border crossings

There are three road crossings between Panama and Costa Rica: in Chiriquí at Paso Canoa on the Carretera Interamericana and at Río Sereno, and in Bocas del Toro between Sixaola, Costa Rica, and Guabito. It is not permitted to take rental cars between Costa Rica and Panama. There are no road crossings between Panama and Colombia.

Driving in Panama

Driving in Panama, particularly in Panama City, is not for the faint of heart. Drivers can be highly aggressive, and many view traffic laws merely as suggestions. Traffic lights and street signs are in short supply, and taking a left turn or merging into traffic can be adventures in themselves. Buses, known as *diablos rojos* ("red devils"), and taxis often drive particularly erratically, and bear close watching for unexpected maneuvers. Pedestrians sometimes seem to have a death wish, and food vendors are unaccountably fond of pushing or driving their carts the wrong way in the fast lane into oncoming traffic. Highly tuned defensive driving skills are virtually a necessity. Offensive driving skills don't hurt, either.

During rush hours (7:00-9:00 AM and 3:30-6:30 PM) and around noontime traffic jams are common. Sunday mornings and holidays have far less traffic than any other time. (However, several national holidays, especially Carnaval and those in November and December, are marked by parades which may close down major arteries such as Vía España, Av. Central, or Av. Balboa.) If you need to drive into the city during the week, 9:00-11:00 is the best time to avoid traffic.

A visiting birder who plans on driving would therefore be well advised to choose a hotel in the Canal Area (Amador, Balboa, Albrook, or farther north), or, if birding eastern Panama, to the east of Panama City, rather than in the middle of the city itself. This will not only minimize the stress of driving, since you will have to deal with less traffic, but also put you closer to the birding areas. Hotels in these areas are also less noisy and have nicer surroundings.

If you prefer to stay in downtown Panama City, it is better to take taxis, which are cheap and plentiful, when going to restaurants or other places within the city, than to attempt to drive. Some birding sites around Panama City, such as Ancón Hill, Metropolitan Nature Park, and Panama Viejo, can easily be reached by taxi. Other sites in the southern part of the Canal Area, such as Summit, Old Gamboa Road, Plantation Road, Gamboa, and Pipeline Road, can be reached by bus, or more expensively by taxi.

Driving outside Panama City is usually not nearly so stressful, except on the Carretera Transístmica to Colón. There are still hazards to be watched for, however. On two-lane roads, trucks and buses often drive a foot or more into the oncoming lane, something that can be quite disconcerting when meeting one rounding a blind curve. On winding roads, drivers will also often attempt to pass on curves. In rural areas, roads may be shared with bicyclists, inattentive pedestrians, and roaming livestock. Because of these hazards, it is not recommended to drive at night outside cities if you can avoid it.

If you are involved in an accident, stay on the scene and do not move the cars until the traffic police (*tránsito*) arrive. They will fill out a report, required for insurance and legal purposes, and tell you what to do. Contact your insurance or rental-car company as soon as possible.

Rental cars

The main rental-car companies in Panama include Alamo, Avis, Barriga, Budget, Dollar, Hertz, National, and Thrifty; there are also a number of smaller ones. The major companies have desks at Tocumen International Airport, as well as branches in downtown Panama City. Several also have desks at the domestic airport at Albrook. Outside of Panama City, there are branches in David (including the airport), Santiago, Chitré, and Colón. There are currently no rental car offices in Bocas del Toro. At present the cost of renting a car is about $310.00-$380.00 per week (including tax and insurance) for a 2WD sedan and $600.00-$680.00 per week for a 4WD SUV.

Buses

In Panama, buses go almost anywhere that can be reached by road. The main Panama City bus terminal is the Gran Terminal de Transportes at Los Pueblos, near Albrook in the Canal Area. Buses from here go to all major cities and towns. There are also small coaster-type buses that run on the main routes along the Carretera Interamericana, Carretera Nacional, and from David to Changuinola in Bocas del Toro. Small buses run from provincial centers to most small towns in the back country. On the more difficult roads, sometimes converted pickup trucks known as *transportes* serve as buses.

The SACA bus line runs to destinations in the Canal Area, including those along Omar Torrijos/Calle Gaillard as far as Gamboa. The terminal is located near the Plaza Cinco de Mayo in the western part of Panama City. See the section on the Canal Area for more details.

Taxis

Taxis are plentiful and cheap in Panama City. They are not metered, but work on a zone system. (Cabs are supposed to display a zone map, but we have not seen one in years.) Most destinations within Panama City will cost no more than $2.00 (not including an extra charge for each additional person). Radio-dispatched taxis are also available, and can be handy if you are going someplace very early in the morning. These cost slightly more. Your hotel will usually be willing to call one for you. Special tourist taxis that wait outside major hotels are more expensive. Taxis can be hired to take you to places outside Panama City in the southern Canal Area, but this can be expensive. Taxis can also be hired by the day or hour. Negotiate a price in advance, and do not pay until you are finished for the day. Taxi drivers do not expect tips.

Smaller cities and towns, even some quite small ones, also have taxis. You may be able to arrange to have one drop you off at a birding area outside of town, and then pick you up at a pre-arranged time later in the day. Your hotel may be able to help you arrange this. Once again, do not pay until your return to make sure your driver comes back.

Hotels and other accommodations

Accommodations in Panama cover a wide range of comfort and price, from extremely cheap bare-bones *pensiones* to very expensive luxury hotels and lodges. Mid-range hotels, renting from about $20.00-$40.00 per night for a single, generally will provide a private bath with hot water, air-conditioning, and TV, and some also offer internet service, usually in the lobby or other public room. Most places in this range will have reasonably pleasant furnishings and decor, and will often have a restaurant on premises. Some or all of these amenities may be available at some places renting for as little as $15.00 per night, depending on the area. Below this level, most places will offer only cold water and fans, and rooms often have shared baths and are usually rather bare of furnishings besides the bed. These cheaper places are known as *pensiones, residenciales,* or *hostales.* They may charge as little as $7.00-$8.00 in some locations, but generally are $10.00-$12.00. Above $40.00 per night you are generally paying for better decor rather than increased amenities, at least such that would be of interest to birders.

Above this range, Panama now offers a number of lodges or ecohotels that are especially dedicated to birding or nature tourism. These are often located inside the forest itself, or are immediately adjacent to prime birding areas. Although relatively expensive, these can be well worth the cost in making it possible to find the greatest number of birds in the shortest time. Many offer packages that include meals (often featuring gourmet cooking) as well as two guided birding trips per day. Some of these may offer simpler accommodations than conventional hotels in the same price range, though this hardly matters to most birders since they spend little time in their rooms anyway. There are also an increasing number of bed-and-breakfast places in popular tourist destinations, often mid-range in terms of cost, that can provide excellent bases for birders.

Many national parks and some other reserves have bunkhouses or bunkrooms where beds can be rented, usually at a cost of around $5.00 per night; a few with better amenities cost more. Beds are usually bunk beds or cots, with foam pads for mattresses. Although some places provide bedding, to be on the safe side it is better to bring your own or a sleeping bag. There are usually cooking facilities, which may include a gas stove, or sometimes a wood-burning one. (Remember to bring your own matches or lighter; in some cases you may need to bring your own gas tank, so check.) Some basic cooking utensils are usually available, although a large group may want to bring their own to make sure there are enough. A few places may even have refrigerators. Check the individual site accounts for the facilities that may be available there.

There are almost no developed campgrounds in Panama. Some national parks have designated camping areas, but these are usually primitive, and offer no more than a place to get water (if that). Camping fees are generally $5.00 per night.

A tax of 10% is added to hotel bills. We have generally not included this in the prices quoted in this guide. We generally give the price of a single room for one person. In some cases the price may be the same for two people (or more) sharing a room, in others additional charges may apply. Check with the hotel in question for their policies. Meals are not included in the prices quoted unless noted.

Restaurants and food

Panama City has a wide range of restaurants, from top-notch places serving excellent international cuisine to cheap diners and cafeterias. Other major cities and towns usually have several decent restaurants, as well as more basic options. Small towns in the countryside will usually have only one or two very basic restaurants, called *fondas,* and these may close soon after dark. The major US fast food chains, such as McDonald's, Burger King, KFC, Pizza Hut, and others, have restaurants in Panama City, and some of these have branches in smaller cities as well, especially those along the Carretera Interamericana. Chinese and Italian food are popular, and Chinese restaurants or pizzerias can often be found even in small towns.

Standard breakfast items in small roadside restaurants include eggs, which may be ordered fried, scrambled, or boiled, bacon, ham, sausage, fried chicken, fried liver, and sometimes *tasajo,* which is like beef jerky but not as tough. There is also a variety of starchy fried items, including *tortillas* (made of corn and smaller and thicker than their Mexican equivalents), *hojaldras* (a kind of fried bread), *yuca* (cassava or manioc), *carimañolas* (cassava-flour fritters with ground meat inside), *empanadas* (flour turnovers filled with meat or cheese), *patacones* (plantain fritters), and a variety of differently-shaped corn fritters. Coffee is usually served black, with an extra charge for milk.

Smaller restaurants often do not have a fixed menu, but instead offer a small selection of dishes from which you may choose. The main item is usually chicken or beef, sometimes pork or fish, either fried (*frito*) or stewed (*guisado*). This will be accompanied by a large pile of white rice, plus beans or lentils, a slice of fried plantain, and cole slaw or a green salad. *Sancocho,* a rich chicken soup, is also frequently on the menu.

Although there are a few 24-hour restaurants in Panama City, in smaller cities and towns it can be difficult to find a place that is open for breakfast before about 6:00-6:30 AM. Your best bet is to look for places near the bus terminal or public market, which are sometimes open as early as 5:30 AM. In this guide we mention restaurants near each site that may be open early in the morning.

Outside Panama City vegetarian food may be hard to find beyond rice, beans, and salads. Some menu items may have hidden meat, since soup stocks may be made from meat broth and many items may be fried in lard. However, fresh fruits and vegetables can often be obtained from markets and roadside stands.

The standard tip in mid-range and better restaurants in Panama is 10%. Tipping is not necessary in small fondas or other cheap restaurants. Some restaurants will apply an automatic 10% service charge (usually marked "servicio," or sometimes "propina") to the bill. No additional tip is necessary in this case. Check your bill to see whether or not this charge has already been included before tipping.

National parks and protected areas

National parks in Panama are quite different in terms of facilities from those Americans, Canadians, and Europeans may be used to. While some national parks have visitor centers, developed trails, and other amenities, many others do not. Some are quite remote, and nearly inaccessible by road.

There is a daily entry fee of $3.50 for non-residents and $1.50 for Panama residents to national parks and other protected areas. (This fee has recently been raised from $3.00 and $1.00 respectively. At the time of this writing most parks were still charging this lower rate, but will eventually raise it.) A few parks have higher admission charges. Upon entering a park, stop at the park headquarters or the ranger post near the entry to pay the fee and register. However, sometimes no one may be present to collect the fee, especially at the early hours birders are likely to enter. In this case, try to pay on the way out. In some of the more remote areas, there may be no one available to collect fees at any time.

Health

Panama is generally a healthy place for the visitor. Food and water are safe in most of the country. While some tropical diseases are present, few are common, and taking a few simple precautions will minimize your risk.

Panama, unlike many other places in the developing world, has tap water that is safe to drink throughout most of the country, including small towns. The exceptions are Bocas del Toro and more remote parts of Darién. Green salads and other fresh vegetables are also generally safe. For those who prefer it, bottled water is available in most small supermarkets and stores. If you do develop traveler's diarrhea, drink plenty of fluids, especially ones containing sugar or salt. Severe cases may require treatment with an antidiarrheal or antibiotics. (Although traveler's diarrhea is less of a problem in Panama than in many other areas, as a precaution in case you do come down with it you may wish to bring antibiotics for treatment with you to avoid having to find a doctor to issue a prescription while traveling.) Of course, you should never drink untreated water from streams. Even if they appear to be pristine, they may have passed through villages or cattle pastures upstream.

Malaria is rare, now being found principally in Darién, Kuna Yala, and Bocas del Toro. Some visitors may wish to take malarial prophylaxis, but this is not really necessary in the Canal Area and major tourist destinations. It may, however, be warranted for those undertaking extended travel in Darién or other remote areas. The main symptom of malaria is periodic high fevers, accompanied by chills, aches, and other symptoms. Severe cases can be fatal. Chloroquine is adequate for Bocas del Toro, but chloroquine-resistant strains are present in eastern Panama, and medications such as mefloquine, doxycycline, or Malarone are necessary. If you plan to take malarial prophylaxis, it is advisable to consult a physician beforehand, since some of these medications, especially mefloquine, can have unpleasant side effects. It is a good idea in any case to avoid being bitten by mosquitoes. See the section on "animal bites" on p. 54 regarding mosquito deterrents.

Dengue fever is another mosquito-borne disease. It is most prevalent in the Panama City metropolitan area, but the average visitor is unlikely to contract it. Symptoms include fever, headache, muscle aches, and joint pains. The main treatment is analgesics to control the symptoms.

Like malaria, yellow fever is considered a risk only in remoter areas of Bocas del Toro, Kuna Yala, eastern Panamá Province, and Darién. It is not a threat in the Canal Area or western Pacific Panama. Yellow fever vaccination is not required for entry into Panama, but if you plan on extended travel in remote areas it would be good to have it.

Leishmaniasis occurs in forested areas in Panama, and is transmitted by the bites of sandflies. The disease causes skin ulcers, and may have more serious systemic effects. The best defense is to avoid being bitten.

Leptospirosis is acquired by exposure to water contaminated by animal urine. Symptoms are flu-like. It is best to avoid going into bodies of freshwater that may be contaminated.

Hepatitis A is a viral disease that can be acquired from contaminated food, water, or ice, or sometimes by direct contact with affected individuals. Symptoms include fever, jaundice, vomiting, and abdominal pain. The vaccine is safe and highly effective, and might be considered by those planning extended travel in Panama (but most residents don't consider it necessary).

Rabies is transmitted by the bite of mammals, the main vector in Panama being vampire bats and feral cats and dogs. Avoid touching or feeding any mammal regardless of how healthy it may appear, and if bitten, obtain medical advice promptly concerning treatment.

Other diseases, such as Chagas disease, hantavirus, and hepatitis B, are present in Panama but are unlikely to be contracted unless you spend extended periods living in rural or remote areas. Cholera has not been reported from Panama recently and vaccination is not recommended.

Sunburn

The tropical sun can be very intense, particularly around midday, and can cause serious sunburn in less than an hour. This can happen even under hazy or overcast skies. If you will be birding in open areas, and particularly if you will spending time in a boat, it is a good idea to wear a long-sleeved shirt, long pants, and a broad-brimmed hat, and use a good sunscreen with a high SPF rating for exposed parts of your body. A bandana can be handy for shading the back of your neck.

Cuts and abrasions

Even small cuts and abrasions can easily become infected under tropical conditions. It is a good idea to bring adhesive bandages and a topical antibiotic ointment to forestall infection in case of minor injuries. Should a cut or thorn puncture wound get infected you should take an antibiotic effective against staphylococcus. As a precaution, you may wish to obtain one through your doctor before traveling.

Dehydration

In tropical heat you can lose a lot of water through perspiration without knowing it, since it may evaporate very quickly. Drink plenty of fluids, and be sure to carry enough water with you when hiking. If you find you are coming down with headaches or feeling fatigued, it could be due to dehydration, and indicate that you have not been drinking enough fluids.

Animal bites

Mosquito bites are not only annoying, they also can potentially carry disease. Mosquitoes are most active at dawn and dusk, but are also active through the night, and in some habitats, during the day as well. Mosquitoes are not a severe problem in most of Panama, the exception being mangroves, swamps, and certain other humid areas. Biting flies can also be a problem. Tiny sandflies known as *chitras* can be pests in beach or seaside areas in many areas along the coast, and also in a few areas in the mountains such as Santa Fe. Some of these small flies may carry leishmaniasis. At certain times of year, biting horseflies or deerflies can be a nuisance in the forest. To avoid being bitten, wear long sleeve shirts, long pants, and shoes and socks (instead of sandals). Liberally apply a repellant containing DEET (N, N-diethyl-3-methylbenzamide); those with a higher concentration will last longer. Do not sleep with the window open, unless it is screened. If you must sleep in unscreened quarters, be sure to use a mosquito net. Sandflies will require a finer mesh size than mosquitoes.

Ticks and chiggers can be nuisances in some areas, in particular those with a high density of mammals (such as pastures with livestock). Ticks are more common in the dry season, and chiggers in the rainy season. Ticks come in three sizes, from tiny "seed ticks" about the size of a pencil point, to juveniles a few millimeters across, to adults about 5 mm (1/4 inch) in diameter. They wait on vegetation for a host to pass by, climb aboard when one brushes past their perch, and then walk around for some time before embedding their mouthparts and beginning to suck blood. After hiking in areas where ticks are prevalent, check yourself over carefully and you will usually be able to detect and remove any before they embed. If they do embed, they can be removed with a gentle tug. Ticks may not cause much itching until they have been embedded for a day or more. Fortunately those in Panama do not generally carry disease. To avoid ticks, spray your boots, socks, and lower legs well with repellent containing DEET and avoid brushing trailside vegetation. Some prefer to tuck their trousers into their socks to keep ticks on the outside of clothing.

Chiggers are even tinier than seed ticks, and are barely visible to the naked eye. When they bite, their saliva causes the formation of a persistent inflamed spot that itches intensely, and which lasts long after the chigger itself is gone. Chiggers particularly favor moist places, and are very common in cattle pastures. Don't sit down on moist ground, leaf litter, or grass. When biting, they gravitate towards places where clothing is tight, such as socks, beltlines, and bra straps. A severe case of chigger bites can cause a long sleepless night. Cortisone lotion helps stop itching once you've been bitten. As with ticks, the best preventive is liberal spraying of boots, socks, and lower legs with repellent.

Many species of ants sting severely, including the giant black *folofa* ant (*Paraponera*), ants that live in bullhorn acacias, army ants, and the tiny *Azteca* ants that build large carton-like nests on many tree trunks. It is best to avoid ants as much as possible, and watch carefully where you place your feet and hands. Africanized honeybees, the so called "killer bees," are present in Panama and are highly aggressive, but rarely cause problems. Avoid disturbing bees, and avoid killing them since they will send out an alarm scent that will attract their hive mates. If attacked, run away as fast as you can in as direct a path as possible. Some species of wasps in Panama can also sting fiercely. Watch for scorpions on logs with loose bark.

There are several kinds of poisonous snakes in Panama, but these are rarely seen, and you are very unlikely to encounter any unless you are actively looking for them. Nevertheless, you should always be careful where you place your feet. The most dangerous are the vipers, especially the large Bushmaster, which is rare, and the more common Fer-de-lance, called *equis* in Panama because of the X-shaped markings on its back. The small Eyelash Viper can be dangerous due to its habit of perching in vegetation. Coral snakes, brightly marked in black in combination with red and/or yellow, are not aggressive but should certainly not be touched. (And bear in mind that rhymes used to distinguish coral snakes from their non-venomous mimics in the US don't work in Panama.) If bitten by any snake, the best course is to find medical assistance as soon as possible.

Jaguars and pumas are present in Panama but avoid contact with humans as much as possible. Avoid any mammal that does not flee from you, since it could be rabid. If bitten, consult a doctor as soon as possible. There are large crocodiles in Lake Gatún and some other places which are potentially dangerous, and swimming in these areas should be avoided.

Panama City, the Canal Area, and Environs

Panama City and its neighboring areas offer an exceptional range of birding opportunities. More than 700 species have been recorded within an easy day's birding trip (a 90-minute drive each way) of downtown Panama City. Many of the region's prime birding sites are easily accessible: Metropolitan Nature Park, with its parrots, toucans, and manakins, is a short cab ride from the city center, and Soberanía National Park, which holds rich lowland tropical forest frequented by curassows and jaguars, is only a half hour away.

Part of the reason for this rich diversity of birds is the wide range of habitats that are available: dry, moist, and wet lowland forest; foothills and cloud forests; savannas and other grasslands; marshes, lakes, and rivers; and coastal habitats bordering two different oceans. It's little wonder that one of the biggest Big Days on record took place in the Canal Area: on 22 March 1997, Doug Robinson and Dan Christian, starting on Pipeline Road in Soberanía National Park and ending at Cerro Azul, recorded 300 species in 24 hours. On the Atlantic side of the Canal Area, the January 1990 Christmas Bird Count recorded 357 species in one day within the 15-mile diameter count circle.

Common species of the Panama Canal Area

The following species are widespread and common in appropriate habitat in the lowlands on both slopes of the Canal Area.

Great Tinamou
Little Tinamou
Blue-winged Teal
Gray-headed Chachalaca
Least Grebe
Pied-billed Grebe
Brown Booby
Brown Pelican
Neotropic Cormorant
Magnificent Frigatebird
Great Egret
Snowy Egret
Little Blue Heron
Tricolored Heron
Cattle Egret
Green Heron
Striated Heron
Yellow-crowned Night-Heron
Black Vulture
Turkey Vulture
Osprey
White-tailed Kite
Double-toothed Kite
Broad-winged Hawk
Swainson's Hawk
Black Hawk-Eagle
White-throated Crake
Gray-necked Wood-Rail
Purple Gallinule
Common Moorhen
Black-bellied Plover
Collared Plover
Semipalmated Plover
Wattled Jacana
Greater Yellowlegs
Lesser Yellowlegs
Willet
Spotted Sandpiper
Whimbrel
Ruddy Turnstone
Semipalmated Sandpiper
Western Sandpiper

Least Sandpiper
Short-billed Dowitcher
Laughing Gull
Royal Tern
Sandwich Tern
Rock Pigeon
Pale-vented Pigeon
Scaled Pigeon
Short-billed Pigeon
Ruddy Ground-Dove
White-tipped Dove
Orange-chinned Parakeet
Blue-headed Parrot
Red-lored Amazon
Squirrel Cuckoo
Striped Cuckoo
Greater Ani
Smooth-billed Ani
Mottled Owl
Lesser Nighthawk
Common Pauraque
Common Potoo
Short-tailed Swift
Band-rumped Swift
Lesser Swallow-tailed Swift
Rufous-breasted Hermit
Long-billed Hermit
Stripe-throated Hermit
White-necked Jacobin
Violet-crowned Woodnymph
Violet-bellied Hummingbird
Sapphire-throated Hummingbird
Rufous-tailed Hummingbird
Violaceous Trogon
Black-throated Trogon
Slaty-tailed Trogon
Blue-crowned Motmot
Ringed Kingfisher
Belted Kingfisher
Green Kingfisher
Collared Aracari
Keel-billed Toucan
Black-cheeked Woodpecker
Red-crowned Woodpecker
Lineated Woodpecker
Crimson-crested Woodpecker
Plain Xenops
Cocoa Woodcreeper
Black-striped Woodcreeper
Fasciated Antshrike
Barred Antshrike
Western Slaty-Antshrike
Checker-throated Antwren
White-flanked Antwren
Dot-winged Antwren
Dusky Antbird
Chestnut-backed Antbird
Spotted Antbird
Black-faced Antthrush
Brown-capped Tyrannulet
Southern Beardless-Tyrannulet
Yellow-crowned Tyrannulet
Yellow-bellied Elaenia
Ochre-bellied Flycatcher
Paltry Tyrannulet
Southern Bentbill
Common Tody-Flycatcher
Yellow-margined Flycatcher
Ruddy-tailed Flycatcher
Eastern Wood-Pewee
Bright-rumped Attila
Dusky-capped Flycatcher
Panama Flycatcher
Great Crested Flycatcher
Lesser Kiskadee
Great Kiskadee
Boat-billed Flycatcher
Rusty-margined Flycatcher
Social Flycatcher
Streaked Flycatcher
Piratic Flycatcher
Tropical Kingbird
Eastern Kingbird
Fork-tailed Flycatcher
Thrush-like Schiffornis
White-winged Becard
Masked Tityra
Purple-throated Fruitcrow
Golden-collared Manakin
Lance-tailed Manakin
Blue-crowned Manakin
Red-capped Manakin
Yellow-throated Vireo
Red-eyed Vireo
Yellow-green Vireo
Scrub Greenlet
Lesser Greenlet
Black-chested Jay
Gray-breasted Martin
Mangrove Swallow
Northern Rough-winged Swallow
Southern Rough-winged Swallow
Barn Swallow
Black-bellied Wren
Buff-breasted Wren
Plain Wren
House Wren
White-breasted Wood-Wren
Song Wren
Long-billed Gnatwren
Tropical Gnatcatcher
Swainson's Thrush
Clay-colored Thrush
Tropical Mockingbird
Tennessee Warbler
Yellow Warbler
Chestnut-sided Warbler
Bay-breasted Warbler
Black-and-white Warbler
Prothonotary Warbler
Northern Waterthrush
Kentucky Warbler
Mourning Warbler
Canada Warbler
Rosy Thrush-Tanager
Gray-headed Tanager
White-shouldered Tanager
Red-throated Ant-Tanager
Summer Tanager
Crimson-backed Tanager
Blue-gray Tanager
Palm Tanager
Plain-colored Tanager
Golden-hooded Tanager
Blue Dacnis
Green Honeycreeper
Red-legged Honeycreeper
Blue-black Grassquit
Variable Seedeater
Yellow-bellied Seedeater
Lesser Seed-Finch
Orange-billed Sparrow
Black-striped Sparrow
Streaked Saltator
Buff-throated Saltator
Rose-breasted Grosbeak
Blue-black Grosbeak
Red-breasted Blackbird
Great-tailed Grackle
Orchard Oriole
Yellow-backed Oriole
Baltimore Oriole
Yellow-billed Cacique
Scarlet-rumped Cacique
Yellow-rumped Cacique
Crested Oropendola
Chestnut-headed Oropendola
Yellow-crowned Euphonia
Thick-billed Euphonia
Fulvous-vented Euphonia

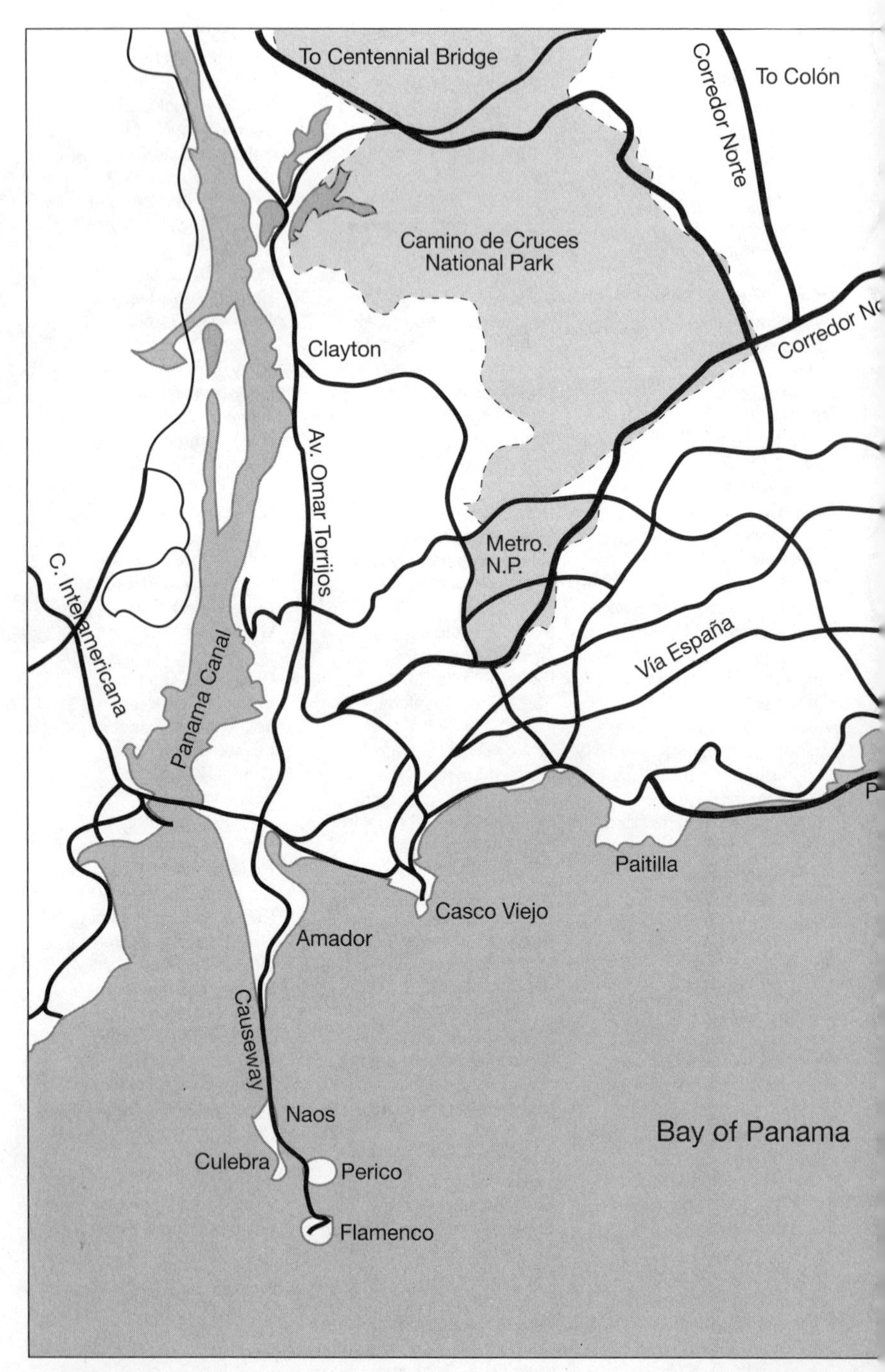

Panama City. Metro N.P. = Metropolitan Nature Park.

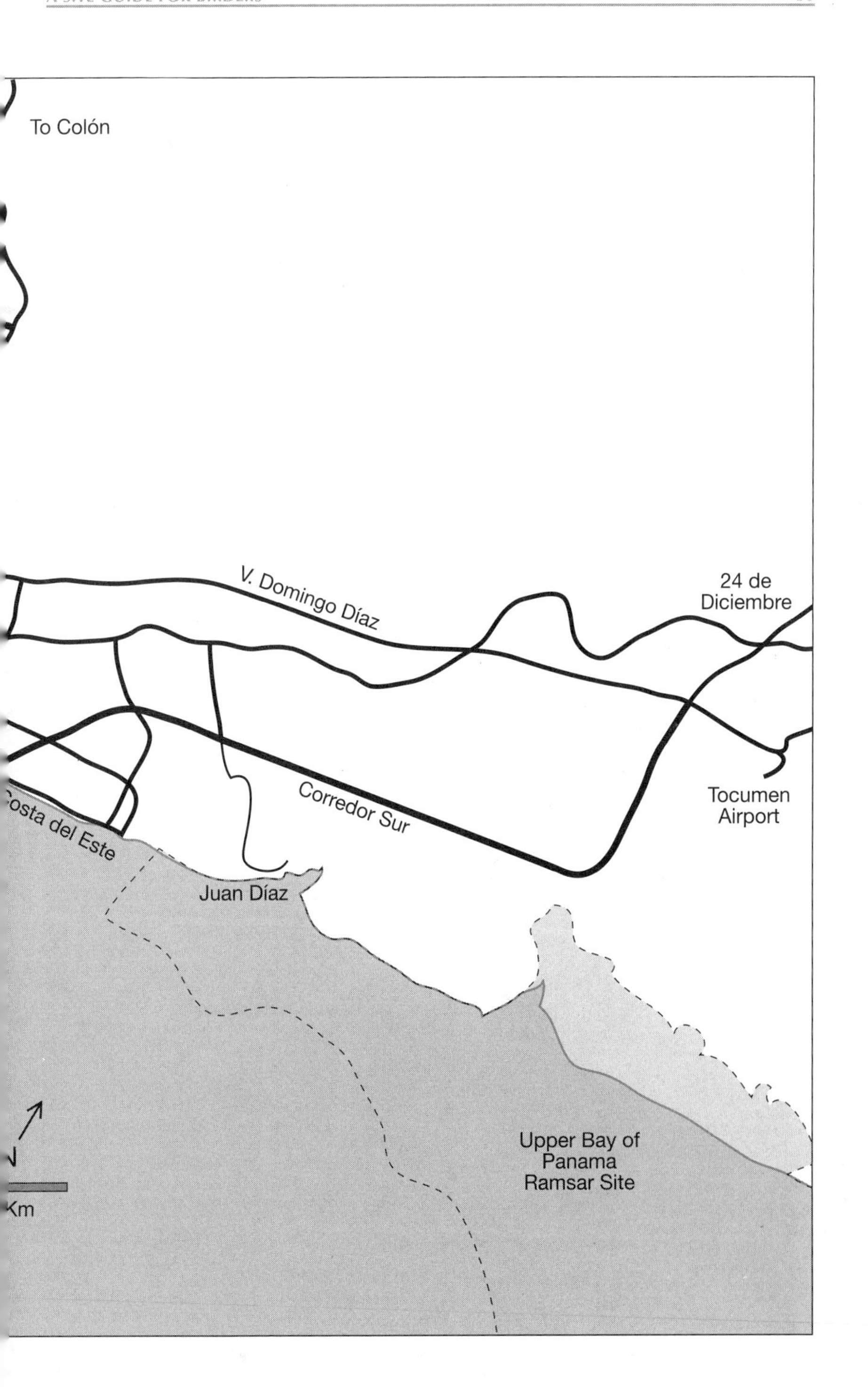
To Colón
V. Domingo Díaz
24 de Diciembre
Corredor Sur
Tocumen Airport
Juan Díaz
Upper Bay of Panama Ramsar Site
Km

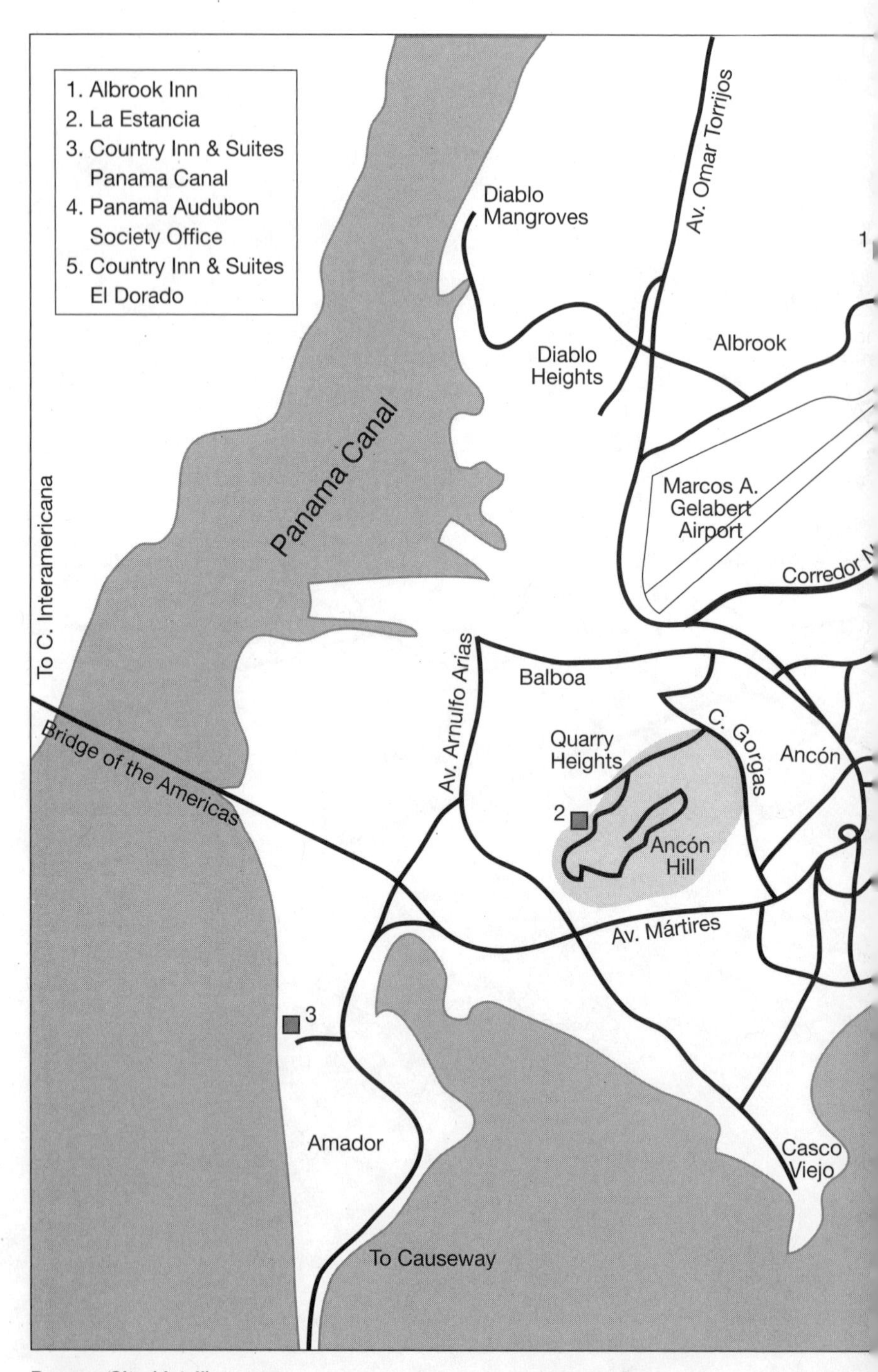

Panama City (detail)

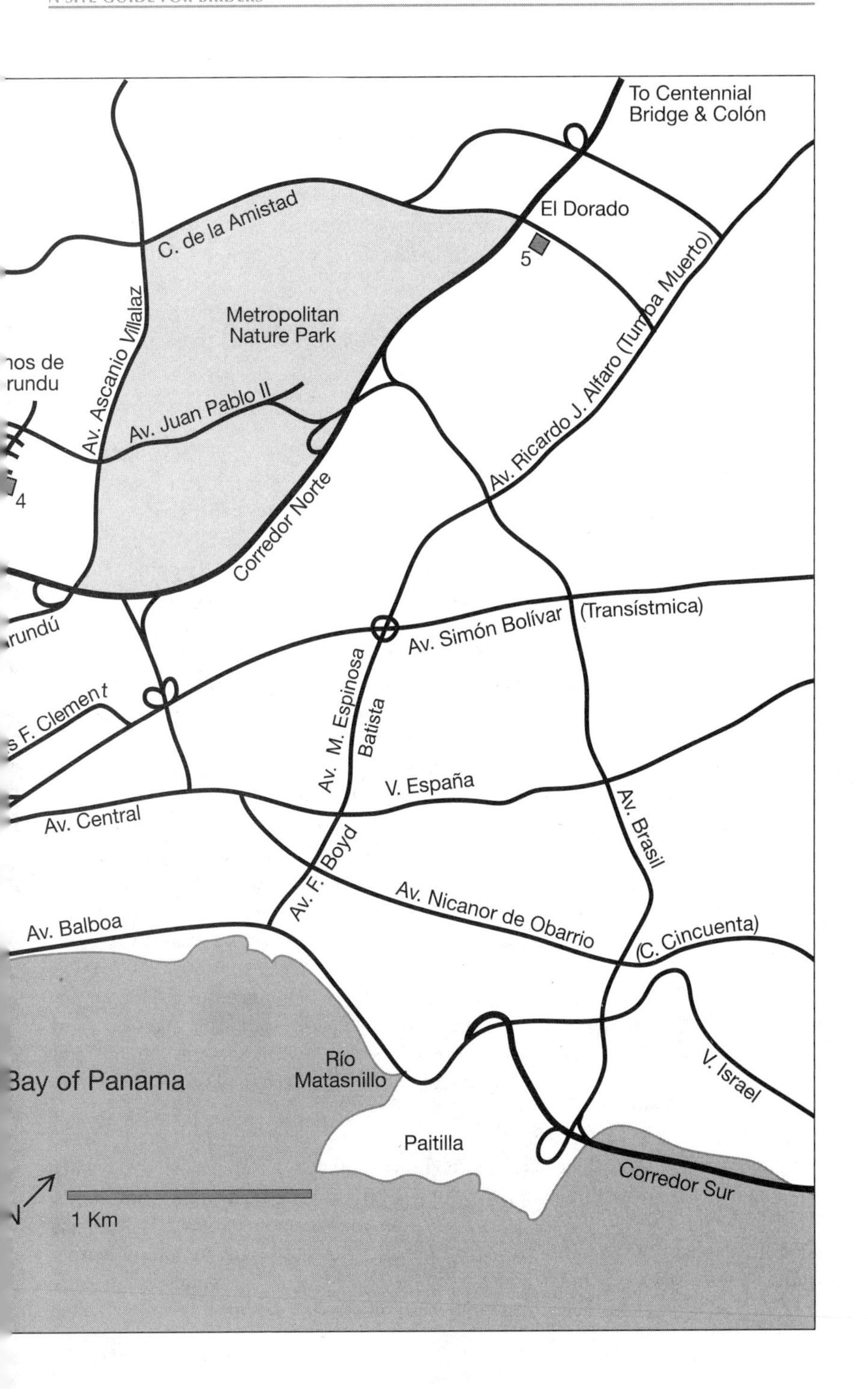

To Centennial Bridge & Colón
El Dorado
5
C. de la Amistad
Metropolitan Nature Park
Av. Ascanio Villalaz
Av. Juan Pablo II
Av. Ricardo J. Alfaro (Tumba Muerto)
Corredor Norte
4
Av. Simón Bolívar (Transístmica)
Av. M. Espinosa Batista
V. España
Av. Central
Av. F. Boyd
Av. Brasil
Av. Nicanor de Obarrio
(C. Cincuenta)
Av. Balboa
Río Matasnillo
Paitilla
V. Israel
Corredor Sur
1 Km

Panama City

Panama City itself offers many species of interest to visiting birders from the temperate zone, especially around city parks and in residential neighborhoods with tree-lined streets. Birds that can often be seen within the city include Yellow-headed Caracara, Ruddy Ground-Dove, Orange-chinned Parakeet, Yellow-crowned Amazon, Garden Emerald, Rufous-tailed Hummingbird, Red-crowned Woodpecker, Barred Antshrike, Tropical Kingbird, Great Kiskadee, Gray-breasted Martin, House Wren, Clay-colored Thrush, Tropical Mockingbird, Crimson-backed, Blue-gray, and Palm Tanagers, Red-legged Honeycreeper, Streaked Saltator, Yellow-crowned Euphonia, and the ubiquitous Great-tailed Grackle and Black and Turkey Vultures.

Among the better areas in and near the city are Parque Recreativo Omar in the San Francisco district, and the Ancón and Balboa areas of the former Canal Zone. Ancón and Balboa feature many back streets planted with native and exotic trees and shrubs. The area around Ancón Hill is particularly good, especially the section of Calle Gorgas (Gorgas Road) that runs from just above Gorgas Hospital and the Palacio de Justicia (Supreme Court Building) to the Canal Administration Building. Ancón Hill is covered with dry forest that hosts many of the more widespread species of this habitat, and also offers the best vantage point for observing the huge flocks of migrating raptors, including Turkey Vulture and Broad-winged and Swainson's Hawks, that pass above the city in October and November.

Another part of the former Canal Zone that offers birding possibilities is the Amador area and the islands at the end of the Causeway (Calzada de Amador) near the mouth of the Panama Canal. Grassy areas of Amador are the best place to find Cattle Tyrant in Panama, and the species has bred for several years around the old post office (more recently ARI headquarters). Elegant Terns and Black Skimmers have sometimes been seen roosting on the shore on the other side of the seawall behind the parking area here. Culebra Island on the Causeway is the site of the Marine Education and Exhibition Center of the Smithsonian Tropical Research Institute. It has a small patch of dry forest where some of the common birds of this habitat can be seen. Many seabirds can be seen from the Causeway and its islands, including Brown Pelican, Neotropic Cormorant, Magnificent Frigatebird, Brown Booby, Laughing Gull, Sandwich Tern, and on rare occasions Blue-footed Booby.

Within the city, Brown Pelicans can often be seen plunge-diving for fish off Avenida Balboa along the waterfront, and Neotropic Cormorants, Magnificent Frigatebirds, Laughing Gulls, and Royal Terns are also common. The end of Avenida Balboa near Paitilla, opposite the Multicentro shopping mall, often has concentrations of herons, gulls, terns, other seabirds, and some shorebirds near the outlet of the Río Matasnillo. Check for Franklin's and Ring-billed Gulls among the much more numerous Laughing Gulls.

Two areas at the edge of the city proper offer even better birding. A good selection of the birds of deciduous and second-growth forest can be found in Metropolitan Nature Park, and Panama Viejo and Costa del Este, on the coast at the eastern edge of the city, are excellent for migratory shorebirds and other coastal species. These sites are discussed in detail on pp. 65-69 and 69-72 respectively.

A note on driving directions for Panama City and the Canal Area. For consistency, directions are given here from the intersection of Vía España with Av. Federico Boyd in downtown Panama City, near the El Carmen church and the El Panama Hotel. See pp. 46-47 in the introduction for a general description of the major streets in Panama City. For the benefit of drivers unfamiliar with Panama City and its traffic, the directions given are those that minimize left turns, crossing uncontrolled intersections, and other difficult driving maneuvers. In some cases there are shorter or more direct routes available to those who know the city. We would, however, caution against attempting to work out your own route based on what seems to be possible on a street map!

Getting to Parque Recreativo Omar. Parque Recreativo Omar is located off Av. Belisario Porras. From the intersection of Vía España with Av. Federico Boyd, take Federico Boyd south for two blocks and take a left onto Av. Nicanor de Obarrio (C. Cincuenta) at the traffic light. Follow Av. Obarrio for 3.5 km, and take a left onto Av. Porras at the sign for that street. The main entrance to the park is in 700 m on the right, at a gate with yellow pillars.

Getting to Ancón and Balboa. To reach Ancón and Balboa from the intersection of Vía España and Av. Federico Boyd, go south on Federico Boyd six blocks until you come to Av. Balboa, opposite the large Miramar Hotel. Take a right onto Balboa. After 2.1 km, near the end of Balboa, turn right just before an overpass, next to a Shell gas station, prominently signposted for "Terminal Nacional Los Pueblos/Albrook." After 300 m (get into the left lane as soon as you can), you will come to a traffic light. Continue straight, and after another short block you will come to a second light. Just beyond this the road divides; take the left lane going up onto a traffic overpass over Av. de los Mártires, signposted for "Puente de las Américas" and "Amador." This is the on-ramp for Av. Mártires headed west. After merging onto Av. Mártires, after 400 m turn right at the first traffic light. This is Calle Gorgas (Gorgas Road). Continue straight past Gorgas Hospital and the Palacio de Justicia (Supreme Court Building), which are the large buildings on your left. The tree-lined part of C. Gorgas begins just past the Palacio de Justicia.

To reach Ancón Hill, continue for 500 m beyond the Palacio de Justicio. At a sharp bend, just as the street begins to descend towards the Canal Administration Building, take a left. After 200 m, you will see a small guard kiosk (usually unmanned), signposted "Quarry Heights." Take a left just past the kiosk, signposted for the conservation organization ANCON. At the top of the street is a T-junction where you go right. (The headquarters of ANCON, Panama's largest conservation organization, is at the left at this intersection.) The road to the lookout at the top of the hill is open from 7:00 AM- 6:00 PM Monday to Friday. A guard near the beginning of the road will advise you when it is safe to go up; traffic is one-way due to the narrowness of the road. The road can also be walked; it is about 20-30 minutes to the top.

Balboa is most easily reached by following the directions for Ancón above, but instead of turning off Av. Mártires at C. Gorgas, stay on it for another 1.0 km, turning off to the right at a prominent intersection onto Avenida Arnulfo Arias Madrid (formerly Balboa Road, not to be confused with Avenida Balboa along the Panama City waterfront). Turning off at either the first or second right, after about 400 m, will bring you onto tree-lined streets.

Getting to Amador and the Causeway. To reach Amador and the Causeway and its islands, follow the directions for Ancón above. However, instead of turning off onto Calle Gorgas, continue on Av. de los Mártires for another 1.4 km past the traffic light, when you will see the signposted exit lane on the right for Amador. (Make sure you are in the right lane well before this point since it can be easy to miss as you come around a curve, and the next chance to turn around is on the other side of the canal.) In 900 km, just past a sign marking the entrance for the Country Inn and Suites and TGI Friday's, you will see a large cream-colored building on the left, which is the old post office. (More recently it has been the headquarters for the Interoceanic Regional Authority, ARI, but since this is closing the building is likely to be converted to something else.) Carry on a short distance to where you can make a U-turn to return to this building. Just beyond the building as you come back is the entrance to the public parking lot. You can park here (you may have to talk the guard into letting you in) to look for roosting sea and shorebirds beyond the seawall, or look for Cattle Tyrants on grassy areas near the building (which you can do from the sidewalk if the guard won't let you in).

To reach the Causeway and the islands, continue on the main road past the old post office instead of making the U-turn. The first island on the Causeway is Naos, which mainly contains marine laboratories of the Smithsonian Tropical Research Institute. Just past these buildings on the right is the sign and entrance to the Smithsonian's Marine Exhibition Center at Culebra (ph 212-8793; puntaculebra@si.edu; www.stri.org/english/visit_us/culebra/). The center is open Tuesday-Friday 1:00 PM to 5:00 PM; Saturday and Sunday 10:00 AM to 6:00 PM; closed Mondays; during school holidays open Tuesday to Sunday 10:00 AM to 6:00 PM. Admission is $2.00 for adults, $0.50 for children under 12, and $1.00 for retirees. The islands beyond Naos and Culebra are now mainly occupied by restaurants, shops, and other tourism developments, but offer ample parking and places to look out for seabirds.

Getting to Avenida Balboa. To reach Av. Balboa from the intersection of Vía España and Federico Boyd, going south on Federico Boyd for six blocks brings you to the intersection with Balboa. For the best viewpoint for birds, at this intersection turn right (a left turn is not permitted) and immediately get into the left lane. In one block you can make a U-turn to get on Balboa heading east. In 1.0 km, at the end of the waterfront promenade, you can turn off to the right for a parking area overlooking the outlet of the Río Matasnillo where birds often congregate.

Metropolitan Nature Park

Metropolitan Nature Park (265 ha/655 acres) is the only tropical forest park located within a major metropolitan area in Latin America. It includes approximately 190 ha (470 acres) of semideciduous lowland tropical forest about 70 years old, as well as areas of grass and younger second growth. Receiving much less rainfall than areas farther north in the Canal Area, many of the trees in the park shed their leaves during the four-month dry season.

The park is an excellent place to find birds of dry and second-growth forest, as well as the more widespread forest species of the Canal Area. Among the dry-forest or Pacific-slope specialties found here are the Panama endemic **Yellow-green Tyrannulet** (uncommon), Yellow-crowned Amazon, Pheasant Cuckoo (rare), Sepia-capped Flycatcher (uncommon), Lance-tailed Manakin, Rufous-breasted and Rufous-and-white Wrens, Rosy Thrush-Tanager, and Red-crowned Ant-Tanager. Gray-headed Chachalaca, Orange-chinned Parakeet, Blue-headed Parrot, Red-lored Amazon, Short-tailed Hawk, Black Hawk-Eagle, Keel-billed Toucan, Collared Aracari, Violaceous, Black-throated, and Slaty-tailed Trogons, White-necked Puffbird, Blue-crowned Motmot, Golden-collared Manakin, Western Slaty-Antshrike, Long-billed Gnatwren, and many others can also be found.

The park has five well-marked trails: Los Momótides (The Motmots), Los Caobos (The Mahogany Trees), El Roble (named after a pink-flowered tree), Mono Tití (Titi Monkey, the local name for Geoffrey's Tamarin, a small monkey that is common in the park), and La Cienaguita (The Little Marsh, named for a small marshy area that is present near the start of the trail during the rainy season). The first three trails start at the park's Visitor Center on Avenida Juan Pablo II. The easy *Los Momótides* (0.9 km/0.6 mi) is a loop, mostly level, through scrubby forest. It begins opposite the entrance to the Visitor Center parking lot across Juan Pablo II. *Los Caobos* (0.9 km/0.6 mi) makes a moderate climb through forest to a lookout, runs along a low ridge for its middle section, and then descends to meet El Roble. It begins a short distance up a gravel road to the left of the Visitor Center. *El Roble* (0.7 km/0.4 mi) parallels the main road past a nursery area and the far end of Los Caobos to reach the start of La Cieneguita and Mono Tití. It begins to the right of the visitor center. Follow the concrete path and the sign marked "sendero."

At the end of El Roble are a small guard booth, a parking area, and a picnic area on Avenida Juan Pablo II. From here the *Mono Tití Trail* (1.1 km/0.7 mi) follows a narrow asphalt road (closed to traffic) to climb through fairly open dry forest to a lookout (*mirador*) offering views of Panama City, the Panama Canal, and Camino de Cruces National Park. *La Cienaguita* (1.1 km/0.7 mi) also ascends through forest, eventually reaching a relatively level ridge and joining Mono Tití near the mirador.

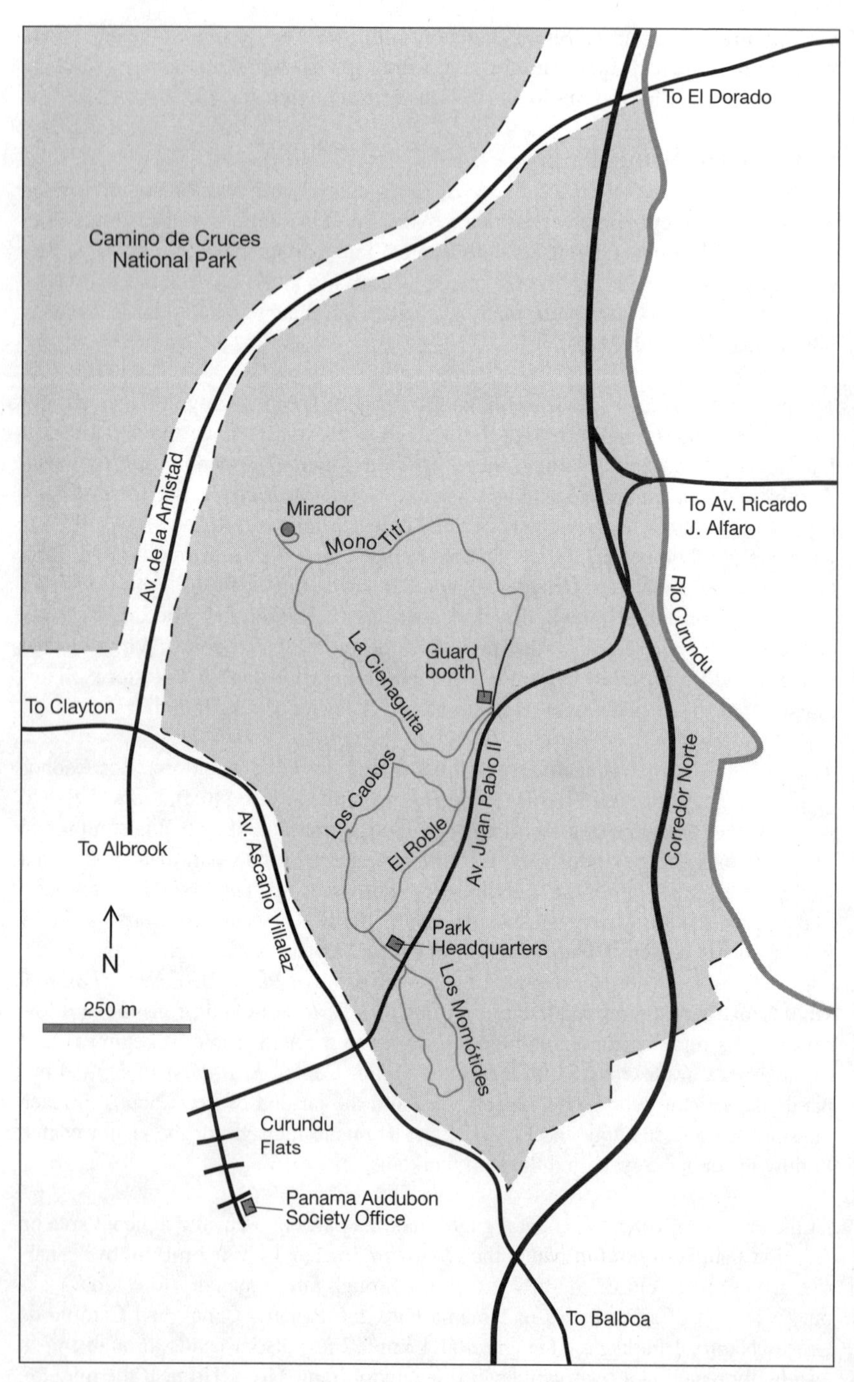

Metropolitan Nature Park

La Cienaguita is a self-guiding nature trail for which a booklet can be purchased at the Visitor Center. These two trails can conveniently be walked as a loop, ascending one and descending by the other. During the migration season in October and November the Mirador can be a good spot from which to watch the massive migrations of Turkey Vulture and Broad-winged and Swainson's Hawk that pass over the city (although Ancón Hill is somewhat better for this). Near the start of Mono Tití is a construction crane used by the Smithsonian Tropical Research Institute to conduct research on the tropical forest canopy, but this area is not open to the public.

Getting to Metropolitan Nature Park. From the intersection of Vía España with Av. Federico Boyd, take Av. Manuel Espinosa Batista (which is the continuation northward of Av. Federico Boyd) north. At 1.0 km from Vía España, you come to a traffic circle at an underpass below Vía Simón Bolívar (Transístmica). Continue straight through the traffic circle; Av. Manuel Espinosa Batista here changes its name and becomes Vía Ricardo J. Alfaro (Tumba Muerto). At 700 m past this traffic circle, just after passing under a pedestrian overpass, take a left at the traffic signal, signposted for Av. Juan Pablo II. After 600 m you will reach the edge of the park; follow the signs to go left toward Curundu. After another 900 m you will come to a sign on the right that says "Mirador," where you can pull off to park at the start of the Mono Tití and La Cienaguita Trails. Continue on another 700 m to find the park's Visitor Center, on the right off Juan Pablo II, and the start of the Los Momótides, Los Caobos, and El Roble Trails. (Go slow since the entrance is concealed by trees and easy to miss.)

The Park can also be reached coming from Av. Omar Torrijos. Follow the directions for Ancón on p. 63 After turning off Av. Balboa, when you come to the second traffic light, instead of going left onto the overpass, bear right, signposted for Albrook and Vía Simón Bolívar, and merge onto Av. Mártires going north. The turn for the park is 500 m past the next underpass; take a right just before the traffic light, signposted for Av. Ascanio Villalaz. Follow this for 1.9 km, and take a right, signposted for Av. Juan Pablo II. The park Visitor Center is on the left 200 m beyond this turn. If you don't have a vehicle, a taxi to the Visitor Center from downtown should cost about $1.50. For your return, taxis can generally be flagged down on Juan Pablo II, though traffic may be a bit sparse on Sundays or other holidays.

Fees to enter the park are $1.00 for Panama residents, $2.00 for foreign visitors, and $3.00 for visitors with a guide. Guides can be arranged by calling the Visitor Center (ph 232-6723, 232-5552) a day in advance. Fees can be paid at the visitor center from 7:00 AM (sometimes earlier) to 4:00 PM. If you enter the park before someone is available to collect the fee, pass by the Visitor Center to pay on your way out.

Bird list for Metropolitan Nature Park. * = vagrant; R = rare.

Little Tinamou
Gray-headed Chachalaca
Brown Pelican
Neotropic Cormorant
Magnificent Frigatebird
Great Egret
Cattle Egret
Yellow-crowned Night-Heron
Wood Stork R
Black Vulture
Turkey Vulture
Osprey
Gray-headed Kite
White-tailed Kite
Double-toothed Kite

Mississippi Kite
Tiny Hawk*
Bicolored Hawk*
Crane Hawk*
Gray Hawk
Mangrove Black-Hawk
Roadside Hawk
Broad-winged Hawk
Short-tailed Hawk
Swainson's Hawk
White-tailed Hawk*
Zone-tailed Hawk
Black Hawk-Eagle
Ornate Hawk-Eagle R
Collared Forest-Falcon
Crested Caracara
Yellow-headed Caracara
American Kestrel
Merlin
Bat Falcon
Peregrine Falcon
Spotted Sandpiper
Rock Pigeon
Pale-vented Pigeon
Scaled Pigeon
Ruddy Ground-Dove
Blue Ground-Dove
White-tipped Dove
Gray-chested Dove
Orange-chinned Parakeet
Brown-hooded Parrot R
Blue-headed Parrot
Red-lored Amazon
Yellow-crowned Amazon
Black-billed Cuckoo
Yellow-billed Cuckoo
Mangrove Cuckoo R
Squirrel Cuckoo
Pheasant Cuckoo R
Smooth-billed Ani
Barn Owl
Tropical Screech-Owl
Vermiculated Screech-Owl
Mottled Owl
Black-and-white Owl
Striped Owl
Lesser Nighthawk
Common Nighthawk
Common Pauraque
Rufous Nightjar
Great Potoo
Common Potoo
Chimney Swift
Short-tailed Swift
Band-rumped Swift
Lesser Swallow-tailed Swift
Band-tailed Barbthroat
Long-billed Hermit
Stripe-throated Hermit
Scaly-breasted Hummingbird
White-necked Jacobin
Black-throated Mango
Garden Emerald
Violet-crowned Woodnymph
Violet-bellied Hummingbird
Sapphire-throated Hummingbird
Snowy-bellied Hummingbird
Rufous-tailed Hummingbird
White-vented Plumeleteer
Purple-crowned Fairy
Violaceous Trogon
Black-throated Trogon
Slaty-tailed Trogon
Blue-crowned Motmot
Ringed Kingfisher
Green Kingfisher
American Pygmy Kingfisher
White-necked Puffbird
Black-breasted Puffbird
Collared Aracari
Keel-billed Toucan
Black-cheeked Woodpecker
Red-crowned Woodpecker
Yellow-bellied Sapsucker R
Lineated Woodpecker
Crimson-crested Woodpecker
Plain Xenops
Scaly-throated Leaftosser R
Plain-brown Woodcreeper
Olivaceous Woodcreeper
Cocoa Woodcreeper
Streak-headed Woodcreeper
Fasciated Antshrike
Barred Antshrike
Western Slaty-Antshrike
Checker-throated Antwren
Dot-winged Antwren
Dusky Antbird
White-bellied Antbird
Chestnut-backed Antbird
Spotted Antbird
Black-faced Antthrush
Brown-capped Tyrannulet
Southern Beardless-Tyrannulet
Mouse-colored Tyrannulet R
Yellow Tyrannulet
Yellow-crowned Tyrannulet
Forest Elaenia
Greenish Elaenia
Yellow-bellied Elaenia
Lesser Elaenia
Ochre-bellied Flycatcher
Sepia-capped Flycatcher
Yellow-green Tyrannulet
Paltry Tyrannulet
Northern Scrub-Flycatcher
Pale-eyed Pygmy-Tyrant
Southern Bentbill
Common Tody-Flycatcher
Olivaceous Flatbill
Yellow-olive Flycatcher
Yellow-margined Flycatcher
Royal Flycatcher
Ruddy-tailed Flycatcher
Olive-sided Flycatcher
Western Wood-Pewee
Eastern Wood-Pewee
Tropical Pewee
Acadian Flycatcher
Bright-rumped Attila
Dusky-capped Flycatcher
Panama Flycatcher
Great Crested Flycatcher
Lesser Kiskadee
Great Kiskadee
Boat-billed Flycatcher
Rusty-margined Flycatcher
Social Flycatcher
Streaked Flycatcher
Sulphur-bellied Flycatcher
Piratic Flycatcher
Tropical Kingbird
Eastern Kingbird
Gray Kingbird
Fork-tailed Flycatcher
White-winged Becard
Masked Tityra
Black-crowned Tityra
Blue Cotinga
Golden-collared Manakin
Lance-tailed Manakin
Red-capped Manakin
Yellow-throated Vireo
Red-eyed Vireo
Yellow-green Vireo
Scrub Greenlet
Golden-fronted Greenlet
Lesser Greenlet
Green Shrike-Vireo
Gray-breasted Martin
Northern Rough-winged Swallow
Southern Rough-winged Swallow
Sand Martin
Cliff Swallow
Barn Swallow
Black-bellied Wren
Rufous-breasted Wren
Rufous-and-white Wren

Buff-breasted Wren
Plain Wren
House Wren
White-breasted Wood-Wren
Southern Nightingale-Wren
Long-billed Gnatwren
Tropical Gnatcatcher
Veery
Gray-cheeked Thrush
Swainson's Thrush
Wood Thrush
Clay-colored Thrush
Gray Catbird
Tropical Mockingbird
Golden-winged Warbler
Tennessee Warbler
Yellow Warbler
Chestnut-sided Warbler
Magnolia Warbler
Cape May Warbler R
Blackburnian Warbler
Bay-breasted Warbler
Blackpoll Warbler R
Cerulean Warbler R
Black-and-white Warbler
American Redstart
Prothonotary Warbler
Worm-eating Warbler R
Ovenbird
Northern Waterthrush
Kentucky Warbler
Mourning Warbler
Canada Warbler
Buff-rumped Warbler
Bananaquit
Rosy Thrush-Tanager
Gray-headed Tanager
White-shouldered Tanager
Red-crowned Ant-Tanager
Red-throated Ant-Tanager
Hepatic Tanager*
Summer Tanager
Scarlet Tanager
Crimson-backed Tanager
Blue-gray Tanager
Palm Tanager
Plain-colored Tanager
Golden-hooded Tanager
Blue Dacnis
Green Honeycreeper
Red-legged Honeycreeper
Blue-black Grassquit
Slate-colored Seedeater R
Variable Seedeater
Yellow-bellied Seedeater
Ruddy-breasted Seedeater
Orange-billed Sparrow
Black-striped Sparrow
Streaked Saltator
Buff-throated Saltator
Rose-breasted Grosbeak
Blue-black Grosbeak
Blue Grosbeak
Great-tailed Grackle
Bronzed Cowbird
Giant Cowbird
Orchard Oriole
Yellow-backed Oriole
Baltimore Oriole
Yellow-billed Cacique
Scarlet-rumped Cacique
Yellow-rumped Cacique
Chestnut-headed Oropendola
Yellow-crowned Euphonia
Thick-billed Euphonia
Fulvous-vented Euphonia
White-vented Euphonia
Lesser Goldfinch
House Sparrow
Tricolored Munia*

Panama Viejo and Costa del Este

Panama Viejo, more properly Panama La Vieja, is the ruins of the original city of Panama, which was founded in 1519 and sacked by the pirate Henry Morgan in 1671. For birders the site offers an easy place to view migratory shorebirds and other coastal species. An interpretive trail starts at the main visitor center and goes east along the coast along the edge of the mudflats and scattered mangroves to the ruins, and is a good vantage point for observing the birds. You can also bird the grassy area that extends from the Cathedral in the middle of the ruins to the shore. Common species include Black-bellied and Semipalmated Plovers, Whimbrel, Willet, Western and Semipalmated Sandpipers, Short-billed Dowitcher, Brown Pelican, Neotropic Cormorant, Laughing Gull, and Royal Tern. Red Knot, Marbled Godwit, Barn Owl, Merlin, and Peregrine Falcon are less common but sometimes seen.

Costa del Este is a new upscale housing development located just east of Panama Viejo. It was originally an area of marshes and mangroves, and later became the site of Panama City's main garbage dump, now closed. The area first became accessible to birders in 1996 when construction of the housing development began. Since then the marshes have been drained and much of the area has been built up, but the mudflats offshore still continue to attract very large numbers of shorebirds and other coastal species.

The best place to observe shorebirds is along the 2-km-long seawall and promenade that fronts the mudflats of Panama Bay. While birds are present along its

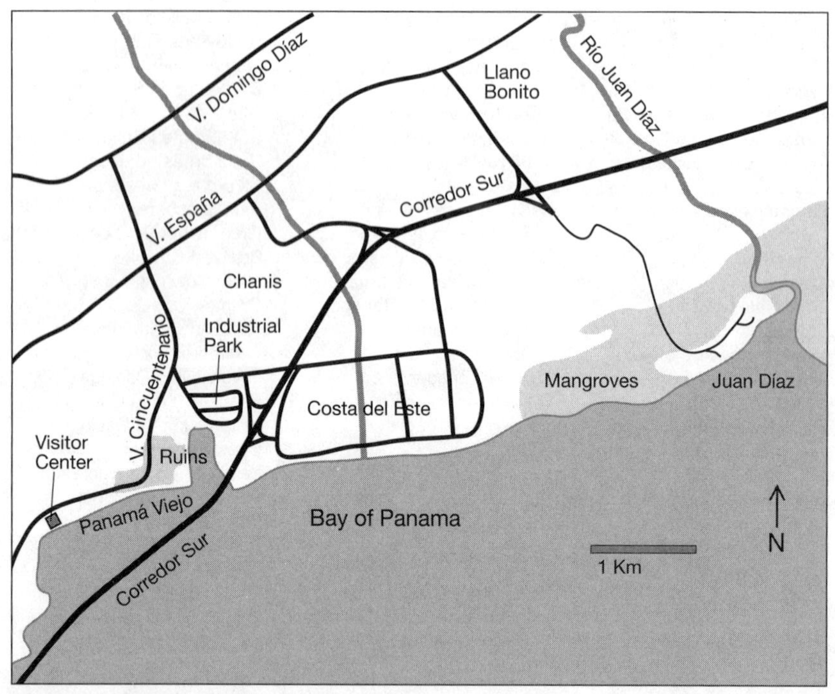

Panama Viejo, Costa del Este, and Juan Díaz

length, the greatest concentrations are found near the mouths of rivers and streams, which are near the eastern and western ends of the promenade, and 1.0 km from the western end. Besides the species mentioned for Panama Viejo, additional species include Cocoi Heron (as well as the more widespread herons), Roseate Spoonbill, Wood Stork, Black-necked Stilt, Collared and Wilson's Plover, Ruddy Turnstone, Least Sandpiper, Greater Yellowlegs, Franklin's, Ring-billed, and Herring Gulls, and Gull-billed, Common, and Least Terns. Rarer vagrants from the north have included Ruff, Lesser Black-backed Gull, and Arctic Tern. Strong El Niño years sometimes bring seabirds from the west coast of South America, including Gray, Gray-hooded, and Kelp Gulls.

A number of grassland-frequenting and marshland species were sometimes seen in the area in the past, but recent construction has caused it to become much less suitable for them. However, it is possible that some of these may still occur at times. These include Yellow-headed Vulture, Glossy Ibis, American Golden-Plover, Buff-breasted Sandpiper, Upland Sandpiper, Baird's Sandpiper, and Pied Water-Tyrant.

Getting to Panama Viejo and Costa del Este. From the intersection of Vía España and Av. Federico Boyd, go south two blocks on Federico Boyd and take a left onto Av. Nicanor de Obarrio (Calle Cincuenta). Continue on Obarrio 3.7 km, turning left onto Vía Cincuentenario, which is 200 m past the well-marked intersection of Av. Obarrio and Vía Porras. Follow Cincuentenario for 1.7 km, where you will see the Panama Viejo Visitor Center on your right just after passing a restored colonial bridge. You can park either at the Visitor Center, or at the main part of the ruins 800 m farther along Vía Cincuentenario where it makes a sharp bend to the left. From either location, you can follow the interpretive trail that connects the Visitor Center with the ruins to see birds along the coast and on the mudflats. Cost to visit the site is $2.00 for adults, and $0.50 for students, which can be paid at the Visitor Center. (It is $3.00 for adults and $0.50 for students for admission to the Visitor Center plus the ruins.)

For Costa del Este, continue on Vía Cincuetenario for another 900 m past the point where the road takes a sharp left in front of the main ruins. Turn to the right at the sign for Costa del Este, across from a Texaco station. Continue for 100 m and turn left (going straight goes into a gated industrial park). After 900 m you will come to an underpass below the Corredor Sur. Take the right immediately past the underpass, signposted "Tocumen Cuota." In 700 m you will come to Av. Paseo del Mar, the seaside promenade. Turning right will bring you to the western end of the promenade in 300 m. You can then work your way back along the promenade to its eastern end.

Bird list for Costa del Este. Most of these species also can be found at Panama Viejo. * = vagrant; R = rare.

Black-bellied Whistling-Duck
Blue-winged Teal
Cinnamon Teal R
Northern Shoveler R
Brown Booby
Brown Pelican
Neotropic Cormorant
Magnificent Frigatebird
Great Blue Heron
Cocoi Heron
Great Egret
Snowy Egret
Little Blue Heron
Tricolored Heron
Green Heron
Cattle Egret
Black-crowned Night-Heron
Yellow-crowned Night-Heron
White Ibis
Glossy Ibis
Roseate Spoonbill R
Wood Stork R
Black Vulture
Turkey Vulture
Lesser Yellow-headed Vulture R
Osprey
White-tailed Kite
Mississippi Kite R
Mangrove Black-Hawk
Savanna Hawk R
Broad-winged Hawk
Swainson's Hawk
Zone-tailed Hawk R
Crested Caracara
Yellow-headed Caracara
American Kestrel
Merlin
Peregrine Falcon
Purple Gallinule
Southern Lapwing
Black-bellied Plover
American Golden-Plover R
Collared Plover
Wilson's Plover
Semipalmated Plover
Killdeer
Black-necked Stilt
Wattled Jacana
Greater Yellowlegs
Lesser Yellowlegs
Solitary Sandpiper
Willet
Spotted Sandpiper
Upland Sandpiper
Whimbrel
Long-billed Curlew R
Marbled Godwit
Ruddy Turnstone
Surfbird
Red Knot
Sanderling
Semipalmated Sandpiper
Western Sandpiper
Least Sandpiper
White-rumped Sandpiper R
Baird's Sandpiper R
Pectoral Sandpiper
Dunlin R
Stilt Sandpiper
Buff-breasted Sandpiper
Ruff*
Short-billed Dowitcher
Wilson's Snipe
Wilson's Phalarope R
Red-necked Phalarope R
Laughing Gull
Franklin's Gull
Gray-hooded Gull*
Gray Gull*
Ring-billed Gull

Herring Gull
Lesser Black-backed Gull*
Kelp Gull*
Sabine's Gull*
Gull-billed Tern
Caspian Tern
Royal Tern
Sandwich Tern
Common Tern
Arctic Tern*
Least Tern
Black Tern
Black Skimmer
Ruddy Ground-Dove
White-tipped Dove
Smooth-billed Ani
Groove-billed Ani R
Barn Owl
Common Nighthawk
Sapphire-throated Hummingbird
Ringed Kingfisher
Pied Water-Tyrant R
Tropical Kingbird
Gray Kingbird
Barn Swallow
Sand Martin
Cliff Swallow
Blue-black Grassquit
Variable Seedeater
Ruddy-breasted Seedeater
Saffron Finch
Red-breasted Blackbird
Great-tailed Grackle
House Sparrow

Juan Díaz Mangroves

A few kilometers east of Costa del Este, near the Panama City suburb of Juan Díaz, there is a small area of mangroves near the coast. Although the area has deteriorated considerably in recent years due to the dumping of trash on the access road and the development of several sand-quarrying operations at the coast, it can still be worth a visit. It is mentioned here especially because it is the easiest spot near Panama City to find Spot-breasted Woodpecker, here near the western limit of its range. (The species is also present at the mangroves immediately east of Costa del Este, but these are no longer easily accessible due to the construction of housing.) Other species include Mangrove Black-Hawk, Peregrine Falcon, Scaly-breasted and Sapphire-throated Hummingbirds, Straight-billed Woodcreeper, Rufous-browed Peppershrike, and the "Mangrove" form of Yellow Warbler, as well as migrant warblers in season. Many species of shorebirds and other coastal birds, similar to those found at Costa del Este, can be found at the shore, but access to the best part of the coast is now blocked by the sand quarries.

From the turnoff to the Embarcadero at the top of the exit ramp from the Corredor Sur (see below) it is about 1.3 km through marshy and grassy areas to where the mangroves start. Unfortunately, piles of trash and construction waste have been dumped along this stretch, although recently there has been a guard posted at the start of the road to prevent this. For the next 700 m or so there are mangroves on either side of the road, and this is the best birding area. The road then bends east next to an antenna, and has mangroves along one side for the next 300 m. In another 300 m the road reaches an area of sand quarries, boat sheds, and other buildings. Next to the entrance for the Arenaria (sand quarry) Balboa there is a gate in a wall where you can reach the shore of the estuary and check for shorebirds. Unfortunately, the best areas right on the coast cannot now be reached.

Be aware that safety can be a concern in this isolated area, which is near several poor communities. A group should not have any problems, but it is not recommended to bird here alone or with only a couple of people. In any case, do not go far from your car, and use caution if you see someone approaching on foot. If there is a guard at the booth at the start of the road, ask him if there have been any problems recently.

Getting to the Juan Díaz Mangroves. From the intersection of Av. Federico Boyd and Vía España, take Federico Boyd two blocks south and take a left at the traffic signal onto Av. Nicanor de Obarrio (Calle Cincuenta). Go 700 m, and turn off to the right at a sign for the Corredor Sur. Follow the signs to get on the Corredor Sur headed for the airport. After about 9 km, take the exit for Llano Bonito-Los Pueblos. At the top of the exit ramp, go right, signposted "Embarcadero." You can also easily get to Juan Díaz from Costa del Este by getting on the Corredor Sur there; there are entrances at the eastern and western ends of Costa del Este. The Llano Bonito-Los Pueblos exit is about 1 or 2 km east of Costa del Este, depending on whether you get on the entrance at the eastern or western end.

Diablo Mangroves

Another area of mangroves that is a bit closer to Panama City than Juan Díaz is on the banks of the Panama Canal near Diablo Heights in the Canal Area. Although the area is small, Straight-billed Woodcreeper can be found here, and it can also be good for hummingbirds and, in season, migrant warblers.

From the parking area after the small bridge mentioned below, you can walk back across the bridge to check the mangroves on both sides of the road. Just beyond the bridge there is a small concrete drainage channel, and during the dry season it's sometimes possible to walk along a path along the channel (on its left as you face inland). The woodcreeper has also been found in the mangroves on the Canal's edge behind the sheds back about 200-300 m along the road. However, with patience and a little luck, you can get the birds without leaving the asphalt.

(Note that this is a different patch of mangroves from the one mentioned in the "Finding Birds in Panama" section of Ridgely. That patch, within Diablo itself, is now so deteriorated that it is no longer recommended.)

Getting to the Diablo Mangroves. To reach the Diablo mangroves from the intersection of Vía España and Av. Federico Boyd, go south on Federico Boyd six blocks until you come to Av. Balboa, opposite the large Miramar Hotel. Take a right onto Balboa. After 2.1 km, near the end of Balboa, turn right just before an overpass, next to a Shell gas station, prominently signposted for "Terminal Nacional Los Pueblos/Albrook." After 300 m (get into the right lane as soon as you can), you will come to a traffic light. Continue straight, and after another short block you will come to a second light. Just beyond this the road divides; bear right, signposted for Albrook and Vía Simón Bolívar. Shortly after you will merge onto Av. Mártires going north. Follow the signs for Av. Omar Torrijos H. At 3.6 km from getting onto Mártires, you will see a large El Rey supermarket sign just past the entrance to the Albrook housing area. Turn left at the traffic signal opposite the entrance to Albrook, where you will immediately cross some railroad tracks. Continue straight, following this road through the town of Diablo. At 1.2 km from the traffic light you'll see the first mangroves along the road to the right. Continue for 300 m, cross a small one-lane bridge and park on the right. You can then walk back across the bridge to bird the mangroves.

Accommodations and meals in the Panama City area. Panama City has a great range of hotel options, ranging from extremely cheap and basic to luxury-class. Restaurants and other eating places can be found for every taste and budget. Hotels and restaurants are especially concentrated in the districts of La Exposición, Bella Vista, and El Cangrejo. A detailed discussion of these is beyond the scope of this guide. If you plan to stay in downtown Panama City, information can be found in standard tourist guides, such as Lonely Planet or Moon Handbooks. Here we mention several hotels on the periphery of the city that provide particularly good bases for birders, being closer to birding areas and mostly in quieter neighborhoods with more birds around.

La Estancia (ph 314-1417, cel 6651-6232; stay@bedandbreakfastpanama.com; www.bedandbreakfastpanama.com) is a small but very attractive bed and breakfast, ideally situated for birders, at the base of Ancón Hill in Quarry Heights. There are seven regular rooms ($45.00-$49.00), plus two suites ($75.00), all with private hot-water bath (although two rooms have their bath across a corridor) and air-conditioning; breakfast is included. Several rooms have balconies and views of the Bridge of the Americas, and there are two shared sitting rooms. The friendly owners, Tammy and Gustavo, provide excellent hospitality. La Estancia is near the beginning of the road that goes up Ancón Hill which provides good birding for the more common species of dry forest, and which is the best vantage point to observe the vast raptor migrations that transit Panama in October and November. It is also convenient to other localities near Panama City and in the southern Canal Area, such as Metropolitan Nature Park and Clayton, and is only about a half-hour to Pipeline Road. It is also a good place to overnight if headed west, since it is close to the Bridge of the Americas. To find La Estancia, follow the directions for Ancón Hill on pp. 63-64. The hotel is the third house, painted peach, after turning right at the T-junction on the road that goes up the hill. If you are coming from the airport and haven't rented a car, it may be best to take advantage of their pick-up service, since the place is a little off the beaten path and some taxi drivers may not know where it is.

Country Inn & Suites Panama Canal (ph 211-4500; fax 211-4501; josalca@unesa.com, www.panamacanalcountry.com/amador/) is located on Amador, next to the entrance to the Panama Canal. Its 159 rooms and suites ($96.00-$225.00, depending on size and amenities; packages and discounts available) all have balconies, all with private hot-water bath and air-conditioning. The location is park-like, and offers great views of ships entering and leaving the Canal. It is convenient for birding sites in Panama City and the southern Canal Area, as well as to overnight if headed west, being near the Bridge of the Americas. To find the hotel, follow the directions for Amador on p. 64. The turn is on the right just before the old post office.

Country Inn & Suites El Dorado (ph 300-3700; fax 300-3701; josalca@unesa.com; www.panamacanalcountry.com/eldorado/) is located in the El Dorado shopping district near Metropolitan Park. It has 83 rooms and suites ($74.00-$81.00), all with private hot-water bath and air-conditioning. The hotel is convenient for Metropolitan

Nature Park, as well as for the Corredor Norte if you are heading for the Atlantic side of the Canal Area. To find it, follow the first set of directions for Metropolitan Park on p. 67. Instead of turning off on Av. Juan Pablo II, continue on Av. Alfaro for another 1.2 km, turning left at a traffic signal, just beyond a McDonald's restaurant on the left. Continue on this street about 500 m, where the hotel will be on the left just before an overpass.

Albrook Inn (ph 315-1789; fax 315-1975; reserva@albrookinn.com; albrookinn panama.com) is located in a quiet suburban residential area on the former Albrook Air Force Base. It has 20 regular rooms and 10 suites ($50.00-$90.00, continental breakfast included), all with private hot-water bath and air-conditioning. The hotel is convenient to Metropolitan Park and to birding sites in the southern Canal Area, as well as to Albrook Airport if you are taking a local flight. To find it, follow the directions to the Diablo mangroves (p. 73). Instead of turning left at the traffic light before the El Rey supermarket, turn right to enter Albrook. After 500 m, turn left at a T-junction. signposted for Av. La Amistad. After 800 m, take a left, signposted for the hotel, which is a short distance beyond this turn.

Riande Aeropuerto Hotel & Resort (ph 290-3333, 266-4066; fax 290-3105; resort@hotelesriande.com; www.hotelesriande.com/hoteles/aerubici.htm) is located near Tocumen International Airport. It has 200 regular rooms and 5 suites ($90.00-$130.00), all with private hot-water bath and air-conditioning. It is of interest mainly to those who may be arriving on a late flight, or leaving very early, since it is five minutes from the airport. It may also be convenient for those who plan to visit Cerro Azul or Tocumen, or going to the Bayano region, Nusagandi, or Burbayar Lodge. To get there from the airport, upon exiting follow the signs for "Panama Centro" as the road curves around to the left. (Be careful not to take the first exit, signposted for the town of 24 de Diciembre.) After the road straightens out it is 1.4 km to the entrance to the hotel on the right, just before the road crosses the Corredor Sur on an overpass. Be careful since the entrance is easy to miss from this direction. To get to the hotel from Panama City, take the Corredor Sur to its end. Instead of turning off for the airport, continue straight past the underpass and the entrance to the hotel is immediately after on the right.

There is an abundance of restaurants in the Panama City area; consult one of the standard tourist guides for recommendations. For those wanting to get an early start, *Niko's* cafeteria-style restaurants can be a good option. The one next to the El Rey supermarket on Vía España, a few blocks west of Av. Federico Boyd, and the one at the back of the El Dorado shopping center off Av. Ricardo Alfaro, are open 24 hours. There is also one in Balboa, but this opens at 6:00 AM.

The Southern Canal Area

The southern part of the Canal Area contains two major national parks, Camino de Cruces and Soberanía. Forest types range from dry semideciduous forest in Camino de Cruces to much wetter forest at the north end of Soberanía. Other habitats include town sites, areas of grass and scrub, and wetland habitats such as small ponds and marshes, the Río Chagres, and the edges of Lake Gatún. Most birding sites in the area can be reached within a 30-45 minute drive from downtown Panama City.

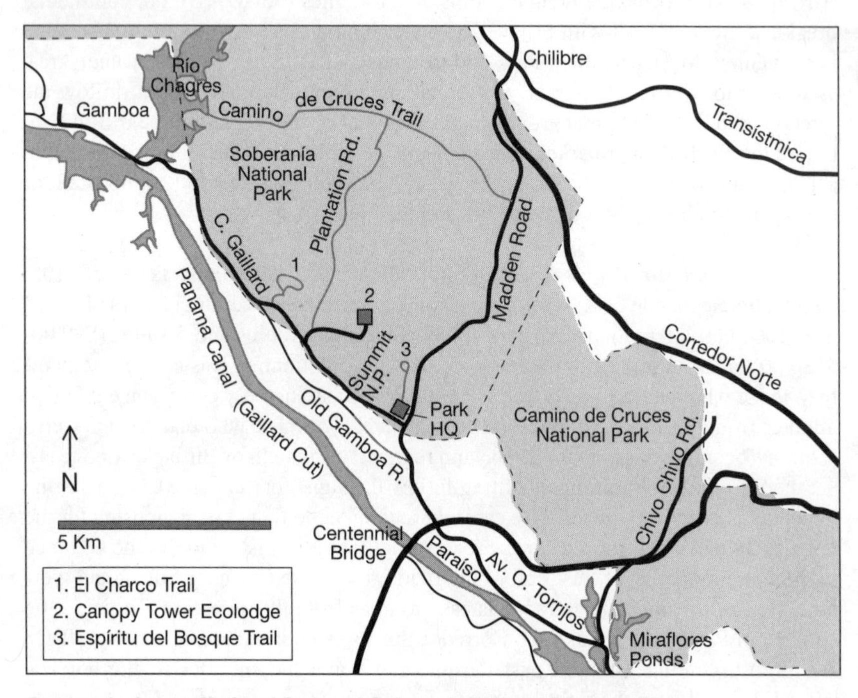

Camino de Cruces National Park and Soberanía National Park (south)

Miraflores Ponds and Camino de Cruces National Park

Camino de Cruces National Park (4,000 ha/9,884 acres) forms a forested link between Metropolitan Nature Park and Soberanía National Park. It protects part of the colonial Camino de Cruces, a trail via which the Spanish transported the treasures of Peru from Panama City to the Río Chagres, then on to the ports of Nombre de Dios and Portobelo for shipment to Spain. The park contains mostly dry semideciduous forest, with some more open areas of grass and scrub. Unfortunately, the park at present has no developed trails, and most of it is not easily accessible, even though bordering parts of Panama City and its suburbs.

The *Miraflores Ponds,* on Av. Omar Torrijos just north of the Panama Canal's Miraflores Visitor Center, and the first part of Chivo Chivo Road next to them, are the only part of the park that is easily visited. There are three ponds, two on the east side (on the right headed north) of Omar Torrijos, and one on the west. The best of the ponds is the second one on the east side. There are usually Wattled Jacanas on the floating vegetation, and the pond can also be scanned for ducks and other aquatic species.

Chivo Chivo Road, bordered by dry woodland, parallels the south side of the second pond on the east. (This is called Chiva Chiva Road in the "Finding Birds in Panama" section of Ridgely.) Along the first 500 m or so there are several vantage points for scanning the pond and the marshy area above it. A number of dry forest species, such as those of Metropolitan Nature Park, can be found here. However, the road is no longer recommended for birding beyond the first fork, about 900 m from Omar Torrijos. The road goes to Panama City's main landfill at Cerro Patacón, and much of it farther out is lined with trash. Birding along the road has also deteriorated due to construction of an access road for the new Centennial Bridge, and the antenna field mentioned in the "Finding Birds in Panama" section in Ridgely is no longer accessible. There are also some safety issues in the area, so use caution if visiting it.

Soberanía National Park

Soberanía National Park (22,104 ha/54,620 acres), with a bird list topping 400 species, is one of the premier birding localities in the New World tropics. The park is easily accessible from Panama City, several of the best birding trails being served by public transportation, and also has two excellent ecotourism hotels located next to or within it. Trails cover a range of habitats, including semideciduous and moist tropical forest, scrub, and grassy areas, and aquatic habitats on the Río Chagres and Lake Gatún can be visited by boat. In the following sections, the main birding areas associated with Soberanía are discussed starting from the south end and going north.

Madden Road. Madden Road runs from the end of Av. Omar Torrijos near Summit to the Carretera Transístmica between Panama City and Colón, passing through tall forest along most of its length. The colonial Camino de Cruces crosses the road. This trail, along sections of which some of the original paving blocks can still be seen, as well as several side roads along Madden Road, can be birded. However, the road can be very busy, and due to the area's proximity to poor communities crime has sometimes been a problem here, so that unless you want to combine birding with historical interest by walking the Camino de Cruces, some of the other sites described in this section provide better options.

Summit Nature Park. Summit Nature Park is small botanical garden and zoo located near the southern end of Soberanía National Park. Many of the more common species that frequent garden or suburban areas can be found here. The park is especially good for hummingbirds, and during the early rainy season may have colonies of oropendolas or caciques at which Giant Cowbird can be seen prospecting for nests to parasitize. There are exhibits containing many local species such as Harpy Eagle and macaws. The park is best visited on weekdays and early in the morning, since it is a popular picnic spot and can become crowded on weekends. The park is open 8:30 AM to 4 PM, and admission is $1.00.

Espíritu del Bosque Trail. This is a new 1.7 km interpretive loop trail located directly behind Soberanía National Park headquarters at the junction of Av. Omar Torrijos and C. Gaillard, which traverses mainly drier hilltop forest. Opened in 2006, it has not been birded much, but would be expected to have some species of dry forest not present farther north in the park.

Old Gamboa Road and Summit Ponds. Although Old Gamboa Road is not actually within the boundaries of Soberanía National Park, it is covered here because it is immediately adjacent to it. This was formerly the main road to Gamboa, which was later replaced by the Gaillard Highway. Its southern section, about 2 km (1.2 mi) long, runs from Av. Omar Torrijos to near Summit Park and traverses areas of dry forest, scrub, and grassland, being the best place in the immediate Canal Area to find some of the species typical of these habitats. Because of its openness, it can be much easier to see birds here than in areas with more closed forest such as Pipeline Road, so that this is a good place for a visiting birder to get an introduction to tropical birds. Much of the asphalt road is still present, although broken up in places, and is very easy for walking. Species include Great Antshrike, Lance-tailed Manakin, Rosy Thrush-Tanager, and at the end near the Summit Golf Club, Jet Antbird. This is also the best place in the immediate Canal Area for Pale-eyed Pygmy-Tyrant and Bran-colored Flycatcher. Summit Ponds are two small ponds located on Old Gamboa Road near Summit Park. Boat-billed Herons nest here, and this nocturnal species can often be seen by arriving just after dawn. Capped Herons also are occasionally present, but irregular. Kingfishers and Rusty-margined Flycatcher are also common at the ponds. Note that the section of the road north of Summit mentioned in the "Finding Birds in Panama" section of Ridgely is no longer open to the public.

Plantation Road and Canopy Tower Ecolodge. Plantation Road, named for a long-overgrown cacao plantation it once provided access to, is 2.5 km beyond Summit Park. The road is no longer open to vehicles, but is maintained as a trail, extending 7 km to its junction with the Camino de Cruces. Since it is just on the north side of the Continental Divide, some species typical of the Atlantic slope begin to appear here, such as Royal Flycatcher, Golden-crowned Spadebill, Blue-crowned Manakin, and Rufous and Broad-billed Motmots. Rare species such as Plumbeous Hawk and Slaty-backed Forest-Falcon have also been seen. The first part of the trail runs above a small stream along which Buff-rumped Warbler can be found. The Canopy Tower Ecolodge, described in more detail in the section on accommodations below, is located at the top of Semaphore Hill just above Plantation Road. The observation deck at the hotel can offer exceptional views of canopy species such as Blue Cotinga, Green Shrike-Vireo, and many honeycreepers and tanagers.

El Charco. El Charco is a short (0.8 km/0.5 mile) and easy interpretive loop trail located on the C. Gaillard 1.1.km north of Plantation Road. It features most of the same birds found there and in the first part of Pipeline Road. While not worth a special stop for birding, there is a shelter next to a pool with a small waterfall which can be a convenient spot for a picnic lunch.

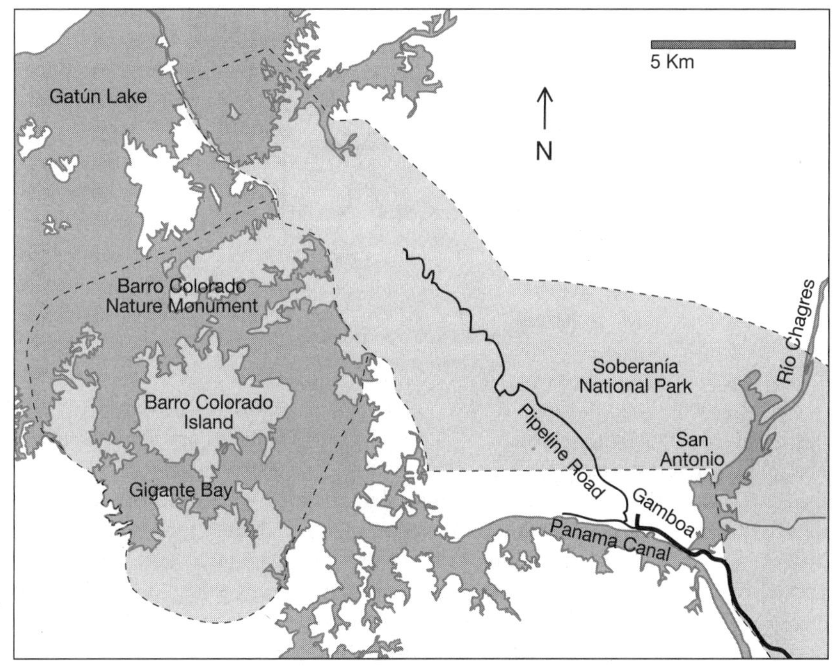

Soberanía National Park (north), Barro Colorado Nature Monument, the Río Chagres, and Lake Gatún

Gamboa and the Gamboa Rainforest Resort. The sleepy small town of Gamboa, at the juncture of the Río Chagres, the Gaillard Cut of the Panama Canal, and Lake Gatún, was founded to serve as headquarters of the Canal's Dredging Division. Today it is the site of the Gamboa Rainforest Resort, a major ecotourism development on the banks of the Chagres, as well as the gateway to the Pipeline Road section of Soberanía Park and to Barro Colorado Island. The *La Laguna Trail* is on the grounds of the Resort and runs for about 600 m (0.4 mi) through a patch of woodland where species such as White-bellied Antbird, Lance-tailed Manakin, and Rosy Thrush-Tanager can be found. To get to it, take a right just after crossing the bridge into Gamboa, signposted for the resort. At 300 m from the turn there is a guard hut for the hotel. Just past this guard station, one end of the trail leaves the road to the left, opposite a wooden sign for the hotel. However, it is easier to enter from the

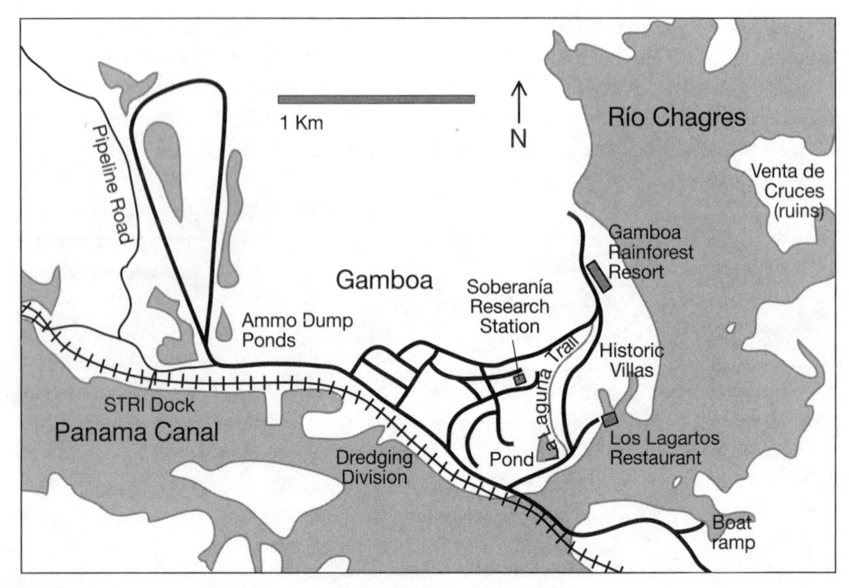

Gamboa

other end, which is better marked. Go left at the sign for the resort, where you will enter the area of historical villas. Continue through this until you get to the entry for the hotel parking lot at the end of the villas. Go left here, signposted for the "centro de actividades," instead of entering the parking lot. In 100 m, at the bottom of the hill, is the signposted entry to the trail on the left. Although there are several unmarked side trails, the patch of woods is so small it's easy enough to retrace your steps if you take a wrong turn. At the end of the trail near the guard hut there is a small pond that in the past sometimes had Rufescent Tiger-Heron and Lesser Kiskadee. It is now so overgrown, however, that it is uncertain that they still occur.

Ammo Dump Ponds. Two other ponds can be found by staying on the main road after crossing the bridge, instead of turning right. In 300 m, the road forks at the beginning of town; bear left here on C. Gaillard to continue straight through town. At 1.0 km from the bridge, at the end of town, there is a "Prohibido entrar sin autorización" sign ("Entry prohibited without authorization"); it is OK to enter to go to the National Park. In another 300 m the paved road ends at a chain-link fence at an area known as the Ammo Dump (signposted *Depósito de Explosivos*) where explosives are stored for use in work on the Canal. You can park here (being sure not to block the gate) and scan the marshy area that surrounds the fence. Purple Gallinule, White-throated Crake (and more rarely Yellow-breasted Crake), Wattled Jacana, and other marsh birds can be found here. A gravel road that goes off to the left where the asphalt road curves sharply right just before the Ammo Dump leads to Pipeline Road, skirting a larger pond which can also be checked. However, this pond and its borders are now so overgrown that visibility is limited.

Pipeline Road. Pipeline Road is one of the best places to see tropical forest birds in the Americas, with a species list exceeding 400. It was originally constructed as an asphalt road (now mostly gravel) to service a pipeline that was built to cross the Isthmus during World War II but never used. Today it extends for 17.5 km through the center of Soberanía, passing through tall secondary and mature tropical forest and crossing many small streams. Pipeline Road is reached by the gravel road that goes to the left just before the Ammo Dump, as described in the section above. This road continues past the pond, curving right and going up hill. At 600 m from the Ammo Dump, take the road to the right, signposted Camino del Oleoducto, which is the beginning of Pipeline.

At the time of this writing there is a locked gate at 200 m in from the start of the road, and you must walk in from there. The Fundación Avifauna Eugene Eisenmann (named in memory of Panama's most important native ornithologist) is in the process of building the Panama Rainforest Discovery Center, which will include a canopy observation tower and interpretive and birding trails, near the Quebrada Juan Grande, 2.0 km from the start of the road. The observation tower is scheduled to open in late 2006. The gate may eventually be moved to the Juan Grande, and visitors may be able to drive in that far, but these plans are still under discussion. To get updated information on the status of the project at Juan Grande and on access to it, call Fundación Avifauna (ph 264-6266; cel 6677-1479). For information on access to Pipeline Road in general, call Soberanía National Park headquarters (ph 276-6370).

The first 7 km (4.2 mi) of Pipeline road is mostly level, passing through tall secondary forest with some scrubbier areas. It then enters hillier and more humid terrain, repeatedly ascending and descending ridges until it terminates at the Río Agua Salud at 17.5 km (10.5 mi). The road is narrow enough so that the forest canopy meets overhead along much of its length, making it relatively easy to see forest interior birds.

One strategy for birding the area is to wade the streams crossed by the road. Species such as Agami Heron, Sunbittern, Dull-mantled Antbird, Sapayoa, Buff-rumped Warbler, and with great good luck, Great Curassow, can be found in this fashion.

Río Chagres and Lake Gatún. The Río Chagres and Lake Gatún are popular for sport fishing, and it is possible to hire a boatman and boat to explore them in search of waterbirds. Species such as Wattled Jacana, Common Moorhen, Purple Gallinule, and Green and Striated Herons are common, and others include Least Grebe, Least Bittern, Black-bellied Whistling, Muscovy, and Masked Ducks, along with migrant ducks, Snail Kite, Yellow-breasted Crake, Limpkin, Greater Ani, American Pygmy, Green, Amazon, and Ringed Kingfishers, Lesser Kiskadee, and Rusty-margined Flycatcher. Sungrebe occurs, but is so shy it can only rarely be seen from a motorboat. The best area for ducks, grebes, and species that prefer floating vegetation is a complex of islands and quiet backwaters about 4 km up the Chagres from Gamboa.

Snail Kite is more common in Gigante Bay south of Barro Colorado Island, especially in the inlets on the south side of the bay, where they nest in mangrove-like clumps of Pond Apple (*Annona glabra*). Limpkin can also sometimes be found here. (Note that entry is prohibited to the inlets on the north side of the bay on Barro Colorado itself, a strict reserve.)

Boats can be hired at the Gamboa boat ramp. There are usually a few boatmen available for hire here early in the morning (some sleep in a shed at the site), with more around on weekends than on weekdays. Prices are about $60.00 per half day (which would be enough time to do both the Chagres and Gigante Bay) for two people, and $70.00 for four. The turnoff for the boat ramp is on C. Gaillard, next to a parking area on the right 9.0 km after you turn onto Gaillard before Summit, and 400 m before the bridge over the Chagres to Gamboa.

San Antonio and Ella Puru. Two small indigenous communities on the Río Chagres near Gamboa, San Antonio, inhabited by Wounaan, and Ella Puru, by Emberá, have recently established birding trails near their villages, offering an interesting cultural experience in combination with birding. A trip to the villages includes a guided walk on the trails, a traditional dance, and a visit to the colonial ruin of Venta de Cruces at the end of the Camino de Cruces on the other side of the river. Birds to be expected here would be similar to those on the first part of Pipeline Road. A trip by boat on the Chagres to see aquatic birds could also be arranged. The cost is $20.00 per person for up to six people, and $15.00 per person for more than six. To arrange a trip you can contact Harmodio Menbache (cel 6637-9503), of San Antonio, or Marcial Caisamo (cel 6502-7292), of Ella Puru. If you have trouble getting in touch with them, you can also call Fundación Avifauna (ph 264-6266; cel 6677-1479).

Getting to Soberanía National Park, Gamboa, and nearby sites. To reach Soberanía National Park and other sites in the southern Canal Area, from the intersection of Vía España and Av. Federico Boyd, go south on Federico Boyd six blocks until you come to Av. Balboa, opposite the large Miramar Hotel. Take a right onto Balboa. After 2.1 km, near the end of Balboa, turn right just before an overpass, next to a Shell gas station, prominently signposted for "Terminal Nacional Los Pueblos/Albrook." After 300 m (get into the right lane as soon as you can), you will come to a traffic light. Continue straight, and after another short block you will come to a second light. Just beyond this the road divides; bear right, signposted for Albrook and Vía Simón Bolívar. Shortly after you will merge onto Av. Mártires going north. Follow the signs for Av. Omar Torrijos H. At 3.6 km from getting onto Mártires, you will pass a large El Rey supermarket sign near the entrance to the Albrook housing area, from which the remaining distances are given.

At 12.5 km from Albrook you pass the entrance to the Summit Golf Course. (The southern end of Old Gamboa Road is marked by bus shelters on both sides of Av. Torrijos 900 m past the golf course entrance. However, it is recommended to begin birding the road at the entrance near Summit, since you are likely to see more birds at Summit Ponds by getting there earlier.) A short distance beyond, the road bends sharply right, passing under an overpass for the railroad which has the date 1929 on its side. Immediately after this overpass, at a bus shelter, C. Gaillard branches off to the left to go to Gamboa. Soberanía National Park Headquarters is at this intersection. You can pay the entrance fee to the park here ($3.50 for foreign visitors and $1.50 for Panama residents). The Espíritu del Bosque Trail is behind the headquarters.

To get to Camino de Cruces, continue straight at this intersection instead of turning onto C. Gaillard. The start of the trail is at a parking area on the left 6.3 km from the intersection.

To get to the remaining sites, take the turn to the left onto C. Gaillard. Summit Nature Park is 1.5 km from the junction, on the right.

To get to Summit Ponds and Old Gamboa Road, turn left opposite the Summit Park entrance just beyond the parking lot. Crossing the railroad tracks, go 300 m, where you meet Old Gamboa Road at a T-junction. Turn left here for Summit Ponds, which are 300 m from this junction. You can park here to walk Old Gamboa Road towards its southern end. Access to Old Gamboa Road going north (right from the T-junction) is now restricted.

The entrance to Plantation Road and to the road up Semaphore Hill to the Canopy Tower Ecolodge is on the right, 3.0 km beyond the start of C. Gaillard. The parking lot for Plantation Road is just after the turnoff, on the left. Just beyond this is a gate on the road that goes up Semaphore Hill to the Canopy Tower. Arrangements must be made in advance to visit the hotel. See the section on accommodations for contact information.

The pullout for the El Charco Trail is on the right 4.2 km from the start of Av. Gaillard, and the entrance to the Gamboa boat ramp is on the right at 9.0 km. The one-lane bridge over the Río Chagres to Gamboa is 9.4 km from the start of Gaillard.

To reach the Gamboa Rainforest Resort, make a sharp right immediately after crossing the bridge and follow the signs. To find the Ammo Dump and Pipeline Road, instead continue straight after crossing the bridge, and then bear left at the first junction, 300 m past the bridge. In another 1.3 m, the paved road ends at the Ammo Dump.

To get to Pipeline Road, take the gravel road that goes to the left where the asphalt road curves sharply right just before the Ammo Dump. Pipeline branches off to the right after 600 m, signposted Camino del Oleoducto.

All of the sites in Soberanía National Park except Camino de Cruces can easily be reached by the orange-and-white SACA buses to Gamboa, which run along Av. Omar Torrijos and C. Gaillard. The SACA bus terminal is just west of the Plaza

Cinco de Mayo in the western part of Panama City, not far from the Av. del los Mártires. The earliest buses currently leave Panama City at 5:00 AM on weekdays and 5:45 AM on weekends, arriving in Gamboa about 45 minutes later. They run at intervals of about 1-2 hours throughout the day. The price to Gamboa is $0.65. Schedules may change, so check with SACA (ph 212-3220) for updates to service.

Bird list for Soberanía National Park and environs. The list includes Lake Gatún and areas adjacent to the park, including Summit Nature Park, Old Gamboa Road, and Gamboa. * = vagrant, R = rare.

Great Tinamou
Little Tinamou
White-faced Whistling-Duck*
Black-bellied Whistling-Duck
Muscovy Duck
Blue-winged Teal
American Wigeon R
Ring-necked Duck R
Lesser Scaup
Masked Duck
Gray-headed Chachalaca
Crested Guan R
Great Curassow R
Marbled Wood-Quail
Tawny-faced Quail R
Least Grebe
Pied-billed Grebe
Brown Booby R
Red-footed Booby*
Brown Pelican
Neotropic Cormorant
Anhinga
Magnificent Frigatebird
Least Bittern
Rufescent Tiger-Heron
Fasciated Tiger-Heron R
Great Blue Heron
Cocoi Heron R
Great Egret
Snowy Egret
Little Blue Heron
Tricolored Heron
Cattle Egret
Green Heron
Striated Heron
Agami Heron R
Black-crowned Night-Heron
Boat-billed Heron
Green Ibis*
Wood Stork R
Black Vulture
Turkey Vulture
King Vulture
Osprey
Gray-headed Kite
Hook-billed Kite
Swallow-tailed Kite
White-tailed Kite
Snail Kite
Double-toothed Kite
Plumbeous Kite
Mississippi Kite
Tiny Hawk
Crane Hawk
Semiplumbeous Hawk
Plumbeous Hawk R
White Hawk
Common Black-Hawk
Great Black-Hawk
Savanna Hawk R
Gray Hawk
Roadside Hawk
Broad-winged Hawk
Short-tailed Hawk
Swainson's Hawk
Zone-tailed Hawk
Crested Eagle*
Harpy Eagle*
Black Hawk-Eagle
Ornate Hawk-Eagle R
Barred Forest-Falcon
Slaty-backed Forest-Falcon R
Collared Forest-Falcon
Red-throated Caracara*
Crested Caracara
Yellow-headed Caracara
American Kestrel
Bat Falcon
Peregrine Falcon
White-throated Crake
Gray-necked Wood-Rail
Uniform Crake R
Yellow-breasted Crake
Purple Gallinule
Common Moorhen
American Coot
Sungrebe
Sunbittern
Limpkin
Northern Jacana*
Wattled Jacana
Collared Plover
Lesser Yellowlegs
Solitary Sandpiper
Spotted Sandpiper
Semipalmated Sandpiper
Western Sandpiper
Least Sandpiper
Wilson's Snipe
Parasitic Jaeger R
Laughing Gull
Franklin's Gull R
Ring-billed Gull R
Herring Gull R
Gull-billed Tern
Royal Tern
Sandwich Tern
Common Tern
Least Tern
Sooty Tern*
Black Tern
Brown Noddy*
Pale-vented Pigeon
Scaled Pigeon
Short-billed Pigeon
Plain-breasted Ground-Dove R
Ruddy Ground-Dove
Blue Ground-Dove
White-tipped Dove
Gray-chested Dove
Violaceous Quail-Dove
Ruddy Quail-Dove
Olive-backed Quail-Dove R
Orange-chinned Parakeet
Brown-hooded Parrot
Blue-headed Parrot
Red-lored Amazon
Yellow-headed Amazon
Mealy Amazon
Yellow-billed Cuckoo
Black-billed Cuckoo
Squirrel Cuckoo
Little Cuckoo R

Striped Cuckoo
Pheasant Cuckoo R
Rufous-vented Ground-Cuckoo R
Greater Ani
Smooth-billed Ani
Vermiculated Screech-Owl
Tropical Screech-Owl
Crested Owl
Spectacled Owl
Central American Pygmy-Owl R
Mottled Owl
Black-and-white Owl
Striped Owl
Oilbird R
Short-tailed Nighthawk
Lesser Nighthawk
Common Pauraque
Rufous Nightjar
Chuck-will's-widow
Great Potoo
Common Potoo
Black Swift R
White-collared Swift R
Chimney Swift
Short-tailed Swift
Band-rumped Swift
Lesser Swallow-tailed Swift
Rufous-breasted Hermit
Band-tailed Barbthroat
Green Hermit*
Long-billed Hermit
Stripe-throated Hermit
White-tipped Sicklebill*
Scaly-breasted Hummingbird
White-necked Jacobin
Black-throated Mango
Violet-headed Hummingbird*
Rufous-crested Coquette R
Garden Emerald
Violet-crowned Woodnymph
Violet-bellied Hummingbird
Sapphire-throated Hummingbird R
Blue-chested Hummingbird
Snowy-bellied Hummingbird
Rufous-tailed Hummingbird
White-vented Plumeleteer
Purple-crowned Fairy
Long-billed Starthroat R
White-tailed Trogon
Violaceous Trogon
Black-throated Trogon
Black-tailed Trogon
Slaty-tailed Trogon
Broad-billed Motmot
Rufous Motmot
Blue-crowned Motmot
Ringed Kingfisher
Belted Kingfisher
Green Kingfisher
Amazon Kingfisher
Green-and-rufous Kingfisher
American Pygmy Kingfisher
Great Jacamar
Barred Puffbird*
White-necked Puffbird
Black-breasted Puffbird
Pied Puffbird
White-whiskered Puffbird
Spot-crowned Barbet R
Collared Aracari
Yellow-eared Toucanet R
Keel-billed Toucan
Chestnut-mandibled Toucan
Stripe-cheeked Woodpecker*
Black-cheeked Woodpecker
Red-crowned Woodpecker
Yellow-bellied Sapsucker R
Cinnamon Woodpecker
Lineated Woodpecker
Crimson-crested Woodpecker
Crimson-bellied Woodpecker R
Slaty-winged Foliage-gleaner R
Buff-throated Foliage-gleaner
Plain Xenops
Tawny-throated Leaftosser
Scaly-throated Leaftosser
Plain-brown Woodcreeper
Ruddy Woodcreeper
Long-tailed Woodcreeper
Wedge-billed Woodcreeper
Northern Barred-Woodcreeper
Cocoa Woodcreeper
Black-striped Woodcreeper
Streak-headed Woodcreeper
Fasciated Antshrike
Great Antshrike
Barred Antshrike
Western Slaty-Antshrike
Russet Antshrike
Spot-crowned Antvireo
Moustached Antwren
Pacific Antwren R
Checker-throated Antwren
White-flanked Antwren
Dot-winged Antwren
Dusky Antbird
Bare-crowned Antbird R
White-bellied Antbird
Chestnut-backed Antbird
Dull-mantled Antbird R
Spotted Antbird
Wing-banded Antbird R
Bicolored Antbird
Ocellated Antbird
Black-faced Antthrush
Streak-chested Antpitta
Black-crowned Antpitta R
Paltry Tyrannulet
Brown-capped Tyrannulet
Southern Beardless-Tyrannulet
Northern Scrub-Flycatcher R
Yellow-crowned Tyrannulet
Forest Elaenia
Gray Elaenia
Greenish Elaenia
Yellow-bellied Elaenia
Lesser Elaenia
Olive-striped Flycatcher R
Ochre-bellied Flycatcher
Yellow Tyrannulet
Yellow-green Tyrannulet
Black-capped Pygmy-Tyrant
Pale-eyed Pygmy-Tyrant
Southern Bentbill
Common Tody-Flycatcher
Slate-headed Tody-Flycatcher
Brownish Twistwing
Olivaceous Flatbill
Yellow-margined Flycatcher
Yellow-olive Flycatcher
Golden-crowned Spadebill
Royal Flycatcher
Ruddy-tailed Flycatcher
Sulphur-rumped Flycatcher
Black-tailed Flycatcher
Bran-colored Flycatcher
Olive-sided Flycatcher
Eastern Wood-Pewee
Western Wood-Pewee
Tropical Pewee
Acadian Flycatcher
Yellow-bellied Flycatcher
Alder Flycatcher
Willow Flycatcher
Pied Water-Tyrant R
Long-tailed Tyrant
Bright-rumped Attila
Speckled Mourner
Sirystes
Rufous Mourner
Dusky-capped Flycatcher

Great Crested Flycatcher
Panama Flycatcher
Lesser Kiskadee
Great Kiskadee
Boat-billed Flycatcher
Rusty-margined Flycatcher
Social Flycatcher
Gray-capped Flycatcher
White-ringed Flycatcher R
Streaked Flycatcher
Sulphur-bellied Flycatcher
Piratic Flycatcher
Tropical Kingbird
Gray Kingbird
Eastern Kingbird
Fork-tailed Flycatcher
Sapayoa R
Thrush-like Schiffornis
Rufous Piha
Cinnamon Becard
White-winged Becard
One-colored Becard*
Masked Tityra
Black-crowned Tityra
Blue Cotinga
Purple-throated Fruitcrow
Three-wattled Bellbird*
Golden-collared Manakin
Lance-tailed Manakin
Blue-crowned Manakin
Red-capped Manakin
Yellow-throated Vireo
Philadelphia Vireo
Red-eyed Vireo
Yellow-green Vireo
Black-whiskered Vireo R
Scrub Greenlet
Tawny-crowned Greenlet
Golden-fronted Greenlet
Lesser Greenlet
Green Shrike-Vireo
Rufous-browed Peppershrike R
Black-chested Jay
Purple Martin
Gray-breasted Martin
Brown-chested Martin
Tree Swallow
Mangrove Swallow
White-thighed Swallow
Blue-and-white Swallow
Northern Rough-winged Swallow
Southern Rough-winged Swallow
Sand Martin
Cliff Swallow
Barn Swallow
White-headed Wren R
Black-bellied Wren
Bay Wren
Rufous-breasted Wren
Rufous-and-white Wren
Buff-breasted Wren
Plain Wren
House Wren
White-breasted Wood-Wren
Southern Nightingale-Wren
Song Wren
Tawny-faced Gnatwren
Long-billed Gnatwren
Tropical Gnatcatcher
Veery
Gray-cheeked Thrush
Swainson's Thrush
Wood Thrush
Clay-colored Thrush
White-throated Thrush R
Gray Catbird
Tropical Mockingbird
Blue-winged Warbler
Golden-winged Warbler
Tennessee Warbler
Yellow Warbler
Chestnut-sided Warbler
Magnolia Warbler
Cape May Warbler R
Black-throated Blue Warbler R
Blackburnian Warbler
Bay-breasted Warbler
Blackpoll Warbler R
Cerulean Warbler
Black-and-white Warbler
American Redstart
Prothonotary Warbler
Worm-eating Warbler R
Swainson's Warbler*
Ovenbird
Northern Waterthrush
Louisiana Waterthrush
Kentucky Warbler
Mourning Warbler
Common Yellowthroat R
Hooded Warbler R
Canada Warbler
Chestnut-capped Warbler
Buff-rumped Warbler
Bananaquit
Rosy Thrush-Tanager
Dusky-faced Tanager
Olive Tanager
Gray-headed Tanager
Sulphur-rumped Tanager
White-shouldered Tanager
Tawny-crested Tanager
White-lined Tanager
Red-throated Ant-Tanager
Summer Tanager
Scarlet Tanager
Crimson-backed Tanager
Flame-rumped Tanager
Plain-colored Tanager
Bay-headed Tanager R
Golden-hooded Tanager
Scarlet-thighed Dacnis
Blue Dacnis
Green Honeycreeper
Shining Honeycreeper
Red-legged Honeycreeper
Blue-gray Tanager
Palm Tanager
Streaked Saltator
Blue-black Grassquit
Slate-colored Seedeater R
Variable Seedeater
Yellow-bellied Seedeater
Ruddy-breasted Seedeater
Lesser Seed-Finch
Yellow-faced Grassquit
Buff-throated Saltator
Black-headed Saltator
Slate-colored Grosbeak
Rose-breasted Grosbeak
Blue-black Grosbeak
Indigo Bunting
Dickcissel
Orange-billed Sparrow
Black-striped Sparrow
Bobolink
Great-tailed Grackle
Giant Cowbird
Orchard Oriole
Yellow-backed Oriole
Orange-crowned Oriole R
Yellow-tailed Oriole
Baltimore Oriole
Yellow-billed Cacique
Scarlet-rumped Cacique
Yellow-rumped Cacique
Crested Oropendola
Chestnut-headed Oropendola
Yellow-crowned Euphonia
Thick-billed Euphonia
Fulvous-vented Euphonia
White-vented Euphonia
Nutmeg Munia*
Tricolored Munia*

Accommodations and meals in the southern Canal Area. Since Gamboa is only about 30 to 45 minutes from Panama City, sites in the southern Canal Area can easily be visited from hotels in town, especially those in the Canal Area such as La Estancia and the Albrook Inn. However, there are two hotels that are especially suitable for birders in the area, as well as a research station/lodge that offers more basic accommodation.

The *Canopy Tower Ecolodge and Nature Observatory* (ph 264-5720, 263-2784, 214-9724, during business hours; cel 6578-5711, 6687-0291 after hours; fax 263-2784; birding@canopytower.com; www.canopytower.com) is one of Panama's premier birding destinations. There are 10 rooms and 2 suites ($100.00-$145.00 May-September; $120.00-$175.00 October-November; $130.00-$200.00 December-April). Prices are per person (double occupancy) and include three meals per day and two daily guided birding tours. Multi-night packages are available. The suites and five upper rooms have private hot-water bath; the lower five rooms share one between them. Rooms have fans instead of air-conditioning due to noise and ecological considerations, but since the site is at about 270 m (900 ft) in elevation and breezy, air-conditioning is not really needed.

The Canopy Tower was once a U.S. government radar installation, and is located on a hilltop within Soberanía National Park and overlooking the Panama Canal. The upper open-air deck, which surrounds the original radar dome, is above the forest canopy and offers superb views of treetop species such as the Green Shrike-Vireo, Blue Cotinga, toucans, tanagers, and honeycreepers, as well as the central part of the Canal. It also provides an excellent vantage point to watch the huge raptor migrations that traverse Panama in October and November; over 230,000 raptors have been counted from here in a single day. The level below this, within the forest canopy itself, contains the dining room and lounge, and the next two levels contain the guest bedrooms. The lounge and guest rooms provide good views of birds and other wildlife within the upper forest levels. Feeders for fruit and nectar attract tanagers and hummingbirds for close observation views. There is good birding on Semaphore Hill Road, which winds up the hill from C. Gaillard, as well as on Plantation Road, which starts at the bottom of the hill. The Canopy Tower also offers guided birding tours, featuring excellent guides, to non-guests for $85.00-$95.00, which includes a meal and an opportunity to use the observation deck.

The *Gamboa Rainforest Resort* (ph 314-9000; fax 314-9020; reservation@gamboaresort.com; www.gamboaresort.com) is a much larger and more elaborate operation than the Canopy Tower. It is located on the Río Chagres in the town of Gamboa, just above the point where the river meets the Panama Canal. The main hotel has 110 rooms and suites ($175.00-$500.00 mid-April to mid-December; $250.00-$650.00 mid-December to January 1; $225.00-$600.00 January 1 to mid-April). There are also 65 rooms in adjacent villas that once housed Panama Canal staff ($135.00-$165.00 mid-April to mid-December; $175-205.00 mid-

December to January 1; $150.00-$180.00 January 1 to mid-April). Rates are single or double occupancy with a $26.00 charge for another person in the room. All rooms have private hot-water baths and air-conditioning. The villas may be more attractive for birders than the main hotel since they are located on a tree-lined street, and some have woodland just in back of them. The hotel has several restaurants and cafes, as well as the separate Los Lagartos Restaurant next to the Chagres.

The hotel has an aerial tram ride that takes you through tropical forest to an observation tower on the top of a nearby hill overlooking the Canal. The tower can be a good spot to observe the raptor migrations that pass through Panama in October and November, although numbers are somewhat lower than closer to the Pacific coast. The resort offers birding tours on trails adjacent to the hotel grounds, as well as to other nearby sites including Pipeline Road, the Las Cruces Trail, and Barro Colorado Nature Monument. There are also tours by boat on the Río Chagres and Lake Gatún. Other non-birding recreational activities are also available, such as sport-fishing and kayaking.

To find the Canopy Tower and the Gamboa Rainforest Resort, follow the directions in the section on "Getting to Soberanía National Park and nearby sites" on pp. 82-83.

The *Soberanía Research Station and Lodge* (ph 228-8704; cel 6676-2466; fax 228-6535; info@advantagepanama.com; www.advantagepanama.com/soberania), offers basic accommodation in the town of Gamboa in quarters formerly used by Panama Canal staff. There are seven units: three regular rooms with a double or two single beds ($35.00); two mini-studio rooms with kitchenette ($55.00), and two apartments with a kitchenette/dining/living area, and a separate room with private bathroom ($70.00). Additional beds in rooms are available at $10.00 per person. Rooms have private hot-water baths and ceiling fans. There are also bunkrooms and a shared kitchen. Reduced rates are available for longer stays, as well as packages including birding excursions on Pipeline, Plantation, Old Gamboa, and Achiote Roads. The place is run by Advantage Tours, which includes some of Panama's best birding guides on its staff; see p. 39. Scientists working on research at the Smithsonian Tropical Research Institute regularly stay here, and it also hosts student groups taking courses on tropical biology.

The Los Lagartos Restaurant of the Gamboa Rainforest Resort (closed on Mondays), located on the bank of the Río Chagres, allows you to watch jacanas and other waterbirds as you eat lunch. Aside from this and the other restaurants at the Rainforest Resort and at the Canopy Tower (prior arrangements must be made to eat at the latter), dining options in the southern Canal Area are few. There are a few small lunch trucks and kiosks near the entrance to the Panama Canal Dredging Division on C. Gaillard in Gamboa, although these are generally only open on weekdays. If you plan to spend the day in the area, and don't want to eat at the hotel restaurants, it's best to pack a lunch.

Barro Colorado Island

Barro Colorado is a 1,500 hectare (3,700 acre) island that was created in 1913 when the Río Chagres was dammed to form Lake Gatún, the central section of the Panama Canal. The island is a major site for the study of tropical forest biology, administered by the Smithsonian Tropical Research Institute (STRI). Barro Colorado is not an exceptionally good place for birding, since many species have become locally extinct on the island since its creation due to the effects of isolation from mainland forests. It is, however, the best place in the Canal Area to see Crested Guan, which is fairly common here and tame due to the island's protection from hunting for more than 80 years. The usually very elusive Great Tinamou is also regularly seen on the trails here. Barro Colorado is an excellent place to see mammals as well, including agoutis, coatis, monkeys, and many others, which are likewise very tame.

STRI operates its own tours of the islands five days a week, which include a guided walk on forest trails for 2½-3 hours, plus an introductory talk on the island's ecology. Several ecotourism companies also offer tours to other parts of Barro Colorado Nature Monument, which includes five mainland peninsulas in addition to the island. If you are particularly interested in seeing Crested Guans or other wildlife, make sure you go on the tour of the island itself rather than another part of the monument. The guan is not found on the mainland in the monument, and wildlife is not as tame there as on the island.

Getting to Barro Colorado Island. A visit to Barro Colorado Island must be arranged through STRI (ph general reception 212-8000; for BCI arrangements 212-8951, 212-8026; ObaldiaA@si.edu; www.stri.org/english/visit_us/barro_colorado/index.php). Visits to the island are heavily booked, so it is best to make arrangements as far in advance as possible. There are sometimes last-minute cancellations, however, so it can be worth enquiring to see if there is space even if you have not been able to book in advance. The cost is $70.00 for foreign visitors; $40.00 for foreign students; $25.00 for Panamanian residents; and $12.00 for Panamanian students. The price includes buffet lunch, a guided tour, and transportation to the island by STRI's launch from Gamboa. To find the STRI dock, follow the directions for Soberanía National Park on pp. 82-83. Once in Gamboa, take the gravel road that goes to the left before the Ammo Dump. The parking lot for the STRI dock is on the left 300 m down this road, on the other side of the railroad tracks. Gamboa can also be reached by SACA bus; see pp. 83-84 for details. On Tuesdays, Wednesdays, and Fridays, the launch leaves the pier at 7:15 AM, and returns from the island at 3:40 PM. On Saturdays and Sundays, it leaves Gamboa at 8:00 AM and leaves the island at 2:30 PM. The boat departs promptly, so if you are late you will miss your tour. The trip to the island through the Canal takes a half hour to an hour depending on the boat used.

Accommodations and meals. Barro Colorado Island can be done as a day trip from Panama City or hotels in the southern Canal Area. Lunch is provided as part of a visit to the island. Bring water and snacks for your hike in the forest.

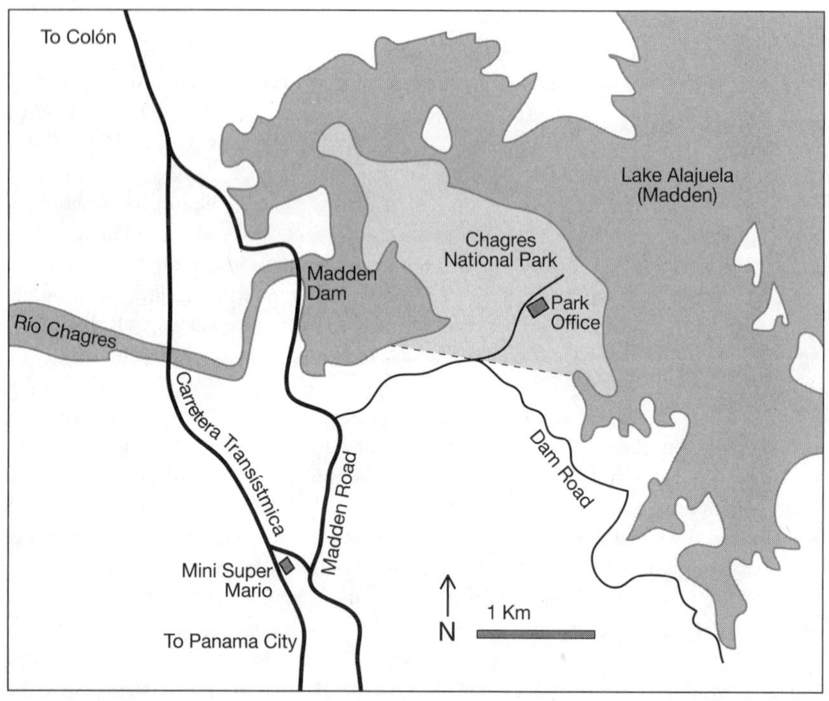

Campo Chagres

Campo Chagres

Campo Chagres is located near the Carretera Transístmica about halfway across the isthmus, and features species typical of both the drier Pacific and the wetter Atlantic sides. It is the headquarters of Chagres National Park, and located near Madden Dam, which was built on the upper Río Chagres to create Lake Alajuela (formerly Lake Madden), a supplemental source of water for the operation of the Panama Canal. Campo Chagres is located on a small peninsula that juts into the lake. The main part of Chagres National Park, across the lake, protects the upper Chagres watershed. Except near Cerro Azul and Cerro Jefe, much of the rest of the park is difficult of access, a true wilderness frequented by jaguars, tapirs, and Harpy Eagles. Species typical of dry forest that occur at Campo Chagres include Lance-tailed Manakin, **Yellow-green Tyrannulet**, Rufous-and-white Wren, and Rufous-breasted Wren. Other species that are sometimes seen include Laughing Falcon, Royal Flycatcher, and Black-headed Saltator.

The entry road into Campo Chagres from the main road runs through fairly scrubby semideciduous forest. The road is broken asphalt and easily driven in 2WD. At 1.4 km from the entrance, the road bends sharply left next to a yellow gate, continuing on through somewhat better semideciduous forest to Park Headquarters at 2.0 km. There are two trails at headquarters. *Las Grutas* is a 2.3 km (1.4 mi) loop interpretive trail, while *El Cuipo,* 900 m (0.5 mi) long, goes down to the lake. (The headquarters is on the site of the "abandoned Boy Scout Camp" referred to in the "Finding Birds in Panama" section of Ridgely.)

The road beyond the yellow gate where the entrance road turns left is a maintenance road for several small dams at the edge of the Lake. (This area is actually outside the park proper.) This road at first runs through humid forest and then into cattle pastures and scrub. From the dams you can scan the lake for waterbirds when its water levels are high. Interestingly, the differing aspect of the forest along the entrance road and this one is not due to differences in rainfall, but rather to geology. The semideciduous forest grows on limestone, which produces very well-drained soils so that trees have less access to water during the dry season. The humid forest is on nonlimestone soil that retains water better.

Getting to Campo Chagres. From the intersection of Vía España with Av. Federico Boyd, take Av. Manuel Espinosa Batista (which is the continuation northward of Av. Federico Boyd) north. At 1.0 from Vía España, you come to a traffic circle at an underpass below Vía Simón Bolívar (Transístmica). Continue straight through the traffic circle; Av. Manuel Espinosa Batista here changes its name and becomes Av. Ricardo J. Alfaro (Tumba Muerto). At 700 m past this traffic circle, just after passing under a pedestrian overpass, take a left at the traffic signal, signposted for the Corredor Norte. After 600 m you will reach the edge of Metropolitan Park; follow the signs to go right for the Corredor Norte and Colón. (You can get on any other entrance of the Corredor Norte as well if more convenient.) Watch carefully for the exit for Colón, which is 5.0 km from the entrance at Metropolitan Nature Park. After 20.5 km, you will exit from the Corredor onto the eastern end of Madden Road; go towards the right. In about 900 m you will reach the entrance ramp to the Transístmica, on the left just before an overpass. At the top of the ramp, go left for Colón. After 6.5 km on the Transístmica, you will see the minisuper Mario and Fonda Cristal on the right. Turn off the highway here. Continue for 100 m, where you turn left just before a small bus shelter. After 1.6 km, you will come to the well-marked entrance for Parque Nacional Chagres on the right. The gate is always kept unlocked. If you arrive early and it is closed, simply open it and drive in (closing it behind you). Continuing on the main road for another 600 m beyond the entrance to the park will bring you to Madden Dam. Do not leave your car unattended along this road, because there have been break-ins. Drive well in from the main road before starting to bird, and use caution if you stop at the overlooks by the dam.

Accommodations and meals. Campo Chagres is easily done as a day trip from Panama City or hotels in the southern Canal Area. If you plan to stay past the morning, it is best to pack a lunch, since there is no convenient place to eat nearby (although there are some small restaurants along the Transístmica).

Canal Area Atlantic side

For those visiting the Panama City area, a foray over to the Atlantic side of the Canal Area can substantially enhance your species list. Because the Atlantic coast here receives almost twice the rainfall of the Panama City area, many species of birds that favor more humid conditions are more common, and sometimes restricted to, this region. San Lorenzo National Park, especially in its southern part along Achiote Road, is one of the most outstanding birding localities in Panama and well worth the trip across the isthmus. Besides San Lorenzo, to the west of the Panama Canal, there are a several other birding localities on the eastern side of the Canal that can be worthwhile for those with sufficient time to spend exploring the area. More than 475 species have been recorded in this region, and the Atlantic Audubon Christmas Bird Count at one time held the record for total species recorded in one day, with 357 in 1990.

The Atlantic side of the Canal Area has changed considerably since it was described in the "Finding Birds in Panama" section of Ridgely in 1989. The former U.S. military bases of Ft. Sherman, Ft. Davis, Galeta Naval Station, and others have reverted to Panamanian control. A major positive step has been the formal protection of much of the forest in these areas. However, in some cases roads and trails that once gave access to good areas have been closed or allowed to become overgrown. The situation has not been as positive outside these protected areas. In the area east of the Canal, mangroves and other habitats have been destroyed by unplanned development, sometimes of dubious legality. It is to be hoped that future projects will take better account of environmental values in the region and of its value for ecotourism.

San Lorenzo National Park

San Lorenzo National Park (9,653 hectares/23,853 acres), established in 2004, occupies much of the former U.S. Army Base of Ft. Sherman, which reverted to Panama at the end of 1999. Most of the area, which stretches along either side of the lower Río Chagres, is hilly and consists of wet lowland tropical forest, but there are also areas of grass, scrub, mangroves, swamps, marshes, and coastline. Paved and gravel roads suitable for 2WD vehicles provide access to most of the better areas. The U.S. Army formerly maintained many additional roads and trails in the area for use in its jungle training operations. Although some of these were excellent for birding, most have not been maintained since the area reverted to Panama, and some have been closed to public access because of proximity to areas once used for firing ranges or for other reasons. In the following accounts we have included some of the better of these roads and trails that are in areas still open to public access, since some can still be followed fairly easily even though somewhat overgrown. As the park is developed, it is possible that some of these sites will be reopened and maintained for birding and tourism.

To get to San Lorenzo, you must cross Gatún Locks at the Atlantic end of the Panama Canal. Vehicles must cross on a one-lane swing bridge, which is closed whenever ships are entering or leaving the locks. There can be delays of a half hour to an hour or even more to cross, so it's best to leave early and be prepared to wait. There are currently no regular restaurants or hotel accommodations in the area of the Park on the east side of the Locks (although arrangements can be made with advance notice to have lunch in the town of Achiote), so you will have to pack a lunch if you plan to make a day of it.

Information on the southern part of the park, south of the Río Chagres, including Achiote Road, is presented first, followed by that on the northern part, including Sherman and Ft. San Lorenzo. To reach the southern part of the park, turn left after crossing Gatún Locks; for the northern sector, go straight.

Southern Sector

Gatún Dam and the Río Chagres. After crossing Gatún Locks, the road to the left first parallels the locks for 500 meters, then swerves right to run along the base of the grass-covered Gatún Dam, at one time the largest earthen dam in the world. The grass on the dam is a good place for several grassland species, such as Red-breasted Blackbird, Bobolink and Dickcissel in season, and several raptors. To the right of the road is a fringe of second-growth woodland where birds such as Blue-crowned Motmot, Great Antshrike, and Rosy Thrush-Tanager can be found. About 2 km from the Locks, the road then crosses the Río Chagres on a one-lane bridge just below the Dam's Spillway. In the late rainy season the Spillway is sometimes opened to release excess water from the lake, which is an impressive spectacle. Just before the bridge, you can pull into the parking lot of the former Tarpon Club restaurant (now closed) to scan the river below the Spillway, which often has herons and kingfishers. After dark, check for Boat-billed Herons at the base of the spillway with a spotlight. Beyond the bridge, the road climbs to run along the crest of the Dam along the shore of Lake Gatún.

Tiger Trail. At the west end of the dam, 1.5 km from the bridge, marked by a yellow-and-green national park sign, is a gravel road that goes to Tiger Trail. The road extends 200 meters to a parking area next to some rusted metal, which is where the trail begins. Although the trail has not been maintained, it can still be followed. Of course, use care to ensure that you can find your way back; break twigs or make other markers wherever your route is in doubt. The first few dozen meters of the trail are the most difficult to find. After that it runs along the gravel bed of a former road and is easier to follow. Because it is within forest, bird activity can be better here later in the day than along the roads. The area is good for forest understory species such as Scaly-throated and Tawny-throated Leaftossers and Streak-chested Antpitta.

Escobal Road. From the far end of the dam the heavily potholed Escobal Road runs southward near Lake Gatún for 15 km before arriving in the small rural community of Escobal. The road is forested for its first 11 km, as far as the turnoff to Achiote, and can provide excellent birding, with most of the same species as Achiote itself. On weekdays traffic can pick up later in the morning and make parts of the road dusty, so try to get there early. All of the side roads that once provided access to for-

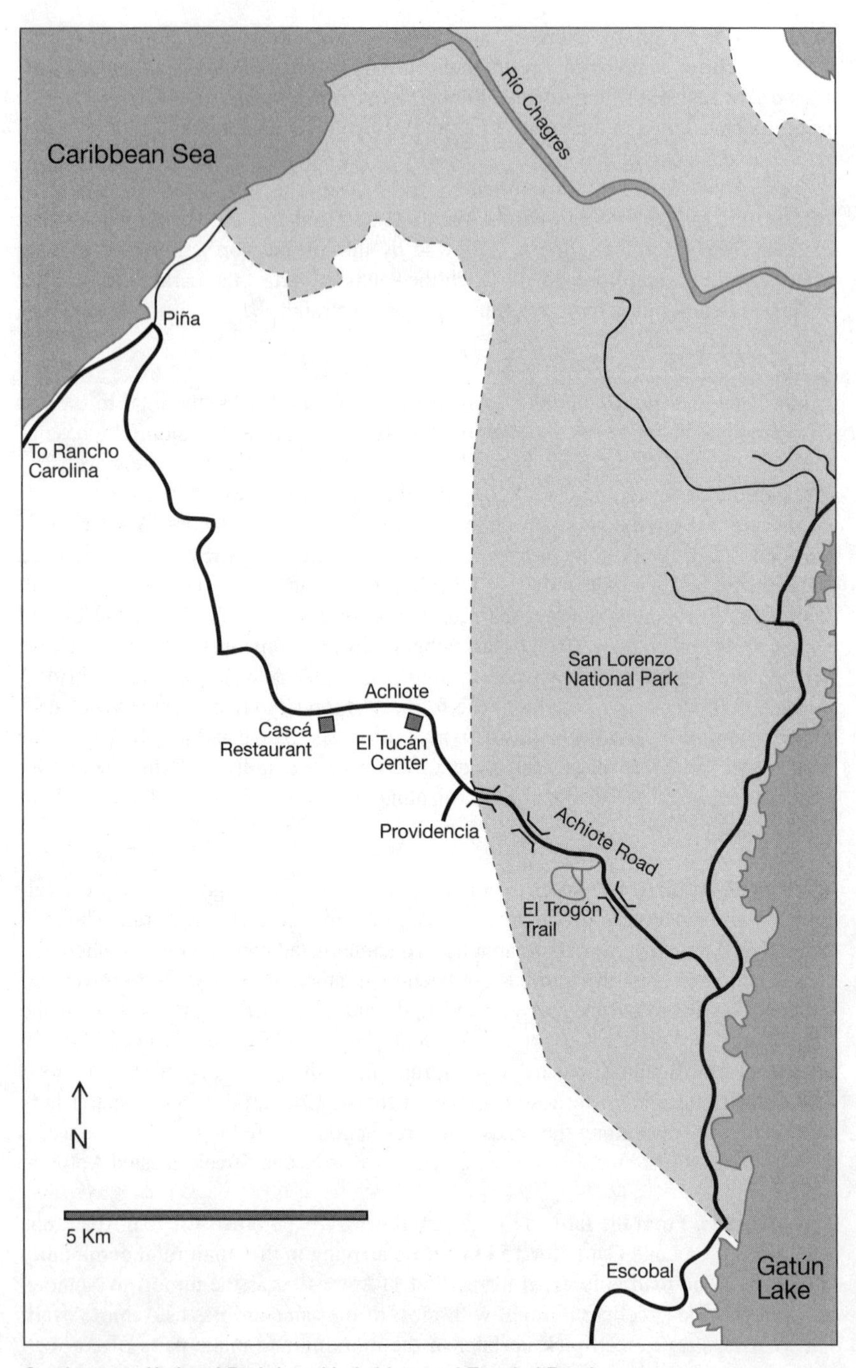

San Lorenzo National Park (south): Achiote and Escobal Roads

est on the right side of Escobal are now closed to the public. The former Piña Firing Range, used for artillery practice by the U.S. military, occupies the right side of the road along the southern part of Escobal Road and the first part of Achiote, and is posted against entry due to the possible presence of unexploded ordinance.

Achiote Road. At 11 km beyond the end of Gatún Dam, just as the Escobal Road leaves the forest, Achiote Road branches off to the right. It runs through forest for its first 5.4 km, at which point it leaves the National Park. The entire road is good, so it is best to drive it slowly while watching for activity, then pulling off the road where possible and birding on foot. Among the species that can be seen along Achiote are Hook-billed Kite, Spot-crowned Barbet, Bare-crowned Antbird, Moustached Antwren, and Stripe-breasted and White-headed Wrens. Agami Herons occur along the streams. Crested Eagle and Black-and-white Hawk-Eagle have also been seen but are very rare. Gray-cheeked Nunlet and **Black-crowned Antpitta** formerly occurred here but have not been seen since the early 1980s.

Streams cross the road at three bridges at 2.1, 4.9, and 5.4 km from its start. Achiote has been the focus of a program by the Panamanian Center for Research and Social Action (CEASPA), which has worked with neighboring communities to make the area friendlier for bird watching. As part of this initiative, the road has been posted with signs asking drivers to slow down, but you should be aware that there is substantial traffic on the road and use caution when birding on foot. In addition, the program established the *El Trogón Trail*, on the left side of the road 3.0 km from its start. White-headed Wren and Agami Heron, among many other species, can be found here. There are signs 200 m on either side alerting you to slow down in advance. There are two loops to the trail. The main loop (600 m/0.4 mi) starts at the parking lot and is level. The second loop (350 m/0.2 mi) begins at the far end of the main loop, crossing the Quebrada Treinticinco, a small stream, to climbs gently to the top of a low ridge. Before the area became a national park, local people cleared small plantations for coffee and other crops a short distance off the road. As part of the park's management plan, these plantations have been allowed to remain but it is not permitted to expand them. The main loop of the El Trogón Trail skirts part of one of these plantations, where visibility of birds can be better due to its relative openness. The cost to use the trail is $5.00 per person, which can be paid at the El Tucán Center in Achiote. (This fee includes use of the facilities at the Center; see below.)

The *El Tucán Visitor Center* (cel 6567-5634), part of the CEASPA project, is in the town of Achiote itself, about 3 km beyond the point where the road leaves the forest. The center is open from 8:00 AM-4:00 PM Monday-Friday (open on weekends by request if you call in advance, or when courses are in progress) has coffee, drinking water, and rest rooms. It also has books, illustrated maps, local arts and crafts, and t-shirts for sale, as well as a bird list for Achiote Road and the El Trogón Trail ($3.00). Behind the center is a short trail through a coffee garden, leading to a wooden deck overlooking the valley. Local guides for Achiote Road and nearby areas can also be contacted through the Center and the Ecotourism Group *Los Rapaces* (cel 6637-8522), and lunch can be obtained at the Restaurant Cascá, about 1 km beyond the Visitor Center, by calling at least a day in advance. See the section on accommodations for information on arranging a home stay in Achiote.

Northern Sector

After crossing the Gatún Locks, drive straight ahead to reach the northern sector of the park, including the former Ft. Sherman headquarters area and colonial Ft. San Lorenzo.

Gatún Drop Zone Road: About 1.7 km from the Locks the road crosses the French Canal. This is a remnant of the excavations for a canal made by a French company in the 1880s but abandoned when it went bankrupt. The later American-built Canal bypassed this short stretch, which can be scanned for kingfishers and other species. About 100 m beyond the bridge, next to a green-and-yellow national park sign, a gravel road takes off to the left. This goes to the Gatún Drop Zone, a large grassy area that was formerly used for parachute practice by the U.S. military. The road extends for 2 km, ending at the edge of the French Canal. Grassland species such as Smooth-billed Ani, Red-breasted Blackbird, and seedeaters can be found here. Gray-breasted Crake occurs but is very elusive. American Pygmy Kingfisher has been seen on the small tidal creek 300 m in from the entrance. The area is particularly good for night birding, with Common and Great Potoo being frequently seen at the edge of the woodland, while Barn Owl and Black-crowned and Yellow-crowned Night-Herons are found in the fields beside the road. Boat-billed Heron can also sometimes be seen on the tidal creek.

Black Tank Road: This gravel road leaves the main road to the left at 2.5 km past the Locks. After 800 m, the road forks. Take the right fork, which continues another 500 m before becoming too overgrown to continue. The left fork ends at two small houses. Sepia-capped Flycatcher has sometimes been seen on the right fork.

Skunk Hollow Road: At 9.1 km from the Locks, the entrance to Skunk Hollow Road is marked by a green and yellow national park sign on the left. The road is overgrown but walkable, The first 500 m passes through forest growing on limestone that resembles the dry Pacific side forest because water drains from the soil so rapidly, then enters more typical wet Atlantic forest. The road extends for 3 km along a ridge.

Sherman and San Lorenzo: About 1 km beyond Skunk Hollow the road enters mangroves where Black-tailed Trogon and Straight-billed Woodcreeper can be found. At 12.3 km from the locks the Sherman gate is reached, where you will be asked to produce identification. Proceed 300 m through a housing area, where there is a sign indicating the turnoff to Ft. San Lorenzo to the left. If you continue straight at this point, you will come to the former airfield and the main buildings of the former base. During the migratory period the grassland around the airfield and the rocky beaches near the cove at its north end can be good for shorebirds. Some of the buildings here are controlled by Panama's Maritime Service, so if you see a guard it is best to inquire where you may be permitted to bird.

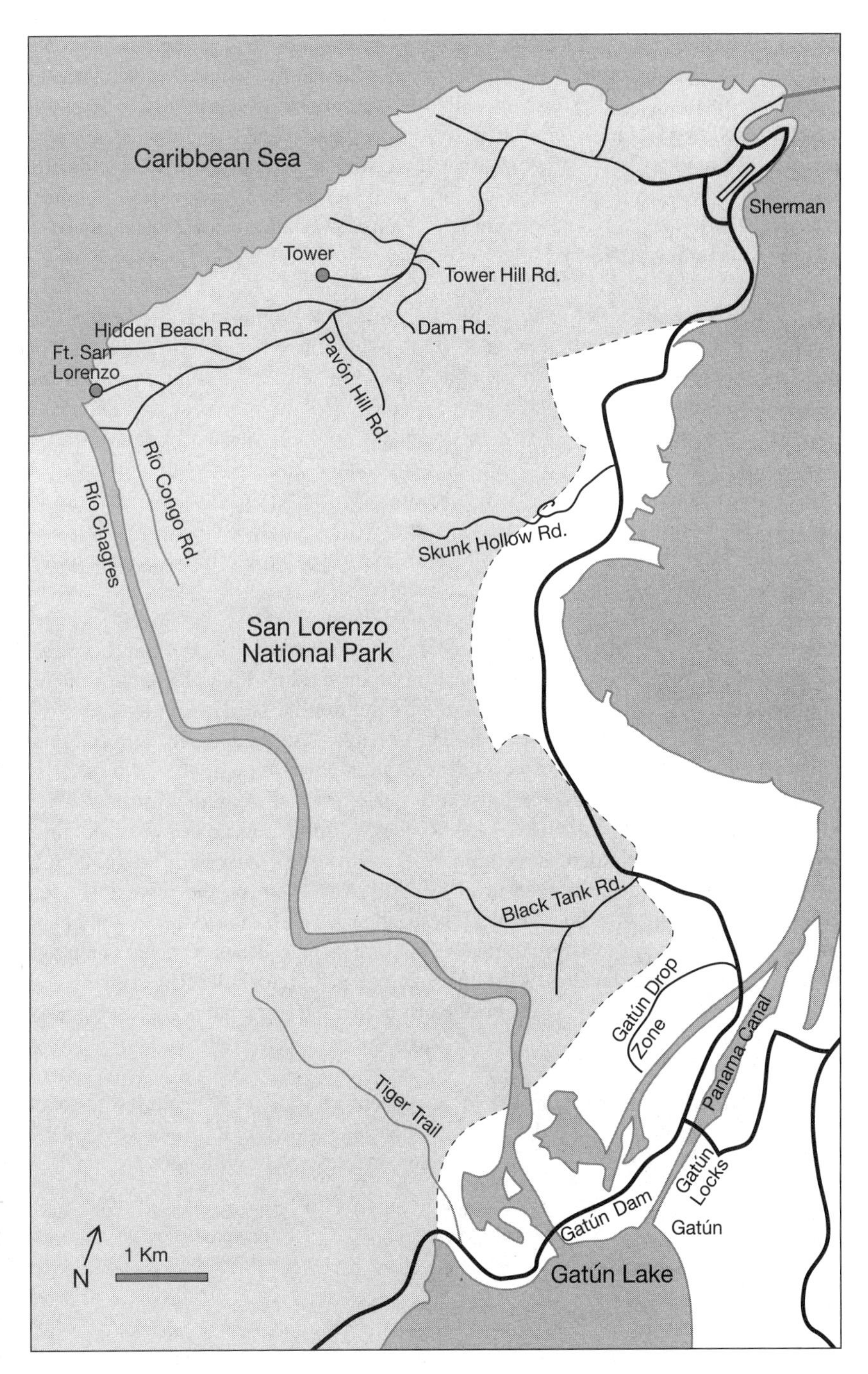

San Lorenzo National Park (north): Sherman and Ft. San Lorenzo

Taking the left turn for San Lorenzo will bring you to another sign and turnoff to the left after 900 m. Distances from this point on are given from this turnoff. Another 1.3 km on this road will bring you to the park administration building, where you will be asked to pay admission ($3.50 for foreign visitors and $1.50 for Panama residents). The road to San Lorenzo is gravel and can be handled with 2WD with care. At 4.4 km the road forks, with the left fork (well signposted) going to San Lorenzo. (The right fork goes to a fenced communications tower after a short distance and is of little interest for birding.)

At 7.8 km the road forks again, with the right fork (signposted) going the San Lorenzo fort, reaching it after another 500 m, while the other goes to the mouth of the Río Chagres. The ruined fort commands magnificent views of the coast and the mouth of the Río Chagres which it once defended. Seabirds and in season Peregrine Falcon can be seen cruising past this vantage point. In the early rainy season colonies of Crested and Chestnut-headed Oropendolas and Yellow-rumped Cacique can be found in the large trees near the parking lot. The fork of the road that goes to the Chagres is one of the better places to find White-thighed Swallow. A trail that leaves the San Lorenzo fork about 300 m from its start on the left descends to the Chagres fork and has good birding.

Sherman side roads: Many gravel roads branch off the main road to San Lorenzo. Most are now overgrown, but some can still be driven in a high-clearance vehicle, and most can easily be walked. Check their condition before attempting them. All distances are given from the turnoff to San Lorenzo described above. The entrance to several of these roads can be found near a national park sign on the main road. At 4.4.km, on the left side of the road just before the turn to the communications tower mentioned above, is the entrance to Dam Road (1.1 km). A short distance before this, on the same side of the road, is the unmarked entrance to Tower Hill Road (200 m). At 5.4 km on the left is the road to Pavón Hill. At 6.8 km on the right is Hidden Beach Road, possibly the best of all of these side roads. The road, still in good condition, extends 1.4 km to a turnaround just before a small beach. Crested Guan has been seen on its outer reaches. After the fork to San Lorenzo, the Río Congo Road is on the left just after the concrete bridge just before the river, but is at present heavily overgrown.

Bird list for San Lorenzo National Park and environs. The list includes adjacent areas, such as Gatún Dam and Spillway, Sherman, and Achiote Road beyond the park. * = vagrant, R = rare, ? = present status uncertain, X = no longer present.

Great Tinamou
Little Tinamou
Blue-winged Teal
Lesser Scaup R
Gray-headed Chachalaca
Crested Guan R
Great Curassow R
Marbled Wood-Quail
Brown Pelican
Neotropic Cormorant
Anhinga
Magnificent Frigatebird
Rufescent Tiger-Heron
Great Blue Heron
Great Egret
Snowy Egret
Little Blue Heron
Tricolored Heron
Reddish Egret*
Cattle Egret
Green Heron
Striated Heron
Agami Heron R
Black-crowned Night-Heron

Yellow-crowned Night-Heron
Boat-billed Heron
White Ibis
Jabiru*
Black Vulture
Turkey Vulture
King Vulture
Osprey
Gray-headed Kite
Hook-billed Kite
Swallow-tailed Kite
Double-toothed Kite
Mississippi Kite
Plumbeous Kite
Bicolored Hawk R
Crane Hawk
Plumbeous Hawk R
Semiplumbeous Hawk
White Hawk
Common Black-Hawk
Great Black-Hawk
Gray Hawk
Roadside Hawk
Broad-winged Hawk
Short-tailed Hawk
Swainson's Hawk
Zone-tailed Hawk
Crested Eagle R
Harpy Eagle*
Black Hawk-Eagle
Barred Forest-Falcon
Slaty-backed Forest-Falcon R
Collared Forest-Falcon
Red-throated Caracara*
Crested Caracara
Yellow-headed Caracara
Laughing Falcon
Bat Falcon
Peregrine Falcon
White-throated Crake
Gray-breasted Crake R
Gray-necked Wood-Rail
Rufous-necked Wood-Rail R
Uniform Crake R
Colombian Crake R
Spotted Rail R
Purple Gallinule
Common Moorhen
Sunbittern R
Limpkin
Black-bellied Plover
Collared Plover
Wilson's Plover
Semipalmated Plover
Wattled Jacana
Greater Yellowlegs
Lesser Yellowlegs
Solitary Sandpiper
Willet
Wandering Tattler*
Spotted Sandpiper
Whimbrel
Long-billed Curlew*
Ruddy Turnstone
Red Knot R
Sanderling R
Semipalmated Sandpiper
Western Sandpiper
Least Sandpiper
Pectoral Sandpiper
Buff-breasted Sandpiper
Short-billed Dowitcher
Wilson's Snipe
Wilson's Phalarope
Laughing Gull
Bonaparte's Gull*
Gull-billed Tern
Royal Tern
Sandwich Tern
Common Tern
Black Tern
Brown Noddy R
Black Skimmer R
Rock Pigeon
Pale-vented Pigeon
Scaled Pigeon
Short-billed Pigeon
Ruddy Ground-Dove
White-tipped Dove
Gray-chested Dove
Olive-backed Quail-Dove R
Violaceous Quail-Dove
Ruddy Quail-Dove
Orange-chinned Parakeet
Brown-hooded Parrot
Blue-headed Parrot
Red-lored Amazon
Mealy Amazon
Black-billed Cuckoo
Yellow-billed Cuckoo
Squirrel Cuckoo
Little Cuckoo
Striped Cuckoo
Pheasant Cuckoo R
Rufous-vented Ground-Cuckoo R
Greater Ani
Smooth-billed Ani
Vermiculated Screech-Owl
Tropical Screech-Owl
Crested Owl
Spectacled Owl
Mottled Owl
Black-and-white Owl
Short-tailed Nighthawk
Lesser Nighthawk
Common Nighthawk
Common Pauraque
Ocellated Poorwill*
White-tailed Nightjar R
Great Potoo
Common Potoo
White-collared Swift
Chimney Swift
Short-tailed Swift
Band-rumped Swift
Lesser Swallow-tailed Swift
Rufous-breasted Hermit
Band-tailed Barbthroat
Green Hermit*
Long-billed Hermit
Stripe-throated Hermit
White-tipped Sicklebill*
White-necked Jacobin
Veraguan Mango*
Black-throated Mango
Rufous-crested Coquette
Garden Emerald
Violet-crowned Woodnymph
Violet-bellied Hummingbird
Sapphire-throated Hummingbird
Blue-chested Hummingbird
Rufous-tailed Hummingbird
White-vented Plumeleteer
Bronze-tailed Plumeleteer
Purple-crowned Fairy
White-tailed Trogon
Violaceous Trogon
Black-throated Trogon
Black-tailed Trogon
Slaty-tailed Trogon
Blue-crowned Motmot
Rufous Motmot
Broad-billed Motmot
Ringed Kingfisher
Belted Kingfisher
Green Kingfisher
Amazon Kingfisher
Green-and-rufous Kingfisher
American Pygmy Kingfisher
White-necked Puffbird
Black-breasted Puffbird
Pied Puffbird
White-whiskered Puffbird
Gray-cheeked Nunlet X
Great Jacamar
Spot-crowned Barbet
Collared Aracari
Yellow-eared Toucanet R

Keel-billed Toucan
Chestnut-mandibled Toucan
Olivaceous Piculet
Black-cheeked Woodpecker
Red-crowned Woodpecker
Yellow-bellied Sapsucker R
Cinnamon Woodpecker
Lineated Woodpecker
Crimson-crested Woodpecker
Slaty Spinetail
Buff-throated Foliage-gleaner
Plain Xenops
Tawny-throated Leaftosser
Scaly-throated Leaftosser
Plain-brown Woodcreeper
Ruddy Woodcreeper
Long-tailed Woodcreeper
Wedge-billed Woodcreeper
Northern Barred-Woodcreeper
Straight-billed Woodcreeper
Cocoa Woodcreeper
Black-striped Woodcreeper
Fasciated Antshrike
Great Antshrike
Barred Antshrike
Western Slaty-Antshrike
Spot-crowned Antvireo
Moustached Antwren
Pacific Antwren
Checker-throated Antwren
White-flanked Antwren
Dot-winged Antwren
Dusky Antbird
Jet Antbird
Bare-crowned Antbird
White-bellied Antbird
Chestnut-backed Antbird
Dull-mantled Antbird R
Immaculate Antbird*
Spotted Antbird
Bicolored Antbird
Ocellated Antbird
Black-faced Antthrush
Black-crowned Antpitta ?
Streak-chested Antpitta
Brown-capped Tyrannulet
Southern Beardless-Tyrannulet
Yellow-crowned Tyrannulet
Forest Elaenia
Greenish Elaenia
Yellow-bellied Elaenia
Lesser Elaenia
Olive-striped Flycatcher R
Ochre-bellied Flycatcher
Sepia-capped Flycatcher R
Yellow Tyrannulet
Yellow-green Tyrannulet R
Paltry Tyrannulet
Black-capped Pygmy-Tyrant
Southern Bentbill
Slate-headed Tody-Flycatcher
Common Tody-Flycatcher
Black-headed Tody-Flycatcher
Brownish Twistwing
Olivaceous Flatbill
Yellow-olive Flycatcher
Yellow-margined Flycatcher
Golden-crowned Spadebill
Royal Flycatcher
Ruddy-tailed Flycatcher
Sulphur-rumped Flycatcher
Black-tailed Flycatcher
Olive-sided Flycatcher
Western Wood-Pewee
Eastern Wood-Pewee
Tropical Pewee
Yellow-bellied Flycatcher
Acadian Flycatcher
Alder Flycatcher
Willow Flycatcher
Vermilion Flycatcher*
Long-tailed Tyrant
Bright-rumped Attila
Sirystes
Rufous Mourner
Dusky-capped Flycatcher
Panama Flycatcher
Great Crested Flycatcher
Lesser Kiskadee
Great Kiskadee
Boat-billed Flycatcher
Rusty-margined Flycatcher
Social Flycatcher
Gray-capped Flycatcher
White-ringed Flycatcher R
Streaked Flycatcher
Sulphur-bellied Flycatcher
Piratic Flycatcher
Tropical Kingbird
Eastern Kingbird
Gray Kingbird
Fork-tailed Flycatcher
Thrush-like Schiffornis
Rufous Piha
Speckled Mourner
Cinnamon Becard
White-winged Becard
Masked Tityra
Black-crowned Tityra
Blue Cotinga
Purple-throated Fruitcrow
Golden-collared Manakin
Blue-crowned Manakin
Red-capped Manakin
White-eyed Vireo*
Blue-headed Vireo*
Yellow-throated Vireo
Red-eyed Vireo
Yellow-green Vireo
Black-whiskered Vireo R
Scrub Greenlet
Tawny-crowned Greenlet
Golden-fronted Greenlet
Lesser Greenlet
Green Shrike-Vireo
Black-chested Jay
Common Raven*
Purple Martin
Gray-breasted Martin
Brown-chested Martin R
Tree Swallow R
Mangrove Swallow
Blue-and-white Swallow R
White-thighed Swallow
Northern Rough-winged Swallow
Southern Rough-winged Swallow
Sand Martin
Cliff Swallow
Barn Swallow
White-headed Wren
Black-bellied Wren
Bay Wren
Stripe-breasted Wren R
Rufous-and-white Wren
Buff-breasted Wren
Plain Wren
House Wren
White-breasted Wood-Wren
Southern Nightingale-Wren
Song Wren
Tawny-faced Gnatwren
Long-billed Gnatwren
Tropical Gnatcatcher
Veery
Gray-cheeked Thrush
Swainson's Thrush
Wood Thrush
Clay-colored Thrush
Gray Catbird
Tropical Mockingbird
European Starling*
Yellowish Pipit R
Golden-winged Warbler

Tennessee Warbler
Northern Parula R
Yellow Warbler
Chestnut-sided Warbler
Magnolia Warbler
Cape May Warbler R
Black-throated Blue Warbler R
Black-throated Green Warbler
Blackburnian Warbler
Yellow-throated Warbler R
Bay-breasted Warbler
Cerulean Warbler R
Black-and-white Warbler
American Redstart
Prothonotary Warbler
Ovenbird
Northern Waterthrush
Louisiana Waterthrush
Kentucky Warbler
Mourning Warbler
Hooded Warbler R
Canada Warbler
Rufous-capped Warbler
Buff-rumped Warbler
Bananaquit
Rosy Thrush-Tanager
Dusky-faced Tanager
Olive Tanager
Gray-headed Tanager
Sulphur-rumped Tanager
White-shouldered Tanager
Tawny-crested Tanager
White-lined Tanager
Red-throated Ant-Tanager
Summer Tanager
Scarlet Tanager
Crimson-backed Tanager
Flame-rumped Tanager
Plain-colored Tanager
Bay-headed Tanager
Golden-hooded Tanager
Scarlet-thighed Dacnis
Blue Dacnis
Green Honeycreeper
Shining Honeycreeper
Red-legged Honeycreeper
Blue-gray Tanager
Palm Tanager
Blue-black Grassquit
Slate-colored Seedeater R
Variable Seedeater
Lined Seedeater*
Yellow-bellied Seedeater
Ruddy-breasted Seedeater
Lesser Seed-Finch
Saffron Finch
Orange-billed Sparrow
Black-striped Sparrow
Dark-eyed Junco*
Streaked Saltator
Buff-throated Saltator
Black-headed Saltator
Slate-colored Grosbeak
Rose-breasted Grosbeak
Blue-black Grosbeak
Blue Grosbeak
Indigo Bunting
Dickcissel
Bobolink
Red-breasted Blackbird
Eastern Meadowlark
Yellow-headed Blackbird*
Great-tailed Grackle
Shiny Cowbird
Giant Cowbird
Orchard Oriole
Yellow-backed Oriole
Yellow-tailed Oriole
Baltimore Oriole
Yellow-billed Cacique
Scarlet-rumped Cacique
Yellow-rumped Cacique
Crested Oropendola
Chestnut-headed Oropendola
Montezuma Oropendola
Yellow-crowned Euphonia
Thick-billed Euphonia
Fulvous-vented Euphonia
White-vented Euphonia
Lesser Goldfinch

Areas east of the Canal

While the best birding on the Atlantic side is to be found in San Lorenzo National Park west of the Canal, a few areas on the east side can be good as well if you are staying in the area and have plenty of time. These include Galeta Island and areas around the town of Margarita and the Mindi Third Locks Excavation.

Galeta Island

Galeta Island is a former US Naval Station that is presently the site of a marine research station and education center operated by the Smithsonian Tropical Research Institute. It is reached by a good asphalt road that takes off from the right side of the north end of Randolph Road (Calle Randolph) east of Colón. The road passes through extensive areas of mangroves where Black-tailed Trogon and Olivaceous, Straight-billed and Streak-headed Woodcreepers can often be found. It's also good for migrant warblers, including on rare occasions Northern Parula. It is 4.5 km from the turnoff from Randolph Road to the Smithsonian facilities. At 2.5 km there is a small ANAM station and a gate that is kept closed at night. For an early visit, it's best to contact the Smithsonian to make sure that the gate will be open. The

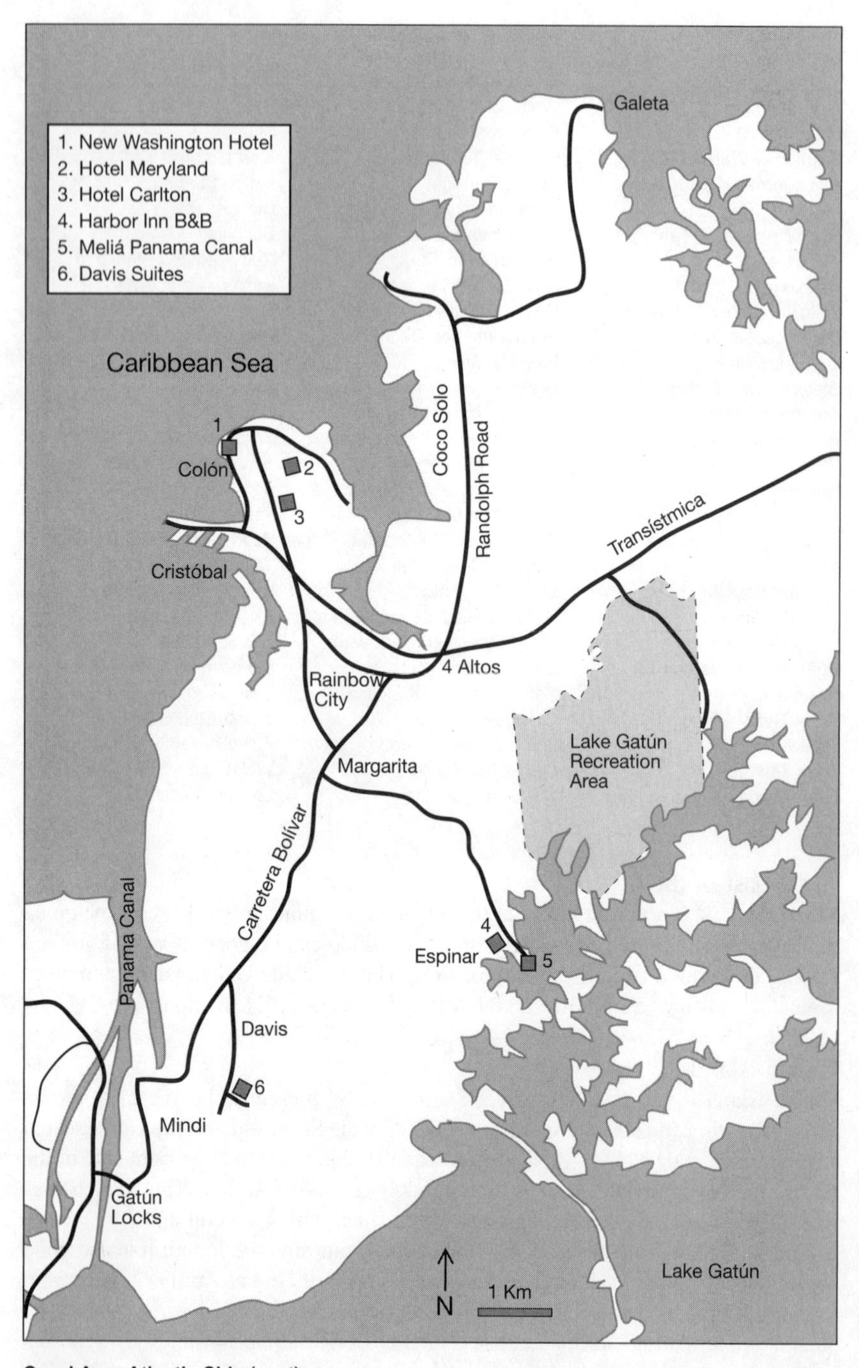

Canal Area Atlantic Side (east)

Smithsonian station (ph 212-8191, 212-8192; galeta@si.edu; www.stri.org/english/visit_us/galeta/) features aquaria, public exhibits, and a boardwalk through the mangroves. To visit the station and its exhibits, prior reservations are necessary. Call the numbers above, or send an e-mail to the address given. The entry fees are adults $2.00; primary or secondary school students $0.50, university students $1.00, retirees $1.50.

Margarita

Margarita is a residential area located south of Colón on the Bolivar Highway en route to San Lorenzo National Park. Its tree-planted streets and lawns can be good for the more common garden birds. Two woodland areas adjacent to Margarita, the Margarita Tank Farm and Rancho Ramos, provide good birding including some forest species as well as those of edge and secondary habitats.

The *Margarita Tank Farm* is an oil storage facility near Margarita, on the north side of the Bolivar Highway and of the diversion canal that parallels it. Turn off the Bolivar Highway to the right (going west in the direction of San Lorenzo) opposite the DUASA (Desarrollo Urbanístico del Atlántico S. A.) building, cross the bridge over the diversion channel and park on the grass near the perimeter fence. Follow the perimeter fence to the left and the drainage ditches through the woodland. This woodland is excellent for birding, including species such as Black-tailed Trogon, Olivaceous Piculet, Jet Antbird, Rosy Thrush-Tanager, and wrens, as well as migrant warblers. The diversion canal has Boat-billed Herons (at night) and American Pygmy Kingfisher.

Rancho Ramos is a patch of woodland just to the west of Margarita that can be good for a few hours birding if you don't have enough time to get over to Achiote or Sherman. It can be found by turning off the Bolivar Highway to the left (going west in the direction of San Lorenzo) at the DUASA building mentioned above. This turnoff is marked by a green-and-white sign that says "Alhambra." After 700 m, turn right at a bus shelter immediately after a large walled housing development. This asphalt road immediately crosses a small yellow bridge, and then in 100 m comes to another small bridge. Just before the bridge is a gravel road to the left with a chain across it where you can walk in. After about 150 m through scattered scrub and grass, the road reaches taller secondary forest. The woodland has some forest species that are usually restricted to larger patches such as Spotted Antbird and Song Wren, as well as species difficult to find on the Atlantic side such as Olivaceous Piculet, Greenish Elaenia, Yellow-olive Flycatcher, and Rufous-and-white Wren. It is also very good for migrant warblers. Continuing on the asphalt road past the bridge, at 800 m from the turnoff another narrower asphalt road goes off to the right. This side road, the Old Gatún Road, extends for about 900 m through woodland and second growth before ending in back of a golf course, and may be good for woodland birds.

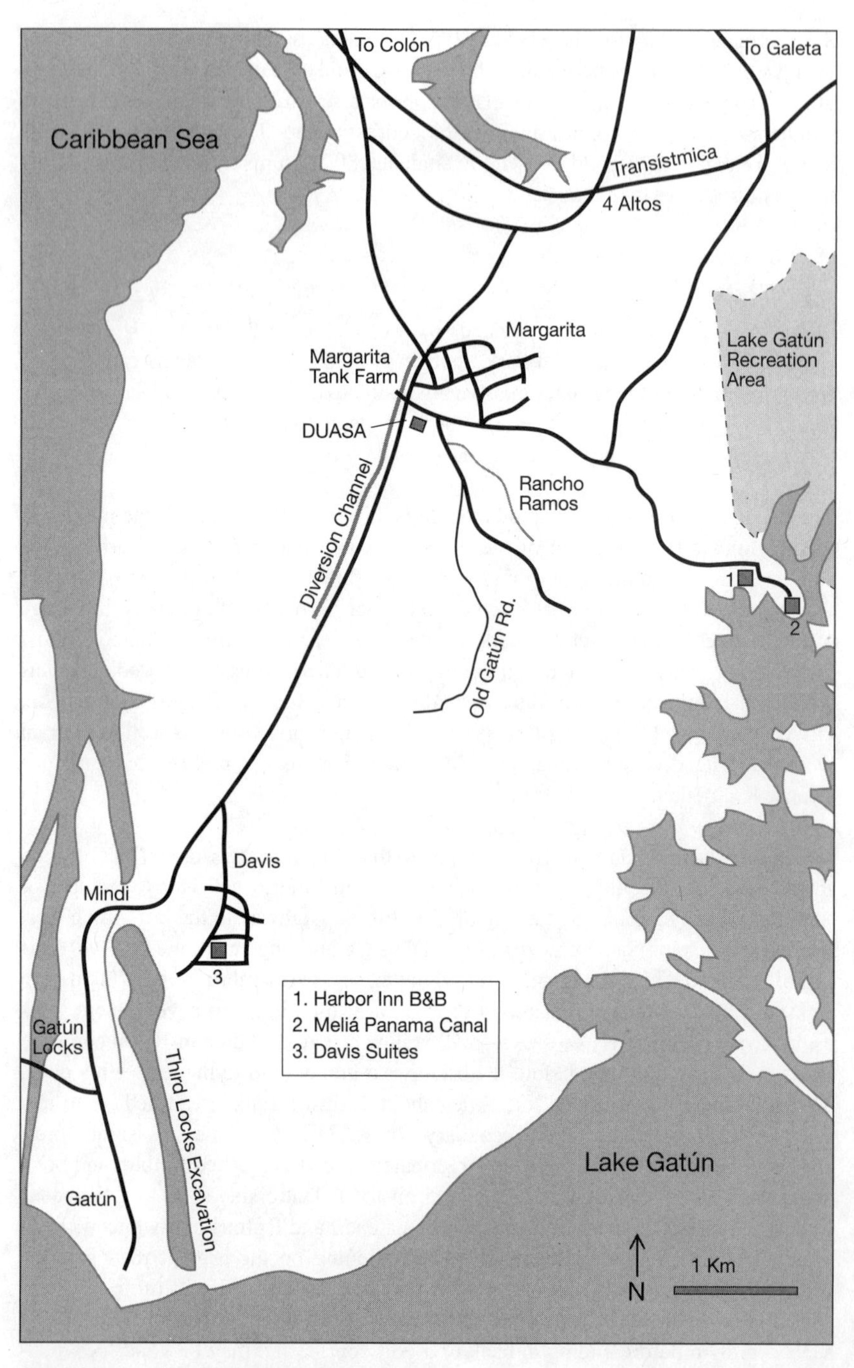

Canal Area Atlantic Side: Margarita and Gatún

Mindi Third Locks Excavation

The Third Locks site was excavated during WW II for the construction of an additional set of locks for the Canal, but the project was never completed. Now filled with water, this elongated pond can be seen to the south of the Bolivar Highway (left heading towards San Lorenzo) a little before it reaches the Gatún Locks. Park off the road, and scan for Snail Kites and, in season, migrant ducks and American Coot. On the western side of the excavation, a small path follows a barbed wire fence at the edge of a pasture, reaching the gravel roadbed of the former route of the Railroad (now relocated to the other side of the cut) after about 50 m. The roadbed can be walked to look for Bran-colored Flycatcher, Rufous-capped Warbler, and Shiny Cowbird, and, in season, migrant Yellow-rumped Warbler.

Getting to San Lorenzo National Park. From the intersection of Vía España with Av. Federico Boyd, take Av. Manuel Espinosa Batista (which is the continuation northward of Av. Federico Boyd) north. At 1.0 from Vía España, you come to a traffic circle at an underpass below Vía Simón Bolívar (Transístmica). Continue straight through the traffic circle; Av. Manuel Espinosa Batista here changes its name and becomes Av. Ricardo J. Alfaro (Tumba Muerto). At 700 m past this traffic circle, just after passing under a pedestrian overpass, take a left at the traffic signal, signposted for the Corredor Norte. After 600 m you will reach the edge of Metropolitan Park; follow the signs to go right for the Corredor Norte and Colón. (Of course, you can get on any other entrance of the Corredor Norte as well if you prefer.) Watch carefully for the exit for Colón, which is 5.0 km from the entrance at Metropolitan Nature Park. The toll for the section that goes north here is $2.40. After 20.5 km, you will exit from the Corredor onto the eastern end of Madden Road; go towards the right. In about 900 m you will reach the entrance ramp to the Transístmica, on the left just before an overpass. At the top of the ramp, go left for Colón. After 49.5 km, you will see a large concrete overpass ahead. Do not take the overpass, but get off the highway on the right side, where you will shortly come to a traffic light. This is the "4 Corners" (*4 Altos* on local signs), a major landmark on the Atlantic side. Take a left at the light. This intersection is supposedly treated as a four-way stop when the signal is not working, which is all the time, but use caution with oncoming traffic. After 800 m on this road, take a left onto Diversion Road. Turn with care since oncoming traffic in two lanes has right-of-way. Watch carefully for this intersection, which is easy to miss; there is a green-and-white sign *Cruce de Ferrocarril* (Railway Crossing) sign just before it, and a sign for the Melia Panama Canal Hotel at its entrance. If you miss it carry on and turn around as soon as you can to catch it from the other direction.

At 1.3 km beyond this turnoff is a somewhat confusing intersection with the Bolivar Highway, marked by a white sign for Margarita and a stop sign. Oncoming traffic has right of way, even when turning across your path. Go straight over this 4-way junction, then bear right and stop again. Turn left onto the highway. After 8.1 km, on the right, is the turnoff for the Gatún Locks, signposted "Sherman" and "Esclusas." In 200 m is another turnoff to the right which goes to the Locks themselves. The

Locks are crossed by a one-lane swing bridge that is only open to traffic when ships are not entering or leaving the chambers. The wait can range from a few minutes to a half-hour, to exceptionally an hour or more. Once across the locks, turn left for the southern sector of San Lorenzo National Park, and stay straight for the northern sector. Follow the directions in each sector for specific localities.

Getting to Galeta Island. Follow the directions for San Lorenzo National Park above as far as the 4 Corners. Turn right at the traffic light onto Randolph Road, and go 4.5 km north, where the road to Galeta turns off to the right just before Randolph ends at a small guard kiosk. The entry is marked by a sign a short distance after the turn. The road is 4.5 km to where it ends at the Smithsonian facilities.

Getting to Margarita Tank Farm and Rancho Ramos. Follow the directions for San Lorenzo National Park above. The Tank Farm and Rancho Ramos are located 1.6 km along the Bolivar Highway from the turnoff from Diversion Road marked by a sign for Margarita. The DUASA building is on the left side of the intersection, which is also marked by a green and white sign pointing left that says "Alhambra." The Tank Farm is to the right at this intersection, and Rancho Ramos to the left. Follow the directions in the site descriptions to reach each site.

Getting to the Mindi Third Locks Excavation. Follow the directions for San Lorenzo National Park above. The Locks Excavation is located 6.6 km along the Bolivar Highway from the turnoff from Diversion Road marked by the Margarita sign, on the left hand just after a railroad crossing.

Accommodations and meals on the Atlantic side of the Canal Area. While the Atlantic side can be visited as a day trip from Panama City, it is about a 90-minute drive from downtown to Achiote Road (somewhat less from hotels in the southern Canal Area), even longer if you have a wait at Gatún Locks. Getting there in time for the best birding in the early morning requires driving the Transístmica in the dark, which is no fun. If possible, it is better to stay on the Atlantic side overnight. Although accommodations are much more limited than on the Pacific side, there are several good places to stay.

The *Sierra Llorona Panama Lodge* (ph 442-8104, cel 6614-8191; info@sierrallorona.com; www.sierrallorona.com) is in Santa Rita Arriba, 4.5 km off the Carretera Transístmica about 11 km before Colón, located in a 200-hectare forested reserve at 180 m (550 ft) elevation. There are two rooms with shared bath, and a suite with private bath; all baths have hot water. The rooms have ceiling fans rather than air-conditioning, which is sufficient at this altitude. Rooms alone cost $68.00-$98.00, including breakfast. All-inclusive packages including three meals and a guided walk are $95.00-$128.00 (single occupancy; two people in a room pay $63.00-$79.00 each). Meals can also be purchased separately. A camping site is also available on the grounds, with a thatched-roof shelter suitable for hammocks and an area outside for tents. There are several trails and lookouts on the property, and over

200 species of birds have been seen in the area. The Lodge provides a convenient and attractive base for independent birding on the Atlantic side of the Canal Area, and also offers half-day and full-day birding tours to sites such as Achiote Road, San Lorenzo and Sherman, Gamboa and Pipeline Road, Plantation Road, Metropolitan Nature Park, and Cerro Azul-Cerro Jefe. The Lodge also offers one or two-week all-inclusive birding packages. Tours are also offered to points of interest such as the Panama Canal, Portobelo, and many other options. Day visits and birding at the lodge are also available for non-guests.

To get to the Lodge, follow the directions for San Lorenzo National Park on pp. 105-106. The turnoff for Santa Rita Arriba and Sierra Llorona is 38 km after the turnoff from Madden Road onto the Transístmica. If you come to the town of Sabanitas, marked by a pedestrian overpass and large El Rey supermarket on the right, you have gone 2 km too far. Turn around at the first opportunity and head back. The road to Santa Rita Arriba is at first potholed asphalt, and then becomes dirt and gravel. The last part of the road may require 4WD or high clearance in the rainy season; check with the Lodge for conditions. At 2.8 km from the Transístmica there is a sharp turn to the left. At 4.5 km you will come to a small yellow house next to a communications antenna. The entrance to the Lodge is just beyond this, where a concrete drive descends steeply on the left. You can arrange to have the Lodge pick you up from Tocumen Airport or other places in the Canal Area. You can also take a bus to Sabanitas and have them pick you up from there.

The *Melia Panama Canal Hotel* (ph 470-1100; fax 470-1200; melia.panama.canal @solmelia.com; www.solmelia.com) is located on a peninsula that juts into Lake Gatún, near Espinar, a former U.S. Army Base. It has 286 rooms and suites ($77.00-$165.00), all with private hot-water bath and air-conditioning. The hotel, which once was the School of the Americas used by the US in training Latin American military officers, has amenities such as an outdoor swimming pool, tennis court, spa, etc., and offers excursions to birding sites. To find it, follow the directions for Rancho Ramos (p. 106). Instead of turning at the bus shelter for Rancho Ramos, continue straight on the main road, and follow the signs for the hotel.

The *Harbor Inn Bed and Breakfast* (ph 470-0640; fax 470-0641; harborinn@ cwpanama.net; www.harborinnpanama.com) is in Espinar near the Melia. It has 24 rooms and 3 suites ($40.75-$81.25, continental breakfast included), all with private hot-water bath and air-conditioning, and has a restaurant on premises. To find it, follow the directions for the Melia above. A little before reaching the Melia, you come to a T-junction. The Harbor Inn is just before the T-junction, on the right.

Davis Suites Panama Canal (ph 473-0639, cel 6628-6454; DavisSuites@yahoo.com; www.davis.suites.iwarp.com) is located in a quiet neighborhood on the former U.S. Army Base of Ft. Davis, and is the closest hotel to San Lorenzo, Sherman, and Achiote Road. It has 10 suites, each with private hot-water bath, air-conditioning, and kitchenette. Suites are $55.00 (single or double occupancy; additional guests per suite are $10.00 each). There are no restaurants nearby (aside from the cafeteria at the Ministerio de Hacienda in Davis; see below), but the hotel can arrange to have meals delivered. If staying here, you may wish to pick up groceries in Panama City or at the El Rey in Sabanitas and cook for yourself. The hotel can arrange tours to Achiote Road and other birding sites. To get there, follow the directions for San Lorenzo National Park (pp. 105-106). After turning on to the Bolivar Highway near Margarita, the turnoff to Davis is at about 4.0 km, on the left. Continue straight on this road, and the Suites will be on the left at 1.3 km from the entrance.

For those who may be interested in getting to know the local culture as well as birding, you can arrange to stay in a private home in the village of *Achiote* through the local ecotourism group, Los Rapaces (cel 6637-8522). Accommodation is basic, without hot water or air-conditioning. The cost varies from $5.00 to $8.00 per night, and may include breakfast. Prior booking is required.

Rancho Carolina (ph 221-6955, cel 6563-6283; agroecoturcarol@hotmail.com) is a new hacienda-style lodge located near the town of Agua Salud on the coast road about 22 km beyond Achiote. The two-story house has seven rooms ($85.00, including all meals) with private baths, some of which have hot water, and air-conditioning. Because the area is distant from population centers, reservations must be made in advance; you can't just show up here. The house itself is located on the coast in an agricultural area, but the hotel owners have a 1,000-hectare ranch about five minutes away. Some of the ranch is used for cattle, but about 40% is forested. It is a bit more than an hour's walk to reach forest through a variety of open habitats including pasture, shrubby areas, ponds, and gallery forest. It is also possible to hire a boat to go up the Río Indio, about 9 km west of the hotel, either through Rancho Carolina or by making arrangements yourself the day before. The cost is about $35.00 per person (two-person minimum). The birds that have been found in the area to date are mostly those of open country and secondary forest, although the forested areas have been birded relatively little so far. The lodge may be of interest, however, to those who want a secluded place away from the main tourist areas. It would also be possible to bird Achiote Road from here, but the road is winding between Achiote and Piña and the drive would take some time. Accommodations on the east side of the Canal may be more convenient despite the possibility of a wait at the Locks.

To get to Rancho Carolina, follow the directions for Achiote Road and continue through to the town of Achiote. About 12 km beyond the start of the town of Achiote you reach the town of Piña on the coast. Take a left here to go towards Palmas Bellas. At about 2 km past the Piña junction you enter the small town of Chagres, and after 7 km you will see the Chagres police station, with a public phone nearby. Right after the police station there is a bridge that marks the beginning of Palmas Bellas. At 3 km west of the Palmas Bellas bridge there is a smaller bridge. Rancho Carolina is the first house on the right after this bridge. The road is asphalt as far as Palmas Bellas, then gravel, but is scheduled eventually to be paved to the Río Indio.

Staying in Colón itself is not recommended because of its very serious crime problem. If for some reason you must stay there, however, we provide information on three of the more suitable hotels. The historic *New Washington Hotel* (ph 441-7133; fax: 441-7397; nwh@eveloz.com) is located in the northwestern part of the city, next to the sea wall, and has a good view of the harbor. It has 124 rooms ($38.50-$110.00). The more modern *Meryland Hotel* (ph 441-7055; fax 441-7105; reservaciones@hotelmeryland.com; www.hotelmeryland.com) is located in a quiet neighborhood in the northeastern part of town, and has 79 rooms ($35.00-$40.00). The *Hotel Carlton* (ph 447-0349; fax 447-0114; oscalo85@hotmail.com; www.elhotelcarlton.com) is located in the center of town, and has 63 rooms ($30.00-$60.00). All three hotels have private hot water baths and air-conditioning, as well as secure parking and a restaurant on premises. If staying in Colón, it is recommended to dine at your hotel and not to leave the premises on foot.

Options for eating on the Atlantic side are remarkably limited, especially since the Tarpon Club, an old standby next to the Gatún Spillway, is now closed. The Sierra Llorona Lodge provides meals if you are staying there, and there are restaurants open to non-guests at the Melia Hotel and the Harbor Inn. Near the Davis Suites, the cafeteria at the Ministerio de Hacienda provides basic meals, and is open to the public 8:30 AM-4:30 PM weekdays. You can find it by continuing on the road past the Davis Suites for another 500 m. Meals can be obtained at the La Cascá Restaurant (cel 6637-8522) in Achiote by making arrangements a day in advance. In Colón, the Colón 2000 development on the east side of town north of the Free Zone has a small restaurant, the Iguana Café, a Subway store, and a supermarket where you can pick up snacks. Because of the difficulty of crossing the locks, if you plan on birding in the San Lorenzo-Sherman area all day, it is best to pack a lunch unless you make arrangements at the La Cascá Restaurant.

Additional Sites in the Panama City area

Besides sites within Panama City and the Canal Area itself, there are several other good birding localities that are easily accessible to visitors staying in this area. Tocumen Marsh features open country and aquatic birds, while Cerro Azul-Cerro Jefe and Altos de Campana National Park have many foothill species not found in the adjacent lowlands. Isla Taboga in Panama Bay holds an important nesting colony of Brown Pelican that can be interesting to visit.

Tocumen Marsh and Cerro Azul-Cerro Jefe are in the same direction and can be done together as part of the same day trip. If possible, visit Tocumen Marsh first, in the morning, since it can get quite hot there in the afternoon. Birds can be active later in the day in the foothills, so Cerro Azul and Cerro Jefe can be visited after Tocumen.

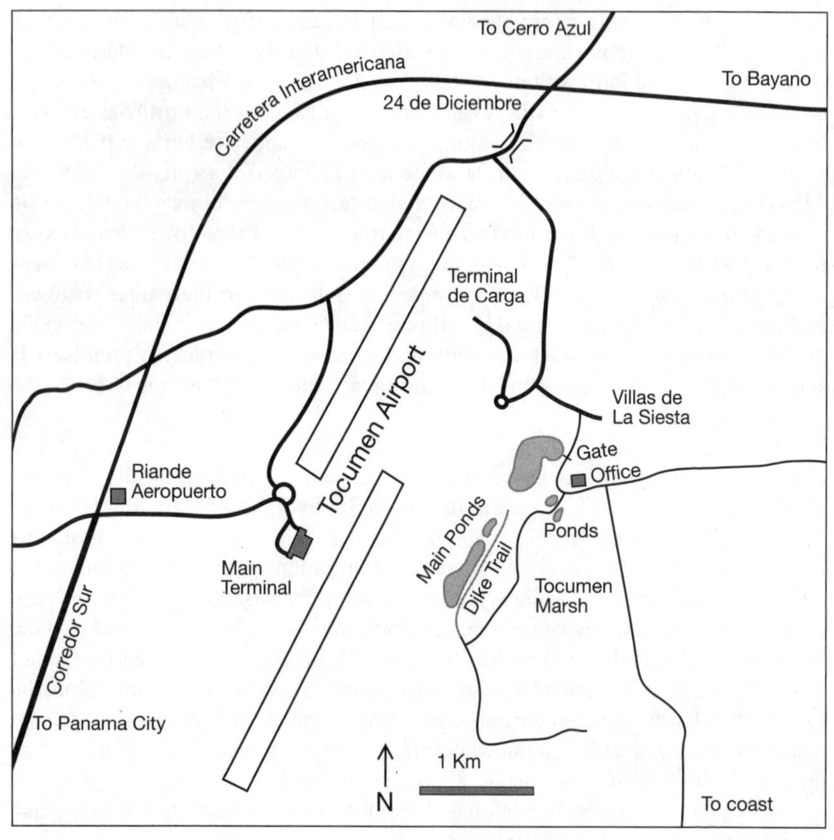

Tocumen Marsh

Tocumen Marsh

Tocumen Marsh is the easiest place near Panama City to see birds of open country, scrub, and freshwater habitats. It once consisted of extensive freshwater wetlands, but these were mostly drained some decades ago to create large rice plantations. However, a few ponds and marshes still remain, and the rice fields themselves can be excellent for some aquatic species at times when they are flooded.

Among the species that are easier to find here than in areas closer to the city are Black-bellied Whistling-Duck, Cocoi and Capped Herons, Glossy Ibis, Lesser Yellow-headed Vulture, Pearl Kite, Savanna Hawk, Laughing Falcon, Striped Owl, Barn Owl, White-tailed Nightjar, Pale-bellied Hermit, Yellow-crowned Amazon, Pale-breasted Spinetail, Pied Water-Tyrant, Yellow Tyrannulet, Mouse-colored Tyrannulet, and Slate-headed Tody-Flycatcher. Bare-throated Tiger-Heron, Least Bittern, and Spot-breasted Woodpecker are rare but possible.

Rails of a great variety of species were commonly seen in the early 1980s, but the habitat has become much less suitable and most of them have seldom been reported in recent years. Rare vagrants that have at times been found in the area have included Long-winged Harrier, Dwarf Cuckoo, Dark-billed Cuckoo, Gray-capped Cuckoo, Lark Sparrow, and Yellow-headed Blackbird.

Tocumen is presently operated by Agropecuaria Palo Grande. To visit, call 295-1175 in advance. Hours for birders are Monday to Friday from 7:00 AM to 4:00 PM, and Saturday from 7:00 AM to 12:00 noon (closed on Sundays). You will need to sign in with the security guard at the front gate, then proceed to the farm headquarters to pay the $5.00 entrance fee. At headquarters, walk around the main building to the left to find the office, which is in the back inside the garage. During the dry season most of the roads may be navigable with 2WD, but 4WD is necessary in the wet season.

From the headquarters, take the road that goes off to the right to get to the main ponds. At about 300 m down this road there are ponds on either side of the road that can be scanned for waterbirds. The one on the right sometimes has nesting Boat-billed Heron and White Ibis; Boat-bills may also roost here during the day when not breeding. At 1.8 km from the headquarters, after driving through extensive rice fields, you come to a T-junction. Park here and walk in on the right, where the road shortly becomes a trail. This trail runs along a dike through scrub, and after about 300 m comes to the first and largest of several ponds. The pond can be checked for Anhinga, Black-bellied Whistling-Duck, and other species. Green-and-rufous and Pygmy Kingfishers can sometimes be found along the ditch on the left side of the dike when it has water in it, and Pale-bellied Hermit along the trail.

If you turn left at the T-junction, the road runs for the next kilometer through scrub before ending 3.7 km from the headquarters. The scrub can be birded for species such as Mouse-colored Tyrannulet and Slate-headed Tody-Flycatcher.

To get to the coast, at headquarters take the road that goes straight beyond it on its right side. At 300 m past headquarters take another road goes off to the right. This reaches the coast in 5.3 km. The first 2 km are through rice fields, and then it enters coastal scrub the rest of the way to the coast, which has extensive mudflats. On the coast many of the species that occur at Costa del Este (pp. 71-72) can be found.

There are several other roads on the property that can be explored. Scan the fields for raptors and other open country birds. After the rains have started, flooded fields may attract ducks, herons, and other waterbirds.

Getting to Tocumen Marsh. From the intersection of Av. Federico Boyd and Vía España, take Federico Boyd two blocks south and take a left at the traffic light onto Av. Nicanor de Obarrio (Calle Cincuenta). Go 700 m, and turn off to the right at a sign for the Corredor Sur. Follow the signs for the airport. When you reach the end of Corredor Sur after 18 km, stay straight on the main road instead of turning off for the airport, until you reach the town of 24 de Diciembre, which is 6.4 km from the airport turnoff. Go through town until you get to a large overpass just beyond the Xtra supermarket (on the left), and turn right at the overpass. At 600 m from this turn you will cross a small bridge, and 100 m past this will come to a small plaza, surrounded by a low blue concrete wall, in the center of an intersection. Take the left fork in front of the plaza. At 2.9 km after this turn, turn left just after the Texaco station, in front of the Supermercado Materiales 288. Stay on this asphalt road as it curves right in front of the Villas de la Siesta housing development. Shortly after this bend, the road becomes dirt. At 900 m from the previous turnoff, you come to the gate for "Tocumen SA" (now Agropecuaria Palo Grande), where you sign in with the security guard. It is another 300 m to the farm headquarters.

Accommodations and meals. Tocumen Marsh is easily done as a day trip from Panama City. The closest hotel is the Riande Aeropuerto Hotel (see p. 75). There are a number of small restaurants and fast-food places along the main street in the town of 24 de Diciembre.

Bird list for Tocumen Marsh. * = vagrant, R = rare, ? = present status uncertain

Little Tinamou
Black-bellied Whistling-Duck
Pied-billed Grebe
Brown Pelican
Neotropic Cormorant
Anhinga
Magnificent Frigatebird
Least Bittern R
Bare-throated Tiger-Heron R
Great Blue Heron
Cocoi Heron
Great Egret
Snowy Egret
Little Blue Heron
Tricolored Heron
Cattle Egret
Green Heron
Striated Heron
Capped Heron
Black-crowned Night-Heron
Boat-billed Heron
White Ibis
Glossy Ibis
Wood Stork
Black Vulture
Turkey Vulture
Lesser Yellow-headed Vulture
Osprey
Pearl Kite
White-tailed Kite
Northern Harrier
Long-winged Harrier*
Crane Hawk
Mangrove Black-Hawk
Great Black-Hawk
Savanna Hawk
Roadside Hawk
Broad-winged Hawk
Short-tailed Hawk
Crested Caracara
Yellow-headed Caracara
Laughing Falcon
American Kestrel
Merlin
Peregrine Falcon
White-throated Crake
Gray-necked Wood-Rail
Sora
Yellow-breasted Crake
Colombian Crake ?
Paint-billed Crake ?
Spotted Rail ?
Purple Gallinule
Southern Lapwing
Black-necked Stilt
Wattled Jacana
Solitary Sandpiper
Willet
Spotted Sandpiper
Whimbrel
Semipalmated Sandpiper
Western Sandpiper
Least Sandpiper
Pectoral Sandpiper
Stilt Sandpiper
Short-billed Dowitcher
Wilson's Phalarope
Laughing Gull
Gray Gull*
Caspian Tern
Pale-vented Pigeon
Plain-breasted Ground-Dove
Ruddy Ground-Dove
White-tipped Dove
Brown-throated Parakeet R
Orange-chinned Parakeet
Blue-headed Parrot
Yellow-crowned Amazon
Dwarf Cuckoo*
Yellow-billed Cuckoo
Dark-billed Cuckoo*
Gray-capped Cuckoo*
Squirrel Cuckoo
Little Cuckoo
Greater Ani
Smooth-billed Ani
Groove-billed Ani
Barn Owl
Tropical Screech-Owl
Striped Owl
Lesser Nighthawk
White-tailed Nightjar
Pale-bellied Hermit
Stripe-throated Hermit
Scaly-breasted Hummingbird
Black-throated Mango
Garden Emerald
Sapphire-throated Hummingbird

Rufous-tailed Hummingbird
Ringed Kingfisher
Amazon Kingfisher
Green Kingfisher
Green-and-rufous Kingfisher
American Pygmy Kingfisher
Olivaceous Piculet
Red-crowned Woodpecker
Pale-breasted Spinetail
Straight-billed Woodcreeper
Great Antshrike
Barred Antshrike
Southern Beardless-Tyrannulet
Mouse-colored Tyrannulet
Yellow-crowned Tyrannulet
Yellow-bellied Elaenia
Yellow Tyrannulet
Pale-eyed Pygmy-Tyrant
Slate-headed Tody-Flycatcher
Common Tody-Flycatcher
Pied Water-Tyrant
Dusky-capped Flycatcher
Panama Flycatcher
Lesser Kiskadee
Great Kiskadee
Boat-billed Flycatcher
Rusty-margined Flycatcher
Social Flycatcher
Tropical Kingbird
Eastern Kingbird
Gray Kingbird
Fork-tailed Flycatcher
Yellow-green Vireo
Golden-fronted Greenlet
Scrub Greenlet
Gray-breasted Martin
Mangrove Swallow
Sand Martin
Cliff Swallow
Cave Swallow R
Barn Swallow
Buff-breasted Wren
Plain Wren
House Wren
Clay-colored Thrush
Tropical Mockingbird
Yellowish Pipit
Yellow Warbler
Prothonotary Warbler
Northern Waterthrush
Common Yellowthroat
Hooded Warbler
Summer Tanager
Crimson-backed Tanager
Blue-gray Tanager
Palm Tanager
Red-legged Honeycreeper
Blue-black Grassquit
Variable Seedeater
Yellow-bellied Seedeater
Ruddy-breasted Seedeater
Black-striped Sparrow
Lark Sparrow*
Streaked Saltator
Rose-breasted Grosbeak
Blue-black Grosbeak
Indigo Bunting
Dickcissel
Bobolink
Red-breasted Blackbird
Yellow-headed Blackbird*
Great-tailed Grackle
Shiny Cowbird
Orchard Oriole
Yellow-crowned Euphonia

Cerro Azul and Cerro Jefe

The mountain ranges of eastern Panama, home to many endemics, extend to just northeast of Panama City. Rising to 1,007 m (3,300 ft) on Cerro Jefe, these cooler highlands hold many foothills species as well, so that about 40 additional species not present in the Canal Area lowlands can be found here. Many Panama City residents maintain weekend homes in the Cerro Azul area, just below Cerro Jefe, to take advantage of the refreshing climate, so that access is relatively easy via paved roads. The area is an easy day trip from Panama City, but also features some convenient places to stay for those who may want to spend more time there.

Weather can be quite different from that in Panama City. Even in the dry season the peaks may be shrouded in cloud and drizzle. On the other hand, bird activity is not nearly as dependent on time of day as it is in the hot lowlands, so that birding almost any time can be productive and it is not so essential to get an early start.

Distances to birding sites in the Cerro Azul and Cerro Jefe area are given from the start of the road to Cerro Azul, at the turnoff described in the section on "Getting to Cerro Azul" following this section.

Monte Fresco. The Monte Fresco area, at about 600 m (2000 ft) elevation, is good for migrants in season and some foothill species all year. The start of the road, on the left at 11.2 km from the start of the Cerro Azul Road, is marked by a very faded "Monte Fresco" sign. At 500 m along this paved road, turn off onto a dirt road to the

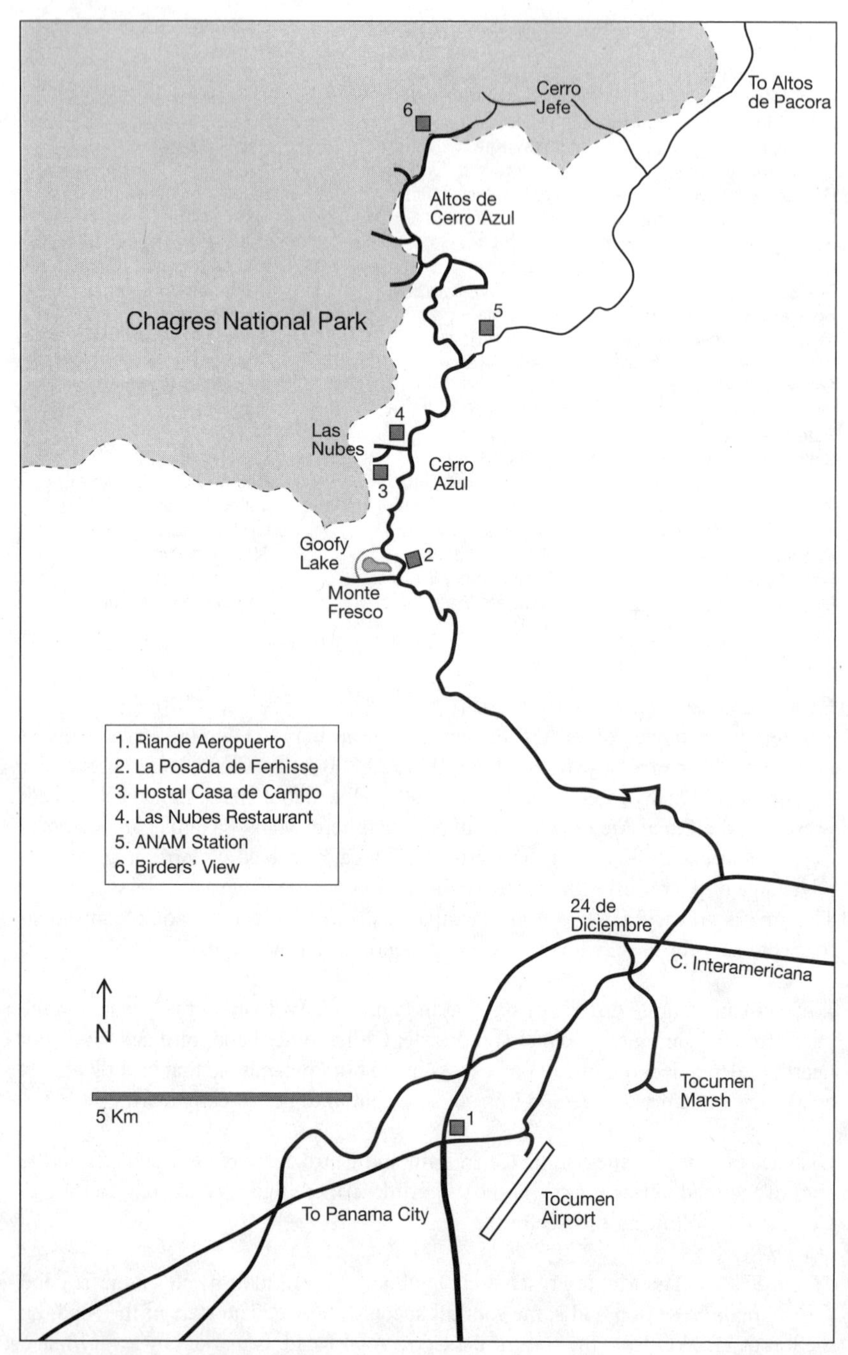

Cerro Azul and Cerro Jefe

right. After 200 m this road comes to an antenna on the right and enters good forest. In another 200 m the road forks; take the right fork. (The left fork here can also be birded, but the road deteriorates after about 500 m.) In less than 100 m along the right fork there is a driveway to the right; keep left here. At 200 m beyond this point, there is a silver wire mesh gate in a barbed wire fence; park near here. Opposite this gate there is an old dirt road that can be walked. At about 400 m along this dirt road there is the concrete shell of an abandoned house on the right which has partly collapsed into the valley below. There is a small trail up to it which offers a beautiful view. Continuing on the road past the silver wire mesh gate will bring you to Lake Cerro Azul, popularly known as "Goofy Lake," although the road is badly eroded and should only be attempted with 4WD. At the bottom of the road bear right to go to Goofy Lake and return to the Cerro Azul Road.

Las Nubes. At 14.8 km along the main road (3.6 km beyond Monte Fresco), the Las Nubes Restaurant is on the left. Take the road to the left just before the restaurant, the Avenida de los Nimbos, the principal avenue of the Las Nubes residential district. After 600 m this paved road turns sharply to the left; a gravel road continues straight here. Park here and walk a short distance down the gravel road. To the right is a fenced property with a concrete driveway ascending the hill. Just before this the traces of an old road go downhill and enter forest. This trail descends steeply to the Río Indio. Birding can be good here but the hike is arduous. White-fronted Nunbird has been reported from this trail.

Altos de Cerro Azul. At 17.1 km on the main road on the left is the entrance to the Altos de Cerro Azul residential community, just before the Policía Nacional office on the right. Just before the entrance is the COMASA store, where you can pick up snacks and basic groceries. Permission is needed to enter Altos de Cerro Azul. Contact Luis Naar (ph 260-4813, cel 6671-4870, lnaar@altosdelmaria.com), Marketing and Sales Manager for Grupo Melo, owner of the development, to request it.

The main road through Altos de Cerro Azul is Paseo Vistamares, marked by a yellow center line. This road and most of the side roads in the development are asphalt. The first few kilometers are through second-growth woods and roadside pine trees, with houses and chicken farms along the way. At 1.4 km you cross the Río Vistamares, a branch of the Río Indio. Check here for Black Phoebe, which is regular near the bridge. There is a viewpoint (*mirador*) for a waterfall at 2.9 km, which often has flocks of foothill tanagers and migrant warblers in the fruiting melastome trees around the platform.

Four kilometers from the entrance you arrive at the Administration Office for Altos de Cerro Azul and a playground. A good birding area is to be found on *Calle Maipo,* the turnoff for which is at Paseo del Himalaya, which goes steeply up on the left 150 m before the Administration Office. After slightly less than 100 m, turn left again, which is the continuation of Paseo del Himalaya. At 600 m from Paseo Vistamares, Paseo del Himalaya turns sharply left. Keep straight on Paseo Chimborazo at this

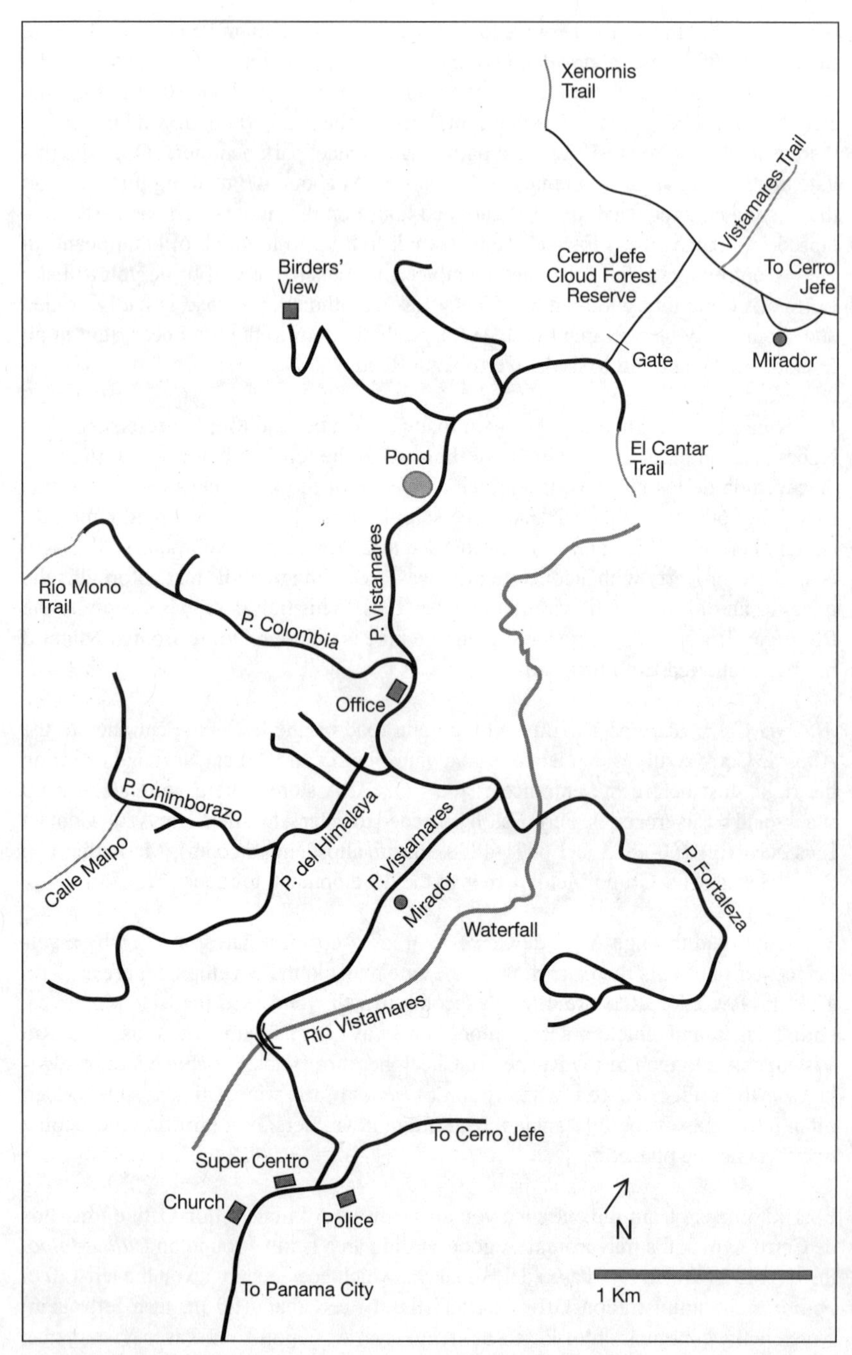

Altos de Cerro Azul

junction, instead of going left to continue on Paseo del Himalaya. (If you continue on Paseo del Himalaya here, it will bring you to Altos del Frente, where the road makes a loop through good forest.) At 1.3 km from Paseo Vistamares, turn left onto Calle Maipo. This is a short grass track, slippery in the wet season. Drive down it (slowly!) and park at the end. In the wet season, if you have a 2WD vehicle, park on the asphalt road and walk from there.

At the end of the grassy track, a wide footpath goes off to the right into good forest. It first descends a steep hill. Look for the nest burrows of leaftossers and Rufous Motmot in the right-hand bank. After that, the trail is undulating but not as steep. Foothill tanagers frequent the fruiting trees, and ant swarms here can yield Barred Forest-Falcon and Rufous-vented Ground-Cuckoo, in addition to the regular antbirds. Other species that can be found here include Black-eared Wood-Quail, Tawny-faced Quail, Yellow-eared Toucanet, **Stripe-cheeked Woodpecker**, and Long-tailed Woodcreeper. Gray-and-gold Tanager, otherwise known mainly from Darién, has also been seen.

Another very good birding area is the *Río Mono Trail*. Just under 400 m beyond the Administration Office turn off on Paseo Colombia, on the left. Continue on Paseo Colombia to the end of the asphalt road, checking for mixed flocks of tanagers in trees along the way, 2.1 km from the Administration Office, where there is a gate. If the gate is open and it is dry season, you can proceed with care by 4WD. Otherwise park here and walk in. The dirt road beyond the gate descends steeply and then branches. The left-hand branch is fairly level for several kilometers, and crosses the Río Mono, while the right-hand fork continues to descend a long way to the Charco de los Monos and twin waterfalls (Romeo y Julieta) with bathing pools. The trails are excellent for forest birds and tanagers, and Tawny-faced Quail, White-tipped Sicklebill, Rufous-vented Ground-Cuckoo, and **Black-crowned Antpitta** can be found. Look for Sunbittern and Fasciated Tiger-Heron on the rivers.

The *Cerro Jefe Conservation Area* also has excellent birding, but 4WD is essential in the wet season, and recommended in the dry season. (However, you can walk in as well, it being a little over 1 km from the gate at the entrance to the start of the Cerro Vistamares Trail, and a little less than 2 km to the Xenornis Trail.) To get there, continue on Paseo Vistamares for another 2 km past Paseo Colombia, following the yellow center line to where it stops at a road junction. Turn left here, where after 100 m there is a gate at the entrance to the Área de Conservación de Cerro Jefe and the asphalt road changes to dirt. (If instead of turning left you continue straight at the road junction, after a short distance the road ends and a signposted walking trail *El Cantar* begins, making a loop through mostly second-growth forest and along streams. This trail may yield foothill tanagers, and, with luck, Sunbittern.)

Just after the dirt road enters the reserve it goes up an incline and then levels out. At 800 m from the start of the dirt road you come to a T-junction. Going right will take you to Cerro Jefe and the Vistamares Trail, while going left brings you to the

Xenornis Trail. The Cerro Jefe area has an interesting elfin forest characterized by scattered palm trees. All along the roads and trails in this forest you may encounter mixed flocks with Tacarcuna Bush-Tanager, **Black-and-yellow Tanager** and Three-striped Warbler. Hummingbirds include **Violet-capped Hummingbird** and Green-crowned Brilliant.

If you turn right at the T-junction, at 300 m look for the *Cerro Vistamares Trail* on the left, which descends fairly steeply and eventually reaches taller forest. Although narrow in places and muddy in others, is has excellent birding. Species that have been found here include **Purplish-backed Quail-Dove**, Brown Violet-ear, Brown-billed Scythebill, Black-headed Antthrush, **Black-crowned Antpitta**, and Stripe-headed Brush-Finch, as well as many other good birds.

Continuing on this road, at about 100 m beyond the entrance to the Cerro Vistamares Trail, a track branches off to the right. This leads to a lookout platform on top of Cerro Vistamares with a magnificent view over most of Chagres National Park. Both the Pacific and Atlantic Oceans can be seen from here on a clear day, hence the name. This platform can give excellent, sometimes even eye-level, views of raptors such as King Vulture, Swallow-tailed Kite, Plumbeous Kite, Black-and-white Hawk-Eagle, and Barred Hawk, as well as migrating swallows and swifts. You can continue on this road as far as Cerro Jefe, marked by an array of communications antennas, but this should only be attempted with 4WD.

Go left at the T-junction for the *Xenornis Trail,* which is the most reliable place to find its namesake bird, the **Speckled (Spiny-faced) Antshrike** *Xenornis setifrons*, other than at Nusagandi. 4WD is required. On the first hundred meters in particular watch out for mixed flocks, especially on fruiting trees, and listen for **Purplish-backed** and (much rarer) **Russet-crowned Quail-Dove**. At 300 m from the junction the track bears left, and shortly after, you come to a very steep downhill section. Only go down if you are sure that you can make it back up again, bearing in mind that rain is frequent (and may occur even in the dry season) and the road can get very slippery, sometimes requiring chains even with 4WD. If in doubt, leave your vehicle here and proceed on foot.

The track ends at 1.1 km from the T-junction. The entrance to the Xenornis Trail is a little difficult to find. At the end of the flat, raised parking area, a wide space for a lookout point has been carved out of the forest to the right. Walk towards this, and about 4 m after leaving the parking area, look for the trail, which goes through a small gap in the scrubby forest to the left. Look for **Violet-capped Hummingbird** at the first patch of red "hot-lips" flowers on the trail, and listen for **Purplish-backed** and **Russet-crowned Quail-Doves** as you descend.

The trail is very steep and is slippery in the wet season, and should only be undertaken by the reasonably fit. It descends first through dwarf cloud forest, which gradually becomes tall forest as you drop down. Once you cross the first stream, the trail levels out considerably, and then crosses two more stream beds (usually dry). After that, it drops down steeply again to another level area, where it ends. All of this area, from the first stream onwards, is the most productive part of the trail for deep forest birds, yielding Great Curassow, Crested Guan, Black-eared Wood-Quail, Yellow-eared Toucanet, Crimson-bellied Woodpecker, Black-headed Antthrush, **Black-crowned** and Scaled Antpittas, Striped Woodhaunter, Slaty-winged and Ruddy Foliage-gleaners, Brown-billed Scythebill, Sapayoa, Stripe-throated Wren, Stripe-headed Brush-Finch, and of course **Speckled Antshrike**.

Cerro Jefe (via the main road). Altos de Cerro Azul provides by far the best birding in the area. However, if for some reason you cannot get permission to enter Altos de Cerro Azul, Cerro Jefe and other cloud forest areas can also be reached via the main road. Continuing past the entrance of Altos de Cerro Azul, after 800 m the asphalt deteriorates. At 1.0 km past Altos de Cerro Azul, on the left, is the yellow-and-green ANAM station that oversees this part of Chagres National Park. A bunkhouse and camping area is available here; see the accommodations section below for details. A trail descends here a few hundred m through second growth forest to the Río Indio.

Between 3.0 and 4.0 km beyond the entrance to Altos de Cerro Azul, the road passes through open fields and then re-enters forest. At 4.9 km the road enters elfin forest; at this point you are at about 850 m (2,800 ft) in elevation. At 5.1 km there is a signposted fork in the road; the left fork goes to Cerro Jefe and the right to Altos de Pacora and the Cabañas 4x4. At 1.5 km on this left fork you will come to the antennas at Cerro Jefe. The road is partly concrete and should be okay for high-clearance 2WD vehicles in the dry season, and perhaps with care in the rainy season. There is also some forest on the road towards Altos de Pacora, but this should only be attempted if you have a good 4WD vehicle; in the wet season, you will probably need chains even with 4WD.

Getting to Cerro Azul and Cerro Jefe. From the intersection of Av. Federico Boyd and Vía España, take Federico Boyd two blocks south and take a left at the traffic signal onto Av. Nicanor de Obarrio (Calle Cincuenta). Go 700 m, and turn off to the right at a sign for the Corredor Sur. Follow the signs for the airport. When you reach the end of Corredor Sur after 18 km, stay straight on the main road instead of turning off for the airport, until you reach the town of 24 de Diciembre, 6.4 km from the airport turnoff. As you come into town, you will cross a bridge just after a shopping center on the left. Immediately after crossing the bridge and just before the large Xtra supermarket (on the left), turn to the left. Take the second right in 200 m, and in another 200 m take a left (all turns are signposted for Cerro Azul). Continue 1.8 km to where you will see a small green sign on the right marking the turnoff on the left to Cerro Azul. From here follow the directions in the site descriptions above.

Accommodations and meals. At a bit over an hour from downtown (depending on traffic, which is often congested in 24 de Diciembre), Cerro Azul and Cerro Jefe can easily be done as a day trip from Panama City. However, there are several places to stay in the area itself.

La Posada de Ferhisse (ph 297-0197, fax 297-0198) is located on the main road near Monte Fresco just past Goofy Lake. It has six rooms ($27.50, including breakfast), all with private hot-water bath and air-conditioning. There is a restaurant on premises.

The *Hostal Casa de Campo Country Inn and Spa* (ph 226-0274 Panama City, fax 226-0336, cel 6677-8993; inform@panamacasadecampo.com; www.panamacasade campo.com) is located amid wooded grounds in Las Nubes. It has 11 rooms ($77.50-$125.00), all different, with private hot-water bath and fan. The hotel offers birding tours, including day trips and nine-day packages. To find it, take the left onto the Avenida de los Nimbos just before the Las Nubes Restaurant, and then take the first left on this street. Take the next right to get to the hotel.

Birders' View (ph 315-1223; info@birdersview.com; www.birdersview.com) in Altos de Cerro Azul is a large private house, surrounded by forest and with a magnificent view overlooking the wildest part of Chagres National Park, which can be rented by birders or other nature enthusiasts for short-term stays. There are four bedrooms (one with private bath, the others sharing one, both hot-water) sleeping a total of seven people, a large sitting room, kitchen, and veranda. There is no air-conditioning, but this is not necessary at this altitude. Birding on the grounds is excellent, including White-tipped Sicklebill, Rufous-crested Coquette, **Violet-capped Hummingbird**, and **Black-crowned Antpitta**; Black-and-white Hawk-Eagle is regularly seen soaring over Chagres National Park from the backyard. Birders' View is convenient for birding the Cerro Jefe Reserve and other sites in Altos de Cerro Azul. Contact Birders' View for further information, including prices and directions.

The *Cabañas 4x4* (cel 6692-0368, 6680-3076; panama4x4@yahoo.com; www.ciu dad.latinol.com/cabanas4x4), are located beyond Cerro Jefe. There are four rustic cabins, lodging from two to eight people each ($45.00 per person including tax), each with shared bath and kitchen. There are also open-sided *bohios* (Indian-style shelters) which sleep 4 people ($15.00 per person). The bohios have a common bath and kitchen area. Meals can be provided at a cost of $20.00 per day. As the name implies, a 4WD vehicle is necessary to reach the cabins. However, transportation can be arranged from the Cerro Azul ANAM station at extra cost. To reach them, follow the directions for Cerro Jefe via the main road on p. 119 and follow the signs for the Cabañas. At the junction with the road to Cerro Jefe, 5.1 km beyond the entrance to Altos de Cerro Azul, take the right fork; after 500 m, go left one; in another 700 m take the right fork (the left fork here goes to Altos de Pacora). Another 1.1 km beyond this junction there is another fork; take the left. The gate for the Cabañas is 200 m down this road. Note that if you stay here, the best birding areas are within Altos de Cerro Azul, so you will still want to get permission to enter this area, and have your own 4WD vehicle available to reach them.

There is a bunkhouse (cold water only) and a camping site at the ANAM station (cel 6664-1562), 1 km past the entrance to Altos de Cerro Azul. The cost is $5.00 per person.

Eating options in the Cerro Azul area are limited. There is a restaurant at the Posada de Ferhisse, near Monte Fresco. The Las Nubes Restaurant, at the entrance to Las Nubes, serves basic meals, as does a small eating place just before ANAM station on the main road.

Bird list for the Cerro Azul-Cerro Jefe area. * = vagrant; R = rare.

Great Tinamou
Little Tinamou
Gray-headed Chachalaca
Crested Guan R
Great Curassow R
Marbled Wood-Quail
Black-eared Wood-Quail
Tawny-faced Quail R
Least Grebe
Neotropic Cormorant
Fasciated Tiger-Heron
Great Blue Heron
Great Egret
Snowy Egret
Little Blue Heron
Tricolored Heron
Cattle Egret
Green Heron
Striated Heron
Black Vulture
Turkey Vulture
King Vulture
Osprey
Gray-headed Kite
Swallow-tailed Kite
Pearl Kite
White-tailed Kite
Double-toothed Kite
Mississippi Kite
Plumbeous Kite
Black-collared Hawk*
Tiny Hawk
Sharp-shinned Hawk
Plumbeous Hawk
Barred Hawk
Semiplumbeous Hawk
White Hawk
Gray Hawk
Solitary Eagle R
Roadside Hawk
Broad-winged Hawk
Short-tailed Hawk
Swainson's Hawk
Red-tailed Hawk
Harpy Eagle R
Black-and-white Hawk-Eagle R
Black Hawk-Eagle
Ornate Hawk-Eagle
Barred Forest-Falcon
Collared Forest-Falcon
American Kestrel
Bat Falcon
Orange-breasted Falcon R
Peregrine Falcon
White-throated Crake
Gray-necked Wood-Rail
Purple Gallinule
Common Moorhen
American Coot R
Sunbittern
Wattled Jacana
Spotted Sandpiper
White-rumped Sandpiper*
Rock Pigeon
Pale-vented Pigeon
Scaled Pigeon
Short-billed Pigeon
Ruddy Ground-Dove
Blue Ground-Dove
White-tipped Dove
Gray-chested Dove
Olive-backed Quail-Dove
Purplish-backed Quail-Dove
Russet-crowned Quail-Dove R
Violaceous Quail-Dove
Ruddy Quail-Dove
Spectacled Parrotlet R
Orange-chinned Parakeet
Blue-fronted Parrotlet
Brown-hooded Parrot
Blue-headed Parrot
Red-lored Amazon
Mealy Amazon
Black-billed Cuckoo
Squirrel Cuckoo
Striped Cuckoo
Rufous-vented Ground-Cuckoo R
Smooth-billed Ani
Tropical Screech-Owl
Vermiculated Screech-Owl
Crested Owl
Spectacled Owl
Mottled Owl
Common Nighthawk
Common Pauraque
Rufous Nightjar
Great Potoo
Common Potoo
Oilbird R
Chestnut-collared Swift
White-collared Swift
Short-tailed Swift
Band-rumped Swift
Band-tailed Barbthroat
Green Hermit
Long-billed Hermit
Stripe-throated Hermit
White-tipped Sicklebill
White-necked Jacobin
Brown Violet-ear R
Violet-headed Hummingbird
Rufous-crested Coquette
Green Thorntail R
Garden Emerald
Violet-crowned Woodnymph
Violet-bellied Hummingbird
Sapphire-throated Hummingbird
Violet-capped Hummingbird
Blue-chested Hummingbird
Snowy-bellied Hummingbird
Rufous-tailed Hummingbird
Bronze-tailed Plumeleteer
Green-crowned Brilliant
Purple-crowned Fairy
Violaceous Trogon
Black-throated Trogon
Black-tailed Trogon
Slaty-tailed Trogon
Tody Motmot R
Rufous Motmot
Broad-billed Motmot
Green Kingfisher
Barred Puffbird R
Black-breasted Puffbird
Pied Puffbird
White-whiskered Puffbird
White-fronted Nunbird R
Rufous-tailed Jacamar R
Great Jacamar
Spot-crowned Barbet
Collared Aracari
Yellow-eared Toucanet

Keel-billed Toucan
Chestnut-mandibled Toucan
Black-cheeked Woodpecker
Red-crowned Woodpecker
Yellow-bellied Sapsucker R
Stripe-cheeked Woodpecker R
Cinnamon Woodpecker
Lineated Woodpecker
Crimson-bellied Woodpecker R
Crimson-crested Woodpecker
Striped Woodhaunter
Slaty-winged Foliage-gleaner
Buff-throated Foliage-gleaner
Ruddy Foliage-gleaner
Plain Xenops
Tawny-throated Leaftosser
Scaly-throated Leaftosser
Plain-brown Woodcreeper
Ruddy Woodcreeper
Olivaceous Woodcreeper
Long-tailed Woodcreeper
Wedge-billed Woodcreeper
Northern Barred-Woodcreeper
Cocoa Woodcreeper
Black-striped Woodcreeper
Spotted Woodcreeper
Streak-headed Woodcreeper
Brown-billed Scythebill
Fasciated Antshrike
Barred Antshrike
Western Slaty-Antshrike
Speckled Antshrike R
Russet Antshrike
Plain Antvireo
Spot-crowned Antvireo
Checker-throated Antwren
White-flanked Antwren
Slaty Antwren
Dot-winged Antwren
Dusky Antbird
Chestnut-backed Antbird
Dull-mantled Antbird R
Spotted Antbird
Bicolored Antbird
Ocellated Antbird
Black-faced Antthrush
Black-headed Antthrush
Black-crowned Antpitta R
Scaled Antpitta R
Streak-chested Antpitta
Brown-capped Tyrannulet
Southern Beardless-Tyrannulet
Yellow Tyrannulet
Yellow-crowned Tyrannulet
Forest Elaenia
Yellow-bellied Elaenia
Lesser Elaenia
Olive-striped Flycatcher
Ochre-bellied Flycatcher
Paltry Tyrannulet
Scale-crested Pygmy-Tyrant
Common Tody-Flycatcher
Brownish Twistwing
Olivaceous Flatbill
Yellow-olive Flycatcher
Yellow-margined Flycatcher
White-throated Spadebill
Golden-crowned Spadebill
Royal Flycatcher
Ruddy-tailed Flycatcher
Sulphur-rumped Flycatcher
Black-tailed Flycatcher
Olive-sided Flycatcher
Western Wood-Pewee
Eastern Wood-Pewee
Tropical Pewee
Acadian Flycatcher
Alder Flycatcher
Willow Flycatcher
Black Phoebe
Long-tailed Tyrant
Bright-rumped Attila
Rufous Mourner
Dusky-capped Flycatcher
Panama Flycatcher
Great Crested Flycatcher
Great Kiskadee
Boat-billed Flycatcher
Rusty-margined Flycatcher
Social Flycatcher
White-ringed Flycatcher
Streaked Flycatcher
Sulphur-bellied Flycatcher
Piratic Flycatcher
Tropical Kingbird
Eastern Kingbird
Fork-tailed Flycatcher
Sapayoa R
Thrush-like Schiffornis
Rufous Piha
Speckled Mourner
Cinnamon Becard
Masked Tityra
Blue Cotinga
Purple-throated Fruitcrow
Golden-collared Manakin
White-ruffed Manakin
Blue-crowned Manakin
Red-capped Manakin
White-eyed Vireo*
Yellow-throated Vireo
Red-eyed Vireo
Yellow-green Vireo
Tawny-crowned Greenlet
Lesser Greenlet
Green Shrike-Vireo
Gray-breasted Martin
Northern Rough-winged Swallow
Southern Rough-winged Swallow
Cliff Swallow
Barn Swallow
Black-bellied Wren
Bay Wren
Stripe-throated Wren R
Buff-breasted Wren
Plain Wren
House Wren
White-breasted Wood-Wren
Gray-breasted Wood-Wren
Southern Nightingale-Wren
Song Wren
Tawny-faced Gnatwren
Long-billed Gnatwren
Tropical Gnatcatcher
Slaty-backed Nightingale-Thrush
Veery
Gray-cheeked Thrush
Swainson's Thrush
Wood Thrush
Pale-vented Thrush
Clay-colored Thrush
Gray Catbird
Cedar Waxwing
Golden-winged Warbler
Tennessee Warbler
Yellow Warbler
Chestnut-sided Warbler
Magnolia Warbler
Black-throated Blue Warbler R
Yellow-rumped Warbler
Black-throated Green Warbler R
Blackburnian Warbler
Yellow-throated Warbler R
Bay-breasted Warbler
Cerulean Warbler
Black-and-white Warbler
American Redstart
Prothonotary Warbler
Worm-eating Warbler R
Ovenbird
Northern Waterthrush
Louisiana Waterthrush
Kentucky Warbler
Mourning Warbler
Wilson's Warbler R
Canada Warbler
Slate-throated Redstart

Rufous-capped Warbler
Three-striped Warbler
Buff-rumped Warbler
Bananaquit
Tacarcuna Bush-Tanager
Yellow-throated Bush-Tanager R
Black-and-yellow Tanager
Rosy Thrush-Tanager
Dusky-faced Tanager
Olive Tanager
Sulphur-rumped Tanager
White-shouldered Tanager
Tawny-crested Tanager
White-lined Tanager
Red-throated Ant-Tanager
Hepatic Tanager
Summer Tanager
Scarlet Tanager
Crimson-backed Tanager
Blue-gray Tanager
Palm Tanager
Blue-and-gold Tanager R
Plain-colored Tanager
Gray-and-gold Tanager R
Emerald Tanager
Silver-throated Tanager
Speckled Tanager
Bay-headed Tanager
Rufous-winged Tanager
Golden-hooded Tanager
Scarlet-thighed Dacnis
Blue Dacnis
Green Honeycreeper
Shining Honeycreeper
Red-legged Honeycreeper
Blue-black Grassquit
Slate-colored Seedeater R
Variable Seedeater
Yellow-bellied Seedeater
Lesser Seed-Finch
Blue Seedeater R
Yellow-faced Grassquit
Stripe-headed Brush-Finch
Orange-billed Sparrow
Black-striped Sparrow
Streaked Saltator
Buff-throated Saltator
Black-headed Saltator
Slate-colored Grosbeak
Rose-breasted Grosbeak
Blue-black Grosbeak
Eastern Meadowlark
Great-tailed Grackle
Orchard Oriole
Yellow-backed Oriole
Yellow-tailed Oriole
Baltimore Oriole
Yellow-billed Cacique
Scarlet-rumped Cacique
Yellow-rumped Cacique
Crested Oropendola
Chestnut-headed Oropendola
Yellow-crowned Euphonia
Thick-billed Euphonia
Fulvous-vented Euphonia
White-vented Euphonia
Tawny-capped Euphonia
Lesser Goldfinch

Cerro Campana

Cerro Campana marks the easternmost extension of the Talamanca Range of Costa Rica and western Panama, and several of the species endemic to this range make it this far. Cerro Campana reaches 1,007 m (3,300 ft), and many foothills species are found in the moist submontane and cloud forests on its slopes. The area features spectacular scenery reminiscent of a Chinese painting. The stumps of eroded volcanoes, the bell-shaped Cerro Campana and three-peaked Cerro Trinidad form precipitous rocky spires often wreathed by mist and cloud.

The area is encompassed by Altos de Campana National Park (4,816 ha/11,900 acres), established in 1966 as Panama's first national park. Unfortunately, much of the area within the park's boundaries has been deforested. Enough still remains, however, so that many primary forest species are still present, and some deforested areas are now regenerating to second growth.

Foothills species such as Scale-crested Pygmy-Tyrant, White-ruffed Manakin, White-throated Thrush, Silver-throated Tanager, and **Black-and-yellow Tanager** can almost be guaranteed, while others such as **Orange-bellied Trogon**, Yellow-eared Toucanet, Plain Antvireo, and Chestnut-capped Brush-Finch are likely.

On the road up to Cerro Campana from the Carretera Interamericana, there are several patches of forest about 2 km from the turnoff which can sometimes be productive for lowland birds. There are several places where you can pull off to park on the left. The road also offers good views of the large area of mangroves in the Bay of

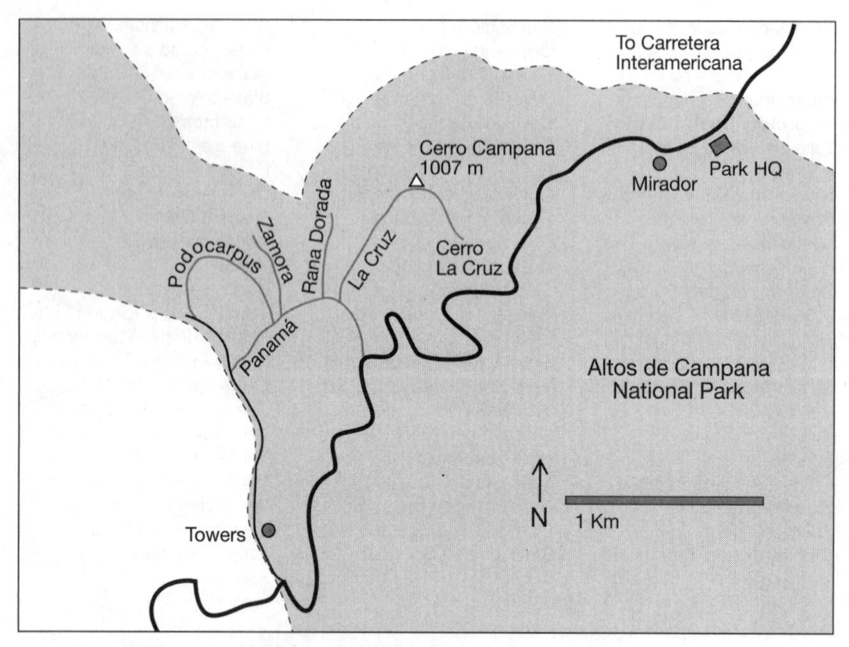

Cerro Campana

Chame and of Punta Chame below. The best place to look for Wedge-tailed Grass-Finch is in the grassy areas around the park headquarters, 4.7 km from the turnoff, and the lookout 200 m beyond it. Scan rocks and small bushes in the grass, where they like to perch to sing their characteristic finch-like song. They are easier to find early in the morning before the winds pick up.

About half a kilometer beyond park headquarters the road begins to pass through woodland below Cerro Campana that can be very good for migrants in season. At 3.3 km from headquarters is the main entrance to the trail system, a road to the right signposted for the Senderos Panamá, Podocarpus, and La Cruz. Turn in here and park. This short road goes up about 100 m to a house, on the right. The end of the Sendero Panamá, which leads to the other trails, is on the left, opposite the house, at a yellow and green interpretive sign. (This is the last numbered sign at the end of the trail, but it is easier to enter the trail system here.)

There are five trails in the park. From the end near the house, the *Sendero Panamá* first climbs gently up through second growth for 300 m and then enters better forest and levels out, becoming a gravel road. About 400 m along it, the signposted *Sendero La Cruz* (Cross Trail) is on the right. This trail climbs from about 820 m (2,700 ft) to the summit of Cerro Campana and to the cross marker on the side of Cerro Campana. Some birds, such as Black-headed Antthrush, can more easily be found near the summit than on the lower trails.

Continuing on the Sendero Panamá, about 100 m beyond the Sendero La Cruz the unmarked *Sendero Rana Dorada* (Golden Frog Trail) goes down on the right. Another 200 m brings you to the Senderos Zamorra and Podocarpus, both signposted, whose entrances are side by side on the right near some benches. The *Sendero Zamorra*, which slopes gently downward, is wide at the start but after about 10 minutes of walking narrows to a foot-path. After another 15 minutes, it crosses a stream, ending about 5 minutes later in an abandoned orange grove.

The *Sendero Podocarpus* is a 1-km (0.6 mi) loop trail. For the first 100 m or so it climbs fairly steeply, then levels out. At a bit over 200 m it reaches a large *Podocarpus* tree, a kind of conifer found mostly in the Southern Hemisphere. The trail here bends sharply to the left. After another 400 m there is a *salida* (exit) sign near two orange and green ANAM buildings, the Refugio Los Pinos. Go down the road at the buildings for about 50 m, where you will come to another gravel road. Go left, where after 250 m you will come to another road, which is the beginning of the Sendero Panamá. Going left here will bring you back to the start of the Sendero Podocarpus in about 100 m.

You can drive to this far end of the trail system and get to the start of any of the trails along the Sendero Panamá by car, but 4WD is required. Instead of turning off at the signs at 3.3 km after the park headquarters, continue another 800 m along the main road, where another road, with a sign showing the way to the Sendero Podocarpus, goes off on the right. This road is at first asphalt but soon becomes broken up and requires high clearance. At 200 m in you will pass some antennas on the right; at 400 m is a fork where you go left. You then pass several houses and enter forest at 700 m. At 900 m another road goes left; the right fork is the beginning of the Sendero Panamá (signposted). The road to the left goes to the Refugio Los Pinos and the far end of the Sendero Podocarpus. You can either park here, or if it is not too muddy, drive in to the start of the Senderos Panamá and Zamora at about 100 m further in, or on to the Sendero La Cruz at about 400 m in.

Getting to Cerro Campana: See the section on "Getting to Western Panama" (p. 130) for directions on getting on to the Carretera Interamericana going west from Panama City. About 33 km from the Bridge of the Americas, just after the point where the Autopista joins the Interamericana, you will start to see the spectacular crags of Cerro Trinidad, in a more remote part of Altos de Campana National Park, in the distance. At 48 km from the Bridge you enter the town of Capira. Near the end of Capira, at 49.3 km, the Quesos Chela cheese store and adjacent minisuper are a convenient place to pick up snacks for the trail. At 52.5 km there is a sign for the turnoff to the town of La Campana – do not confuse this with the entrance to the national park, which is farther along. At 54.3 km the signposted turnoff for the park from the Interamericana is on the right at a blue bus shelter. From here it is 4.7 km on a good asphalt road to park headquarters, where you should pay the entrance fee ($3.50 for foreign visitors and $1.50 for Panama residents). Driving time is about 50 minutes from the Bridge of the Americas.

On your return from the park, you cannot take a left back onto the divided Carretera Interamericana. Instead go right and in 400 m you can turn off to the right and then cross back over the highway to return. (Do not try to make a direct U-turn here, which is prohibited.) Try not to miss this turn, since the next turn-around spot is in another 3.2 km at the bottom of the hill.

If you are coming from the west, the turnoff to Altos de Campana is at the top of a long, 3-km hill where the road climbs up in steep curves. As you crest this hill, you will see the blue bus shelter that marks the turnoff to the park on your left on the other side of the highway. You cannot turn here, but instead carry on for another 800 m, where you can make a turn from the left lane to head back to the turnoff.

Accommodations and meals. Since it is less than an hour's drive away, Cerro Campana is easily done as a day trip from the Panama City area. If you plan to stay all day, pack a lunch since there is no place to eat near the park itself.

There is a cabin available at park headquarters if you wish to stay overnight. It has two rooms with a pair of bunk beds in each (total of 8 beds) and cooking facilities. Gas for the stove can be provided for an additional fee; Panama residents or large parties may prefer to bring their own tank, which requires a Panagas fitting. The cost is $10.00 per person for Panama residents and $12.00 for visitors. For reservations call the ANAM regional office in La Chorrera (ph 254-2848).

Bird list for Altos de Campana National Park and environs. * = vagrant; R = rare; ? = present status uncertain, X = no longer present.

Great Tinamou
Little Tinamou
Gray-headed Chachalaca
Black Guan X
Crested Bobwhite R
Black-eared Wood-Quail
Cattle Egret
Black Vulture
Turkey Vulture
King Vulture
Osprey
Gray-headed Kite
Hook-billed Kite R
Swallow-tailed Kite
White-tailed Kite
Double-toothed Kite
Mississippi Kite
Plumbeous Kite
Bicolored Hawk R
White Hawk
Gray Hawk
Great Black-Hawk
Roadside Hawk
Broad-winged Hawk
Short-tailed Hawk
Swainson's Hawk
White-tailed Hawk
Zone-tailed Hawk
Black Hawk-Eagle
Barred Forest-Falcon
Collared Forest-Falcon
American Kestrel
Aplomado Falcon R
Bat Falcon
Peregrine Falcon
Gray-necked Wood-Rail
Rock Pigeon
Pale-vented Pigeon
Scaled Pigeon
Short-billed Pigeon
Ruddy Ground-Dove
Blue Ground-Dove
Maroon-chested Ground-Dove ?
White-tipped Dove
Gray-chested Dove
Purplish-backed Quail-Dove R
Ruddy Quail-Dove
Orange-chinned Parakeet
Blue-headed Parrot
Black-billed Cuckoo
Squirrel Cuckoo
Striped Cuckoo
Rufous-vented Ground-Cuckoo R
Smooth-billed Ani
Groove-billed Ani
Tropical Screech-Owl
Mottled Owl
Common Nighthawk
Common Pauraque
Rufous Nightjar
Common Potoo
Chestnut-collared Swift
White-collared Swift
Band-rumped Swift
Rufous-breasted Hermit
Band-tailed Barbthroat
Green Hermit
Long-billed Hermit
Stripe-throated Hermit
White-tipped Sicklebill R

White-necked Jacobin
Brown Violet-ear R
Violet-headed Hummingbird
Rufous-crested Coquette
Green Thorntail R
Garden Emerald
Violet-crowned Woodnymph
Violet-bellied Hummingbird
Snowy-bellied Hummingbird
Rufous-tailed Hummingbird
Snowcap R
White-vented Plumeleteer
Bronze-tailed Plumeleteer
Purple-crowned Fairy
Violaceous Trogon
Orange-bellied Trogon
Black-throated Trogon
Slaty-tailed Trogon
Blue-crowned Motmot
Rufous Motmot
Broad-billed Motmot
White-whiskered Puffbird
Great Jacamar
Blue-throated Toucanet
Collared Aracari
Yellow-eared Toucanet
Keel-billed Toucan
Black-cheeked Woodpecker
Red-crowned Woodpecker
Yellow-bellied Sapsucker R
Lineated Woodpecker
Crimson-crested Woodpecker
Spotted Barbtail
Buff-throated Foliage-gleaner
Plain Xenops
Scaly-throated Leaftosser
Plain-brown Woodcreeper
Ruddy Woodcreeper
Olivaceous Woodcreeper
Long-tailed Woodcreeper
Northern Barred-Woodcreeper
Cocoa Woodcreeper
Black-striped Woodcreeper
Spotted Woodcreeper
Fasciated Antshrike
Barred Antshrike
Western Slaty-Antshrike
Russet Antshrike
Plain Antvireo
Spot-crowned Antvireo
Checker-throated Antwren
White-flanked Antwren
Slaty Antwren
Dusky Antbird
White-bellied Antbird
Dull-mantled Antbird R
Spotted Antbird
Bicolored Antbird
Ocellated Antbird
Black-faced Antthrush
Black-headed Antthrush R
Black-crowned Antpitta R
Brown-capped Tyrannulet
Southern Beardless-Tyrannulet
Yellow Tyrannulet
Yellow-crowned Tyrannulet
Yellow-bellied Elaenia
Lesser Elaenia
Olive-striped Flycatcher
Ochre-bellied Flycatcher
Paltry Tyrannulet
Scale-crested Pygmy-Tyrant
Common Tody-Flycatcher
Black-headed Tody-Flycatcher
Eye-ringed Flatbill R
Olivaceous Flatbill
Yellow-olive Flycatcher
Yellow-margined Flycatcher
White-throated Spadebill
Golden-crowned Spadebill
Royal Flycatcher
Ruddy-tailed Flycatcher
Sulphur-rumped Flycatcher
Common Tufted-Flycatcher
Olive-sided Flycatcher
Western Wood-Pewee
Eastern Wood-Pewee
Tropical Pewee
Yellow-bellied Flycatcher
Acadian Flycatcher
Alder Flycatcher
Willow Flycatcher
Bright-rumped Attila
Rufous Mourner
Dusky-capped Flycatcher
Panama Flycatcher
Great Crested Flycatcher
Great Kiskadee
Boat-billed Flycatcher
Social Flycatcher
Streaked Flycatcher
Sulphur-bellied Flycatcher
Piratic Flycatcher
Tropical Kingbird
Eastern Kingbird
Fork-tailed Flycatcher
Thrush-like Schiffornis
Rufous Piha
Cinnamon Becard
Masked Tityra
Golden-collared Manakin
White-ruffed Manakin
Lance-tailed Manakin
Red-capped Manakin
Yellow-throated Vireo
Red-eyed Vireo
Yellow-green Vireo
Scrub Greenlet
Lesser Greenlet
Green Shrike-Vireo
Gray-breasted Martin
Violet-green Swallow R
Blue-and-white Swallow R
Northern Rough-winged Swallow
Southern Rough-winged Swallow
Cliff Swallow
Barn Swallow
Black-bellied Wren
Bay Wren
Stripe-breasted Wren R
Rufous-breasted Wren
Rufous-and-white Wren
Plain Wren
House Wren
Ochraceous Wren
White-breasted Wood-Wren
Gray-breasted Wood-Wren
Southern Nightingale-Wren
Song Wren
Tawny-faced Gnatwren
Long-billed Gnatwren
Tropical Gnatcatcher
Veery
Gray-cheeked Thrush
Swainson's Thrush
Wood Thrush
Pale-vented Thrush
Clay-colored Thrush
White-throated Thrush
Cedar Waxwing R
Golden-winged Warbler
Tennessee Warbler
Yellow Warbler
Chestnut-sided Warbler
Magnolia Warbler
Black-throated Blue Warbler R
Yellow-rumped Warbler
Black-throated Green Warbler
Blackburnian Warbler
Yellow-throated Warbler R
Bay-breasted Warbler
Cerulean Warbler
Black-and-white Warbler
American Redstart
Prothonotary Warbler
Worm-eating Warbler
Ovenbird

Northern Waterthrush
Louisiana Waterthrush
Kentucky Warbler
Mourning Warbler
Wilson's Warbler R
Canada Warbler
Rufous-capped Warbler
Buff-rumped Warbler
Yellow-breasted Chat R
Bananaquit
Black-and-yellow Tanager
Rosy Thrush-Tanager
White-shouldered Tanager
Tawny-crested Tanager
Red-crowned Ant-Tanager
Red-throated Ant-Tanager
Hepatic Tanager
Summer Tanager
Scarlet Tanager
Crimson-backed Tanager
Blue-gray Tanager
Palm Tanager
Plain-colored Tanager
Silver-throated Tanager
Bay-headed Tanager
Rufous-winged Tanager
Golden-hooded Tanager
Scarlet-thighed Dacnis
Blue Dacnis
Green Honeycreeper
Shining Honeycreeper
Red-legged Honeycreeper
Blue-black Grassquit
Slate-colored Seedeater R
Variable Seedeater
Yellow-bellied Seedeater
Lesser Seed-Finch
Blue Seedeater R
Yellow-faced Grassquit
Wedge-tailed Grass-Finch
Chestnut-capped Brush-Finch
Orange-billed Sparrow
Black-striped Sparrow
Rufous-collared Sparrow
Streaked Saltator
Buff-throated Saltator
Slate-colored Grosbeak
Rose-breasted Grosbeak
Blue-black Grosbeak
Eastern Meadowlark
Great-tailed Grackle
Orchard Oriole
Yellow-backed Oriole
Baltimore Oriole
Scarlet-rumped Cacique
Chestnut-headed Oropendola
Yellow-crowned Euphonia
Thick-billed Euphonia
Fulvous-vented Euphonia
White-vented Euphonia
Tawny-capped Euphonia
Lesser Goldfinch

Isla Taboga

The charming island of Taboga is a popular tourist destination near Panama City. The island hosts one of the largest colonies of Brown Pelican in Central America, with over 1,000 pairs, which is protected as a wildlife refuge. At the height of the breeding season the colony is a fascinating spectacle, with flocks of adults flying in from distant fishing areas with food for their raucously begging young. Aside from the pelican colony, however, the island is of limited interest for birding, having mostly species of garden and scrub that can easily be found elsewhere.

The small village of Taboga occupies the north coast of the island, on the side closest to Panama City. The pelicans nest in trees on the forested west and south sides of the island, and on the island of Uravá, adjacent to Taboga on the east. Nesting usually begins in January and is mostly over by late June, with the height of activity in March and April. There is also a small nesting colony of Brown Booby (and perhaps sometimes Blue-footed Booby as well) on Farallón Rock, a tiny islet about 500 m south of Isla Taboguilla, the large island 2 km northeast of Taboga and Uravá.

Taboga can easily be visited as a day-trip from Panama City. See below for details on ferries to the island. The colony is best viewed by taking a small boat around the island; there are no good viewpoints on the island itself that offer a vantage to see much of the nesting area. After arriving on the ferry, ask around for a boatman to take you around the island. Boatmen are easiest to find on weekends, but some should always be available. If you have trouble locating one, ask at the Autoridad Maritima office to the right of the wharf. The basic price is $25.00 for a circuit of the island.

If you have more time on the island and want to do some birding outside the town, there are two ways up through woodland to the island's highpoint at 307 m (1,007 ft), one via a trail and the other by a dirt road. To find them, walk left from the wharf along the promenade above the beach. Near its end, the street divides. To get to the trail, go up the upper fork and continue to the plaza in front of the church. A concrete path next to a phone booth at the right of the plaza leads to a trail that goes steeply up, eventually coming out on the road a short distance from the top. To find the road, continue on the lower fork of the street, and carry on past the end of town. About 15 minutes beyond the end of town, the dirt track curves uphill. At this point there is an ANAM sign for the wildlife refuge; take the right fork here. From here it is roughly 2 km to the peak.

Getting to Isla Taboga. The Calypso line (ph 232-5736) operates ferries to Taboga from Panama City. They leave Panama City at 8:00 AM and return from Taboga at 4:00 PM; on weekends and holidays there is an additional boat to Taboga at 10:30 AM and an additional one returning at 3:00 PM. The trip takes about one hour each way. The price is $5.00 each way. The ferries leave from the La Playita wharf on Naos Island on the Amador Causeway. Follow the directions on p. 64 for the Amador Causeway to get there. Turn off the Causeway at the sign for the Smithsonian's Marine Exhibition Center at Culebra, next to the Mi Ranchito Restaurant. The entrance to the wharf is on the left next to the gate for the Exhibition Center. Taxi fare to the wharf from downtown Panama City is $4.00; taxis wait to meet returning ferries.

Accommodations and meals. While Taboga is easily done as a day-trip, if you want to spend more time on the island there are two places to stay. The pleasant *Vereda Tropical* (ph 250-2154; veredatropicalhotel@hotmail.com), located up from the main street a short walk left from the wharf, has 12 rooms ($55.00-$65.00 with fan, $75.00 with air-conditioning), all with private cold-water bath. There is a restaurant on premises. The *Cool Hostal* (ph 690-2545; luisveron@hotmail.com), up a concrete path opposite the wharf, offers more basic accommodations. The hostal has a bunkroom with beds for $10.00 per person, plus a room with two beds that rents for $25.00 for the room. There is a shared hot-water bath, and kitchen facilities are available. In addition to the restaurant at the Vereda Tropical, there are several other restaurants and snack bars in town. For more information on Taboga, see www.taboga.panamanow.com.

Western Panama

The rugged topography of western Panama has produced a wide range of habitats in close proximity. This region contains parts of three different areas of endemism, one of them, the Talamanca Range, having the second highest number of regional endemics of any mainland area in the world. These mountains, ranging up to 3,474 m (11,297 ft) on Volcán Barú, run along the entire length of the Isthmus from the Costa Rican border to the Canal Area, separating the dry savannas and woodlands of the Pacific slope from the ever-wet, evergreen tropical forests of the Atlantic side. Islands off each coast, Coiba on the Pacific and the Bocas del Toro archipelago on the Caribbean, host specialties of their own.

Western Panama may be visited either by driving from Panama City, or by flying to David or Bocas del Toro and exploring from there. In the following section sites are presented by province from east to west, as they would be visited by a traveler traveling by road from Panama City.

Getting to Western Panama. From the intersection of Vía España and Av. Federico Boyd in downtown Panama City, go south on Federico Boyd six blocks until you come to Av. Balboa, opposite the large Miramar Hotel. Take a right onto Balboa. After 2.1 km, near the end of Balboa, turn right just before an overpass, next to a Shell gas station, prominently signposted for "Terminal Nacional Los Pueblos/Albrook." After 300 m (get into the left lane as soon as you can), you will come to a traffic light. Continue straight, and after another short block you will come to a second light. Just beyond this the road divides; take the left lane going up onto a traffic overpass over Av. de los Mártires, signposted for "Puente de las Américas" and "Amador." This is the on-ramp for Av. Mártires headed west. About 1.7 km after merging onto Mártires, stay in the left lanes, which are signposted for La Chorrera. (There is an exit on the right here for Amador here; make sure you do not take it.) This will shortly bring you to the Bridge of the Americas over the Panama Canal. For the next 10 km or so the four-lane highway is narrow and winding with a concrete barrier in the middle, and traffic is often heavy. Drive with care, and try to avoid traveling this stretch in the dark.

At around 13 km from the start of the Bridge, shortly after passing the town of Arraiján, you will cross an overpass. Just beyond it take the left exit, signposted for

"Autopista a La Chorrera." (Make sure you do not take the right here, confusingly signposted "Carretera Panamericana" and "La Chorrera." This will take you into the center of La Chorrera.) The Autopista is a 19-km toll road that bypasses La Chorrera; the toll is $0.50 for cars and $1.00 for light trucks. At 32.8 km from the start of the Bridge of the Americas, just beyond where the Autopista ends, you come to an intersection. Continue straight across it to merge onto the Carretera Interamericana headed west.

The recently-constructed Centennial Bridge (*Puente Centenario*), which crosses the Canal north of the Bridge of the Americas, provides an alternate route for heading west from Panama City. From Panama City, the distance is somewhat longer than that by the Bridge of the Americas, but the road is better (bypassing the winding stretch before Arraiján). This route would definitely be preferred if you are coming from Gamboa or the Canopy Tower. To get to the Centennial Bridge from Panama City, follow the directions for Soberanía National Park on p. 82. The entrance ramp for the bridge off Av. Omar Torrijos H. is at 11.0 km north of the El Rey supermarket at the entrance to Albrook, just beyond the town of Paraiso. Follow the signs for Arraiján to go west. (If you are coming from the direction of Gamboa, this entrance is 2.8 km south of the intersection of C. Gaillard and Av. Omar Torrijos H. near Summit.) The Bridge can also be reached from the Corredor Norte. When returning from western Panama, follow signs for "Paraiso" (for Av. Omar Torrijos H.) and "Bethania." (for the Corredor Norte).

The Carretera Interamericana is now four-lane and in mostly good condition from Panama City to Santiago, and in Chiriquí from David west to the Costa Rican border. There is a police checkpoint 96 km west of Santiago where the driver and passengers may be asked for passports or other identification. The most difficult stretch is in the hills between Santiago and eastern Chiriquí, where the road is still two-lane and narrow and often pot-holed. (It has recently been improved, however, by the addition of passing lanes on the steeper hills.) It is not recommended to drive this section at night, especially since sections under repair may be poorly marked. Hotels and better restaurants are mainly restricted to Penonomé, Coclé (pp. 143-144), Santiago, Veraguas (pp. 166-167), and David, Chiriquí (p. 201), so plan your trip accordingly.

To reach sites in Herrera and Los Santos Provinces on the Carretera Nacional, the turnoff is at the small town of Divisa, about 208 km from the Bridge of the Americas and 65 km west of Penonomé. Bocas del Toro can be reached by car via the Oleoducto Road that leaves the Carretera Interamericana at the small town of Chiriquí, 178 km west of Santiago and 29 km west of San Lorenzo, and passes through the Fortuna Forest Reserve. It is also not advised to drive this road at night, since it is narrow, winding, often fog-bound, and frequented by large semitrailer trucks. At 98 km from the Carretera Interamericana, at Punta Peña, is the turnoff to Almirante, reached in another 68 km, and Changuinola, in 87 km. See the respective sections for more details.

Driving times from Panama City are about 2 hours to Penonomé; 3-3½ hours to Santiago; 6-7 hours to David; and 7-8 hours to Volcán. It is about 4 hours to Chitré, and 9-10 hours to Changuinola. Traffic on the Interamericana near Panama City can be particularly heavy on long holiday weekends, outbound on Friday afternoon and inbound at the end of the holiday. Inbound traffic can also be heavy on Sunday afternoons due to people returning from beach areas in eastern Panamá Province. Highway police are particularly vigilant for speeders and other traffic violators at these times. Avoid traveling then if you can.

If you prefer, it is also possible to fly to David in Chiriquí province and rent a vehicle there to visit the Western Highlands and other sites. Although you can fly to Changuinola and Bocas del Toro Town, at present there are no car-rental offices in either place. Both Aeroperlas and AirPanama have daily flights to David, Changuinola, and Bocas del Toro Town.

Western Panamá Province

The Talamanca Range, which begins in Costa Rica, reaches its end in western Panamá Province. Besides Cerro Campana (pp. 123-128), Altos del María, just east of El Valle, has recently been discovered to be a good place to find some of the endemic species of this range.

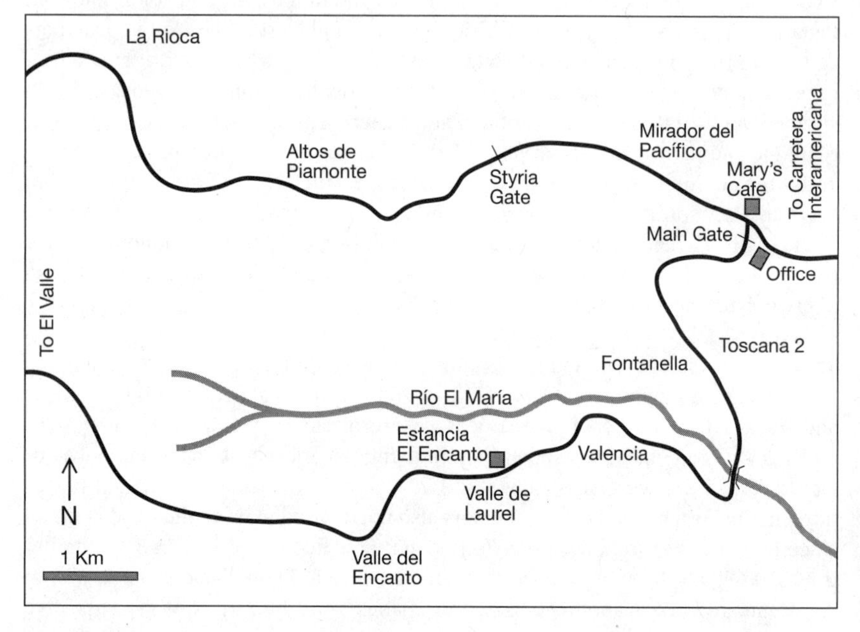

Altos del María

Altos del María

Altos del María is a new housing development in the foothills of western Panamá Province at about 500-1,150 m (1,650-3,800 ft) elevation. It consists of 2,400 ha (6,000 acres) of former cattle pasture and some forest, owned by the Melo Group, which also operates Altos de Cerro Azul east of Panama City. Lots are being sold in the deforested parts for weekend houses in a gated community. To enter the community and pass gates within it, you must obtain permission from the Altos del María company and present identification. A map of the development (found by clicking on "Project Urbanizations") and details on obtaining permission is available at the company's website; see the next section. Most of the area is accessible by paved road, or by short walks on trails which are generally in good condition. On some trails there are steep sections and stairs.

The avifauna is quite similar to that of El Valle (see next section). In recent years many rare and interesting species have been found here, some of them at their easternmost limit. These include Plumbeous Hawk, **Purplish-backed Quail-Dove**, Rufous-vented Ground-Cuckoo, White-tipped Sicklebill, Brown Violet-ear, Violet Sabrewing, Green Thorntail, **White-tailed Emerald**, Snowcap, **Purple-throated Mountain-gem**, **Orange-bellied Trogon**, Red-faced Spinetail, Spotted Barbtail, Brown-billed Scythebill, **Black-crowned Antpitta**, Rufous-browed Tyrannulet, **Ochraceous Wren**, and Pale-vented Thrush. No doubt many other species will be found here as the area is birded more regularly.

There are two main areas for birding, Altos de la Rioca and the road to El Valle. As you arrive at the community on the road from Sorá, you will first see the sales office on the left. Shortly beyond the office, on the left, is the gate to the main area of the development, Valle del María. Continuing straight here, without entering this main gate, brings you to Altos de Rioca; going in through the gate will bring you to the road to El Valle. These two areas will eventually be linked directly by a new road wholly within the development, but at present to go from one to the other you have to come back to the main gate.

Altos de la Rioca, extending up to the Continental Divide at about 1,100 m, is the higher of the two areas. To get there, after you reach the office, go straight on up the ridge without entering the main gate to Valle del María, passing Mary's Cafe on the right. Continue straight through a second gate at 2.0 km past the main gate. You will reach broken forest at about 4.0 km, where you can start birding. The best area is at 5.6 km, in the Altos de la Rioca development itself, where Calle Coruña goes off to the right. Leave the vehicle at this point and walk along Calle Coruña a few meters where you will find a paved pathway going up through good wet forest on both left and right. You can bird this trail and the roads in the development to find birds of the wet foothills (and some of high elevations) of western Panama.

The *El Valle Road* runs for approximately 14 km to the town of El Valle, via Mata Ahogada, coming out by the Hotel Campestre. Beyond the Altos del María sales office, turn left to enter the main gate. At 2.3 km from the gate the road crosses the Río El María. Paths run along the river bank both upstream and down for 2-3 km, but since the area is very deforested it is better to continue on. At 4.2 km, on the right hand side, is the Estancia del Encanto Bed and Breakfast (see section on accommodations below). At 5.4 km the paved road ends, although there are plans to eventually pave it beyond this point. This road is in good condition and passable all year round, although the steeper parts would require 4WD in the wet season.

At 9.8 km you enter forested areas. Being on the Pacific slope, this forest is drier than that at Altos de la Rioca on the Continental Divide. At 12.6 km the road reaches its highest point at 1050 m (3,465 ft). At this point, you can park on the left and walk a good signposted trail to the top of Cerro de la Gaita (1,178 m/3,887 ft). The trail affords magnificent panoramic views of the Pacific coast, the Laguna de San Carlos, El Valle and part of the Atlantic slope, as well as the general area of Altos del María. This hill is forested, while surrounding areas up to the Continental Divide are partially forested. Beyond this point the road drops down quickly to deforested areas, and eventually reaches El Valle. (Note: Beyond Cerro de la Gaita, there is a gate on the road, and it is often closed. Anyone planning to go on to El Valle should find out the current opening hours of the gate beforehand. At present, it is only regularly open when it is manned during dry season weekends.)

Getting to Altos del María. From Panama City, take the Carretera Interamericana west. The turnoff is in the small town of Bejuco, about 65 km from the city, at an archway on the right signposted Altos del María, next to the Restaurante Pío Pío. Enter through the archway, from which the route is signposted. Take the road at the back of the parking lot for 100 m, then turn right, and after another 100 m turn right again. Continue for 18.4 km to the village of Sorá. Turn sharp right just past a Chinese gateway on your left. Continue for 1.5 km where you will see the sales office for Altos del María on your left. From there follow the directions in the section above.

As mentioned above, although birders are welcome to visit, permission is needed to enter the Altos del María development, and you must make an appointment. See their website (www.altosdelmaria.com) for a registration form and other information. You can also contact Luís Naar (ph 260-4813, cel 6671-4870, lnaar@altosdelmaria.com), Marketing and Sales Manager, for further information.

Accommodations and meals

The *La Estancia del Encanto Bed and Breakfast* (cel 6671-6449; hslanham@aol.com), an upscale bed-and-breakfast, is inside Altos del María, and is owned and run by an American couple, Honey and Larry Dodge. There are two double rooms and one more for up to four guests ($75.00 single, with an additional charge of $5.00 for an extra person in the room), which have hot-water baths. It is eventually planned to open a restaurant here as well. It is 4.2 km from the main gate on the road to El Valle.

Cabins are also for rent in the area (ph 231-7292, 208-4011; cel 6672-0801; fraukeschnellmunoz@yahoo.com) for $70.00 per night on weekdays, or $160.00 for two weekend nights, with cheaper prices by the week or month.

In Altos del María itself, the only restaurant at present is *Mary's Cafe* (cel 6612-1372, 6631-4648) near the main gate, serving US-style food and pizza, and open for lunch and dinner Thursday-Sunday and on Monday holidays. Arrangements can be made for groups visiting on other days.

There are a couple of small eating places in Sorá, and in Bejuco on the Interamericana, which are open all week. The Pío Pío fast food restaurant in Bejuco is open from 7:00 AM.

See the Altos del María website listed in the previous section for additional information on accommodations and places to eat.

Coclé Province

The mountains of northern Coclé are the easternmost place where a large contingent of the endemics of the Talamanca Range can be found. The two main foothills birding sites in the province are El Valle and Omar Torrijos National Park (known commonly as El Copé). El Valle is much easier to get to, but El Copé has a wider range of birds. In contrast, southern Coclé features extensive grasslands, pastures, and savannas where birds of open country may be found. On the coast, Aguadulce holds mangroves and dry scrub, and is a good place for migratory shorebirds in season.

El Valle

El Valle de Antón, usually called simply El Valle, is a picturesque mountain town located in the caldera of an extinct volcano. Only a two-hour drive from Panama City, it is a popular tourist destination because of its handicrafts market, and also a weekend getaway spot for Panama City residents. Because of this it is well-furnished with good hotels, restaurants, and other amenities. Unlike many places in Panama away from the Canal Area, birding at El Valle does not require a car. Most of the best birding localities can easily be reached from the center of town on foot or by bus or taxi.

Near the Continental Divide at nearly 600 m (2,000 ft) elevation, El Valle features an interesting mix of species. A number of species of the western foothills and highlands can be found on the mountain peaks that form the crater rim, with some as low as the valley itself, including **Black Guan**, Snowcap, **White-tailed Emerald**, **Purple-throated Mountain-gem**, **Orange-bellied Trogon**, Red-faced Spinetail, and Orange-billed Nightingale-Thrush. The forests at the edge of the crater around the town host species of both the drier Pacific slope and the wet Atlantic side, depending on microclimate. Species typical of the Pacific slope include Blue-throat-

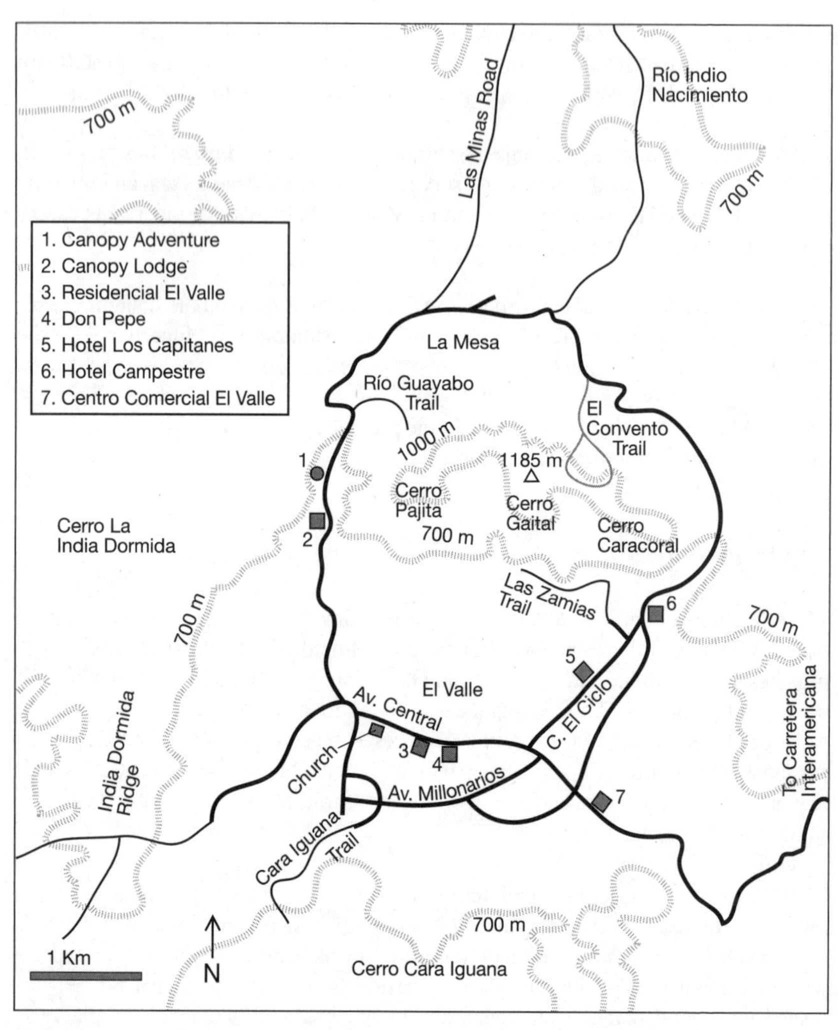

El Valle

ed Goldentail, Sepia-capped Flycatcher, Pale-eyed Pygmy-Tyrant, Lance-tailed Manakin, Rufous-and-white Wren, and Rosy Thrush-Tanager, while species that favor humid forest include Rufous-vented Ground-Cuckoo, Tody Motmot, Dull-mantled Antbird, and Dusky-faced and Tawny-crested Tanagers. Other sought-after species that can be found here include Barred Hawk, Sunbittern, **Purplish-backed Quail-Dove**, White-tipped Sicklebill, Green Thorntail, Green-crowned Brilliant, Long-billed Starthroat, Black-headed Antthrush, Scaled Antpitta, Rufous-browed Tyrannulet, Black-and-white Becard, **Ochraceous Wren**, and Blue Seedeater, along with many commoner species.

The road to El Valle from the Carretera Interamericana enters the crater on its eastern side, descending to the valley floor and becoming Avenida Central as it passes through town. Most hotels and other businesses are found along this thoroughfare. Birding areas are discussed from east to west, according to their access from Avenida Central. Much of the information on the birds and birding areas of El Valle was provided by Danilo Rodríguez, birding guide for the Canopy Lodge and Canopy Adventure, and by Raúl Arias de Para, their proprietor. We thank them both for their assistance.

Las Zamias Trail. This trail runs along the base of Cerro Gaital, which forms the north rim of the crater, and is good for foothills and wet-forest birds. To get there, take Calle de Ciclo, which is on the right about 700 m beyond the Centro Comercial El Valle near the beginning of town. The intersection is just before the minisuper Hong Kong, a white building with green balconies, and is well signposted for the Hotels Campestre and Los Capitanes and the Casa de Lourdes Restaurant. Go 1.1 km on Calle de Ciclo, and turn left on a paved road just before a blue house, signposted for the Casa de Lourdes. This road is paved for 200 m to the Instituto Profesional y Técnico, then continues for another 300 m as a gravel road (past the turnoff to the Casa de Lourdes). At the end of the road, a dirt track takes off at a right angle up the hill to the right. This can be driven if you have 4WD or high clearance; otherwise park off the side of gravel road and hike up. After about 400 m on this track you will reach a yellow-and-green ANAM building. The Los Zamias Trail goes off to the left, in front of an iron gate. The first 300 m or so are through secondary forest with an open understory where species such as Blue-crowned Motmot, Sepia-capped Flycatcher, and Golden-collared Manakin can be found. It then enters primary forest good for wetter forest species; even Rufous-vented Ground-Cuckoo has been seen here. After doing this trail, it may be worth checking out the grounds of the Hotel Campestre at the end of Calle de Ciclo, where Snowcap has been seen recently.

Cara Iguana Trail. Cerro Cara Iguana forms the south rim of the El Valle crater. The Cara Iguana Trail is a gravel road that runs along its base where birds of second growth and Pacific-slope forest can be found. To get to it, continue through town on Av. Central past the market and the church, both on the left. Av. Central ends about 100 m beyond the church. At this intersection, take the left turn onto Av. Los Pozos, signposted for the Pozos Termales (thermal pools). The pavement ends after 600 m; at this point turn onto the paved road to the left. Continue on this road 300 m and take the first road to the right. This road almost immediately crosses a small bridge, and shortly afterward comes to an intersection with another road. Continue straight through the intersection. The road goes uphill and curves to the right. The asphalt ends about 200 m after the intersection, and about 300 m after that is a fork where you should stay left. After about 600 m more the road becomes steeper and more difficult. The road can be birded on foot anywhere from the fork on for species such as such as Little Tinamou, Gray-headed Chachalaca, White-bellied Antbird, Lance-tailed Manakin, Sepia-capped Flycatcher, Pale-eyed Pygmy-Tyrant, Rufous-and-white Wren, and Rosy Thrush-Tanager.

Road to Cerro India Dormida ridge. Cerro India Dormida (Sleeping Indian Woman) forms the west side of the crater wall. This road crosses grassland habitat where species such as Wedge-tailed Grass-Finch and Crested Bobwhite can be found. To find it, go to the intersection at the end of Av. Central, 100 m beyond the church, and turn right. Within a short distance the road crosses a yellow bridge; turn left here. The road is asphalt for 1.7 km, where it turns sharply to the right at a red bus shelter to go uphill. At this point the road becomes very rough and rocky and 4WD is necessary; otherwise you can walk up from here. In another 1.2 km, at an elevation of a bit over 700 m (2,300 ft), the road forks. If you drove up, park about 100 m down the left hand fork on a level stretch next to a concrete cross above the road to the left. (There is a nice view of El Valle from next to the cross.) The finch can be found in the grassland along this road and the other fork, easiest to see when singing from the tops of the rocks that jut out of the grass.

Road to La Mesa and Cerro Gaital. La Mesa is an agricultural area on relatively level land on the north side of Cerro Gaital. At over 800 m (2,600 ft), many foothills species not found in the valley below can be seen here. The road is mostly asphalt, and although it is heavily potholed in places, an ordinary car should not have any trouble. To get to it, go to the intersection at the end of Av. Central, 100 m beyond the church, and turn right. Within a short distance the road crosses a yellow bridge; turn right here again.

At 1.7 km on this road, on the left, is the entrance for the *Canopy Lodge,* and 500 m beyond that, also on the left, is the Canopy Adventure. The Lodge is a hotel that specially caters to birders. See the section on accommodations below for details.

The *Canopy Adventure,* operated by the same owner as the Canopy Lodge and connected to it by trails, is an attraction that features a descent through the forest canopy on a wire cable using a harness. The area also has a number of trails good for wet-forest and foothills species; for a description see the website for the Canopy Lodge (www.canopylodge.com), under "activities." Admission is charged for the trails, $2.50 for the short trail to the Chorro de Macho waterfall, and $15.00 for the rest of the trail system. Guided birding trips can also be arranged.

At 2.6 km along the road is the *Río Guayabo Trail.* Just after a small bridge, pull off on the right next to an ANAM sign for the Cerro Gaital Nature Monument, almost obscured by vegetation. A small trail takes off here, following the Río Guayabo through second growth and small banana plantings for about a kilometer. Species that can be found here include Sunbittern, Band-tailed Barbthroat, Dull-mantled Antbird, White-throated Spadebill, and Southern Nightingale-Wren.

At about 4.7 km, the road enters an area with several chicken farms operated by the Toledano Company. At 4.8 km, on the right, there is a sign for Toledano's Finca Cecilia. Opposite it, on the left, is the *Las Minas Trail,* a dirt track that runs for about 2 km through patches of forest and pasture. Yellow-eared Toucanet, otherwise scarce in the area, has been seen along this road, and Eastern Meadowlark can be found in the pastures.

Continuing along the main road, a fork is reached at 5.3 km. Take the right fork to continue on to Cerro Gaital. (There is an almost-illegible moss-covered wooden signpost here for the Nature Monument.) Bird the patches of forest along this stretch for species such as **Blue-throated Toucanet**, Spotted Woodcreeper, and Silver-throated and Bay-headed Tanagers. The left fork here can also be birded for a few hundred meters before it arrives at Finca Macarena; the road deteriorates after 500 m.

At 6.6 km, at Finca Alemi, the main road bends sharply to the right to continue to Cerro Gaital (the turn is marked by a wooden signpost). At this point another road takes off to the left, running level for about 500 m to a lookout point over the headwaters of the Río Indio on the Atlantic slope. This lookout is a good place to scan for raptors such as Gray-headed Kite and White and Short-tailed Hawks. Beyond this point the road becomes very bad, descending steeply another 2.3 km to the village of *Río Indio Nacimiento* at about 650 m (2,145 ft), and should not be attempted except with a good 4WD vehicle. Species such as Spot-crowned Barbet and Long-tailed Tyrant can be found at the lower elevations around Río Indio Nacimiento. The area is not well known for birds and may repay investigation.

The main road arrives at the entrance to the *Cerro Gaital Nature Monument,* well marked by signs, at 7.0 km from the yellow bridge. The Nature Monument (335 ha/828 acres) was established in 2001 to protect the forests of Cerro Gaital (1,185 m/3,910 ft) and its flanking peaks of Cerro Pajita and Cerro Caracoral. There is a small ANAM building at the entrance which is in theory manned from 8:00 AM to 5:00 PM. The entrance fee is $3.50 for foreign visitors and $1.50 for Panama residents. If there is no one there, enter through the wire gate and pay on the way out.

The trail first follows an abandoned road for about 400 m before entering the forest to the left. The grassy area at this point has flowers favored by **White-tailed Emerald**. After entering the forest, the trail goes up a short set of steps to join the *El Convento Trail*. This loop trail, about 2.5 km (1.5 mi) long, goes up to the ridgetop at a bit over 1000 m (3,300 ft). It is best to go up the left hand fork of the trail since this way is not as steep as the other side. The top section of the trail runs fairly level through cloud forest for several hundred meters. **Black Guan** can be seen along the upper part of the El Convento loop, although the species is rare and wary here. Scaled Antpitta has been found nesting a little beyond the turnoff for the Mirador. Other species to be looked for here include **White-tailed Emerald**, **Purple-throated Mountain-gem**, Green-crowned Brilliant, and Blue Seedeater (in areas with bamboo). At trail station 3 ("Las Heliconias") there is a turnoff to the left for a lookout (*mirador*) which is about 100 m down the trail in elfin forest. When the weather is clear, the lookout provides a magnificent view of El Valle and the countryside below down to the Pacific Ocean, and the Caribbean Sea can be seen on the opposite side. The lookout is a good place to look for Barred Hawk and other raptors. This trail continues on to the peak of Cerro Gaital and then descends to the valley, emerging near the Sendero Las Zamias, but is very steep in places and recommended only for the fit and adventurous. The peak of Cerro Gaital has not been birded very extensively, and some surprises may still await there.

The La Mesa Road and the Cerro Gaital Nature Monument can be reached by local bus for about $1.00. Ask around the market for the schedule. You can also get there by one of the taxis that ply Av. Central for about $3.00.

Getting to El Valle. The signposted turnoff to El Valle from the Carretera Interamericana is 95 km from the Bridge of the Americas, in the small community of Las Uvas about 4 km west of San Carlos. The road is asphalt and is in good condition, with no forks until you reach El Valle in 28 km. Driving time from Panama City is about 2-2 1/2 hours.

Accommodations and meals. As a tourist town, El Valle has a wide variety of accommodations and restaurants ranging from very basic to luxury.

The recently-opened *Canopy Lodge* (ph 264-5720, 263-2784, 214-9724 during office hours; after office hours call Carlos Bethancourt, cel 6578-5711, or Yaritza Medina, cel 6687-0291; birding@canopytower.com; www.canopylodge.com) on the La Mesa Road (see directions above), operated by the same owner as the well-known Canopy Tower in the Canal Area, is especially designed for birders and birding. Amenities include well-appointed, comfortable rooms and excellent food. There is a good trail system, connecting with that of the nearby Canopy Adventure, and the grounds themselves are very birdy, with feeders and plantings of flowering and fruiting shrubs and trees selected to attract hummingbirds, tanagers, honeycreepers, and other species. The dining area, next to the garden, is open-air, allowing you to continue birding through meals, and the rooms also offer good views of the grounds. There is also a small observation platform built into the hotel that allows observation of the forested hillside behind the hotel from various levels. Sunbittern has been seen on the stream that winds past the garden, and Tody Motmot just behind the hotel. The Lodge primarily offers multi-day birding packages, including meals and guided birding trips to sites in El Valle and nearby sites; no walk-ins. Contact the Lodge for details on prices and itineraries.

The *Hotel Campestre* (ph 983-6146; fax 983-6460; ricardoarango@hotmail.com; www.elcampestre.com) has 32 rooms plus 3 cabins ($22.00 Sun-Thu, $32.75 Fri-Sat) with private hot-water bath; some have air-conditioning. It is located at the end of Calle de Ciclo.

The *Hotel Los Capitanes* (ph 983-6080; cel 6687-8819; fax 983-6505; capitanes@cwp.net.pa; www.loscapitanes.com) has rooms and suites ($33.00-$85.00 including breakfast) with private hot-water baths, and has a restaurant on premises. It is located partway up Calle de Ciclo on the left.

The *Residencial El Valle* (ph 983-6536) has 22 rooms ($25.00 Sun-Thu, $33.00 Fri-Sat; "backpacker's rooms" for $16.50 Sun-Thu, $22.00 Fri-Sat) with private hot-water bath. It is located on Av. Central near the market. There is a restaurant on the ground floor.

The *Hotel Don Pepe* (ph 983-6425; ph./fax: 983-6935; hoteldonpepe@hotmail.com) has 11 rooms ($25.00 Sun-Thu, $30.00 Fri-Sat) with private hot-water bath. It is next to the Residencial El Valle on Av. Central near the market. There is a restaurant on the ground floor.

The restaurant at the Hotel Los Capitanes has excellent German, French, and Italian food. The restaurants at the Residencial El Valle and the Hotel Don Pepe area are good for typical Panamanian fare.

There are a number of more expensive hotels located in town and around the valley, as well as cheaper hostals and pensions, and there are a variety of other restaurants along Av. Central and elsewhere. See one of the standard tourist guides for additional choices.

Bird list for El Valle. R = rare.

Great Tinamou
Little Tinamou
Gray-headed Chachalaca
Crested Guan R
Black Guan R
Crested Bobwhite
Neotropic Cormorant
Great Blue Heron
Great Egret
Cattle Egret
Green Heron
Black Vulture
Turkey Vulture
Osprey
Gray-headed Kite
Swallow-tailed Kite
White-tailed Kite
Plumbeous Kite
Tiny Hawk R
Sharp-shinned Hawk R
Bicolored Hawk R
Barred Hawk R
White Hawk
Great Black-Hawk
Gray Hawk
Roadside Hawk
Broad-winged Hawk
Short-tailed Hawk
White-tailed Hawk
Zone-tailed Hawk
Black Hawk-Eagle
Barred Forest-Falcon
Crested Caracara
Yellow-headed Caracara
Merlin
Bat Falcon
White-throated Crake
Gray-necked Wood-Rail
Sunbittern
Wattled Jacana
Southern Lapwing
Spotted Sandpiper
Scaled Pigeon
Pale-vented Pigeon
Ruddy Ground-Dove
Blue Ground-Dove
White-Tipped Dove
Gray-chested Dove
Purplish-backed Quail-Dove
Brown-throated Parakeet
Orange-chinned Parakeet
Brown-hooded Parrot
Blue-headed Parrot
Yellow-billed Cuckoo
Squirrel Cuckoo
Smooth-billed Ani
Groove-billed Ani
Striped Cuckoo
Rufous-vented Ground-Cuckoo R
Barn Owl
Tropical Screech-Owl
Mottled Owl
Black-and-white Owl
Crested Owl
Common Potoo
Common Nighthawk
Common Pauraque
Rufous Nightjar
Chestnut-collared Swift
White-collared Swift
Band-rumped Swift
White-tipped Sicklebill
Rufous-breasted Hermit
Band-tailed Barbthroat
Green Hermit
Long-billed Hermit
Stripe-throated Hermit
White-necked Jacobin
Black-throated Mango
Violet-headed Hummingbird
Rufous-crested Coquette
Green Thorntail
Garden Emerald
White-tailed Emerald
Snowcap R
Violet-crowned Woodnymph
Sapphire-throated Hummingbird
Blue-throated Goldentail
Rufous-tailed Hummingbird
Blue-chested Hummingbird
Snowy-bellied Hummingbird
White-vented Plumeleteer
Bronze-tailed Plumeleteer
Purple-throated Mountain-gem
Green-crowned Brilliant
Purple-crowned Fairy
Long-billed Starthroat
Violaceous Trogon
Orange-bellied Trogon
Black-throated Trogon
Slaty-tailed Trogon
Ringed Kingfisher
Amazon Kingfisher
Green Kingfisher
Tody Motmot R
Blue-crowned Motmot
Rufous Motmot
Broad-billed Motmot
White-whiskered Puffbird

Spot-crowned Barbet
Blue-throated Toucanet
Collared Aracari
Yellow-eared Toucanet
Keel-billed Toucan
Red-crowned Woodpecker
Lineated Woodpecker
Crimson-crested Woodpecker
Red-faced Spinetail R
Spotted Barbtail R
Plain Xenops
Buff-throated Foliage-gleaner
Plain-brown Woodcreeper
Olivaceous Woodcreeper
Wedge-billed Woodcreeper
Northern Barred-Woodcreeper
Cocoa Woodcreeper
Spotted Woodcreeper
Fasciated Antshrike
Great Antshrike
Barred Antshrike
Western Slaty-Antshrike
Russet Antshrike
Plain Antvireo
Spot-crowned Antvireo
Checker-throated Antwren
White-flanked Antwren
Slaty Antwren
Dot-winged Antwren
Dusky Antbird
White-bellied Antbird
Chestnut-backed Antbird
Dull-mantled Antbird R
Bicolored Antbird
Spotted Antbird
Black-headed Antthrush
Black-faced Antthrush
Black-crowned Antpitta R
Scaled Antpitta R
Southern Beardless-Tyrannulet
Yellow Tyrannulet
Yellow-bellied Elaenia
Lesser Elaenia
Ochre-bellied Flycatcher
Olive-striped Flycatcher
Sepia-capped Flycatcher
Slaty-capped Flycatcher
Rufous-browed Tyrannulet R
Paltry Tyrannulet
Black-capped Pygmy-Tyrant
Scale-crested Pygmy-Tyrant
Pale-eyed Pygmy-Tyrant
Common Tody-Flycatcher
Black-headed Tody-Flycatcher
Brownish Twistwing
Eye-ringed Flatbill R
Olivaceous Flatbill
Yellow-olive Flycatcher
Yellow-margined Flycatcher
White-throated Spadebill
Golden-crowned Spadebill
Ruddy-tailed Flycatcher
Black-tailed Flycatcher
Bran-colored Flycatcher
Common Tufted-Flycatcher R
Olive-sided Flycatcher
Western Wood-Pewee
Eastern Wood-Pewee
Tropical Pewee
Long-tailed Tyrant
Bright-rumped Attila
Rufous Mourner
Dusky-capped Flycatcher
Panama Flycatcher
Great Crested Flycatcher
Lesser Kiskadee
Great Kiskadee
Boat-billed Flycatcher
Rusty-margined Flycatcher
Social Flycatcher
Streaked Flycatcher
Sulphur-bellied Flycatcher
Piratic Flycatcher
Tropical Kingbird
Eastern Kingbird
Fork-tailed Flycatcher
Thrush-like Schiffornis
Cinnamon Becard
White-winged Becard
Black-and-white Becard R
Masked Tityra
Golden-collared Manakin
White-ruffed Manakin
Lance-tailed Manakin
Blue-crowned Manakin
Red-capped Manakin
Yellow-throated Vireo
Philadelphia Vireo
Red-eyed Vireo
Yellow-green Vireo
Scrub Greenlet
Lesser Greenlet
Rufous-browed Peppershrike
Black-chested Jay
Gray-breasted Martin
Blue-and-white Swallow R
White-thighed Swallow
Southern Rough-winged Swallow
Bay Wren
Rufous-breasted Wren
Rufous-and-white Wren
Plain Wren
House Wren
Ochraceous Wren
White-breasted Wood-Wren
Gray-breasted Wood-Wren
Southern Nightingale-Wren
Orange-billed Nightingale-Thrush
Gray-cheeked Thrush
Swainson's Thrush
Wood Thrush
Pale-vented Thrush
Clay-colored Thrush
White-throated Thrush
Tawny-faced Gnatwren
Long-billed Gnatwren
Tropical Gnatcatcher
Golden-winged Warbler
Tennessee Warbler
Yellow Warbler
Chestnut-sided Warbler
Magnolia Warbler
Blackburnian Warbler
Palm Warbler R
Bay-breasted Warbler
Black-and-white Warbler
American Redstart
Prothonotary Warbler
Ovenbird
Northern Waterthrush
Louisiana Waterthrush
Kentucky Warbler
Mourning Warbler
Canada Warbler
Rufous-capped Warbler
Buff-rumped Warbler
Bananaquit
Common Bush-Tanager
Black-and-yellow Tanager
Rosy Thrush-Tanager
Dusky-faced Tanager
Olive Tanager
White-shouldered Tanager
Tawny-crested Tanager
White-lined Tanager
Red-crowned Ant-Tanager
Red-throated Ant-Tanager
Hepatic Tanager
Scarlet Tanager
Summer Tanager
Crimson-backed Tanager
Flame-rumped Tanager
Blue-gray Tanager
Palm Tanager
Plain-colored Tanager
Emerald Tanager
Silver-throated Tanager

Bay-headed Tanager
Golden-hooded Tanager
Scarlet-thighed Dacnis
Green Honeycreeper
Red-legged Honeycreeper
Blue-black Grassquit
Variable Seedeater
Yellow-bellied Seedeater
Ruddy-breasted Seedeater
Lesser Seed-Finch
Blue Seedeater R
Yellow-faced Grassquit
Wedge-tailed Grass-Finch
Sooty-faced Finch
Chestnut-capped Brush-Finch
Orange-billed Sparrow
Black-striped Sparrow
Streaked Saltator
Buff-throated Saltator
Black-headed Saltator
Slate-colored Grosbeak
Black-faced Grosbeak
Rose-breasted Grosbeak
Blue-black Grosbeak
Indigo Bunting
Eastern Meadowlark
Great-tailed Grackle
Bronzed Cowbird
Giant Cowbird
Yellow-backed Oriole
Baltimore Oriole
Yellow-billed Cacique
Yellow-rumped Cacique
Scarlet-rumped Cacique
Crested Oropendola
Chestnut-headed Oropendola
Yellow-crowned Euphonia
Thick-billed Euphonia
Tawny-capped Euphonia
Lesser Goldfinch

Penonomé

Penonomé is the largest town in Coclé province. On the Carretera Interamericana about two hours drive west of Panama City and an hour and a half east of Santiago, it has several good hotels and restaurants and is a convenient place to overnight en route between Panama City and the western highlands. It can also serve as a good base for exploring El Copé, Aguadulce, and the grasslands of central Panama.

Accommodations and meals

The *Hotel Las Fuentes* (ph 991-0508, 991-0509) is on the north side of the Interamericana a little before entering the main part of Penonomé coming from the east. It has 20 rooms in a motel-type layout ($18.00), all with private hot-water bath and air-conditioning, and has a restaurant on premises.

The *Hotel Guacamaya* (ph 991-0117, 991-0360, 991-1010; fax 991-0960) is on the north side of the highway just after entering town. It has 40 rooms ($15.40-$33.00), all with private hot-water bath and air-conditioning. The hotel restaurant serves Chinese food.

The *Hotel Dos Continentes* (ph 997-9325) is on the north side of the Interamericana just beyond the Hotel Guacamaya, opposite a large gas station, at the junction between the highway and the main street that goes to the center of town. It has 61 rooms ($22.00-$25.30), all with private hot-water bath and air-conditioning, and a restaurant on premises.

The *Hotel La Pradera* (ph 991-0106, 991-0107) is on the north side of the highway at the western end of town. It has 30 rooms ($15.00-$26.40), all with private hot-water bath and air-conditioning, and a restaurant on premises.

The *Hostal Villa Esperanza* (ph 997-8055; cel 6680-7546; hostal1villaesperanza@yahoo.com; www.hostalvillaesperanza.com) is a bed and breakfast with six rooms ($21.00 Monday-Thursday, $26.00 Thursday-Sunday including breakfast) with hot-water bath and air-conditioning. To find it, coming from the east, take a right off the Interamericana just past the pedestrian overpass beyond the Hotel Dos Continentes at the Colegio Angel María Herrera. Go one block and take a right at the minisuper Jonathan. The hostal is on the right on the second block after this turn.

Outside of town, *Finca La Peregüeta* (cel 6680-5616; info@laperegueta.com; www.laperegueta.com) is a working ranch about 20 minutes drive west of Penonomé near the small community of Cerro Gordo. A guest house is available with sleeping accommodations for up to eight people and a bathroom; cold water only. Electricity is provided by generator from 6:30 PM to 9:00 PM. Camping under shelters is also available. Most of the birds listed for the Coclé grasslands (see next section) can be found on or near the property, and it can also provide a base for birding El Copé and Aguadulce. It may be of particular interest to those who wish to learn more about rural life in Panama.

Just west of the Hotel Dos Continentes there are two restaurants on the Interamericana that are convenient places to grab a bite if you are in transit. The *Parrillada El Gigante,* on the north side of the highway, serves grilled meats, Lebanese food, and pizza, as well as typical Panamanian food. *Las Tinajas,* a cafeteria opposite El Gigante on the south side of the highway, is the best place to get a quick meal if you are in a hurry. The restaurants at the main hotels on the Interamericana, the Las Fuentes, Guacamaya, Dos Continentes, and La Pradera, are also options. There are several other places to eat in Penonomé off the main road.

Note that Penonomé is a popular destination during Carnaval, and rooms are likely to be hard to come by during this period.

Coclé grasslands

The grasslands of Coclé province are a good place to see birds of the open country of the western Pacific lowlands, especially the scarce Grassland Yellow-Finch. Other species of interest include White-tailed Hawk, Aplomado Falcon, and **Veraguan Mango**. An endemic subspecies of Grasshopper Sparrow also once occurred here, but has not been seen since the 1960s.

Getting to the Coclé grasslands. To get to good grassland areas from Penonomé, turn south (left if you are heading west) off the Interamericana next to the pedestrian overpass just west of the Hotel Dos Continentes. A little over 4 km south of Penonomé the road divides, the left fork going to El Coco and the right to El Gago. Either one is good for grassland birds. Both forks go for about 9-10 km before reaching fields near mangroves near the coast. A couple of kilometers west of El Gago is Cerro Cerrezuela, which contains the largest remaining patch of dry forest in the region, although it is highly disturbed.

Accommodations and meals. The Coclé grasslands are best done as a day trip from Penonomé.

Bird list for the Coclé grasslands. The list includes species of grassland and of dry scrub. Aquatic species can also be found in appropriate habitats within the grasslands. R = rare; ? = present status uncertain.

Gray-headed Chachalaca
Crested Bobwhite
Cattle Egret
Black Vulture
Turkey Vulture
Lesser Yellow-headed Vulture
Pearl Kite
White-tailed Kite
Roadside Hawk
Short-tailed Hawk
White-tailed Hawk
Zone-tailed Hawk
Crested Caracara
Yellow-headed Caracara
Aplomado Falcon
Bat Falcon
Gray-necked Wood-Rail
Southern Lapwing
Mourning Dove
Plain-breasted Ground-Dove
Ruddy Ground-Dove
Blue Ground-Dove
White-tipped Dove
Brown-throated Parakeet
Orange-chinned Parakeet
Yellow-crowned Amazon
Squirrel Cuckoo
Striped Cuckoo
Smooth-billed Ani
Groove-billed Ani
Barn Owl
Tropical Screech-Owl
Ferruginous Pygmy-Owl
Striped Owl
Lesser Nighthawk
Common Pauraque
Rufous Nightjar
Common Potoo
Scaly-breasted Hummingbird
Veraguan Mango
Garden Emerald
Sapphire-throated Hummingbird
Snowy-bellied Hummingbird
Rufous-tailed Hummingbird
Blue-crowned Motmot
Red-crowned Woodpecker
Pale-breasted Spinetail
Barred Antshrike
Southern Beardless-Tyrannulet
Mouse-colored Tyrannulet
Yellow-crowned Tyrannulet
Greenish Elaenia
Yellow-bellied Elaenia
Lesser Elaenia
Yellow Tyrannulet
Pale-eyed Pygmy-Tyrant
Slate-headed Tody-Flycatcher
Common Tody-Flycatcher
Yellow-olive Flycatcher
Bran-colored Flycatcher
Tropical Pewee
Panama Flycatcher
Great Kiskadee
Social Flycatcher
Streaked Flycatcher
Tropical Kingbird
Fork-tailed Flycatcher
Lance-tailed Manakin
Yellow-green Vireo
Scrub Greenlet
Golden-fronted Greenlet
Rufous-browed Peppershrike
Gray-breasted Martin
Barn Swallow
House Wren
Tropical Gnatcatcher
Clay-colored Thrush
Yellowish Pipit
Tennessee Warbler
Yellow Warbler
Rufous-capped Warbler
Red-legged Honeycreeper
Summer Tanager
Blue-gray Tanager
Palm Tanager
Blue-black Grassquit
Variable Seedeater
Yellow-bellied Seedeater
Ruddy-breasted Seedeater
Lesser Seed-Finch
Grassland Yellow-Finch R
Black-striped Sparrow
Grasshopper Sparrow ?
Streaked Saltator
Buff-throated Saltator
Red-breasted Blackbird
Eastern Meadowlark
Great-tailed Grackle
Yellow-backed Oriole
Yellow-crowned Euphonia
Lesser Goldfinch

El Copé (General de División Omar Torrijos Herrera National Park)

The mountains above the small town of El Copé are the easternmost locality for many of the endemics of the western highlands. This is the most accessible part of General de División Omar Torrijos Herrera National Park (25,257 ha/62,411 acres), often referred to as "El Copé" after the town, which protects montane and cloud forests along the Central Cordillera in the province of Coclé. Though quite rare, Three-wattled Bellbird and **Bare-necked Umbrellabird** can be found here seasonally. It is the best locality in Panama for Snowcap, and other species that can be seen here include **Black Guan**, **Purplish-backed Quail Dove**, White-tipped Sicklebill, Brown Violet-ear, Green Thorntail, Green-crowned Brilliant, **Purple-throated Mountain-gem**, **Red-fronted Parrotlet**, Red-headed Barbet, Immaculate Antbird, Stripe-breasted Wren, **Black-faced Solitaire**, **Blue-and-gold Tanager**, Yellow-

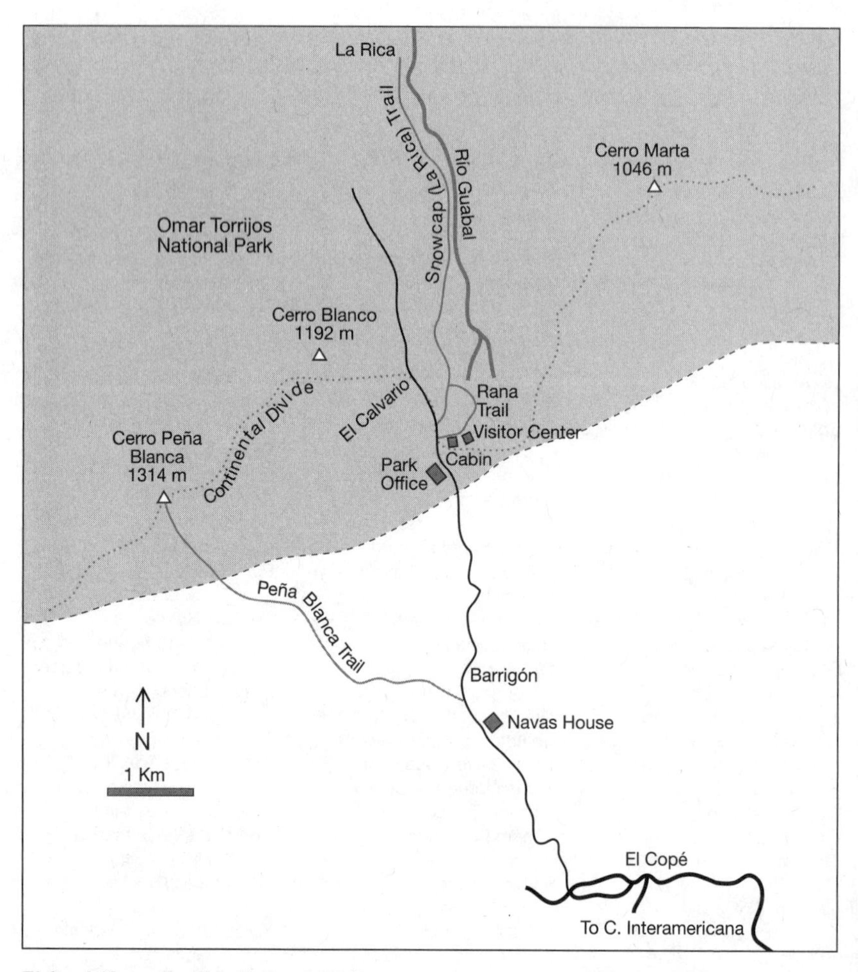

El Copé (Omar Torrijos National Park)

throated Bush-Tanager, **Black-thighed Grosbeak**, and White-throated Shrike-Tanager. At lower elevations, the road that goes to El Copé passes through cattle and agricultural land which can yield Crested Bobwhite, Mourning Dove, and Mouse-colored Tyrannulet, and with luck Aplomado Falcon or White-tailed Hawk.

Weather can be a problem in the area. In the earlier part of the dry season the trade winds may keep it foggy and drizzly all morning or even all day. March has the reputation of being the time most likely to produce good weather. In rainy season there is no wind and clear sunny days often occur, but there is also the chance of all-day rain. If rained out, a good plan can be to check out the Aguadulce area for open country birds and try again the next day.

There are two main birding trails within the park, the Snowcap, or La Rica, Trail, and the Rana ("Frog") Trail. Both of these are near the Visitor's Cabin and Visitor Center near the main park office. The road itself can also be birded to El Calvario, a cross on the Continental Divide. In addition, there is a short, 500-m interpretive loop trail, Los Helechos ("The Ferns"), behind the Visitor Center.

Snowcap (La Rica) Trail. This is not an official park trail but instead a former road that goes to the small community of La Rica and other villages on the Atlantic slope within the park. Its namesake species is regularly seen along it. It takes off from the main gravel road on the right, about 200 m beyond the Visitor's Cabin, which is at about 715 m (2,360 ft) elevation. Because it is regularly used by villagers and their livestock it can be muddy in places. The trail descends somewhat steeply for the first 500 m (0.3 mi), where there is a junction with the Rana Trail on the right, then levels out for several hundred m at about 700 m elevation. It first goes through second growth, then through taller forest with occasional open areas on old landslides, before reaching the first agricultural clearings about 1.5 km (0.9 mi) from its start, at about 600 m (2,000 ft) elevation. Beyond this point patches of forest and second growth alternate with fields.

Rana Trail. To find this new trail, recently constructed by the US Peace Corps, take the gravel path to the right of the Visitor's Cabin, which leads to the Visitor Center. The entrance to the trail is to the right of the Visitor Center. This trail goes down into a small drainage behind the Center, then loops back up to join the Snowcap Trail about 500 m from its beginning. The first part of the trail is steep in some places, but has steps. At the first stream you come to near the bottom there is a new trail on the right, not completed at the time of writing, that will eventually connect to the main road near the park office. A few hundred meters beyond this you reach a larger stream, and turn to the left at a yellow sign. There is a steep slippery uphill bit after this, but the trail then becomes easier again. At the junction with the Snowcap Trail, take a left at the "salida" sign to return to the main road and the Visitor's Cabin.

El Calvario. The road beyond the Visitor Center can also be birded. Because it is steep and can be slippery it is probably better to walk than to drive up. About 500 m beyond the entrance to the Snowcap Trail is El Calvario at 900 m (2,970 ft), a lookout point with a large cross, from which both oceans can be seen on a rare clear day.

Besides the trails near the Visitor Center, the more adventurous can visit more remote areas of the park with the aid of a guide. Cerro Peña Blanca (1,314 m/4,336 ft), a few kilometers west of the Visitor Center, reaches higher elevations and has a number of additional highland species not found in lower areas. The trail starts in the town of Barrigón, about 3 km before the park entrance, but is difficult to follow due to many side trails. Santos Navas (ph 983-9130) and members of his family, who live in Barrigón, can serve as guides; the cost is $12.00 per day. In addition, Santos operates a rustic lodge near the town of La Rica, on the Atlantic side of the park at about 400 m (1,300 ft) elevation. From there Cerro Marta and other areas can be reached by trail.

See the section on accommodations for more information. Although we have not visited the La Rica area, it should be interesting for species of the wet Atlantic slope. Santos' house, signposted "Albergues Navas," is about 100 m before the Buena Vista Bridge in Barrigón, behind a small green building on the right side of the road. See the section below for instructions on how to get to Barrigón.

Getting to El Copé. From Panama City, follow the directions for "Getting to Western Panama" (pp. 130-132). The turnoff to El Copé from the Carretera Interamericana is 19.0 km west of Penonomé. The un-signposted turnoff is just before a pedestrian overpass, on the right at a small collection of stores including the green Farmacia Vega #3. This asphalt road is in good condition all the way to El Copé, at 26 km from the Highway. Upon reaching El Copé, an asphalt road goes left; continue straight on a narrower asphalt road. After about 400 m, there is another asphalt road to the left, while a gravel road, signposted for "Parque Nacional Gral. Omar Torrijos H. – 7.5 km" continues straight. This road is currently in fair condition but requires 4WD or at least high clearance in places, especially past Barrigón. At 2.4 km, near the beginning of the town of Barrigón, you will see a small bridge with benches, the Puente Buena Vista, on the left off the main road. Santos Navas' house is 100 m before this bridge, on the right. Continuing on the main road, you reach the Barrigón school at 3.1 km. The park entrance is at 6.3 km, and the park office, where you can pay the entrance fee ($3.50 for foreign visitors and $1.50 for Panama residents), is 200 m beyond that. Another 500 m brings you to the Visitor's Cabins and the trails.

If you do not have a vehicle, El Copé can be reached by bus from Penonomé, and from there a minibus can be taken as far as Barrigón. From Barrigón it is about a 3.5 km (2.1 mi) hike to the park office. You could also arrange to have Santos Navas take you up to the park from Barrigón by car.

Accommodations and meals. El Copé can be visited as a day trip from Penonomé, which has the closest mid-level hotels. See pp. 143-144 for information on staying in Penonomé.

The town of El Copé has a small and very basic pension, the *Hospedaje Vankiria* (ph 938-9137). There are five rooms ($12.00), several with private bath; fan and cold water only. To find it, take the asphalt road that goes to the left, where the gravel road starts to go to National Park. Go 300 m to where the asphalt ends next to the minisuper Del Pueblo. Go left on the gravel road next to the church; the hospedaje is 100 m down this road. There are three small restaurants in El Copé. If you stay overnight, be aware that they may close soon after dark.

The *Visitor's Cabin* in the park is comfortable and attractive, with a large living room, bunk room, and kitchen. There are four bunks, but larger groups can be accommodated in the living room and a large loft if they bring cots or air mattresses. The cabin has electricity from solar panels, a gas stove, basic cookware, and a refrigerator (but the latter was not in service on a recent visit). The cost is $13.00 per night.

Santos Navas and family (ph 983-9130) rent rooms in their house at Barrigón. They also operate the *Albergue Navas*, a rustic cabin at 400 m near the town of La Rica on the Atlantic side of the park. There is no road to the cabins; you must either walk in or go by horseback. The cabins have no electricity. It is about 1 1/2 hours on foot from the closest point that you can reach by vehicle. The cost is $30.00 per day ($24.00 per person for a group of three), which includes all meals and a guide.

Bird list for El Copé. R = rare. PB = species mainly found at higher elevations on Cerro Peña Blanca and other peaks, above the level of the main trails.

Great Tinamou
Little Tinamou
Gray-headed Chachalaca
Black Guan PB
Crested Guan R
Black-breasted Wood-Quail PB
Gray-necked Wood-Rail
Black Vulture
Turkey Vulture
Hook-billed Kite
Swallow-tailed Kite
Double-toothed Kite
Plumbeous Kite
Barred Hawk
Semiplumbeous Hawk
Gray Hawk
Roadside Hawk
Broad-winged Hawk
Short-tailed Hawk
Swainson's Hawk
Crested Eagle R
Black Hawk-Eagle
Barred Forest-Falcon
Bat Falcon
Sunbittern
Pale-vented Pigeon
Scaled Pigeon
Band-tailed Pigeon PB
Short-billed Pigeon
Ruddy Pigeon PB
Ruddy Ground-Dove
Blue Ground-Dove
White-tipped Dove
Chiriqui Quail-Dove PB
Purplish-backed Quail-Dove
Orange-chinned Parakeet
Red-fronted Parrotlet R
Brown-hooded Parrot
Blue-headed Parrot
Squirrel Cuckoo
Striped Cuckoo
Smooth-billed Ani
Groove-billed Ani
Tropical Screech-Owl
Crested Owl
Ferruginous Pygmy-Owl R
Mottled Owl
Common Pauraque
Chuck-will's-widow
Great Potoo
White-collared Swift
Chimney Swift
Vaux's Swift
Band-rumped Swift
Lesser Swallow-tailed Swift
Rufous-breasted Hermit
Band-tailed Barbthroat
Green Hermit
Long-billed Hermit
Stripe-throated Hermit
White-tipped Sicklebill
Green-fronted Lancebill PB
Violet Sabrewing PB
White-necked Jacobin
Brown Violet-ear R
Violet-headed Hummingbird
Rufous-crested Coquette
Green Thorntail R
Garden Emerald
Violet-crowned Woodnymph
Violet-bellied Hummingbird
Blue-chested Hummingbird
Snowy-bellied Hummingbird
Rufous-tailed Hummingbird
Black-bellied Hummingbird PB R
White-tailed Emerald PB
Snowcap
Bronze-tailed Plumeleteer
Purple-throated Mountain-gem PB
Green-crowned Brilliant
Purple-crowned Fairy
Orange-bellied Trogon
Black-throated Trogon
Slaty-tailed Trogon
Lattice-tailed Trogon R
Blue-crowned Motmot
Rufous Motmot
Broad-billed Motmot
Spot-crowned Barbet
Red-headed Barbet
Prong-billed Barbet PB
Blue-throated Toucanet
Collared Aracari
Yellow-eared Toucanet
Keel-billed Toucan
Black-cheeked Woodpecker
Red-crowned Woodpecker
Golden-olive Woodpecker
Cinnamon Woodpecker
Lineated Woodpecker
Spotted Barbtail PB
Ruddy Treerunner PB
Buffy Tuftedcheek PB
Striped Woodhaunter R
Lineated Foliage-gleaner PB
Slaty-winged Foliage-gleaner
Buff-throated Foliage-gleaner
Ruddy Foliage-gleaner
Streak-breasted Treehunter PB
Plain Xenops
Streaked Xenops R
Tawny-throated Leaftosser
Plain-brown Woodcreeper
Ruddy Woodcreeper
Long-tailed Woodcreeper
Wedge-billed Woodcreeper
Strong-billed Woodcreeper R
Cocoa Woodcreeper
Black-striped Woodcreeper
Spotted Woodcreeper
Brown-billed Scythebill
Barred Antshrike
Western Slaty-Antshrike
Russet Antshrike
Plain Antvireo
Spot-crowned Antvireo
Pacific Antwren
Checker-throated Antwren
White-flanked Antwren
Slaty Antwren
Dot-winged Antwren

White-bellied Antbird
Bare-crowned Antbird
Chestnut-backed Antbird
Dull-mantled Antbird
Immaculate Antbird
Spotted Antbird
Bicolored Antbird
Black-faced Antthrush
Black-headed Antthrush
Black-crowned Antpitta R
Streak-chested Antpitta
Silvery-fronted Tapaculo PB
Paltry Tyrannulet
Brown-capped Tyrannulet
Yellow-crowned Tyrannulet
Forest Elaenia
Yellow-bellied Elaenia
Lesser Elaenia
Olive-striped Flycatcher
Ochre-bellied Flycatcher
Slaty-capped Flycatcher
Rufous-browed Tyrannulet R
Scale-crested Pygmy-Tyrant
Slate-headed Tody-Flycatcher
Common Tody-Flycatcher
Eye-ringed Flatbill
Yellow-margined Flycatcher
White-throated Spadebill
Golden-crowned Spadebill
Ruddy-tailed Flycatcher
Sulphur-rumped Flycatcher
Common Tufted-Flycatcher
Olive-sided Flycatcher
Western Wood-Pewee
Eastern Wood-Pewee
Yellow-bellied Flycatcher
Acadian Flycatcher
Long-tailed Tyrant
Bright-rumped Attila
Rufous Mourner
Dusky-capped Flycatcher
Great Kiskadee
Boat-billed Flycatcher
Social Flycatcher
Golden-bellied Flycatcher PB
Sulphur-bellied Flycatcher
Tropical Kingbird
Eastern Kingbird
Speckled Mourner
Cinnamon Becard
Masked Tityra
Rufous Piha
Thrush-like Schiffornis
Bare-necked Umbrellabird R
Three-wattled Bellbird R
Golden-collared Manakin
White-ruffed Manakin
White-crowned Manakin R
Blue-crowned Manakin
Red-capped Manakin
Philadelphia Vireo
Red-eyed Vireo
Tawny-crowned Greenlet
Golden-fronted Greenlet
Lesser Greenlet
Green Shrike-Vireo
Black-chested Jay
Gray-breasted Martin
Blue-and-white Swallow R
White-thighed Swallow
Northern Rough-winged Swallow
Southern Rough-winged Swallow
Cliff Swallow
Barn Swallow
Bay Wren
Stripe-breasted Wren
Rufous-breasted Wren
Rufous-and-white Wren
Plain Wren
House Wren
White-breasted Wood-Wren
Gray-breasted Wood-Wren
Southern Nightingale-Wren
Song Wren
Tawny-faced Gnatwren
Long-billed Gnatwren
Tropical Gnatcatcher
Black-faced Solitaire
Orange-billed Nightingale-Thrush
Slaty-backed Nightingale-Thrush
Gray-cheeked Thrush
Swainson's Thrush
Pale-vented Thrush
Clay-colored Thrush
White-throated Thrush
Gray Catbird
Golden-winged Warbler
Tennessee Warbler
Chestnut-sided Warbler
Magnolia Warbler
Black-throated Green Warbler R
Blackburnian Warbler
Bay-breasted Warbler
Cerulean Warbler
Black-and-white Warbler
American Redstart
Northern Waterthrush
Louisiana Waterthrush
Mourning Warbler
Wilson's Warbler R
Canada Warbler
Slate-throated Redstart
Collared Redstart PB
Rufous-capped Warbler
Three-striped Warbler
Buff-rumped Warbler
Wrenthrush PB
Bananaquit
Common Bush-Tanager
Yellow-throated Bush-Tanager
Black-and-yellow Tanager
Dusky-faced Tanager
Olive Tanager
Gray-headed Tanager
White-throated Shrike-Tanager R
Tawny-crested Tanager
White-lined Tanager
Red-crowned Ant-Tanager
Red-throated Ant-Tanager
Hepatic Tanager
Summer Tanager
Scarlet Tanager
Crimson-collared Tanager R
Crimson-backed Tanager
Flame-rumped Tanager
Blue-gray Tanager
Palm Tanager
Blue-and-gold Tanager
Plain-colored Tanager
Emerald Tanager
Silver-throated Tanager
Speckled Tanager
Bay-headed Tanager
Rufous-winged Tanager
Golden-hooded Tanager
Spangle-cheeked Tanager PB
Scarlet-thighed Dacnis
Blue Dacnis
Green Honeycreeper
Shining Honeycreeper
Red-legged Honeycreeper
Blue-black Grassquit
Variable Seedeater
Lesser Seed-Finch
Blue Seedeater R
Yellow-faced Grassquit
Sooty-faced Finch PB
Chestnut-capped Brush-Finch
Orange-billed Sparrow
Black-striped Sparrow
Streaked Saltator
Buff-throated Saltator
Black-headed Saltator
Slate-colored Grosbeak
Black-faced Grosbeak R

Black-thighed Grosbeak PB
Rose-breasted Grosbeak
Blue-black Grosbeak
Bronzed Cowbird
Yellow-backed Oriole
Yellow-tailed Oriole
Baltimore Oriole
Yellow-billed Cacique
Scarlet-rumped Cacique
Chestnut-headed Oropendola
Golden-browed Chlorophonia PB
Yellow-crowned Euphonia
Thick-billed Euphonia
Fulvous-vented Euphonia
White-vented Euphonia
Tawny-capped Euphonia

Aguadulce salinas

The coastal area east of Aguadulce can be a good place to see migrant shorebirds and other aquatic species, as well as some species of open country. At one time evaporation ponds here, used for manufacturing salt and known as *salinas*, were a major high-tide roosting and foraging area for many waterbirds, but they are now much reduced from their former extent. Several years ago tariffs on imported salt were removed and most of the salt ponds were taken out of production. While some are still in use much of the area now consists of bare soil. However, the area can still be worth a visit. During migration and the northern winter many shorebirds and terns can be found on and near the coastal mudflats. Non-aquatic species that can be found in the area include Mangrove Black-Hawk, Plain-breasted Ground-Dove, White-winged Dove, Northern Scrub-Flycatcher, and, more rarely, Aplomado Falcon.

The salt ponds (and former salt ponds) extend along the last 3 km of the asphalt road that goes from Aguadulce to the coast. At the end of the salt ponds the road bends sharply to the right where it reaches the shore. Here you can park on the left and walk over a small rise to the beach and coastal mudflats that edge the Bay of Parita. The area is best birded at high tide, when birds are forced closer to shore. At low tide they are often too far out on the mudflats to see well.

Getting to the Aguadulce salinas. Leaving from Panama City, follow the directions for "Getting to Western Panama" (pp. 130-132). Aguadulce is about 43 km beyond Penonomé heading west, and 11 km beyond Natá. Finding your way through town to the salt flats is a bit complicated. Coming from the direction of Penonomé, take a left off the Carretera Interamericana just beyond the first pedestrian overpass and the large Importadora Americana building. After 700 m you will come to an intersection opposite a park with statues and white benches. Go straight across, and after 500 m more you will come to the main city plaza and its church, which are on your left. Past the plaza continue straight down the street that parallels the plaza's right side. After 100 m this street comes to a small park with benches. Turn left here, then take the second right, opposite a building with a sign "Bilingüe La Colmena." Take the next left, signposted for the Turiscentro El Gallo. Immediately after this you will cross a small concrete bridge. At 300 m beyond this turn you will come to a small park with a gazebo and a statue of a salt worker. Bear left where the road forks just before the park to get to the salt flats and the coast. From the statue it is about 4 km to the first salt ponds and 7.9 km to the coast.

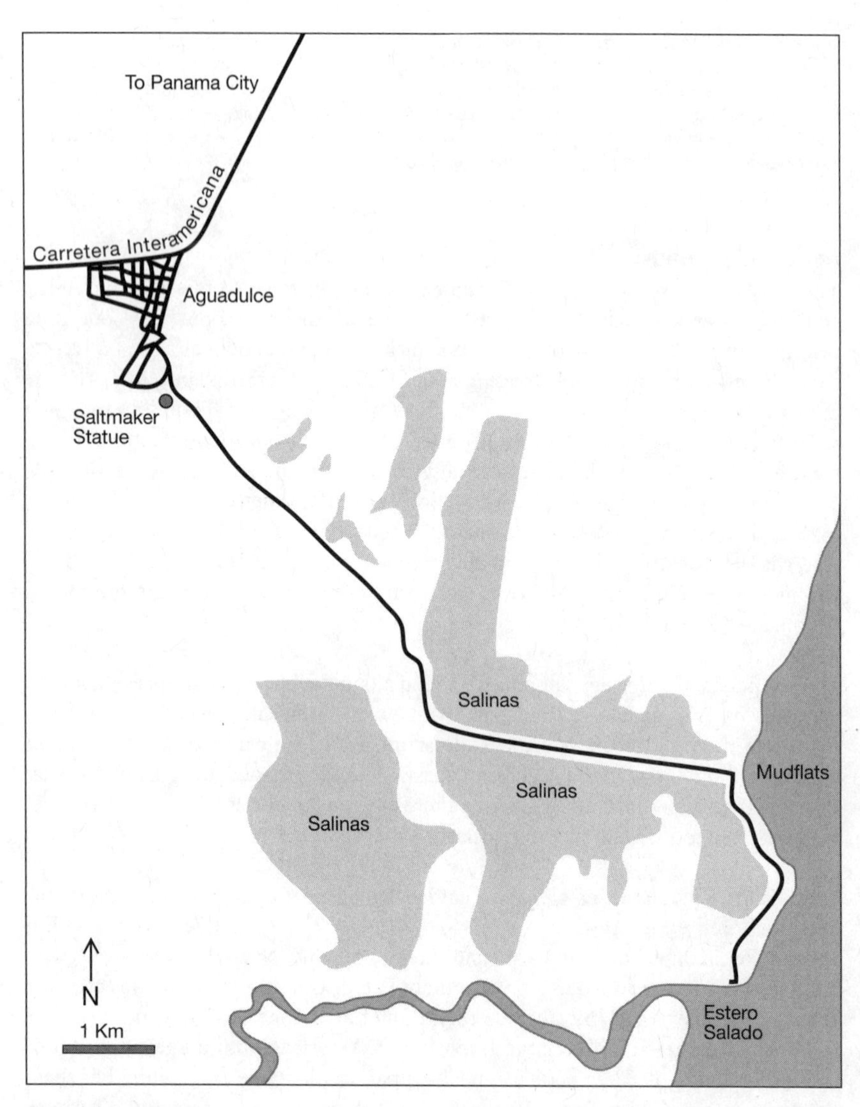

Aguadulce salinas

Accommodations and meals. Since it takes only an hour or two to bird the Aguadulce salinas, they are best done as a stop off en route between other sites to the east, west, or south. In particular, Aguadulce can be a convenient complement to a trip to El Copé. It is preferable to stay in Penonomé, Santiago, or Chitré (depending on your direction of travel), which have a wider selection of hotels and restaurants than Aguadulce itself. However, there are several hotels and restaurants along the Carretera Interamericana and near the downtown area if you prefer to stay here.

Bird list for Aguadulce and environs. * = vagrant; R = rare.

Black-bellied Whistling-Duck
Brown Pelican
Neotropic Cormorant
Magnificent Frigatebird
Bare-throated Tiger-Heron
Great Blue Heron
Great Egret
Snowy Egret
Little Blue Heron
Reddish Egret*
Tricolored Heron
Cattle Egret
Green Heron
Black-crowned Night-Heron
Yellow-crowned Night-Heron
White Ibis
Roseate Spoonbill R
Wood Stork
Black Vulture
Turkey Vulture
Lesser Yellow-headed Vulture
Osprey
White-tailed Kite
Mangrove Black-Hawk
Savanna Hawk
Roadside Hawk
Short-tailed Hawk
Swainson's Hawk
White-tailed Hawk R
Crested Caracara
Yellow-headed Caracara
Aplomado Falcon R
Crested Bobwhite
Black-bellied Plover
American Golden-Plover
Collared Plover
Wilson's Plover
Semipalmated Plover
Killdeer
American Oystercatcher
Black-necked Stilt
American Avocet*
Wattled Jacana
Greater Yellowlegs
Lesser Yellowlegs
Solitary Sandpiper
Willet
Spotted Sandpiper
Whimbrel
Long-billed Curlew*
Marbled Godwit
Ruddy Turnstone
Surfbird
Red Knot
Sanderling
Semipalmated Sandpiper
Western Sandpiper
Least Sandpiper
White-rumped Sandpiper R
Baird's Sandpiper R
Dunlin*
Stilt Sandpiper
Buff-breasted Sandpiper
Short-billed Dowitcher
Wilson's Phalarope R
Red-necked Phalarope R
Laughing Gull
Ring-billed Gull
Herring Gull
Gull-billed Tern
Caspian Tern
Royal Tern
Elegant Tern
Common Tern
Least Tern
Yellow-billed Tern*
Large-billed Tern*
Black Tern
Rock Pigeon
White-winged Dove
Mourning Dove
Common Ground-Dove
Plain-breasted Ground-Dove
Ruddy Ground-Dove
Blue Ground-Dove
White-tipped Dove
Brown-throated Parakeet
Orange-chinned Parakeet
Yellow-crowned Amazon
Black-billed Cuckoo
Yellow-billed Cuckoo
Smooth-billed Ani
Groove-billed Ani
Barn Owl
Lesser Nighthawk
Common Nighthawk
Common Pauraque
Short-tailed Swift
Scaly-breasted Hummingbird
Garden Emerald
Sapphire-throated Hummingbird
Blue-crowned Motmot
Ringed Kingfisher
Green Kingfisher
Amazon Kingfisher
Red-crowned Woodpecker
Pale-breasted Spinetail
Straight-billed Woodcreeper
Barred Antshrike
Southern Beardless-Tyrannulet
Mouse-colored Tyrannulet
Northern Scrub-Flycatcher
Greenish Elaenia
Yellow-bellied Elaenia
Lesser Elaenia
Pale-eyed Pygmy-Tyrant
Common Tody-Flycatcher
Panama Flycatcher
Great Kiskadee
Boat-billed Flycatcher
Rusty-margined Flycatcher
Social Flycatcher
Tropical Kingbird
Gray Kingbird
Scissor-tailed Flycatcher R
Fork-tailed Flycatcher
Yellow-green Vireo
Scrub Greenlet
Rufous-browed Peppershrike
Gray-breasted Martin
Mangrove Swallow
Cliff Swallow
Barn Swallow
Rufous-breasted Wren
Rufous-and-white Wren
Plain Wren
House Wren
Yellow Warbler (migrant forms)
Yellow (Mangrove) Warbler
Yellow-rumped Warbler
Prothonotary Warbler
Northern Waterthrush
Mourning Warbler
Rufous-capped Warbler
White-lined Tanager
Summer Tanager
Blue-gray Tanager
Palm Tanager
Golden-hooded Tanager
Blue-black Grassquit
Variable Seedeater
Yellow-bellied Seedeater
Ruddy-breasted Seedeater
Black-striped Sparrow
Dickcissel
Red-breasted Blackbird
Eastern Meadowlark
Great-tailed Grackle
Bronzed Cowbird
Orchard Oriole
Baltimore Oriole
Yellow-crowned Euphonia
Lesser Goldfinch
House Sparrow

Herrera Province

Herrera Province, in the Azuero Peninsula, along with Los Santos is the heartland of Panama's traditional culture. Although much of the province has been deforested and converted to cattle pasture, some interesting wetland and coastal habitats can be found here. Las Macanas Marsh is the largest freshwater wetland in the region, which is Panama's driest. The Cenegón del Mangle Wildlife Refuge protects an extensive area of mangroves, as well as dry scrub woodland. Coastal areas near Chitré are the best place to find species such as White-winged Dove and Common Ground-Dove. Inland, the El Montuoso Forest Reserve protects the largest remnant of foothills forest still remaining in the province.

Las Macanas Marsh

Las Macanas Marsh is good for many aquatic species, as well as for species of open country and scrub of this region. Las Macanas is a shallow lake, much of the surface of which is covered by thick mats of floating Water Hyacinth and *Pistia* water ferns, surrounded by boggy areas. It is protected as a Multiple Use Area (2,000 ha/4,942 acres), which allows grazing and fishing. Among the wetland species to be found here are Black-bellied and Fulvous Whistling-Ducks, Snail Kite, Limpkin, Glossy Ibis, and Wattled Jacana. In the surrounding fields and scrub can be found Lesser Yellow-headed Vulture, Pearl Kite, Aplomado Falcon, Ferruginous Pygmy-Owl, Mouse-colored Tyrannulet, and many other species. The best vantage point for seeing aquatic species is an observation tower near the shore of the marsh. Scrub species can be found in the small patch of dry woodland around the tower, where there is a short trail, the *Sendero Marao,* or else by driving the roads near the marsh. Open country species can often be seen over fields or perched on power lines along the roads on the way in to the site. Bring water and sunscreen, since the area can be blistering hot, especially at mid-day.

Getting to Las Macanas Marsh. From Panama City, follow the directions for "Getting to Western Panama" (pp. 130-132). Turn south off the Carretera Interamericana at Divisa, which is 65 km west of Penonomé and about 40 km east of Santiago, to go south on the Carretera Nacional. After 2.5 km on this road you will pass through the town of Santa María. The turnoff for Las Macanas is 7.4 km from Divisa, at a blue bus shelter where a tree-lined road goes off to the left. From the turn it is 2.1 km the town of El Rincón, where the road turns sharply right at a church. After 300 m, the asphalt ends and the road divides in two; take the right fork here. (There is another road that goes off at a right angle to the right where the main road divides; take the right fork on the main road, not this one. On the left fork of the main road, the local ANAM office is the first building on the right.) After another 200 m, the road forks again; take the left, signposted "Ciénaga Las Macanas 2.5 km." After 2.2 km on this road, take a turn to the right on a side road, signposted for Las Macanas. After 1.0 km on this road, you will come to a parking area next to a fence and a "Bienvenidos" sign. Go in through the entrance in the fence and follow the gravel path to get to the observation tower and the Sendero Marao.

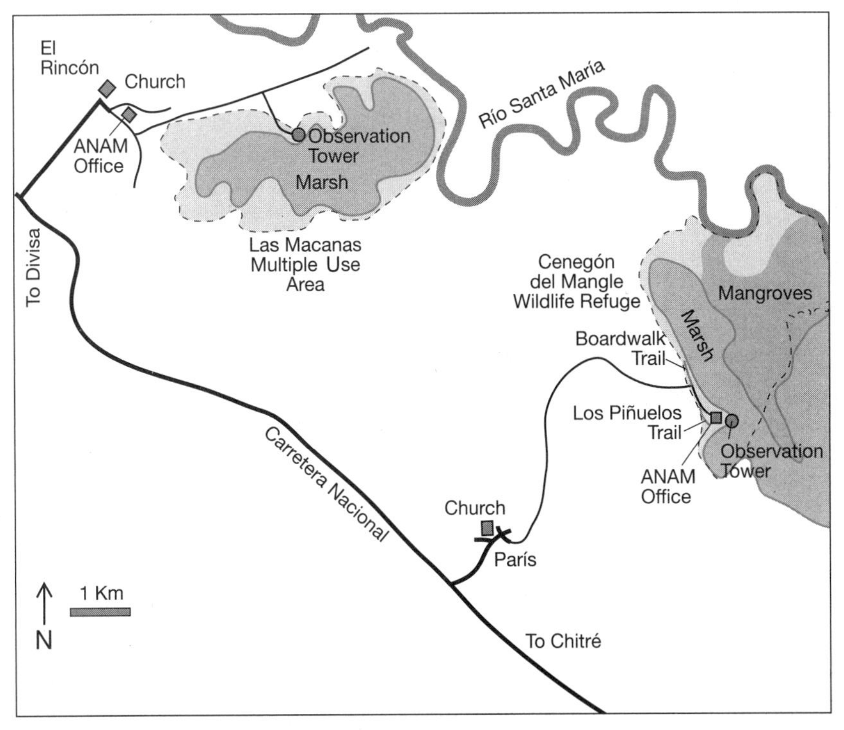

Las Macanas Marsh and Cenegón del Mangle Wildlife Refuge

Accommodations and meals. Since it takes only a short time to bird and is close to the Carretera Interamericana, Las Macanas can make a good detour en route between Panama City and the western highlands. It can also be done as a side trip in combination with a trip to El Copé. The closest accommodations can be found in Penonomé or Santiago along the Carretera Interamericana, or in Chitré if you plan to continue farther down the Azuero Peninsula. There are several small restaurants in Santa María.

Bird list for Las Macanas and environs. * = vagrant; R = rare.

Black-bellied Whistling-Duck
Fulvous Whistling-Duck
Muscovy Duck
Blue-winged Teal
Lesser Scaup
Masked Duck R
Crested Bobwhite
Least Grebe
Pied-billed Grebe
Neotropic Cormorant
Anhinga
Magnificent Frigatebird
Least Bittern
Bare-throated Tiger-Heron R
Great Blue Heron
Cocoi Heron*
Great Egret
Snowy Egret
Little Blue Heron
Tricolored Heron
Cattle Egret
Green Heron
Black-crowned Night-Heron
White Ibis
Glossy Ibis
Roseate Spoonbill R
Wood Stork
Black Vulture
Turkey Vulture
Lesser Yellow-headed Vulture
Osprey
Pearl Kite
White-tailed Kite

Snail Kite
Mangrove Black-Hawk
Savanna Hawk
Roadside Hawk
Short-tailed Hawk
Swainson's Hawk
Crested Caracara
Yellow-headed Caracara
Aplomado Falcon
Peregrine Falcon
Purple Gallinule
Common Moorhen
Limpkin
Southern Lapwing
Black-bellied Plover
Killdeer
Black-necked Stilt
Wattled Jacana
Greater Yellowlegs
Solitary Sandpiper
Pectoral Sandpiper
Short-billed Dowitcher
Caspian Tern
Royal Tern
Pale-vented Pigeon
Mourning Dove
Common Ground-Dove
Plain-breasted Ground-Dove
Ruddy Ground-Dove
White-tipped Dove
Brown-throated Parakeet
Orange-chinned Parakeet
Yellow-crowned Amazon
Mangrove Cuckoo R
Squirrel Cuckoo
Smooth-billed Ani
Groove-billed Ani
Ferruginous Pygmy-Owl
Garden Emerald
Sapphire-throated Hummingbird
Rufous-tailed Hummingbird
Ringed Kingfisher
Green Kingfisher
Amazon Kingfisher
Red-crowned Woodpecker
Pale-breasted Spinetail
Mouse-colored Tyrannulet
Yellow-bellied Elaenia
Common Tody-Flycatcher
Panama Flycatcher
Great Kiskadee
Boat-billed Flycatcher
Social Flycatcher
Streaked Flycatcher
Tropical Kingbird
Eastern Kingbird
Gray Kingbird
Fork-tailed Flycatcher
Red-eyed Vireo
Rufous-browed Peppershrike
Sand Martin
Cliff Swallow
Barn Swallow
House Wren
Tropical Gnatcatcher
Clay-colored Thrush
Yellow Warbler
Black-and-white Warbler
Prothonotary Warbler
Bananaquit
Blue-gray Tanager
Palm Tanager
Blue-black Grassquit
Red-breasted Blackbird
Eastern Meadowlark
Great-tailed Grackle
Yellow-crowned Euphonia

Cenegón del Mangle

Cenegón del Mangle is a Wildlife Refuge (835 ha/2,063 acres) that protects a large area of mangroves and wetlands as well as one of the largest heron colonies in Panama. Unfortunately, an extensive boardwalk that once gave access to the mangroves and the heron colony is now in disrepair, and in any case many of the herons have moved their nesting area to a nearby pond outside the refuge. Although the refuge has not been birded extensively, in addition to mangroves it also has dry forest scrub in which some of the regional specialties, such as Common Ground-Dove, White-winged Dove, and Northern Scrub-Flycatcher, might be expected to occur, and it could be worth exploration if you are in the area.

There are a couple of short trails within the reserve. At the entrance to the refuge there is a barbed-wire fence with a wire gate on the left. Go in here and go up the gravel road to the edge of the mangroves. (There was a small building under construction here in early 2006.) At the point where the road enters the mangroves a gravel path goes off to the left, and can be followed about 350 m through the mangroves to the start of the boardwalk. This boardwalk goes hundreds of meters into the mangroves to the area of the heron colony. However, it has not been maintained and some of the floorboards are now so rotted they give way easily, making it very hazardous to walk. It is recommended not to try using it unless it is clear it has been repaired.

About 400 m beyond the entrance, there is a short trail, *Los Piñuelos,* through dry forest scrub, signposted on the right. Just beyond this a road goes off the main road to the right, signposted "Casa de Visitantes – Cenegón del Mangle." The refuge office and an observation tower are 200 m down this road, along with some small thermal pools. You can pay the entry fee here, which is $3.50 for foreign visitors and $1.50 for Panama residents.

Getting to Cenegón del Mangle. From Panama City, follow the directions for "Getting to Western Panama" (pp. 130-132). Turn south off the Carretera Interamericana at Divisa, which is 65 km west of Penonomé and 40 km west of Santiago, to go south on the Carretera Nacional. The turnoff to Cenegón del Mangle is at 18.6 km from Divisa, on the left next to a yellow bus shelter and a red-and-white gas station, signposted for the town of París and for the Refuge. (París, incidentally, is named for an Indian chief who once ruled the area, and has nothing whatever in common with the city in France aside from its name.) At 1.0 km from the turnoff you come to a church, where you turn right. At 200 m past this turn you come to a four-way intersection; turn right again here. The asphalt ends after 100 m. Continue on this road 4.0, where you come to signs for "Los Positos de París" and "Cenegón del Mangle – Entrada 1.88 km"; turn right here. It is 1.9 km to the Refuge entrance, where there is a barbed wire fence and gate on the left. Go in here to find the trail to the mangroves and the boardwalk. At 2.3 km from the entrance is the Los Piñuelos Trail, and the side road to the right that leads to the Refuge office.

Accommodations and meals. Chitré (p. 159) is the closest town from which to visit Cenegón del Mangle, although it can also easily be done from Penonomé (pp. 143-144).

Chitré (El Agallito Beach)

Chitré is the capital of Herrera province and the largest city in the Azuero Peninsula. Scrub woodland and coastal habitats near the town are the best place in Panama to find Common Ground-Dove and White-winged Dove, and species such as Mouse-colored Tyrannulet and Northern Scrub-Flycatcher can also be found here. The coast itself features many migratory shorebirds and other species in season, and is one of the best places in Panama to find Roseate Spoonbill, Wood Stork, and American Oystercatcher.

The best birding area around Chitré is at El Agallito Beach and surrounding areas. From downtown Chitré (see directions in the next section) on the road to El Agallito it is 2.8 km to the airport. Look for Yellowish Pipit in the grass between the parking lot and the runway and to the left of the runway. At 3.5 km from Chitré the road forks; stay left and continue 1.7 km to the Agallito Beach recreational complex. The road forks again just before reaching the complex, which is to the right. Take the left fork, which parallels the beach and is lined with houses. After 500 m the asphalt ends. Park here and walk in on the dirt road. (Unfortunately there has been a considerable amount of unsightly trash dumped here in recent years.) The road ends in about 100 m and a trail begins that runs along a grassy embankment between the mangroves and the salt flats. The trail and embankment can be followed for about 200 m before petering out. Common Ground-Dove can be found in the scrub along

Chitré and El Agallito Beach

the dirt road and the trail, along with both Ruddy and Plain-breasted Ground-Doves, as well as White-winged and Mourning Doves.

The coastal mudflats here, as well as the salt flats at the end of the paved road, are good for shorebirds, terns, and gulls in season, as well as for Roseate Spoonbill, Wood Stork, and American Oystercatcher. Viewing conditions are best at high tide when the birds are closer to the beach instead of far out on the mudflats.

Getting to Chitré and El Agallito Beach. Coming from Panama City, follow the directions for "Getting to Western Panama" (pp. 130-132). Turn off the Carretera Interamericana at Divisa to go south on the Carretera Nacional. Chitré is 36 km south of Divisa. As you enter the center of town, the concrete divider along the middle of the Carretera Nacional ends at a column with a plaque for the Club de Leones. About 200 m beyond this, you come to a small plaza with a fountain and plantings on the right, and the street becomes one-way. Two streets beyond the end of the plaza you come to the intersection with Av. Herrera, with the Hotel Santa Rita on your right. Turn left here for El Agallito Beach. It is 2.8 km to the airport and 5.2 km to the beach complex. If you are coming from the south on the Carretera Nacional, stay straight when you come to the cathedral, which will be on your left. This will put you on Av. Herrera headed toward El Agallito. (The Hotel Santa Rita and the intersection with the Carretera Nacional are one block after the cathedral.) On your return from El Agallito, if you want to go north, continue one block past the Hotel Santa Rita and take a right in front of the cathedral. Continue straight on this street for several blocks until you come to a stop sign. Turn right here, and after one block you will come to the Carretera Nacional, where you turn left to go north. If you want to go south, continue straight on Av. Herrera, and when you come to the cathedral go around it to the left. Continuing straight beyond the cathedral will put you on the Carretera Nacional headed south.

There is frequent bus service from Chitré to El Agallito Beach, and it can also be reached by taxi.

Accommodations and meals.
The *Hotel Versalles* (ph 996-4422, 996-4563; fax 996-2090; reservaciones@hotelversalles.com; www.hotelversalles.com) is on the Carretera Nacional (called here the Paseo Enrique Geenzier), on the right if you are heading south, about 2 km before you come into the center of town. It has 61 rooms and suites ($23.00 and up), all with private hot-water bath and air-conditioning, and has a restaurant on premises.

The *Hotel Hong Kong* (ph 996-4483, 996-9180) is on the Carretera Nacional, on the right if you are heading south, about 2 km south of the center of town. It has 28 rooms ($22.00), all with private hot-water bath and air-conditioning. The hotel restaurant serves Chinese as well as Panamanian food.

The *Hotel Los Guayacanes* (ph 996-9758) is on the Vía Circunvalación, the bypass that parallels the Carretera Nacional to the south. It has 64 rooms ($38.50), all with private hot-water bath and air-conditioning. Located on a hilltop, the hotel's restaurant has a nice view over town.

Besides the restaurants at the hotels, there are a variety of other restaurants and fast-food places on the Carretera Nacional both north and south of town and around the town center. See one of the standard tourist guides for more information.

Bird list for El Agallito Beach and the Chitré area. * = vagrant, R = rare.

Black-bellied Whistling-Duck
Blue-winged Teal
Least Grebe
American White Pelican*
Brown Pelican
Neotropic Cormorant
Anhinga
Magnificent Frigatebird
Bare-throated Tiger-Heron
Great Blue Heron
Great Egret
Snowy Egret
Little Blue Heron
Tricolored Heron
Cattle Egret
Green Heron
Striated Heron
Black-crowned Night-Heron
Yellow-crowned Night-Heron
Boat-billed Heron
White Ibis
Roseate Spoonbill
Wood Stork
Black Vulture
Turkey Vulture
Lesser Yellow-headed Vulture
Osprey
Pearl Kite
White-tailed Kite
Mangrove Black-Hawk

Savanna Hawk
Roadside Hawk
Short-tailed Hawk
Swainson's Hawk
Zone-tailed Hawk
Crested Caracara
Yellow-headed Caracara
American Kestrel
Merlin
Peregrine Falcon
Gray-headed Chachalaca
Crested Bobwhite
White-throated Crake
Gray-necked Wood-Rail
Purple Gallinule
Common Moorhen
Southern Lapwing
Black-bellied Plover
Wilson's Plover
Semipalmated Plover
Killdeer
American Oystercatcher
Black-necked Stilt
Wattled Jacana
Greater Yellowlegs
Lesser Yellowlegs
Solitary Sandpiper
Willet
Spotted Sandpiper
Whimbrel
Marbled Godwit
Ruddy Turnstone
Red Knot
Sanderling
Semipalmated Sandpiper
Western Sandpiper
Least Sandpiper
Baird's Sandpiper
Pectoral Sandpiper
Stilt Sandpiper
Short-billed Dowitcher
Wilson's Phalarope
Laughing Gull
Gull-billed Tern
Royal Tern
Sandwich Tern
Forster's Tern*
Rock Pigeon
White-winged Dove
Mourning Dove
Common Ground-Dove
Plain-breasted Ground-Dove
Ruddy Ground-Dove
Blue Ground-Dove
White-tipped Dove
Brown-throated Parakeet
Orange-chinned Parakeet
Yellow-crowned Amazon
Black-billed Cuckoo
Yellow-billed Cuckoo
Mangrove Cuckoo R
Squirrel Cuckoo
Striped Cuckoo
Smooth-billed Ani
Groove-billed Ani
Barn Owl
Ferruginous Pygmy-Owl
Common Nighthawk
Common Pauraque
Veraguan Mango
Garden Emerald
Sapphire-throated Hummingbird
Snowy-bellied Hummingbird
Rufous-tailed Hummingbird
Ringed Kingfisher
Green Kingfisher
Amazon Kingfisher
Red-crowned Woodpecker
Straight-billed Woodcreeper
Barred Antshrike
Southern Beardless-Tyrannulet
Mouse-colored Tyrannulet
Northern Scrub-Flycatcher
Yellow-bellied Elaenia
Lesser Elaenia
Common Tody-Flycatcher
Yellow-olive Flycatcher
Eastern Wood-Pewee
Acadian Flycatcher
Panama Flycatcher
Great Kiskadee
Social Flycatcher
Tropical Kingbird
Fork-tailed Flycatcher
Cinnamon Becard
Yellow-throated Vireo
Red-eyed Vireo
Yellow-green Vireo
Scrub Greenlet
Rufous-browed Peppershrike
Gray-breasted Martin
Mangrove Swallow
Sand Martin
Cliff Swallow
Barn Swallow
Plain Wren
House Wren
Swainson's Thrush
Clay-colored Thrush
Yellowish Pipit
Tennessee Warbler
Yellow Warbler (migrant forms)
Yellow (Mangrove) Warbler
Black-and-white Warbler
Prothonotary Warbler
Northern Waterthrush
Rufous-capped Warbler
Bananaquit
Red-legged Honeycreeper
Blue-gray Tanager
Summer Tanager
Blue-black Grassquit
Variable Seedeater
Ruddy-breasted Seedeater
Red-breasted Blackbird
Eastern Meadowlark
Great-tailed Grackle
Orchard Oriole
Baltimore Oriole
Yellow-crowned Euphonia
House Sparrow

El Montuoso Forest Reserve

The El Montuoso Forest Reserve (10,375 ha/25,637 acres) contains a remnant patch of foothills forest at the northern end of the mountains that form the spine of the Azuero Peninsula. Although off the beaten track, it is the easiest place on the mainland to find the Panama national endemic **Brown-backed Dove**, which is also found on Isla Coiba. El Montuoso also has a number of the endemics and specialties of the western Pacific lowlands, including **Orange-collared Manakin**, Blue-throated Goldentail, and Sepia-capped Flycatcher, as well as a few highland species such as Green Thorntail and Slaty Antwren.

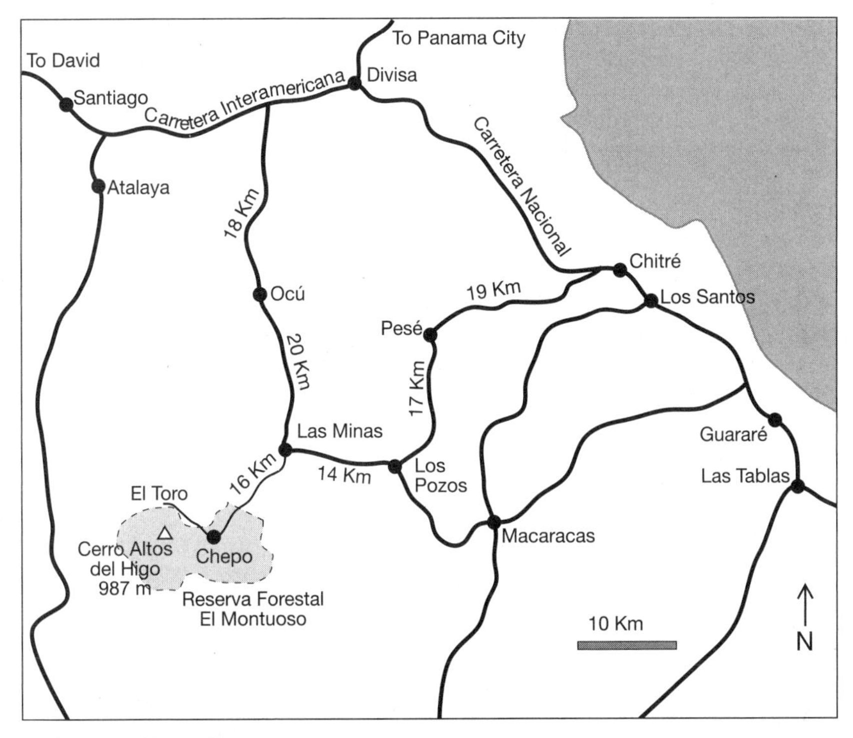

El Montuoso Forest Reserve

Although it is a Forest Reserve, much of the forest within the reserve's boundaries has been cut. The main patch of remaining forest surrounds the summit of Cerro Altos del Higo (987 m/3,238 ft). Two trails go to the summit from near the village of Chepo (not to be confused with the larger and better-known town in eastern Panama province). The easier of the two is the Caras Pintadas Trail, which starts at a higher elevation and has a gentler climb to the top. A second trail starts at the Chepo school.

For directions to the village of Chepo see the next section. A little after entering the town, you will see a small church on the right. For the trail that begins at the school, continue 1.7 km beyond the church, where there is a sign on a concrete slab for the Centro de Educación de Chepo. Take the steep downhill road to the left, which requires 4WD or high clearance. Passing through pine trees, the road reaches the school complex after about 500 m. Go into the complex, where you can ask permission to park. At the far end of the complex you will see a gravel road exiting to the left. Just where this road leaves the complex next to the last building, there is a disused gravel road going downhill. Walk down this road, where after 50 m you will come to a stream where there is a sign for the El Montuoso Reserve on the other side. Ford the stream, and go up the path to the right. The trail first passes through second growth and then reaches taller forest en route to the top.

The *Caras Pintadas ("Painted Faces") Trail* is named for a rock with pre-Columbian carvings near its beginning. To find it, continue on the main road 700 m past the sign for the Centro de Educación, where a narrow dirt track, requiring 4WD, goes steeply downhill on the left. After 300 m, you will come to a wire gate. Go through (leaving the gate the way you found it), and after another 200 m you will come to a house on the right amid pine trees. Although the road can be driven for another 300 m, it gets worse near the end, so it is convenient to park here and walk up. The *caras pintadas* rocks are just beyond where the road ends at a water tank. Just past them, take the uphill fork to the right. About 100 m beyond this there is an old clearing with a partly burned wooden sign where the trail divides. The original trail is to the right; the left fork goes through the clearing to bypass a heavily eroded section but reunites with it after about 100 m, and the trail then enters good forest. The first 400 m of the trail are steep and eroded in parts, but after that it broadens and climbs more gently for the next 700 m, staying at around 850 m (2,800 ft) elevation. The trail was improved with signs and small cable bridges about six or seven years ago, but these have not been maintained and some have rotted away. At about 1.1 km (0.7 mi) from the beginning it reaches a camping area, signposted "Area de Acampar." It then climbs more steeply again, after 150 m coming to a sign for a trail to the left that goes to the peak at Altos de Higo. (The peak itself is a communications area and restricted.) Shortly after this the trail reaches its high point at 920 m (3,000 ft) and begins to descend steeply. We have not followed it beyond this point.

Getting to El Montuoso Forest Reserve. The El Montuoso Forest Reserve is reached by way of the towns of Las Minas and Chepo. Las Minas can be reached either via Pesé, from the Carretera Nacional near Chitré, or via Ocú, from the Carretera Interamericana near Santiago. The turnoff to Pesé from the Carretera Nacional is about 33 km south of Divisa, or a little over 3 km north of Chitré. From the turnoff it is 19 km to Pesé. Shortly after you enter Pesé, you will come to a T-junction. Turn right, and then take an immediate left, signposted for Los Pozos and Las Minas. At Los Pozos, 17 km from Pesé, continue on the main road as it bends right as it enters town. Las Minas, marked by an abandoned gas station, the Servicentro El Montuoso, is 13.7 km beyond Los Pozos. If instead you are coming from the Carretera Interamericana, the turnoff to Ocú is approximately 10 km west of Divisa and 30 km east of Santiago. It is about 18 km from the highway to Ocú and an additional 20 km to Las Minas. All these roads as far as Las Minas are asphalt, and mostly in good condition.

Whichever direction you come from, once you get to Las Minas, turn up the road next to the Servicentro El Montuoso. Continue straight for 300 m until you reach the plaza, which has a church at its far end. Take a left at the end of the plaza you came in on, then another left immediately after, at the minisuper Melisa. A few dozen meters after this turn, take a right onto a small gravel road that goes right at the minisuper Vicente. After 200 m, there is a sign for the El Montuoso Reserve (although the actual boundary is at La Peña, 11 km from Las Minas). The road is mostly gravel, but has asphalt sections on the steeper hills. In the dry season it may be passable to 2WD drive, but even a little rain could be problematic for some sections, so 4WD would be preferred. At 4.7 km from Las Minas there is a fork where you stay right. At 16.4 km there is a faded "Bienvenidos a Chepo" sign on the left, and shortly after you come to the restaurant Brisas de Montuoso. At 17.0 km you come to a small church on the right. From here, follow the directions in the previous section to find the trails.

Accommodations and meals. The closest towns with a range of better hotels and restaurants are Santiago (pp. 166-167) and Chitré (p. 159). There is a small *hospedaje* in Las Minas; it is the white house with red doors opposite the minisuper Melisa on the plaza. The owner was not there when we visited so we have no further information on it, but it is sure to be very basic. There are also two small restaurants on the plaza in Las Minas. In Chepo, you can get a meal at the Brisas de Montuoso near the start of town. Surprisingly for a place so remote, a 12-room hotel was under construction next door when we visited in March 2006, and was planned to be open within a few months.

Bird list for El Montuoso Forest Reserve.

Great Tinamou
Little Tinamou
Black Vulture
Turkey Vulture
Swallow-tailed Kite
White Hawk
Broad-winged Hawk
Short-tailed Hawk
Black Hawk-Eagle
Gray-headed Chachalaca
Blue Ground-Dove
White-tipped Dove
Brown-backed Dove
Crimson-fronted Parakeet
Brown-throated Parakeet
Squirrel Cuckoo
Long-billed Hermit
Stripe-throated Hermit
Violet Sabrewing
Garden Emerald
Blue-throated Goldentail
Snowy-bellied Hummingbird
Violaceous Trogon
Black-throated Trogon
Slaty-tailed Trogon
Blue-crowned Motmot
Keel-billed Toucan
Red-crowned Woodpecker
Lineated Woodpecker
Plain Xenops
Olivaceous Woodcreeper
Cocoa Woodcreeper
Slaty Antwren
Chestnut-backed Antbird
Paltry Tyrannulet
Sepia-capped Flycatcher
Ochre-bellied Flycatcher
Eye-ringed Flatbill
Yellow-olive Flycatcher
White-throated Spadebill
Royal Flycatcher
Ruddy-tailed Flycatcher
Tropical Pewee
Dusky-capped Flycatcher
Panama Flycatcher
Boat-billed Flycatcher
Thrush-like Schiffornis
Orange-collared Manakin
White-ruffed Manakin
Lance-tailed Manakin
Red-capped Manakin
Yellow-throated Vireo
Philadelphia Vireo
Yellow-green Vireo
Lesser Greenlet
Black-chested Jay
Rufous-breasted Wren
Rufous-and-white Wren
House Wren
Long-billed Gnatwren
Orange-billed Nightingale-Thrush
Swainson's Thrush
White-throated Thrush
Golden-winged Warbler
Tennessee Warbler
Tropical Parula
Chestnut-sided Warbler
Black-throated Green Warbler
Bay-breasted Warbler
Black-and-white Warbler
Golden-crowned Warbler
Rufous-capped Warbler
Rosy Thrush-Tanager
Red-crowned Ant-Tanager
White-winged Tanager
Bay-headed Tanager
Blue-gray Tanager
Streaked Saltator
Buff-throated Saltator
Orange-billed Sparrow
Black-striped Sparrow
Yellow-faced Grassquit
Yellow-crowned Euphonia
Thick-billed Euphonia
Lesser Goldfinch

Los Santos Province

Los Santos Province, in the southeastern part of the Azuero Peninsula, like Herrera is almost entirely deforested. While the standard open-country and scrub birds can be found here, it is of interest mainly for islands off its coast that are important breeding sites for seabirds, Isla Iguana and the Islas Frailes del Sur. Pelagic birding off Punta Mala is also an option for those who might be interested. Cerro Hoya National Park, which is mostly in Veraguas, can be reached through Los Santos. See the account on pp. 181-183 for details.

Isla Iguana

Isla Iguana, off the southeastern tip of the Azuero Peninsula near Punta Mala, is the site of the largest and most accessible nesting colony of Magnificent Frigatebird in Panama, numbering over 1,000 pairs. During courtship, a male Frigatebird inflates his large bright-red throat pouch until it looks like a balloon nearly ready to burst. Many males displaying together make a spectacular show. Males sometimes fly about with the throat patch inflated. This display is most likely to be seen in the earlier stages of the breeding season, during the dry season from January to April. The nesting period covers 11 months in all, and adults and juveniles are present on the island through most of the year.

Isla Iguana is low, flat, and sandy, covered mainly in scrub and grass, and is little over 4 km offshore. It is also a popular location for snorkeling and scuba diving to view the fringing coral reefs that surround the island. The island is a Wildlife Refuge and has an attractive visitor center, but on a recent visit there were no park rangers on the island and the visitor center was closed. The entrance fee is $3.50 for foreign visitors and $1.50 for Panama residents, if you can find someone to pay.

The best way to see the colony is to have your boatman first take you on a circuit of the island. The nests are mostly on the southeastern side of the island, in spiny vegetation and scrub. Unfortunately, there is no observation tower or other viewpoint on the island itself that permits an overall view of the colony. However, to get a closer view of some of the birds, you can land on the beach on the western side of the island next to the visitor center. From here a short 200-m trail goes from the water pump next to the picnic shelter across to a small beach, La Playita, on the other side of the island next to a lighthouse. From La Playita about 50 to 100 nests may be visible in the scrub north of the beach. Please do not approach the nests more closely than the rocks at the left end of the beach, since the adults are likely to leave the nest, making eggs or small chicks vulnerable to scavenging Yellow-headed Caracaras or Great-tailed Grackles. Although some maps of the refuge show trails into the southern part near the colony, these are now overgrown. Do not attempt to enter the main part of the colony, which in any case is protected by a rocky shoreline and dense spiny vegetation.

Getting to Isla Iguana. Isla Iguana can be most easily reached via the town of Pedasí in the southern Azuero Peninsula. Coming from Panama City, follow the directions for "Getting to Western Panama" (pp. 130-132). Turn off the Carretera Interamericana at Divisa to go south on the Carretera Nacional. You can bypass downtown Chitré by taking the Vía Circunvalación, signposted for Las Tablas, on the right at 32.8 km south of Divisa and about 100 m after the turnoff to Pesé. (If you miss the turn, continue into Chitré and follow the directions for El Agallito Beach on pp. 158-159, but take a right at the Hotel Santa Rita). It is approximately 25 km from Chitré to Las Tablas. Entering Las Tablas, take a left at the end of the main town plaza to get to Pedasí, which is 42 km south.

In Pedasí, the best place to get a boat for Isla Iguana is at Playa El Arenal. The road to El Arenal is on the left next to the Accel gas at the north end of town; it is 3 km on this asphalt road to the beach. Probably the easiest way to arrange a boat if you are staying in Pedasí overnight is to ask your hotel to contact a boatman. Otherwise you can try to find a boatman at the beach itself, or ask at the Accel station; the IPAT office two buildings down from the Accel station on the El Arenal road may also be able to provide phone numbers. The price is $40.00 for the round trip, regardless of the number of people. It is 6 km from the beach to the island, and the trip across usually takes 20 minutes, but can take longer if a strong onshore wind is blowing. It is therefore best to go in the morning, when there is less wind, especially in the dry season.

Accommodations and meals. Pedasí is a very small town, and has three places to stay, all on the Carretera Nacional. For a trip to Isla Iguana, it would also be possible to overnight in Chitré (p. 159), which is only a bit more than an hour's drive away. There are several small restaurants in town, all very basic.

The *Residencial Pedasí* (ph 995-2322) is on the left as you come into town. It has 16 rooms ($15.00).

Dim's Hostal (ph 995-2303) is a bed-and-breakfast on the right a few blocks south of the Residencial Pedasí. There are five rooms ($15.00, including breakfast), all with private cold-water bath and air-conditioning.

The *Residencial Moscoso* (ph 995-2203) is on the left a few blocks further on. Rooms with air-conditioning and private cold-water bath are $14.00, those with fan and shared bath are cheaper.

Islas Frailes del Sur

These two rocky stacks, one much larger than the other, are the only nesting site in Panama for Sooty and Bridled Terns and Brown Noddy. Sooty Tern at times nests in the thousands here, while there are perhaps 50 pairs each of Bridled Tern and Brown Noddy. The islets are about 8 km off the southern coast of the Azuero, and 15 km west of Punta Mala. The islets are so steep-sided and the ocean swells are so large that it is very difficult to land, but the birds, which nest on the flat tops of the islands and on the tops of the cliffs, can be seen as they swirl about the colony. The breeding cycle at this colony is poorly known, and the birds may not nest every year.

Getting to the Islas Frailes del Sur. The Frailes del Sur can be visited by hiring a boat from Pedasí, although the trip is much longer and more expensive than that to Isla Iguana. You may also be able to hire a boat from around Playa Venao (or Venado), which is on the mainland north of the islands, or from elsewhere in the vicinity of Punta Mala. You can make arrangements at the La Playita Resort or Villa Marina near Playa Venao; see the section on accommodations below. To get to Playa Venao go to Pedasí and continue south on the main road. Playa Venao is about 31 km southwest of Pedasí.

Accommodations and meals. There are two places to stay near Playa Venao.

The *La Playita Resort* (ph 996-5491) has five rooms and two large cabañas ($45.00 and up) with private hot-water bath. The beach at this place is popular on weekends and it can be noisy then.

The *Villa Marina* (ph 211-2277; cel 6673-9445; information@playavenado.com; www.playavenado.com) has 3 rooms with hot water bath and air-conditioning. Rooms with shared bath are $95.00; with private bath $125.00; breakfast included. Three meals per day can be provided for $40.00. The entrance to Villa Marina is the next one beyond the entrance to La Playita.

Pelagic birding near Punta Mala

The continental shelf is very close to shore near Punta Mala and the southern coast of the Azuero Peninsula. Pelagic birding is likely to be productive here at certain times of year, but has been little tried. Those who wish to investigate it can hire boats in Pedasí or other localities near Punta Mala, or make arrangements through the hotels mentioned above; see the sections on Isla Iguana and Isla Frailes del Sur above.

Veraguas Province

Veraguas is Panama's only province that borders on both oceans. In the highlands, the small mountain town of Santa Fe features many of the specialties of the western highlands, as well as some of those of the wet Atlantic slope foothills. Isla Coiba, off the Pacific coast, has Panama's best remaining population of Scarlet Macaw, as well as two national endemics, Brown-backed Dove and Coiba Spinetail. Cerro Hoya National Park, at the southern tip of the Azuero Peninsula, is the best locality for the nationally endemic Azuero Parakeet, and also has Brown-backed Dove and several regional endemics.

Santiago

Santiago is the largest town in Veraguas province. On the Carretera Interamericana about 3-3½ hours west of Panama City and 2½ hours east of David, it has several decent hotels and restaurants and can be a convenient place to overnight if driving between Panama City and Chiriquí and Bocas del Toro. You may also wish to overnight here before a trip to Coiba National Park. Santiago is also the route to Santa Fe in the Veraguas highlands.

Accommodations and meals.
The *Hotel Plaza Gran David* (ph 998-3433; fax 998-2553) is on the north side of the Carretera Interamericana (right side coming from the east). Laid out motel style, it has 32 rooms ($19.00), all with private hot-water bath and air-conditioning, and has a restaurant on premises. (Do not confuse this with the Hotel Gran David listed below.)

The *Hotel Piramidal* (ph 998-3123; fax 998-4511), marked by its many pyramidal red roofs, is on the on the north side of the Interamericana, a little over 1 km past the Plaza Gran David, opposite where Av. Central forks off the highway to go the center of town. It has 62 rooms ($25.00), all with private hot-water bath and air-conditioning, and has a restaurant and outdoor cafeteria on premises.

The *Hotel Galería* (ph 958-7950, 958-7955; fax 958-7954; hotelgaleria@cwpanama.net) is on the north side of the Interamericana a few hundred meters past the Piramidal. It has 40 rooms ($38.50), all with private hot-water bath and air-conditioning, and a restaurant on premises.

The *Hotel Gran David* (ph 998-4510; fax 998-1866) is on the north side of the Interamericana a few hundred meters past the Galería. It has 74 rooms ($18.70), all with private hot-water bath and air-conditioning, and a restaurant on premises.

The *Hotel La Hacienda* (ph 958-8580; fax 958-8579; reservahacienda@hotmail.com), on the south side of the Interamericana about a kilometer west of town, is the newest and fanciest hotel in the area and is decorated in a Mexican theme. It has 30 rooms ($38.50), all with private hot-water bath and air-conditioning, and a restaurant on premises.

The best place to eat in Santiago is the *Restaurante Delicias del Mar,* specializing in Peruvian-style seafood. It is on the south side of the Interamericana, where Av. Central forks off to go to the center of town, almost opposite the hotel Piramidal.

The *Restaurante Los Tucanes* is a cafeteria at the end of a complex of shops and gas station on the north side of the Interamericana just west of the Hotel Gran David, and is a convenient place to grab a quick meal or take a break when driving between Panama City and Chiriquí. It can be crowded, however, when a couple of long-distance buses stop off here at the same time.

Like Los Tucanes, the cafeteria at the Piramidal Hotel is a popular stop for long-distance buses and can get crowded. The restaurants at the other hotels on the Interamericana are also options for meals. There are several fast-food places and other small restaurants on the Interamericana as well.

Santa Fe

Santa Fe, at about 400 m (1,300 ft) elevation, is a picturesque mountain town in the foothills of Veraguas and the gateway to the national park of the same name that protects the rugged mountains above it. Birding here can be spectacular, with many endemics and rarities of the western highlands, plus several specialties of the Atlantic slope that barely reach the Pacific side here. Highland species that can be found here include **Black Guan**, Snowcap, Immaculate Antbird, Slaty-capped Flycatcher, Eye-ringed Flatbill, Three-wattled Bellbird, **Bare-necked Umbrellabird**, **Black-faced Solitaire**, Slaty-backed Nightingale-Thrush, Three-striped and Golden-crowned Warblers, and many others. Some Atlantic-slope species of the area include **Lattice-tailed Trogon**, Crimson-collared Tanager, and Yellow-throated Bush-Tanager, and rarities such as Solitary Eagle, Rufous-winged Woodpecker, Lanceolated Monklet, and Black-and-white Becard have also been found. **Glow-throated Hummingbird** has been recorded, but there are no recent reports. **Yellow-green Finch** should occur by range on the higher peaks but has not yet been definitely reported.

Santa Fe National Park (72,636 ha/179,487 acres) was established in 2001, but still remains a "paper park." There are no facilities or infrastructure, and no established trails for visitors. Parts of the park are unfortunately being subjected to colonization and deforestation. Nonetheless, the park contains huge areas of nearly pristine montane forest rising to over 1,500 m (4,900 ft), and has the potential to be one of the crown jewels of Panama's protected area system if it can be preserved.

One of the best contacts for information about birding and birding sites in the Santa Fe area is Berta de Castrellón (ph 954-0910). She and her brother are both birding guides, although since Berta was recently elected town mayor she says she has less time available to devote to this. She also has an extensive collection of orchids at her house which she enjoys showing to visitors. To find her place, take the dirt road that goes to the right just before the "Bienvenidos a Santa Fe" sign at the entrance to town. Take the second right off this, which is the road which goes to the Río Mulabá and El Pantano. Berta's house is about 200 m down this road on the right, marked by a sign "Orquideario y Cultivos Las Fragancias de Santa Fe." Bear in mind this is a private home, so do not try to visit before 9:00 AM or after 5:00 PM. If the gate is closed, call out from the road, and do not enter if there is no response.

In the dry season the area tends to be foggier and often rainier but less muddy. During the rainy season the weather is similar to elsewhere in the western highlands, being clear to cloudy in the mornings with light misty rain beginning around noon. During the late rainy season, however, it possible to get a completely rainless day.

Alto de Piedra. The best birding in the area is to be found above the town of Santa Fe itself, along the road to Alto de Piedra. This gravel/dirt road was repaired in early 2006, but 4WD is recommended since there are several steep sections. To get to Alto de Piedra, take the left fork at the "Bienvenidos a Santa Fe" sign at the entrance to town. After 200 m take a left just past a school; there is a sign there for Alto de Piedra. After another 400 m, you will come to a Cable and Wireless building with an antenna; turn left here. There is a sign saying "Alto de Piedra 5 km" just past the turn, next to a MIDA building. This dirt road crosses a small cement bridge at 300 m, and a more substantial new bridge at 1 km. At 3.9 km from the antenna there is a sign saying "Bienvenidos a Alto de Piedra." Cerro Mariposa, the highest point on the Cerro Tute ridge, can be seen in the distance beyond this sign. At 4.3 km you reach the blue-and-white Alto de Piedra school complex.

Continuing past the school, at 4.9 km a road to the left passes along the base of the Tute ridge at about 850 m (2,800 ft) elevation. Although the forest along the road is mainly second growth, birding can be excellent here. Crimson-collared Tanager, **Bare-necked Umbrellabird**, and the rare Rufous-winged Woodpecker have been found, and Bellbirds can be seen in season. This 4WD road is in good condition and can be driven for several kilometers, but for birding it is better to walk. At 300 m in, a gated road goes uphill to the right to private property. Just beyond is a convenient spot to park. Several footpaths lead off to the right about 400 m past the gated road.

Continuing on the main road, a little beyond this side road a trail ascends the Tute ridge, but it is little difficult to find. At 5.5 km from Santa Fe there is a citrus orchard on the left; on the right is a small red house with a steeply pitched roof (like most roofs in this very rainy area). There is a wire gate in the barbed wire fence on the left just before the house. On our most recent visit this was tied shut with barbed wire, but you can undo it and refasten it after passing through. (We were told by the neighbors that access is permitted through the orchard.) Take the trail up through the orchard; after about 300 m it passes by a concrete water tank and shortly beyond that parallels the edge of the forest, reaching a gate in a barbed wire fence 200 m beyond the tank. Instead of going through the gate, take a small trail that parallels the fence uphill to the left. After 30 m, you will reach another wider trail that follows the contour around the hill (referred to as the "contour trail" below). To the right, through a gate, this trail enters pasture before going up into the forest again for a good distance. We have not birded it but it could be productive.

To get to the trail to the ridge, go left along the contour trail, which is at about 850 m (2,800 ft) elevation. After about 350 m it reaches a gate in another barbed wire fence. Passing through the gate, you will see a steep trail going uphill next to the fence on your right. This is the trail to the ridge. It goes up very steeply along the fence for about 80 m before leveling out somewhat. Use caution along this stretch since the trail is slippery (and the barbed wire is right next to it!) The trail continues for about 5 km (3 mi), eventually attaining an altitude of over 1,200 m (4,000 ft) on Cerro Mariposa. Although we have not followed it so far, species that have been reported from higher elevations on the trail include **Black Guan**, **Black-faced Solitaire**, Slaty-backed Nightingale-Thrush, and Three-striped Warbler. **Glow-throated Hummingbird** and perhaps **Yellow-green Finch** would also be possibilities.

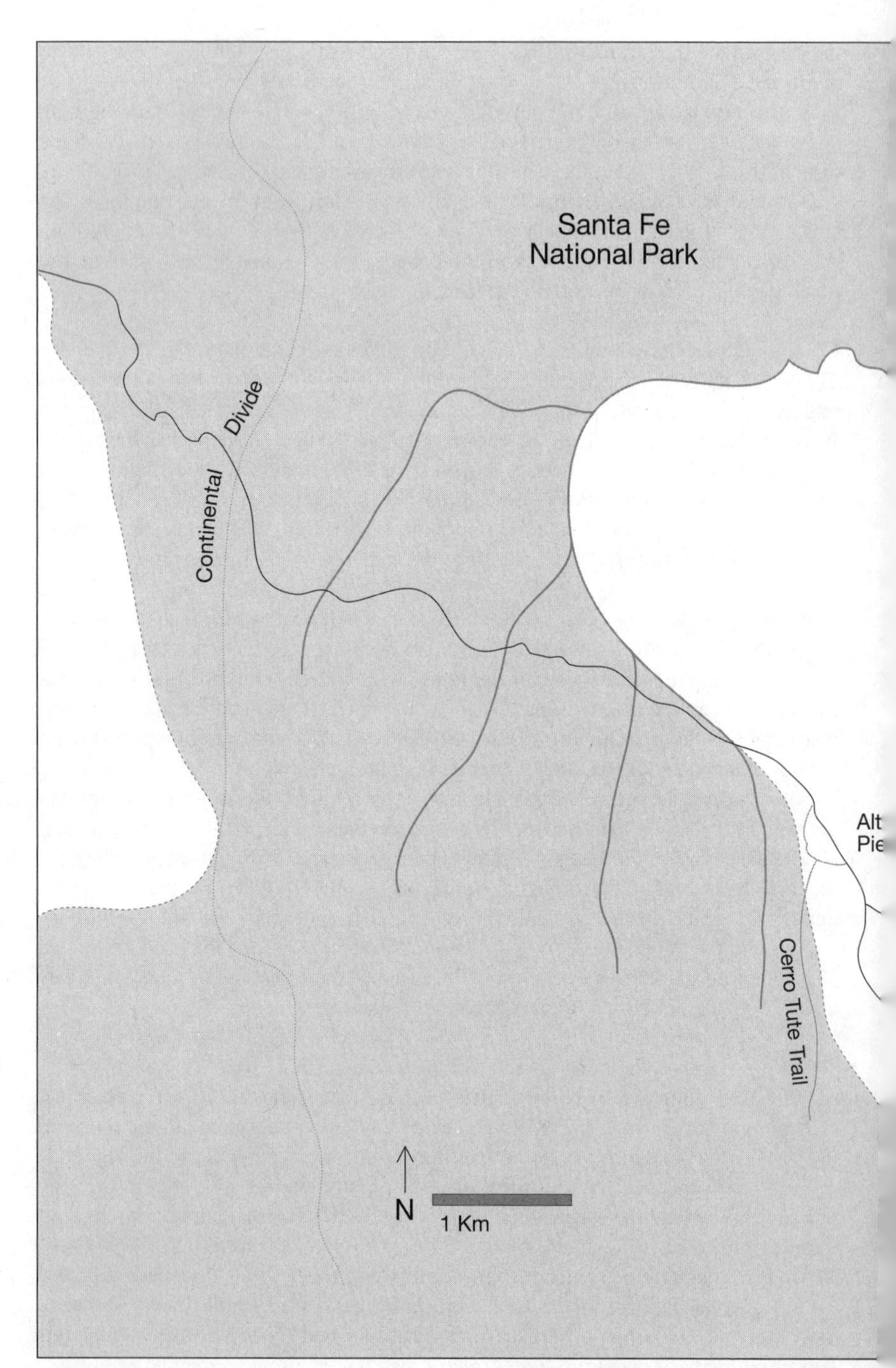
Santa Fe
National Park
Continental
Divide
Cerro Tute Trail
N
1 Km

Santa Fe

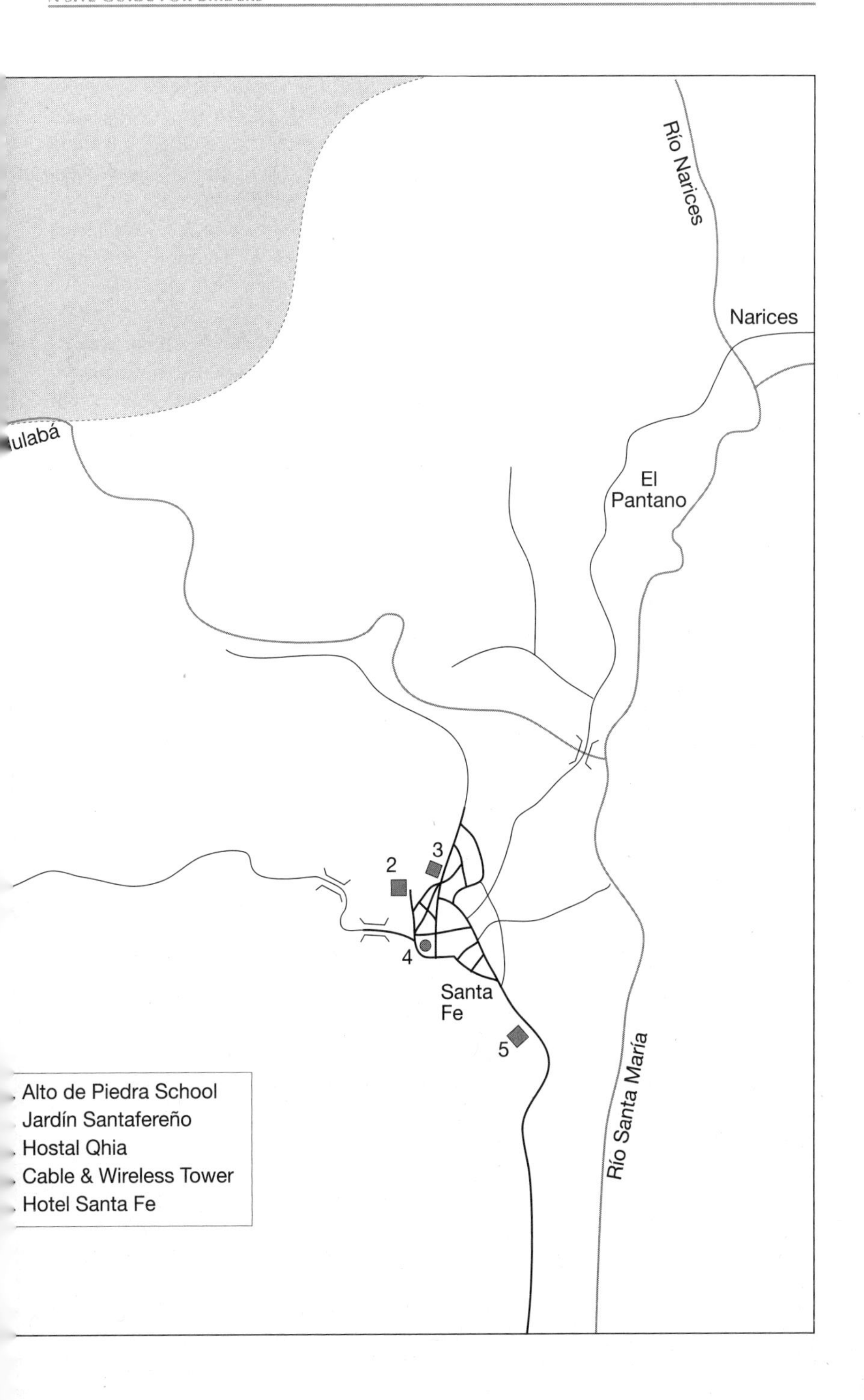
Río Narices
Narices
ulabá
El Pantano
2
3
4
Santa Fe
5
Río Santa María
Alto de Piedra School
Jardín Santafereño
Hostal Qhia
Cable & Wireless Tower
Hotel Santa Fe

About 50 m beyond this ridge trail the contour trail reaches the edge of a pasture. Descending along the fence that parallels the trail brings you to a locked and chained silver metal gate back on the main road. This gate is at about 5.3 km along the Alto de Piedra Road, and is next to a red metal gate. If you find this gate open, or can locate someone who can let you in (or don't mind clambering over the locked gate), this would provide easier and more direct access to the contour and ridge trails. We have been told by the neighbors that access to the trail is permitted through this pasture.

The birding along the contour trail and the edge of the forest can be superb. Species found here include Snowcap, **Lattice-tailed Trogon**, Yellow-eared Toucanet, Immaculate Antbird, Slaty-capped Flycatcher, Eye-ringed Flatbill, both Red-crowned and Red-throated Ant-Tanagers, Crimson-collared Tanager, Yellow-throated Bush-Tanager, Orange-billed Nightingale-Thrush, and Golden-crowned Warbler.

A few meters down the road from the gate that leads to the citrus grove, opposite the small red house, there is a gap in the fence where a trail begins that goes steeply down to a waterfall, Salto de Alto de Piedra. We have not birded it but it could be worth a look.

A short distance beyond this point the road goes steeply down for 1 km through pastures to the Río Mulabá (called the Río Bulabá by locals), where Lanceolated Monklet has been seen. It then crosses the river and continues as far as El Guabal, at about 200 m (660 ft) elevation. In July 2006 the road was drivable for at least 2 km past the river, and here has good forest on either side. However, in the past this road has often not been kept in good repair and sometimes has been impassable, even with 4WD, from the steep downhill just beyond the Salto de Alto de Piedra onwards. If the road is in bad condition, it is worth birding on foot down to the river and beyond.

The road continues at about 800 m (2,600 ft), dipping to 600 m (2,000 ft) in valleys, for the next 3 km beyond the river as it crosses three ridges between the upper tributaries of the Río Mulabá before descending steeply again beyond the Continental Divide. White-throated Shrike-Tanager has been seen at an altitude of 550 m (1,800 m) along this road. We have never been able to go as far as El Guabal, but if the road is passable this area could yield some interesting species of the wet Atlantic slope.

El Pantano. El Pantano is an area of small farms, woodland, and patches of forest at about 400 m (1,300 ft) elevation to the northeast of Santa Fe itself. Although we have not birded the area ourselves, it is recommended by Berta de Castrellón. Although it would be expected to lack the higher elevation species found at Alto de Piedra, it could be a good alternative if the weather is bad up above. To get there, take the dirt road that goes right just before the "Bienvenidos a Santa Fe" sign, then take the second right. This is the Río Mulabá road, which runs by Berta's house, as described above. After 1.2 km on this road you will cross a girder bridge over the Río Mulabá. At 200 m beyond the bridge the road forks; take the right (downhill) branch. The next 2 or 3 km run through the farms and broken woodland of the El Pantano area. There are trails that enter the national park to the north of El Pantano and the Narices area beyond it, but these are best explored with the aid of a guide.

Getting to Santa Fe. From Panama City, follow the directions for "Getting to Western Panama." Santiago is about 250 km from the Bridge of the Americas. To get to Santa Fe from Santiago, take the right turn off the Carretera Interamericana 1.1 km west of the Hotel Piramidal. At 17 km on this road you will reach the town of San Francisco. Stay left at the police station at the entrance of town. The road continues through San Juan at 26 km and San José at 33 km, entering Santa Fe at 56 km from the Interamericana. There is a prominent white "Bienvenidos a Santa Fe" sign at the entrance to town where the road divides. The right paved branch goes to the center of town.

Accommodations and meals. Although Santa Fe can be visited as a day trip from Santiago, it's a long drive and it is better to overnight in Santa Fe itself. There are three places to stay in town, all of them fairly basic: the Hotel Santa Fe, the new Hostal Qhia, and the Jardín Santafereño. The first two places can arrange guides to the area.

The *Hotel Santa Fe* (ph 954-0941) is on the main road, on the left 600 m before the "Bienvenidos a Santa Fe" sign. Look for a blue "Hotel" sign on the right side of the main road as you approach town; the place is easy to pass by. The hotel has 20 rooms. Smaller rooms without air-conditioning or fan are $13.00; somewhat larger rooms with air-conditioning are $25.00. (Neither fan nor air-conditioning are usually necessary at this altitude.) All rooms have cold water only. There is a restaurant on premises.

The new *Hostal Qhia* (yes, that's how it's spelled, pronounced approximately Kee-hee-ah; ph 954-0903; hostal_laqhia@yahoo.es), run by Horacio and Stephanie Toma, an Argentine-French couple, has a backpacker feel. There are two rooms, the smaller of which is $15.00 and the larger $25.00. Extra beds in these rooms are $5.00 each. There is also a dorm room with six bunks at $8.00 each. There are two shared bathrooms, one with hot water (the only place to stay in town that has it). There is a nice second-story veranda and a gazebo with hammocks. The Hostal also serves food. To find it, take the right fork at the Bienvenidos sign. After 600 m, turn right at a red-and-white building onto a dirt road. Follow this road as it curves around to the right. Just beyond a restaurant with a blue roof, turn uphill to the left. At the next intersection, go right to find the hostal. (There are small yellow arrows for the hostal at the intersections, but some are hard to see.)

The *Jardín Santafereño* (ph 954-0866) has four small, very basic cabins which rent for $11.00 (cold water only) near a small beer-garden/restaurant. The cabins are on a hill above town and have a nice view. To find them, take the left fork at the Bienvenidos sign, then after 200 m take a left just past a school. After another 400 m, you come to a Cable and Wireless antenna. Continue straight for 100 m more, then turn left uphill where the cabins will be found after another 100 m.

Besides the restaurants at the hotels, there are a few basic restaurants near the center of town, as well as ones down both the right and left forks of the main street just north of the center of town.

Bird list for Santa Fe. * = vagrant; R = rare; ? = present status uncertain.

Great Tinamou
Little Tinamou
Gray-headed Chachalaca
Black Guan
Black-eared Wood-Quail
Black-breasted Wood-Quail
Crested Bobwhite
Little Blue Heron
Cattle Egret
Black Vulture
Turkey Vulture
Swallow-tailed Kite
White-tailed Kite
Double-toothed Kite
Plumbeous Hawk R
Barred Hawk
White Hawk
Great Black-Hawk
Harris's Hawk*
Solitary Eagle R
Gray Hawk
Roadside Hawk
Broad-winged Hawk
Short-tailed Hawk
Swainson's Hawk
White-tailed Hawk
Black Hawk-Eagle
Barred Forest-Falcon
Collared Forest-Falcon
Crested Caracara
Yellow-headed Caracara
American Kestrel
Bat Falcon
Gray-necked Wood-Rail
Sunbittern
Common Snipe
Short-billed Pigeon
Ruddy Ground-Dove
Blue Ground-Dove
White-tipped Dove
Gray-chested Dove
Purplish-backed Quail-Dove R
Brown-throated Parakeet
Sulphur-winged Parakeet
Orange-chinned Parakeet
Brown-hooded Parrot
Blue-headed Parrot
Squirrel Cuckoo
Striped Cuckoo
Rufous-vented Ground-Cuckoo R
Smooth-billed Ani
Groove-billed Ani
Vermiculated Screech-Owl
Tropical Screech-Owl
Crested Owl
Ferruginous Pygmy-Owl
Short-tailed Nighthawk
Common Pauraque
Chuck-will's-widow
White-collared Swift
Vaux's Swift
Band-rumped Swift
Band-tailed Barbthroat
Green Hermit
Long-billed Hermit
Stripe-throated Hermit
White-tipped Sicklebill
Violet Sabrewing
White-necked Jacobin
Brown Violet-ear R
Violet-headed Hummingbird
Green Thorntail R
Garden Emerald
Violet-crowned Woodnymph
Violet-bellied Hummingbird
Snowy-bellied Hummingbird
Rufous-tailed Hummingbird
Stripe-tailed Hummingbird
Black-bellied Hummingbird
White-tailed Emerald
Snowcap R
Bronze-tailed Plumeleteer
White-bellied Mountain-gem
Purple-throated Mountain-gem
Green-crowned Brilliant
Purple-crowned Fairy
Glow-throated Hummingbird ?
Violaceous Trogon
Orange-bellied Trogon
Lattice-tailed Trogon R
Rufous Motmot
Broad-billed Motmot
Ringed Kingfisher
Green Kingfisher
Lanceolated Monklet R
Red-headed Barbet
Blue-throated Toucanet
Collared Aracari
Yellow-eared Toucanet
Keel-billed Toucan
Chestnut-mandibled Toucan
Black-cheeked Woodpecker
Red-crowned Woodpecker
Smoky-brown Woodpecker
Rufous-winged Woodpecker R
Golden-olive Woodpecker
Cinnamon Woodpecker
Lineated Woodpecker
Crimson-crested Woodpecker
Red-faced Spinetail
Spotted Barbtail
Striped Woodhaunter
Lineated Foliage-gleaner
Slaty-winged Foliage-gleaner
Plain Xenops
Plain-brown Woodcreeper

Olivaceous Woodcreeper
Wedge-billed Woodcreeper
Cocoa Woodcreeper
Black-striped Woodcreeper
Spotted Woodcreeper
Streak-headed Woodcreeper
Brown-billed Scythebill
Fasciated Antshrike
Russet Antshrike
Plain Antvireo
Checker-throated Antwren
White-flanked Antwren
Slaty Antwren
Dot-winged Antwren
Dusky Antbird
Chestnut-backed Antbird
Dull-mantled Antbird R
Immaculate Antbird
Bicolored Antbird
Black-headed Antthrush
Rufous-breasted Antthrush
Black-crowned Antpitta R
Scaled Antpitta R
Streak-breasted Antpitta
Paltry Tyrannulet
Southern Beardless-Tyrannulet
Yellow-bellied Elaenia
Mountain Elaenia
Torrent Tyrannulet
Sepia-capped Flycatcher
Olive-striped Flycatcher
Ochre-bellied Flycatcher
Slaty-capped Flycatcher
Rufous-browed Tyrannulet R
Scale-crested Pygmy-Tyrant
Brownish Twistwing
Eye-ringed Flatbill
Olivaceous Flatbill
Yellow-olive Flycatcher
Yellow-margined Flycatcher
White-throated Spadebill
Golden-crowned Spadebill
Royal Flycatcher
Ruddy-tailed Flycatcher
Sulphur-rumped Flycatcher
Black-tailed Flycatcher
Bran-colored Flycatcher
Common Tufted-Flycatcher
Olive-sided Flycatcher
Western Wood-Pewee
Eastern Wood-Pewee
Tropical Pewee
Long-tailed Tyrant
Bright-rumped Attila
Rufous Mourner
Dusky-capped Flycatcher
Panama Flycatcher
Great Crested Flycatcher
Boat-billed Flycatcher
Social Flycatcher
Golden-bellied Flycatcher
Streaked Flycatcher
Tropical Kingbird
Eastern Kingbird
Fork-tailed Flycatcher
Thrush-like Mourner
Speckled Mourner
Cinnamon Becard
Black-and-white Becard R
Masked Tityra
Purple-throated Fruitcrow
Three-wattled Bellbird
Bare-necked Umbrellabird R
Golden-collared Manakin
White-ruffed Manakin
Lance-tailed Manakin
White-crowned Manakin
Red-capped Manakin
Yellow-throated Vireo
Philadelphia Vireo
Red-eyed Vireo
Scrub Greenlet
Tawny-crowned Greenlet
Lesser Greenlet
Green Shrike-Vireo
Rufous-browed Peppershrike
Black-chested Jay
Blue-and-white Swallow
Northern Rough-winged Swallow
Southern Rough-winged Swallow
Sand Martin
Cliff Swallow
Barn Swallow
Band-backed Wren
Bay Wren
Stripe-breasted Wren
Rufous-breasted Wren
Rufous-and-white Wren
Plain Wren
House Wren
Ochraceous Wren
White-breasted Wood-Wren
Gray-breasted Wood-Wren
Southern Nightingale-Wren
American Dipper
Tawny-faced Gnatwren
Long-billed Gnatwren
Tropical Gnatcatcher
Black-faced Solitaire
Orange-billed Nightingale-Thrush
Slaty-backed Nightingale-Thrush
Black-headed Nightingale-Thrush R
Veery
Swainson's Thrush
Wood Thrush
Pale-vented Thrush
Clay-colored Thrush
Cedar Waxwing R
White-throated Thrush
Golden-winged Warbler
Tennessee Warbler
Tropical Parula
Yellow Warbler
Chestnut-sided Warbler
Black-throated Green Warbler
Blackburnian Warbler
Palm Warbler R
Bay-breasted Warbler
Cerulean Warbler
Black-and-white Warbler
Northern Waterthrush
Kentucky Warbler
Mourning Warbler
Common Yellowthroat
Wilson's Warbler
Canada Warbler
Slate-throated Redstart
Collared Redstart
Golden-crowned Warbler
Rufous-capped Warbler
Three-striped Warbler
Buff-rumped Warbler
Bananaquit
Common Bush-Tanager
Yellow-throated Bush-Tanager
Black-and-yellow Tanager
Rosy Thrush-Tanager
Dusky-faced Tanager
Olive Tanager
White-throated Shrike-Tanager R
White-shouldered Tanager
Tawny-crested Tanager
White-lined Tanager
Red-crowned Ant-Tanager
Red-throated Ant-Tanager
Hepatic Tanager
Summer Tanager
Flame-colored Tanager
Crimson-collared Tanager R

Crimson-backed Tanager
Flame-rumped Tanager
Blue-gray Tanager
Palm Tanager
Blue-and-gold Tanager
Plain-colored Tanager
Emerald Tanager
Silver-throated Tanager
Speckled Tanager
Bay-headed Tanager
Golden-hooded Tanager
Spangle-cheeked Tanager
Blue Dacnis
Green Honeycreeper
Shining Honeycreeper
Blue-black Grassquit
Variable Seedeater
Lesser Seed-Finch
Blue Seedeater R
Yellow-faced Grassquit
Sooty-faced Finch
White-naped Brush-Finch
Chestnut-capped Brush-Finch
Orange-billed Sparrow
Black-striped Sparrow
Streaked Saltator
Buff-throated Saltator
Black-headed Saltator
Slate-colored Grosbeak
Black-faced Grosbeak
Black-thighed Grosbeak
Blue-black Grosbeak
Red-breasted Blackbird
Eastern Meadowlark
Great-tailed Grackle
Bronzed Cowbird
Baltimore Oriole
Yellow-billed Cacique
Scarlet-rumped Cacique
Crested Oropendola
Chestnut-headed Oropendola
Golden-browed Chlorophonia
Yellow-crowned Euphonia
Thick-billed Euphonia
Elegant Euphonia
White-vented Euphonia
Tawny-capped Euphonia

Coiba National Park

Isla Coiba (50,314 ha/124,328 acres; 34 km long by 21 km wide) is the largest island off the Pacific coast of Central America and the centerpiece of one of Panama's most spectacular national parks. The park also includes many smaller islands nearby as well as the surrounding marine waters. The island's isolation has resulted in the evolution of a host of endemic forms, including the **Coiba Spinetail**, found only on the island, and 19 endemic subspecies of birds. This is the only place in Panama where Scarlet Macaw can be found in significant numbers, and the easiest place to see the nationally endemic **Brown-backed Dove**. The island has a number of Pacific-slope species that can be difficult to find on the mainland, including Bare-throated Tiger-Heron, Blue-throated Goldentail, and Sepia-capped Flycatcher. The island was formerly a penal colony, Panama's equivalent of Devil's Island, but the last prisoners were removed several years ago. When the island was a penal colony parts of it were logged, and there were cattle ranches and plantations at work camps scattered around the island's perimeter. However, most of the island away from the former prison facilities is still covered in luxuriant tropical forest, and the former cultivated areas are beginning to re-grow. The offshore waters contain the most extensive coral reefs on Panama's Pacific coast, and are popular for snorkeling and scuba diving. Humpback Whales can be seen regularly offshore.

To visit the island it is necessary to hire a boat. (There is an airstrip on the island but there are no regular flights.) See the section on the "How to get to Coiba National Park" for details on making arrangements. The park headquarters, called La Aguja, and cabins for visitors are on a small peninsula near the north end of the island. The site is a former fishing lodge and unfortunately doesn't have much forest nearby, nor trails that connect it to the rest of the island. Although there are a couple of short trails here, the better ones are found in other parts of the island reachable only by boat. For this reason it is necessary to have a boat available for the duration of your stay.

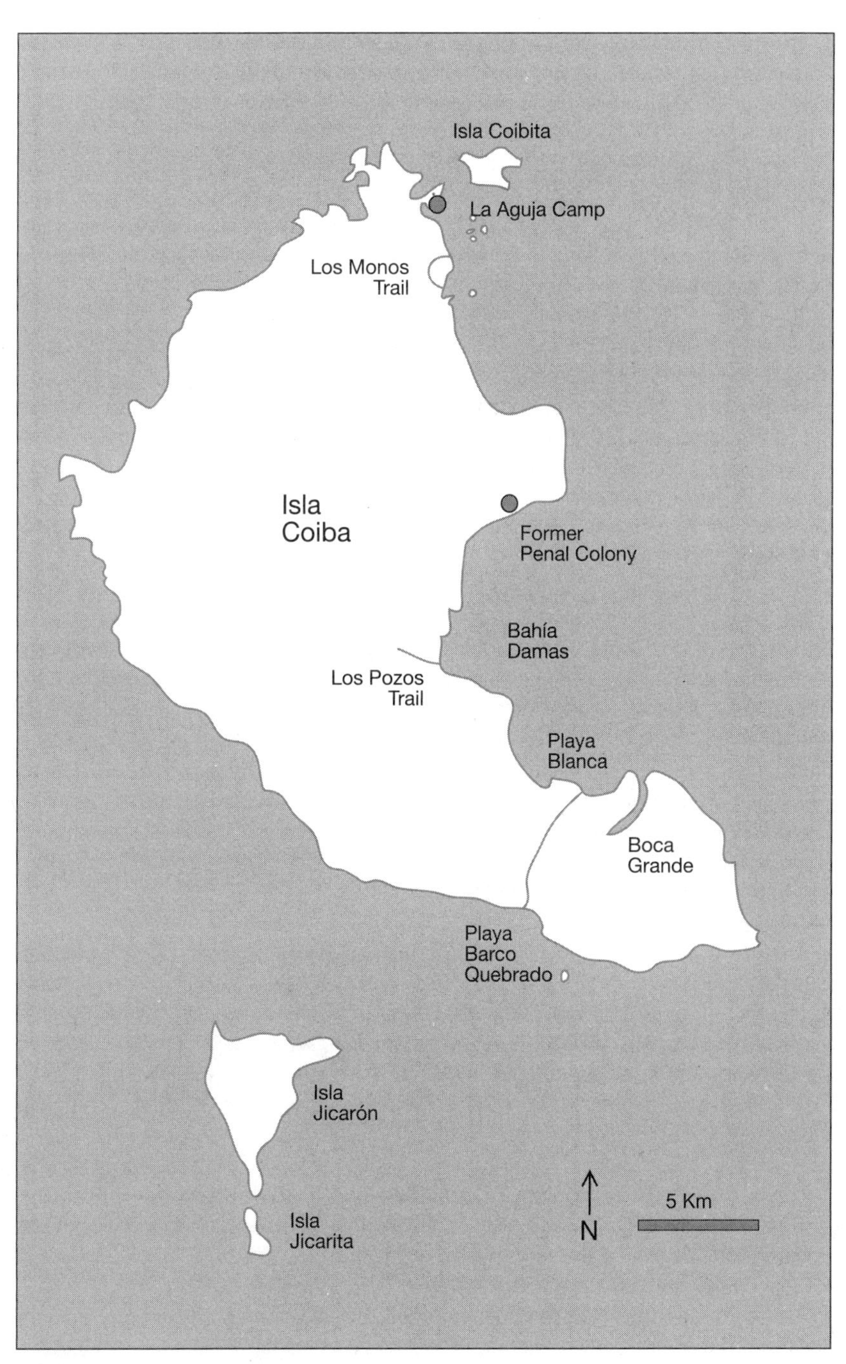
Isla Coibita
La Aguja Camp
Los Monos
Trail
Isla
Coiba
Former
Penal Colony
Bahía
Damas
Los Pozos
Trail
Playa
Blanca
Boca
Grande
Playa
Barco
Quebrado
Isla
Jicarón
Isla
Jicarita
N
5 Km

Isla Coiba

There are two short trails at La Aguja. The *Sendero del Observatorio* (Observatory Trail) starts behind cabin 6 and is a 15-minute walk to a lookout point. Another short trail behind the kitchen leads up to a second lookout point on a small hill.

The *Sendero de los Monos* (Monkey Trail) is about 3 km south of the La Aguja station (about 7 minutes by boat), near an islet called Granito de Oro, a popular snorkeling spot. This 1 km (0.6 mi) trail curves inland from one beach to another. **Brown-Backed Dove**, Blue-throated Goldentail, Lance-tailed Manakin, Tropical Parula, and other species can be seen here.

Presently the best available trail for birding is the one at the *Pozos Termales* (thermal pools), about 20 km south of La Aguja (30 minutes by fast boat). It is better to arrive at mid to high tide, since otherwise you have to walk for about 25 minutes on the rocky shoreline to get to the trail from the landing spot. The trail is only 750 m (0.5 mi) long, and leads to a few small thermal pools through mangroves, scrub, and secondary forest. Scarlet Macaws are sometimes been seen near the beginning of this trail, and King Vulture, **Brown-backed Dove**, Blue-throated Goldentail, **Coiba Spinetail**, Lance-tailed Manakin, Rufous-browed Peppershrike, and White-Throated Thrush can also be found here.

Traditionally the best place for Scarlet Macaws has been considered to be near Playa Barco Quebrado on the southern side of the island, where they are supposed to nest. A trail, about 5 km (3 mi) one-way, once ran from Playa Blanca on the south side of Bahía Damas across to Playa Barco Quebrado, but it has not been maintained and at the time of writing it was closed. It is to be hoped that as the park infrastructure is developed eventually this trail will be reopened. Playa Blanca is about 25 km from La Aguja (40 minutes by fast boat). The trail, which initially traverses regenerating cattle pasture and then enters better forest, is somewhat hilly especially at the start, and it takes at least a full morning for a round trip.

Getting to Coiba National Park. Because the logistics of getting to Coiba are quite complicated, most visitors will find it more convenient to make arrangements through a tour operator rather than trying to do it on their own. To make a trip economical, you will also probably need to organize a group. Advantage Tours, Ancon Expeditions, and EcoCircuitos (see p. 39) all have experience in arranging trips to Coiba. (Other companies offer tours as well, but focus on snorkeling, fishing, or other marine activities rather than birding.)

You can get to Coiba from either Puerto Mutis in the Gulf of Montijo near Santiago, or from Playa Santa Catalina, a popular destination for surfers, on the coast south of Soná. Puerto Mutis is a shorter drive but a longer trip by boat.

To get to Puerto Mutis from Santiago, turn left off the Carretera Interamericana onto Av. Central just past the Hotel Piramidal (marked by its red pyramidal roofs). After 1.8 km you will reach a large church. Go around the left side of the church, and take a left onto the street directly behind it (Calle Segunda Norte). The Hotel Santiago will be on your left after this turn. Take the second right, at the minisuper El Progreso, and bear left at the intersection just beyond. (Although there are two huge signs for Puerto Mutis on Av. Central, there are no signs at all at the intersections where they are needed!) From here it is 24 km to Puerto Mutis, with no turns. Once in Puerto Mutis, which consists of a few houses, basic restaurants, and the pier, ask at the police post for a secure place to leave your car.

To get to Santa Catalina, follow the directions for Puerto Mutis, but in back of the church make a right turn instead of left. After 200 m you get to an intersection with a small store on the left. Turn left here for Soná, at a blue sign that says "Soná 45 km." At 5 km from this point there is a fork; continue straight on the left one. At 43 km from the blue sign for Soná, just before getting into town, there is a gas station where you take a left turn for Santa Catalina. From here you go another 45 km until you get to a turnoff at Tigrè, where you make another left turn. After 9 km more you come to the village of Hicaco, where you make a right turn at the police station. In another 7 km you will arrive at Santa Catalina. The road, formerly very bad, is now asphalt all the way to Santa Catalina. It takes about 45 minutes from Santiago to Soná and another hour to Santa Catalina.

If you have not hired a tour operator, to arrange a boat to the island you can ask the ANAM office in Santiago for advice, or else try to arrange one in either Puerto Mutis or Santa Catalina. From Santa Catalina, a high speed boat with a 200 HP motor can get to Coiba in about an hour, a less powerful one will take 1½ hour or more. From Puerto Mutis it will likely take 2 hours or more. In Santa Catalina, the Catalina Punta Brava Lodge has boats available that can make the trip, and they can also be arranged through the Cabañas Rolo. Bear in mind that to get to the better trails and have much time to bird on the island you will have to either leave very early or stay overnight and keep the boat with you.

Accommodations and meals. The park headquarters at La Aguja has six rustic cabins, each of which has two rooms with four bunk beds each and shared cold-water showers ($18.00 per person per night). The rooms have air-conditioning, but generator power is provided only until 9:00 PM. If you want air-conditioning all night you will have to bring enough diesel fuel for the generator with you, about 10-15 gallons per night. You also need to bring all your own food and a gas tank for the stove; check with ANAM for the kind of fitting needed. As in other coastal areas, sandflies (*chitras*) can be a problem here. To make reservations for the cabins, call the ANAM office in Santiago (ph 998-4271). They may also be able to suggest how to make arrangements for getting to the island if you have not hired a tour operator. There is also a $10.00 park entry fee, which must be paid even by day visitors.

If you are leaving from Puerto Mutis, the best place to overnight is Santiago (pp. 166-167). Places to stay in Santa Catalina include the nicer Catalina Punta Brava Lodge and the more basic Cabañas Rolo.

The *Catalina Punta Brava Lodge* (ph 202-5505; cel 6614-3868; jose@puntabrava.com; www.puntabrava.com) is an oceanfront lodge with terraces that offer a panoramic view of the Pacific. There are 12 rooms ($30.00 for two persons), all with air-conditioning and private bath, and some with hot water. There is a restaurant that has good international cooking; meals are $16.00 a day. The lodge has two boats available for trips to Coiba. To get to it once in Santa Catalina, go past the school and take the road that goes left where there is a store next to a restaurant. After 300 m you pass another restaurant on the left, and 100 m after that come to a fork where you keep left. At 400 m from this fork there is a second fork where you go right, and after another 100 m come to the entrance for the lodge.

The *Cabañas Rolo* (ph 998-8600; info@rolocabins.com; reservations@rolocabins.com) are in Santa Catalina and have five rooms ($10.00 per person) with fan and shared bath. Meals can be had here for about $3:00 per meal. They also offer trips to Coiba.

Bird list for Isla Coiba and surrounding waters. * = vagrant; R = rare.

Blue-winged Teal
Lesser Scaup
Muscovy Duck
Least Grebe
Wedge-rumped Storm-Petrel
Brown Pelican
Brown Booby
Neotropic Cormorant
Magnificent Frigatebird
Great Blue Heron
Great Egret
Snowy Egret
Reddish Egret*
Little Blue Heron
Green Heron
Yellow-crowned Night-Heron
Bare-throated Tiger-Heron
White Ibis
Black Vulture
Turkey Vulture
King Vulture
Osprey
Double-toothed Kite
Bicolored Hawk
Broad-winged Hawk
Roadside Hawk
White Hawk
Mangrove Black-Hawk
Ornate Hawk-Eagle
Yellow-headed Caracara
Bat Falcon
Peregrine Falcon
American Kestrel
Wattled Jacana
Sora
Gray-necked Wood-Rail
White-throated Crake
Gray-breasted Crake R
Yellow-breasted Crake R
Purple Gallinule
American Oystercatcher
Black-bellied Plover
Semipalmated Plover
Wilson's Plover
Ruddy Turnstone
Semipalmated Sandpiper
Western Sandpiper
Least Sandpiper
Willet
Wandering Tattler R
Sanderling
Lesser Yellowlegs
Spotted Sandpiper
Whimbrel
Laughing Gull
Ring-billed Gull R
Sandwich Tern
Royal Tern
Pale-vented Pigeon
Ruddy Ground-Dove
Blue Ground-Dove
Brown-backed Dove
White-tipped Dove
Ruddy Quail-Dove
Scarlet Macaw
Orange-chinned Parakeet
Blue-headed Parrot
Mealy Amazon
Red-lored Amazon
Smooth-billed Ani
Great Horned Owl*
Rufous Nightjar
White-collared Swift
White-chinned Swift
Vaux's Swift
Lesser Swallow-tailed Swift
Scaly-breasted Hummingbird
Garden Emerald
Sapphire-throated Hummingbird
Blue-throated Goldentail
Snowy-breasted Hummingbird
Rufous-tailed Hummingbird
Ringed Kingfisher
Green Kingfisher
American Pygmy Kingfisher
Red-crowned Woodpecker
Red-rumped Woodpecker
Coiba Spinetail
Barred Antshrike
Tropical Kingbird
Fork-tailed Flycatcher
Streaked Flycatcher
Boat-billed Flycatcher
Great Kiskadee
Panama Flycatcher
Great Crested Flycatcher
Tropical Pewee
Common Tody-Flycatcher
Yellow Tyrannulet
Yellow-bellied Elaenia
Lesser Elaenia
Greenish Elaenia
Scrub Flycatcher

Southern Beardless-Tyrannulet
Sepia-capped Flycatcher
Ochre-bellied Flycatcher
Bright-rumped Attila
Masked Tityra
Three-wattled Bellbird
Lance-tailed Manakin
Red-eyed Vireo
Yellow-green Vireo
Philadelphia Vireo
Scrub Greenlet
Rufous-browed Peppershrike
Gray-breasted Martin
Mangrove Swallow
Barn Swallow
House Wren
Swainson's Thrush
White-throated Thrush
Tropical Gnatcatcher
Tropical Parula
Prothonotary Warbler
Tennessee Warbler
Yellow-rumped Warbler
Bay-breasted Warbler
Yellow Warbler
Chestnut-sided Warbler
Hooded Warbler
Northern Waterthrush
Kentucky Warbler
American Redstart
Rufous-capped Warbler
Red-legged Honeycreeper
Bananaquit
Summer Tanager
Scarlet Tanager
Crimson-backed Tanager
Blue-gray Tanager
Blue-black Grassquit
Variable Seedeater
Yellow-faced Grassquit
Lesser Seed-Finch
Black-striped Sparrow
Streaked Saltator
Rose-breasted Grosbeak
Dickcissel
Great-tailed Grackle
Thick-billed Euphonia

Cerro Hoya National Park

The mountains of the southwestern Azuero Peninsula are isolated from Panama's other ranges and are of considerable biogeographic interest. The **Azuero Parakeet** (known locally as the *Perico Carato*), a Panama national endemic, is found only here. It was originally described as a subspecies of Painted Parakeet, but recently has been recommended for full species status. The **Brown-backed Dove**, another Panama national endemic, occurs only here and on Islas Coiba and Cébaco. Two other regional endemics of the western Pacific lowlands are also found, **Black-hooded Antshrike** and **Orange-collared Manakin**. A number of species typical of the western highlands, including Three-wattled Bellbird, occur at higher elevations, and a few Great Green and Scarlet Macaws still hang on here.

Cerro Hoya National Park (32,577 ha/80,500 acres) nominally protects part of these Azuero mountains, but the area continues to be threatened by deforestation. The park is almost completely undeveloped and difficult to visit. Road access is poor, and there are no good hotels or other facilities for tourism close to the park. Seeing the **Azuero Parakeet** or the other species found in the area almost requires mounting an expedition. This is not a place easily visited by the casual birder.

Getting to Cerro Hoya National Park: Cerro Hoya National Park may be visited either from the east, through Los Santos Province, or from the west by driving down the western side of the Azuero Peninsula through Veraguas.

Eastern side. Coming from Panama City, follow the directions for "Getting to Western Panama" (pp. 130-132). Turn off the Carretera Interamericana at Divisa to go south on the Carretera Nacional. You can bypass downtown Chitré by taking the Via Circunvalación, signposted for Las Tablas, on the right at 32.8 km south of Divisa and 100 m after the turnoff to Pesé. (If you miss the turn, continue into Chitré and follow the directions for El Agallito Beach on pp. 158-159, but take a right at the Hotel Santa Rita). It is about 25 km from Chitré to Las Tablas, where you turn

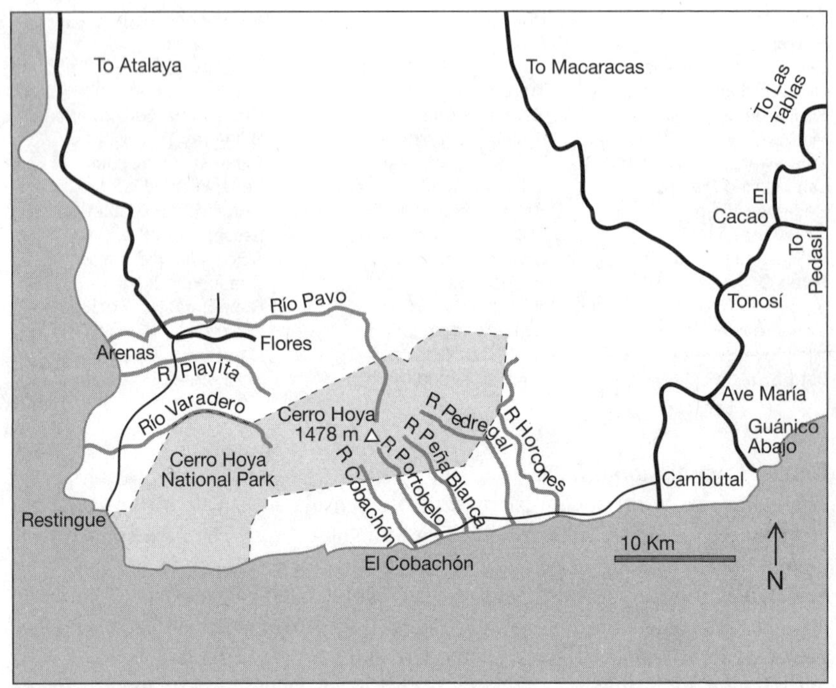

Cerro Hoya National Park

off from the Carretera Nacional for Tonosí. From Las Tablas, it is 58 km via a winding asphalt road through the hills to Tonosí. When you come to a T-junction in the center of Tonosí, take a right. In 100 m, the road forks; go left at the Centro Comercial Tonosí. From here it is 25 km on a badly potholed asphalt and gravel road to Cambutal. At 11.3 km on this road is a fork where you go right; at 15.7 is another fork where you go left (the right is signposted for Guanico).

A little after entering Cambutal, a dirt road goes right at the minisuper Hnos. Valdéz. This road continues 22 km to the village of El Cobachón, from which you can hike or hire horses to go up to the park, which is above it. The road requires 4WD, and is even with it is passable only in the dry season. However, it is not difficult if it is dry. The first 10 km are fairly level. After that the road climbs up and down steep hills to get around headlands, and several rivers must be forded. The drive is very scenic, passing through some remnant patches of forest with huge *cuipo* trees, and with some excellent views of the blue Pacific with white surf pounding in on black sand beaches. There are only a few houses in El Cobachón. Daniel Saénz, who lives there, is an excellent guide for the area. To find his house, when you arrive at the El Cobachón school, walk down the edge of the clearing in front of the school to the top of the beach. Go left on the path here and his is the second house. Daniel can help hire horses and make other arrangements. Great

Green Macaws nest in the area, and **Azuero Parakeet** can be found in the forest above. The hills above El Cobachón are deforested up to about 600 m (2,000 ft), but above this there is still forest.

Accommodations and meals (eastern side). The closest place to the eastern side of Cerro Hoya National Park that has hotels and restaurants is Tonosí, where the following places can be found. In Cambutal, there are also some basic cabins near the beach.

The *Residencial Mar y Selva* (ph 995-8003) has 10 basic rooms ($20.00) with private hot-water bath and air-conditioning. There is a restaurant on the ground floor. It is on the right just as you come into Tonosí from the direction of Las Tablas or Pedasí.

The *Pensión Boamy* (ph 995-8142) has 11 rooms ($18.70) with private hot-water bath and air-conditioning and a restaurant on premises. To find it, take a right at the T-junction in the center of town, and then a left at the Centro Comercial Tonosí for the road to Cambutal. The Boamy is a short distance down this road on the left.

The best place to eat is *El Charcón* (the sign on the road actually says Restaurante Peroniles), an outdoor place specializing in grilled meat. It is on the road to Cambutal, a little over 1 km south of Tonosí, on the left just after a bridge.

Western side. The western side of Cerro Hoya National Park can be reached by asphalt road from Santiago. Take the turn-off for Atalaya, approximately 4 km east of Santiago on the Carretera Interamericana. At 3.7 km from the Interamericana, before reaching Atalaya, take a right at the minisuper El Cruce, and at 24.6 km, take the right fork, signposted for Mariato. You will reach Mariato, the largest town en route, at 55 km, where there are a few stores and restaurants. Mangroves can be found along or near the road at Morillo, at 87 km, and at 92 km along a dirt road to the right. These can be worth checking for Wood Stork and possibly **Yellow-billed Cotinga**. (Although there are as yet no records of the latter from here, it has been found a bit farther north along this coast on the Gulf of Montijo.) At 100 km from the Interamericana there is a police station and ANAM office on the outskirts of the village of Arenas. At 600 m past these, the road crosses the rather misnamed Río Grande, with the El Cruce Restaurant just past it on the left. From here the road turns eastward to parallel the north side of the Cerro Hoya range. The asphalt road ends in the town of Flores at about 107 km from the Interamericana. Flores is about 3 km north of where forest starts. This forest is outside the park proper but is contiguous with it.

The forest can be reached by driving across pastures on private land. Permission to enter can be obtained with the assistance of local guides. Two residents of the area who can serve as guides are Ricaurte Moreno ("Cauca"), formerly an ANAM park guard, and Isidro Barría Castro ("Chiro"), who has lived in the area for many years and is very familiar with the forest. To get in touch with Ricaurte, call the public phone at 999-8143, and ask the person who answers to tell Ricaurte that you will call again in 15 minutes. Ricaurte's house is the first one on the right-hand side when you reach the village of Flores. If Ricaurte cannot act as guide, he will contact Isidro.

There are several trails in the area, including one into the valley of the Río Playita that first passes through secondary forest and then partially-logged primary forest, and gives access to a trail that goes up into the foothills, and another that ascends more steeply to over 1,000 m (3,300 ft). Great Green Macaw, **Black-hooded Antshrike**, and **Orange-collared Manakin** can be found on the Río Playita, and according to local residents, **Azuero Parakeet** can also sometimes be seen on both trails. We would be very interested in reports from anyone who visits this area.

Another road which can be explored with 4WD goes down the side of the Jardín El Cruce in Arenas (see below), fords the Río Grande, and continues through gallery forest, light woodland, and cattle pasture before making an impressive crossing of the very wide rocky bed of the Río Pavo. Another dirt road, which goes right just after the bridge in Arenas, goes to Restingue at the edge of Cerro Hoya National Park at the southwestern tip of the peninsula. Although there was formerly an ANAM bunkhouse here, it no longer exists. While the road is scenic in places, there is very little forest along it, and according to locals the **Azuero Parakeet** is not found in the forest near Restingue. The road is not passable past the wide Río Varadero in the wet season.

If you don't have a car, Flores and Arenas can be reached by bus from Santiago.

Accommodations and meals (western side)

The most convenient place to stay or eat near the park is the *Jardín y Restaurante El Cruce* in Arenas, on the left just after you cross the bridge over the Río Grande. There are two double rooms with air-conditioning and private cold-water bath ($18.00) and two without air-conditioning ($13.00). The restaurant serves breakfast, lunch, and dinner. The place may be noisy on weekends and public holidays, but is otherwise very quiet.

The *Cabañas Torio* (US ph (253) 843-6583; torioreservations@torioresort.com; www.torioresort.com) has eight cabins ($26.00) with private bath and fans. It is on the main road 70 km from the Interamericana and 37 km north of Flores. There is a restaurant on premises that serves breakfast and dinner (but not lunch).

Farther up the Azuero, the *Río Negro Sportsfishing Lodge* (US ph (912) 786-5926; panamasportsman@earthlink.net; www.panamasportsman.com) is located near Mariato. Although as its name implies it is mostly dedicated to sport fishing, the lodge has also developed a package for birders. **Yellow-billed Cotinga** has been seen in mangroves in Montijo Bay not too far from the lodge, and could occur here as well.

Chiriquí Province

After the Canal Area, the western highlands of Chiriquí are Panama's most popular birding destination – and little wonder. More than 50 regional endemics can be found here, as well as many other species that occur in Panama only in this region. This is one of the easiest places in the world to see Resplendent Quetzal, and many other spectacular species, such as Three-wattled Bellbird, **Bare-necked Umbrellabird**, and a great diversity of brilliantly-colored hummingbirds and tanagers can also be found. The climate is spring-like year round, and the area has many charming hotels and lodges well suited to serve as bases for birding.

The best-known birding areas are on the flanks of Volcán Barú, Panama's highest peak, including Boquete on the mountain's eastern side and Volcán and Cerro Punta to its west. Fortuna, east of Boquete, features many species that favor wetter forests, while the Santa Clara area, near the Costa Rican border, has several species that are hard to find elsewhere. Cerro Santiago, in eastern Chiriquí, is more difficult to get to, but is the best place to find the Panama national endemics **Glow-throated Hummingbird** and **Yellow-green Finch**.

The lowlands of Chiriquí also have a number of regional endemics, including **Charming Hummingbird**, **Fiery-billed Aracari**, **Black-hooded Antshrike**, **Yellow-billed Cotinga**, **Orange-collared Manakin**, and others. The best area for most of these species is the Burica Peninsula around Puerto Armuelles, but several can be found farther east, at Chorcha Abajo, Cerro Batipa, and Las Lajas Beach.

Cerro Santiago (Cerro Colorado)

The central part of the western highlands of Panama, the Tabasará range, is home to two Panama national endemics found nowhere else, **Glow-throated Hummingbird** and **Yellow-green Finch**. The Cerro Santiago area, above Cerro Colorado, the site of a major copper deposit and proposed mine, is the most accessible place where both these species can be found. (Although the finch has been found at higher elevations at Fortuna, and the hummingbird above Santa Fe, they are more difficult to see at these localities.) A variety of more widespread of endemics of the western highlands can also be found here, some of them at lower elevations than in areas farther west. These include **Ruddy Treerunner**, **Silvery-fronted Tapaculo**, **Black-faced Solitaire**, **Black-and-yellow Silky-flycatcher**, **Collared Redstart**, **Spangle-cheeked Tanager**, **Sooty-capped Bush-Tanager**, and **Slaty Flowerpiercer**.

The Cerro Santiago region is in the heart of the recently established Ngöbe-Buglé Comarca (indigenous homeland), and nearly all of the residents are members of this group (also known as Guaymí). At the time of writing, the area was not easy to get to. Development of the Cerro Colorado mine was halted several years ago due to falling copper prices, and the road has deteriorated severely due to lack of maintenance. The first 30 km of the road, from San Félix to Hato Chamí at about 1,050 m (3,465 ft), at present require a strong 4WD vehicle with high clearance, and even at

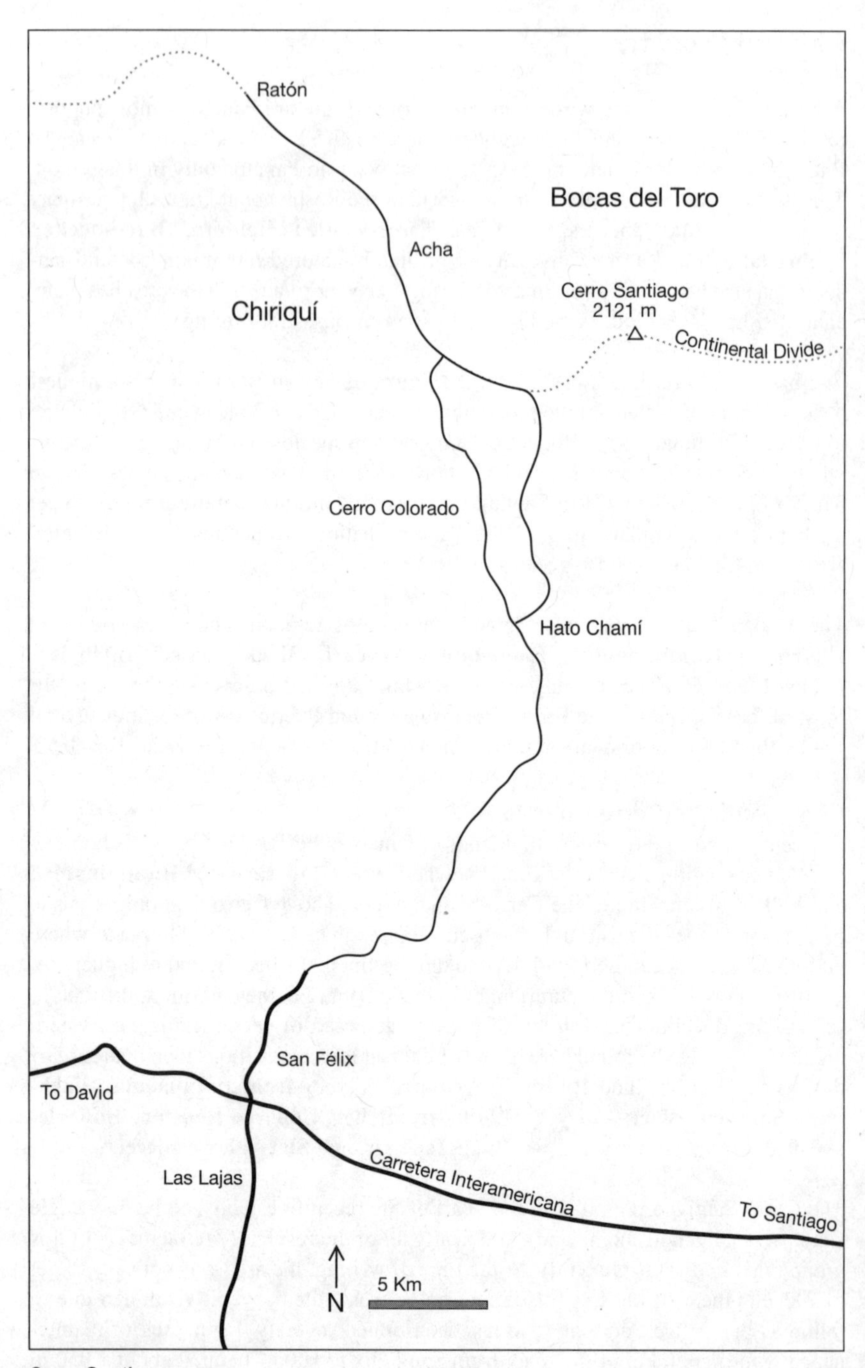

Cerro Santiago

that is only passable during the dry season. Above Hato Chamí the road is less traveled and in better condition, though there are still many rough spots. However, in early 2006 the road was under repair, and is scheduled to be improved all the way to Ratón (see below). While this will make the area easier to get to, it is likely to accelerate the already rapid deforestation occurring at higher altitudes.

Most of the lower part of the road is deforested, except for a couple of small patches of forest at about 900 m (3,000 ft). The first forest appears about 3 km beyond Hato Chamí, at about 1,150 m (3,800 ft) elevation. Both **Glow-throated Hummingbird** and **Yellow-green Finch** have been seen here, but are more easily found somewhat higher up. The finch is generally quite common in both forest and scrubby areas. The hummingbird is more erratic, and its abundance seems to vary seasonally together with the availability of its preferred flowers, which include those of a low heath-like shrub found on steeper windswept slopes. One of the better areas for the hummer is around and beyond an antenna about 10 km past Hato Chamí, where there are extensive areas of this heath-like shrub at about 1,700 m (5,600 ft).

From about 6 km beyond Hato Chamí the road runs along the Continental Divide at elevations between 1,600-1,700 m (5,300-5,600 ft), with Chiriquí on the left and Bocas del Toro on the right. The ridge can be windy, with blowing mist and rain, especially in the dry season. The road passes mainly through cloud and elfin forest, low scrub, and patches of taller forest in the more sheltered areas. There are, however, scattered areas of pasture along the road, and the small community of Acha is at about 14 km above Hato Chamí. About 4 km beyond Acha the forest gives way to pastures; the road ends near the community of Ratón about 3 km beyond that.

Given the difficulty of getting to the area, it is preferable to camp there for at least one night in order to have adequate birding time. Look for a sheltered area out of the wind where there is space to pull off the road and where there is room for a tent. There is little running water on the ridgetop, so bring your own supply.

Getting to Cerro Santiago. The road to Cerro Colorado and Cerro Santiago takes off from the Carretera Interamericana at San Félix, which is about 118 km west of Santiago and 72 km east of David. The turnoff is marked by the El Cruce gas station and restaurant. (The road to Las Lajas goes south from the Interamericana here.) You will first pass through the town of San Félix, which is set back from the highway a short distance. In March 2006, the asphalt ended about 9 km from the Interamericana, but as mentioned above the road was under repair and scheduled to be improved as far as Ratón. Hato Chamí is 30 km from the Interamericana. Go right at the fork at the top end of town; the left fork goes down to Cerro Colorado. After about 100 m the road forks again; take the fork to the left, which goes uphill. Forest is reached at about 3 km beyond this fork.

If you don't have a suitable 4WD vehicle, the Cerro Santiago area can be reached by public transportation from San Félix. High-clearance pick-up *transportes* currently serve as buses to Hato Chamí and areas above it. These transportes leave from the supermarket Kevin, at the north end of San Félix 3.7 km from the Interamericana. Service is most frequent to Hato Chamí, but there are daily transportes as far as Acha, and sometimes beyond, leaving as early as 6:00 AM. It would be best to find one going as far as Acha, and then get off at a likely area to bird (or camp if you decide to do so). It's not cheap; fare to Hato Chamí is $5.00 and Acha $12.00 (one-way). Once the road is improved less-expensive minibuses may start to make the trip.

Accommodations and meals. The closest accommodations to the Cerro Santiago area are the *Cabañas Carrizal* (ph 727-0021) near San Félix. The cabins are located on the Carretera Interamericana about 700 m west of the turnoff for San Félix, directly behind the red San Félix gas station (not to be confused with the El Cruce station at the intersection itself). The cabins are basic, and rent for $15.00 with fan and $25.00 for air-conditioning; cold water only. There is a restaurant on premises that serves breakfast, lunch, and dinner.

The *El Cruce* restaurant at the intersection is rather dismal even for a Panamanian greasy spoon. There are also likely to be a few basic restaurants in San Félix, but we have not investigated them.

There are also cabins at Las Lajas Beach (see p. 189), but these are likely to be too noisy to be suitable for birders, at least during the high season during the dry season. The area can also be visited as a day trip from either David or Santiago, which are about 1 hour or 1½ hours away respectively from San Félix. For accommodations and restaurants in these cities see p. 201 for David and pp. 166-167 for Santiago. However, this would require you to leave either city well before first light in order to get up to birding areas by mid-morning.

Bird list for Cerro Santiago and vicinity.

Black Vulture
Swallow-tailed Kite
Broad-winged Hawk
Band-tailed Pigeon
Bare-shanked Screech-Owl
Dusky Nightjar
White-collared Swift
Green Hermit
Green Violet-ear
Garden Emerald
Rufous-tailed Hummingbird
Stripe-tailed Hummingbird
White-tailed Emerald
Snowcap
White-bellied Mountain-gem
Purple-throated Mountain-gem
Magnificent Hummingbird
Glow-throated Hummingbird
Scintillant Hummingbird
Orange-bellied Trogon
Prong-billed Barbet
Blue-throated Toucanet
Olivaceous Piculet
Yellow-bellied Sapsucker
Hairy Woodpecker
Silvery-fronted Tapaculo
Ruddy Treerunner
Spotted Barbtail
Ruddy Treerunner
Buffy Tuftedcheek
Scaly-throated Foliage-gleaner
Streak-breasted Treehunter
Ruddy Woodcreeper
Olivaceous Woodcreeper
Streak-headed Woodcreeper
Spot-crowned Woodcreeper
Silvery-fronted Tapaculo
Yellow-bellied Elaenia
Mountain Elaenia
Olive-striped Flycatcher
Eye-ringed Flatbill
White-throated Spadebill
Yellow-bellied Flycatcher
Yellowish Flycatcher
Bright-rumped Attila
Black-and-white Becard
Three-wattled Bellbird
White-ruffed Manakin
Brown-capped Vireo
Rufous-and-white Wren
Gray-breasted Wood-Wren
Ochraceous Wren
Blue-and-white Swallow
Tawny-faced Gnatwren
Black-faced Solitaire
Black-billed Nightingale-Thrush

Ruddy-capped Nightingale-Thrush
Mountain Thrush
White-throated Thrush
Black-and-yellow Silky-flycatcher
Tennessee Warbler
Wilson's Warbler
Slate-throated Redstart
Collared Redstart
Three-striped Warbler
Common Bush-Tanager
Sooty-capped Bush-Tanager
Silver-throated Tanager
Spangle-cheeked Tanager
Yellow-faced Grassquit
Sooty-faced Finch
White-naped Brush-Finch
Chestnut-capped Brush-Finch
Yellow-green Finch
Slaty Flowerpiercer
Wedge-tailed Grass-Finch
Rufous-collared Sparrow
Golden-browed Chlorophonia
Yellow-crowned Euphonia
Elegant Euphonia
Tawny-capped Euphonia

Las Lajas Beach

The area around Las Lajas Beach in eastern Chiriquí features some of the species of open country and scrub of the western lowlands. White-collared Seedeater breeds here, and **Black-hooded Antshrike** can be found in the scrub. Grassland Yellow-Finch has also been seen once in the area, far from its more usual range in the Coclé grasslands. Migrant Scissor-tailed Flycatcher and Dickcissel have also been found. A marsh near the beach features ducks and wading birds (sometimes including Wood Stork and Roseate Spoonbill). The area can make a good stop for an hour or two en route between Santiago and David.

From the turnoff south from the Carretera Interamericana at San Félix, it is 2.3 km to a fork just before the town of Las Lajas, where you turn right at a small yellow building (signposted "Playa Las Lajas 15 km"). The road is asphalt and in mostly good condition all the way to the coast, although potholed in places. The center of the town of Las Lajas is at 3.4 km. For the next 9 km or so the road runs through pasture but is lined with shade trees. Look for raptors, including Laughing Falcon and Pearl Kite, in the pastures, and waterfowl in marshy places. At about 12 km from the Interamericana, the road curves around a small marsh on the left. Check the marsh for ducks, waders, shorebirds, and other species, and the nearby pastures for White-collared Seedeater and Yellow Tyrannulet. After a few hundred meters, the road turns sharply right at a small patch of woodland. **Black-hooded Antshrike** has been found here, though you may need a tape to bring it into view. Seasonally there is sometimes a colony of Crested Oropendola in the pasture on the right just before this woodland.

Las Lajas Beach is reached at 13 km from the Interamericana. About 200 m before reaching the beach area, there is a gravel road to the left, signposted "Hostal Playa Las Lajas." This road passes through pastures for about 1 km before reaching the gate for the hostal. These pastures are where Scissor-tailed Flycatcher has most regularly been seen. If you continue on the main road all the way to the beach, there is another road to the right that passes for about 1.4 km through similar grassy habitat that also could be checked (although we have not seen the species here).

The beach is a popular destination and can be quite crowded during the dry season. In March 2006 there was a police checkpoint about 4 km south of Las Lajas town, where a $1.00 fee was being collected for "security" for visitors to the beach. We are not certain whether this fee will continue to be collected during the off season.

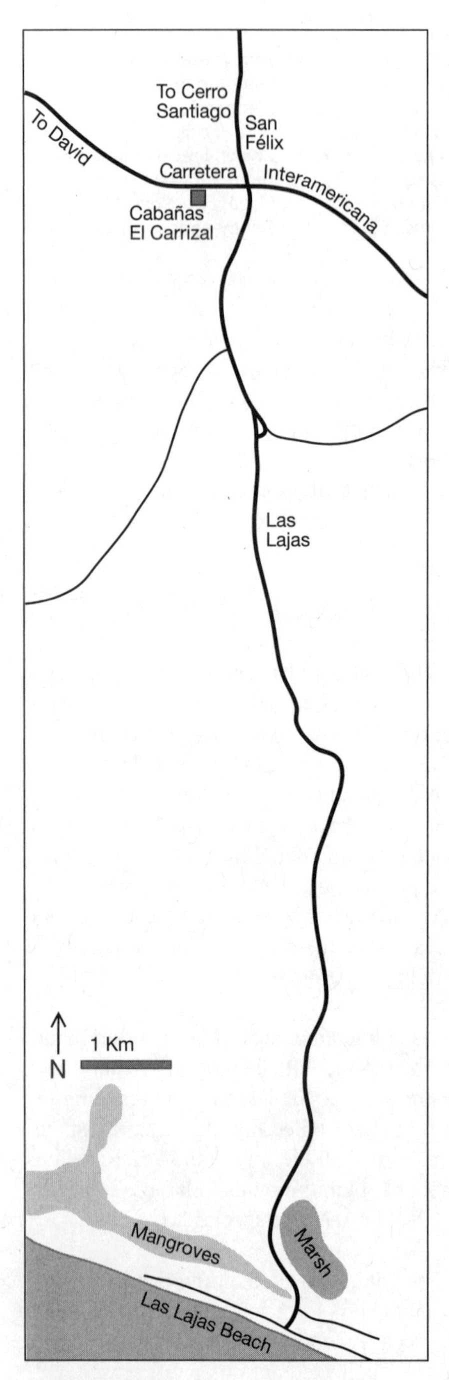

Las Lajas Beach

How to get to Las Lajas Beach. From the Carretera Interamericana, the signposted turnoff to Las Lajas is on the south at the town of San Félix, about 118 km west of Santiago and 72 km east of David. It is marked by the El Cruce gas station and restaurant. The turnoff is opposite for that for San Félix and Cerro Colorado/Cerro Santiago on the north side of the highway.

Accommodations and meals. Since it only takes an hour or two to bird, Las Lajas Beach is best done as a stop-off en route between Santiago and David. If you do want to stay in the area, the Cabañas Carrizal are on the Interamericana 700 m west of the San Félix/Las Lajas intersection. See the previous section on Cerro Santiago/Cerro Colorado for information these cabins and for other options in the area. There are several small eating places at the beach.

Bird list for Las Lajas. R = rare.

Black-bellied Whistling-Duck
Neotropic Cormorant
Magnificent Frigatebird
Least Bittern R
Great Blue Heron
Great Egret
Snowy Egret
Little Blue Heron
Tricolored Heron
Cattle Egret
Green Heron
White Ibis
Glossy Ibis R
Roseate Spoonbill R
Black Vulture
Turkey Vulture
Osprey
Mangrove Black-Hawk
Savannah Hawk
Roadside Hawk
Crested Caracara

Laughing Falcon
Collared Plover
Black-necked Stilt
Northern Jacana
Greater Yellowlegs
Willet
Least Sandpiper
Short-billed Dowitcher
Laughing Gull
Royal Tern
Sandwich Tern
Plain-breasted Ground-Dove
Ruddy Ground-Dove
Brown-throated Parakeet
Orange-chinned Parakeet
Ringed Kingfisher
Red-crowned Woodpecker
Black-hooded Antshrike
Yellow Tyrannulet
Yellow-crowned Tyrannulet
Yellow-bellied Elaenia
Common Tody-Flycatcher
Tropical Pewee
Panama Flycatcher
Great Kiskadee
Tropical Kingbird
Scissor-tailed Flycatcher R
Fork-tailed Flycatcher
Cliff Swallow
Barn Swallow
House Wren
Yellow Warbler
Crimson-backed Tanager
Blue-gray Tanager
Red-legged Honeycreeper
Blue-black Grassquit
White-collared Seedeater
Ruddy-breasted Seedeater
Grassland Yellow-Finch R
Dickcissel
Red-breasted Blackbird
Crested Oropendola
Lesser Goldfinch

Cerro Batipa

Cerro Batipa contains the largest patch of lowland forest remaining in Chiriquí east of Puerto Armuelles. There are about 330 hectares (815 acres) of older forest on Cerro Batipa itself and two smaller hills nearby which are maintained as a private reserve. Located on a peninsula surrounded by mangroves southeast of David, it is the best place in Panama to find **Yellow-billed Cotinga**. Other endemics of the western Pacific slope that can be found here include **Fiery-billed Aracari**, **Black-hooded Antshrike**, **Orange-collared Manakin**, and **Cherrie's Tanager**. Additional species of interest include Bronzy Hermit, Blue-throated Goldentail, Sepia-capped Flycatcher, Scissor-tailed Flycatcher, and Gray-crowned Yellowthroat.

Cerro Batipa is on a working cattle ranch. At the time of this writing, plans were in development to open facilities here to accommodate birders and others interested in nature tourism. Those interested in visiting the site should contact the owner, Luis Ríos, who lives in David (office ph 775-2512 , home ph 775-3968; lare@fertica.com.pa), for current information on visiting the site.

How to get to Cerro Batipa. The turnoff to Cerro Batipa from the Carretera Interamericana is 16.4 km west of San Lorenzo and 24.5 km east of David. The gate is locked and permission to enter must be obtained by contacting the owner.

Fortuna Forest Reserve

Fortuna Forest Reserve (19,500 ha/48,185 acres) was created to protect the nearly pristine montane and cloud forests that surround the Fortuna hydroelectric project. Many regional endemics of the western highlands are found here, as well as some Atlantic-slope specialties such as **Lattice-tailed Trogon** which here reach the Pacific side. Fortuna is the best place in Panama to find **Bare-necked Umbrellabird** (although even here it is scarce). Other highland specialties and endemics include **Black Guan**, Violet Sabrewing, **Black-bellied Hummingbird**, **Bare-shanked Screech-Owl**, Red-headed and **Prong-billed Barbets**, Black-banded Woodcreeper, Rufous-rumped Antwren, Rufous-breasted Antthrush, **Silvery-fronted Tapaculo**, Rufous-browed Tyrannulet, Black-and-white Becard, Three-wattled Bellbird,

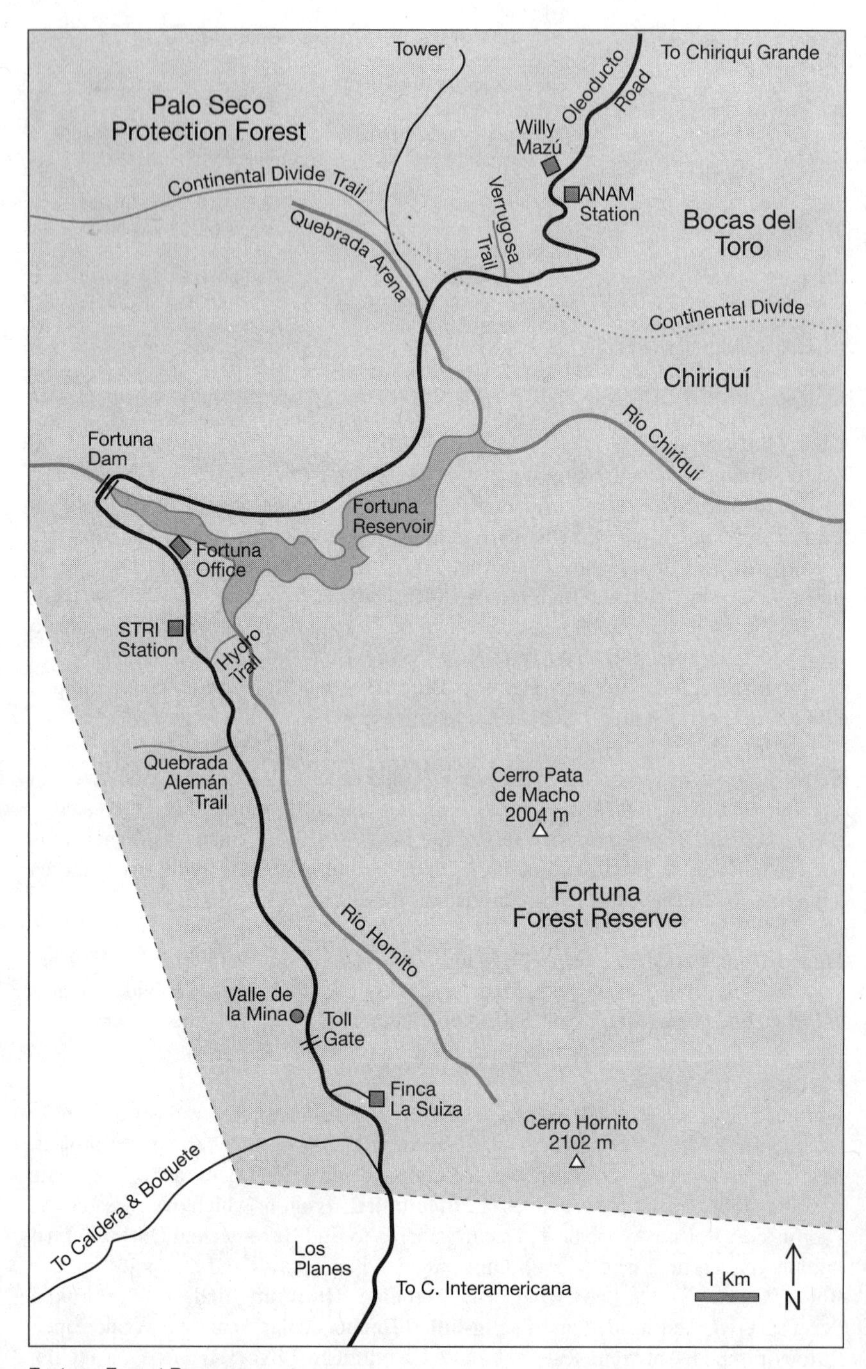

Fortuna Forest Reserve

Azure-hooded Jay, **Black-and-yellow Silky-flycatcher**, Slaty-backed Nightingale-Thrush, **Wrenthrush**, and **Sooty-faced Finch**. Fortuna is the westernmost locality known for the nationally endemic **Yellow-green Finch**.

The reserve is traversed by the Oleoducto Road (Camino Oleoducto), a two-lane asphalt highway that was built in conjunction with an oil pipeline from Puerto Armuelles on the Pacific to the port of Chiriquí Grande in Bocas del Toro on the Caribbean coast. The 16 km of the road within the reserve provides access to a number of excellent birding trails. As you drive along, keep an eye out for raptors and swifts, or for patches of flowers that may attract hummingbirds. The many swift streams crossed by the road are worth checking out for Torrent Tyrannulet and American Dipper. In many places there are also extensive level stretches that can be walked to look for mixed flocks of tanagers and other species. (If you stop along the road, be sure to pull off completely, and use care when walking along it, since heavy trucks use it regularly.)

The reserve is just south of a low point in the central cordillera which provides a gap for moisture-laden winds to surge through from the Atlantic slope. Weather in the area can be very unpredictable. During the dry season, January to April, the trade winds blow steadily from the north. It is often foggy and rainy, but frequently one can find a protected area where conditions are relatively clear somewhere along the road. South of the reservoir it is quite windy all year long. During the rainy season, May to December, the basin may be socked in, or there can be several sunny, rainless days in succession. Temperature can range from 60-80°F (15-27°C), and it can be quite chilly at night or dawn or dusk, especially when there is wind and rain. Within the reserve the Oleoducto Road runs mostly between 1,100-1,200 m (3,600-4,000 ft), dropping to 1000 m (3,300 ft) where it crosses the reservoir, but the trails reach substantially higher. It is best to dress for cool weather but be prepared to shed outer layers on warm days; and to always be prepared for rain even if the day starts out sunny. The best chance for good weather is the rainy season, but the best time for Three-wattled Bellbird and **Bare-necked Umbrellabird** is March to September since they migrate to lower elevations during the non-breeding season. If conditions are bad within the Reserve, you can continue on into Bocas del Toro and bird the lowlands of the Caribbean coast.

The entrance to the reserve is marked by a "Bienvenidos Reserva Forestal Fortuna" sign about 22 km north of Gualaca (and 38 km from the Interamericana). Distances to trails and other localities within the reserve are given from this sign. (There are also roadside markers indicating kilometers from the Interamericana.)

Finca La Suiza, a small tourist lodge located about 1.1 km past the Fortuna entrance sign, has a complex of trails open to non-guests on payment of an entrance fee. The entrance is marked by a wooden sign in a small meadow to the right of the road. You may enter between 7:00-11:00 AM, but must leave the property by 4:00 PM in the dry season and 2:00 PM in the rainy season. The lodge and its trail system are closed during June, September, and October, the rainiest months. Drive in through the gate and park in the area indicated. Walk up the gravel road a short distance to a chain barrier, where someone from the small white house by the entrance will meet you to collect the $8.00 entrance fee and give you a trail map. It is about 400 m up the gravel road to the trail entrance. There are four well-maintained and well-marked interconnected loop trails, which range between 1,150-1,600 m (3,800-5,300 ft) in elevation and offer 20 hours of hiking. See the section on accommodations below for more information on the lodge.

There is a toll gate for trucks at 2.4 km; cars are not charged. Just beyond is Valle de La Mina, a collection of four or five restaurants and shops that cater to the bus trade.

At 7 km the road crosses an open windswept saddle with drop-offs on each side to enter the watershed of the Fortuna Reservoir. At 9.3 km the road crosses the Quebrada Alemán on a concrete bridge, where the *Quebrada Alemán Trail* starts. At 10.7 km is the start of the *Hydro Trail.* [4]

At 11.3 km on the left is a small field research station operated by the Smithsonian Tropical Research Institute (STRI), the Centro de Investigaciones Tropicales Jorge L. Araúz. There is a loop trail behind the cabin, but advance permission from STRI is necessary to enter. **Bare-shanked Screech-Owl** can sometimes be found along the road here.

Offices for the hydro project are on the right side of the road a couple of kilometers past the STRI station. At 15.5 km the road crosses the Fortuna Dam. Park here and check for swifts and swallows over the reservoir and river. At 1.3 km beyond the dam, there is a bridge over a stream worth checking for Torrent Tyrannulet, American Dipper, and swifts.

The stretch of road just to the north of the reservoir is the area referred to as Umbrellabird Road in the "Finding Birds in Panama" section of Ridgely (the layout of the roads in the area has changed since this was written). Look for **Umbrellabirds** anywhere along the road to the north of the reservoir (there is only one record in the reserve to the south of it), although sightings at best are rare.

At 24.2 km, the road reaches the Continental Divide at an altitude of 1,100 m (3,600 ft), marked with a sign and a guard post. A few hundred meters before reaching the Divide, the road crosses the *Quebrada Arena.* Check the stream from the bridge for Fasciated Tiger-Heron, which can also be found by wading the stream.

Not far beyond the Quebrada Arena, about 20 m before the guard post at the Continental Divide, is a gravel side road to the left that leads to the *Continental Divide*

[4] At the time of this writing, Fortuna S.A., the company that operates the hydroelectric project, has restricted public access to these and other trails within the reserve while a policy on visitation by tourists is under development. For current information on access to trails in the area, contact the Panama Audubon Society.

Trail. The road (4WD preferred but usually passable with 2WD with care) first descends towards the stream, then turns right and ascends a straight stretch through scrubby vegetation. After 1.3 km the road turns sharply to the right between two small hills where you will see two utility poles, one numbered 112, on the left. The trail starts here, climbing the steep bank just to the left of the poles. The trail is steep for the first 100 m or so but then rises more gradually, with periodic dips. The trail traverses moss-cloaked cloud forest along the Continental Divide as it runs between the provinces of Chiriquí and Bocas del Toro for several kilometers. This trail can be superb for highland specialties and endemics when the weather cooperates. Even in clear weather it is often muddy, however, so rubber boots are recommended.

Beyond the entrance to the Continental Divide Trail the gravel road continues another 3.4 km to a transmission tower. (Here it is on the Atlantic slope and in Palo Seco Protection Forest in the province of Bocas del Toro.) There is good forest on either side, although it is sometimes over 30 m away because the road follows the buried oil pipeline. This can be a good place to look for soaring raptors such as Barred, Great Black, and Red-tailed Hawks and Ornate Hawk-Eagle. The road can be driven (parts are steep and require 4WD) or walked looking for mixed species flocks and hummingbirds. This is the best place to find **White-bellied Mountain-gem**, and Gray-breasted Crake has been seen and heard in the roadside grass. At 900 m beyond the entrance to the Continental Divide Trail (2.2 km from the Oleoducto Road), a trail goes down on the left behind a dilapidated wooden shed with a corrugated iron roof. It first descends steeply for a short distance, then levels out for several hundred meters as it goes through good fairly open forest with tall trees. Species such as Rufous-rumped Antwren, Rufous-browed Tyrannulet, Black-and-white Becard, Ashy-throated Bush-Tanager, and **Golden-browed Chlorophonia** have been found here.

The Oleoducto Road continues past the Continental Divide to descend steeply through the Palo Seco Protection Forest and the Atlantic foothills of Bocas del Toro to the town of Chiriquí Grande on the coast (35 road miles from the divide). See the section on Bocas del Toro Province (pp. 233-235) for descriptions on birding sites along the road past the Divide.

Getting to Fortuna. Driving time from Panama City on the Interamericana to the turnoff to Fortuna is about 5-6 hours, and it is about another 1½ hours to the Reserve itself. The turnoff is at the village of Chiriquí, 29 km west of San Lorenzo (and 178 km west of Santiago) and 12 km east of David. On the north side of the highway, the turnoff is signposted for Gualaca and Chiriquí Grande; there is a large gas station on the northwest corner of the intersection and the Centro Comercial El Progreso is on the south side. There are no gas stations between Gualaca and the Caribbean coast, so it would be wise to fill up before heading on to the Reserve. The town of Gualaca is at 17 km from the turn off. A road forks off to the left here; take this road to the left instead of continuing straight, which goes into Gualaca proper. (There are a couple of confusing signs here that suggest that Gualaca is to the right and Chiriquí Grande is straight; ignore them and go left.) At 1.4 km beyond the fork, there is a T-junction by a gas station; take the right turn, signposted for Bocas del Toro. In 600 m you will cross a bridge, after which the road turns left, passing a hydroelectric station on the left. (On your return, look for the hydroelectric station, the right turn after it, and the bridge, because it is easy to miss the left turn at the gas station to get back to the Interamericana.) At 36 km from the Interamericana, you pass Los Planes, which formerly housed workers on the hydroelectric project but which is now mostly empty. At about 38 km beyond the Interamericana there is a bus shelter and a gravel road to the left signposted for Chiriquicito. (This is a short cut to Boquete if that site is also on your itinerary.) The "Bienvenidos Reserva Forestal Fortuna" sign is 200 m beyond this turnoff. For the trails within the Reserve, follow the directions in the account above.

There are frequent buses from David to Chiriquí Grande, Almirante, and Changuinola that pass through Fortuna en route, and they will stop at any of the trail heads, accommodations and restaurants mentioned in this section, and the section on Chiriquí Grande and the Oleoducto Road (see pp. 233-235). It is quite feasible to get around in this area by bus if you do not have a car.

Accommodations and meals. The only public accommodations available within the reserve are at Finca La Suiza. However, the area can also be birded as a day trip from David (p. 201), or from the Rancho Ecológico Willy Mazú or Chiriquí Grande (p. 236) on the Bocas del Toro side.

Finca La Suiza (cel 6615-3774; call between 7:00 PM and 9:00 PM; fincalasuiza @hotmail.com; www.panama.net.tc), though small, is one of Panama's most attractive hotels for birders. Operated by a Swiss couple, Herbert Brullmann and Monika Kohler, the lodge has three nicely appointed bedrooms, each with two double beds and with private hot-water bath ($31.00 single and $40.00 double per night for a minimum stay of two nights; higher for one night; credit cards and checks not accepted). Gourmet meals are provided, at $4.00 for breakfast and $11.00 dinner. (Lunch and snacks are not provided, so you may wish to bring your own supplies.) The lodge, at 1,300 m (4,300 ft), provides panoramic views from its terraces, and has a system of well-maintained trails offering 20 hours of hiking (see the section above). Guests pay a single $8.00 trail fee for the length of their stay. The lodge is closed during June, September, and October, the rainiest months. The signposted entrance is at a small meadow 1.1 km beyond the "Bienvenidos Reserva Forestal Fortuna" sign (see above). If they know you are coming the gate will be left open for you, but you must arrive by 6:30 PM since it is locked then. If you have 4WD you may drive up to the lodge. If you have 2WD or arrive by bus you may leave your luggage in the car or at the small house near the entrance and walk up (it is about 800 m) and it will be brought up later. The lodge can easily be reached by bus from David or Changuinola; the fare from David is about $3.00.

There are four or five small restaurants at Valle de la Mina, just past the tollgate 2.4 km above the "Bienvenidos" sign. Meals here are generally very basic standard Panamanian fare.

Bird list for Fortuna and the Oleoducto Road. * = vagrant; R = rare; P = species recorded only from Pacific slope (Fortuna); A = species recorded only from Atlantic slope (Oleoducto Road). Oleoducto Road records include only those from the foothill zone.

Highland Tinamou R
Little Tinamou
Black Guan
Crested Guan R
Great Curassow R
Black-breasted Wood-Quail
Spotted Wood-Quail
Pied-billed Grebe
Brown Pelican
Neotropic Cormorant
Anhinga R
Fasciated Tiger-Heron R
Cattle Egret
Black Vulture
Turkey Vulture
King Vulture
Osprey
Hook-billed Kite A
Swallow-tailed Kite
Double-toothed Kite
Tiny Hawk R
Sharp-shinned Hawk
Cooper's Hawk* P
Bicolored Hawk R
Barred Hawk
White Hawk
Great Black-Hawk
Gray Hawk
Roadside Hawk
Broad-winged Hawk
Short-tailed Hawk
Swainson's Hawk
White-tailed Hawk P R
Red-tailed Hawk
Black Hawk-Eagle
Ornate Hawk-Eagle
Yellow-headed Caracara
Barred Forest-Falcon
American Kestrel
Bat Falcon
White-throated Crake
Gray-breasted Crake A R
Sunbittern
Spotted Sandpiper
Pale-vented Pigeon
Band-tailed Pigeon
Ruddy Pigeon
Ruddy Ground-Dove
White-tipped Dove
Chiriqui Quail-Dove
Buff-fronted Quail-Dove
Purplish-backed Quail-Dove A
Sulphur-winged Parakeet
Crimson-fronted Parakeet
Orange-chinned Parakeet
Red-fronted Parrotlet R
Blue-headed Parrot
White-crowned Parrot A
Squirrel Cuckoo
Striped Cuckoo
Groove-billed Ani
Tropical Screech-Owl
Bare-shanked Screech-Owl
Costa Rican Pygmy-Owl R
Mottled Owl
Black-and-white Owl
Common Pauraque
White-tailed Nightjar P
Black Swift R A
Chestnut-collared Swift
White-collared Swift
Chimney Swift
Vaux's Swift
Gray-rumped Swift A
Band-tailed Barbthroat
Green Hermit
Stripe-throated Hermit
White-tipped Sicklebill
Green-fronted Lancebill
Violet Sabrewing
Green Violet-ear
Violet-headed Hummingbird A
Green Thorntail R
Garden Emerald
Violet-crowned Woodnymph A
Fiery-throated Hummingbird
Charming Hummingbird P
Snowy-bellied Hummingbird
Rufous-tailed Hummingbird
Stripe-tailed Hummingbird
Black-bellied Hummingbird R
White-tailed Emerald
White-bellied Mountain-gem
Purple-throated Mountain-gem
Green-crowned Brilliant
Magnificent Hummingbird
Purple-crowned Fairy A
Magenta-throated Woodstar R
Scintillant Hummingbird
Collared Trogon
Orange-bellied Trogon
Black-throated Trogon
Slaty-tailed Trogon A
Lattice-tailed Trogon
Resplendent Quetzal R
Broad-billed Motmot
Amazon Kingfisher
Green Kingfisher
Lanceolated Monklet R
Rufous-tailed Jacamar A
Red-headed Barbet
Prong-billed Barbet
Blue-throated Toucanet
Collared Aracari A
Keel-billed Toucan
Chestnut-mandibled Toucan A
Olivaceous Piculet
Acorn Woodpecker
Black-cheeked Woodpecker
Red-crowned Woodpecker
Yellow-bellied Sapsucker
Smoky-brown Woodpecker
Rufous-winged Woodpecker R
Golden-olive Woodpecker
Cinnamon Woodpecker A
Lineated Woodpecker A
Pale-breasted Spinetail P
Red-faced Spinetail
Spotted Barbtail
Ruddy Treerunner
Buffy Tuftedcheek
Striped Woodhaunter
Lineated Foliage-gleaner
Scaly-throated Foliage-gleaner
Buff-throated Foliage-gleaner
Ruddy Foliage-gleaner
Streak-breasted Treehunter
Tawny-throated Leaftosser
Plain-brown Woodcreeper
Olivaceous Woodcreeper
Wedge-billed Woodcreeper

Strong-billed Woodcreeper R
Northern Barred-Woodcreeper
Black-banded Woodcreeper R
Cocoa Woodcreeper
Spotted Woodcreeper
Streak-headed Woodcreeper
Spot-crowned Woodcreeper
Brown-billed Scythebill
Fasciated Antshrike
Barred Antshrike
Russet Antshrike
Plain Antvireo
Slaty Antwren
Rufous-rumped Antwren
Immaculate Antbird
Black-headed Antthrush
Rufous-breasted Antthrush
Scaled Antpitta R
Ochre-breasted Antpitta R
Silvery-fronted Tapaculo
White-fronted Tyrannulet
Paltry Tyrannulet
Yellow-bellied Elaenia
Lesser Elaenia
Mountain Elaenia
Torrent Tyrannulet
Olive-striped Flycatcher
Ochre-bellied Flycatcher
Slaty-capped Flycatcher
Rufous-browed Tyrannulet R
Black-capped Pygmy-Tyrant A
Scale-crested Pygmy-Tyrant
Pale-eyed Pygmy-Tyrant P
Common Tody-Flycatcher
Eye-ringed Flatbill
Yellow-margined Flycatcher A
White-throated Spadebill
Sulphur-rumped Flycatcher
Bran-colored Flycatcher
Common Tufted-Flycatcher
Dark Pewee
Olive-sided Flycatcher
Western Wood-Pewee
Eastern Wood-Pewee
Tropical Pewee
Yellow-bellied Flycatcher
Yellowish Flycatcher
Willow/Alder Flycatcher
Black Phoebe
Long-tailed Tyrant
Bright-rumped Attila
Rufous Mourner
Dusky-capped Flycatcher
Great Crested Flycatcher A
Boat-billed Flycatcher
Gray-capped Flycatcher
Golden-bellied Flycatcher
Streaked Flycatcher
Sulphur-bellied Flycatcher
Piratic Flycatcher
Tropical Kingbird
Thrush-like Schiffornis
Rufous Piha
Cinnamon Becard P
White-winged Becard
Black-and-white Becard R
Masked Tityra
Snowy Cotinga R A
Purple-throated Fruitcrow A
Bare-necked Umbrellabird R
Three-wattled Bellbird
White-ruffed Manakin
Lance-tailed Manakin P
White-crowned Manakin
Sharpbill A
Yellow-throated Vireo
Warbling Vireo* A
Yellow-winged Vireo
Brown-capped Vireo
Philadelphia Vireo
Red-eyed Vireo
Yellow-green Vireo
Tawny-crowned Greenlet
Lesser Greenlet
Rufous-browed Peppershrike
Azure-hooded Jay
Gray-breasted Martin
Blue-and-white Swallow
Southern Rough-winged Swallow
Barn Swallow
Band-backed Wren A
Black-throated Wren A
Stripe-breasted Wren A
Rufous-breasted Wren P
Rufous-and-white Wren P
Plain Wren P
House Wren
Ochraceous Wren
White-breasted Wood-Wren
Gray-breasted Wood-Wren
Southern Nightingale-Wren
American Dipper
Long-billed Gnatwren
Black-faced Solitaire
Orange-billed Nightingale-Thrush
Slaty-backed Nightingale-Thrush
Black-headed Nightingale-Thrush A
Swainson's Thrush
Wood Thrush
Mountain Thrush
Pale-vented Thrush
Clay-colored Thrush
White-throated Thrush
Black-and-yellow Silky-flycatcher
Long-tailed Silky-flycatcher
Golden-winged Warbler
Tennessee Warbler
Tropical Parula
Yellow Warbler
Chestnut-sided Warbler
Magnolia Warbler A
Yellow-rumped Warbler
Black-throated Green Warbler
Townsend's Warbler* A
Blackburnian Warbler
Palm Warbler R
Bay-breasted Warbler A
Black-and-white Warbler
American Redstart A
Louisiana Waterthrush
Kentucky Warbler
Mourning Warbler
MacGillivray's Warbler A R
Olive-crowned Yellowthroat A
Gray-crowned Yellowthroat P
Hooded Warbler R
Wilson's Warbler
Canada Warbler
Slate-throated Redstart
Collared Redstart
Golden-crowned Warbler
Rufous-capped Warbler
Black-cheeked Warbler
Three-striped Warbler
Buff-rumped Warbler
Wrenthrush R
Bananaquit
Common Bush-Tanager
Sooty-capped Bush-Tanager
Ashy-throated Bush-Tanager A
Black-and-yellow Tanager
Rosy Thrush-Tanager
Dusky-faced Tanager A
Olive Tanager A
White-shouldered Tanager A
White-lined Tanager A
Red-crowned Ant-Tanager
Hepatic Tanager
Summer Tanager
Scarlet Tanager
Flame-colored Tanager
White-winged Tanager
Crimson-collared Tanager A
Passerini's Tanager A
Flame-rumped Tanager A
Blue-gray Tanager

Palm Tanager A
Blue-and-gold Tanager
Plain-colored Tanager A
Emerald Tanager
Silver-throated Tanager
Speckled Tanager
Bay-headed Tanager
Rufous-winged Tanager
Golden-hooded Tanager
Spangle-cheeked Tanager
Scarlet-thighed Dacnis
Green Honeycreeper A
Red-legged Honeycreeper
Blue-black Grassquit
Variable Seedeater
Yellow-bellied Seedeater
Slate-colored Seedeater
Lesser Seed-Finch
Yellow-faced Grassquit
Slaty Finch
Slaty Flowerpiercer
Sooty-faced Finch
Yellow-green Finch R
White-naped Brush-Finch
Chestnut-capped Brush-Finch
Stripe-headed Brush-Finch
Orange-billed Sparrow A
Black-striped Sparrow
Rufous-collared Sparrow
Streaked Saltator
Buff-throated Saltator
Black-headed Saltator A
Slate-colored Grosbeak A
Black-faced Grosbeak
Black-thighed Grosbeak
Rose-breasted Grosbeak
Eastern Meadowlark P
Yellow-tailed Oriole A
Baltimore Oriole
Chestnut-headed Oropendola
Montezuma Oropendola A
Golden-browed Chlorophonia
Thick-billed Euphonia
Elegant Euphonia
Olive-backed Euphonia A
White-vented Euphonia
Tawny-capped Euphonia
Yellow-bellied Siskin
Lesser Goldfinch

Chorcha Abajo

Most of Panama's western Pacific lowlands have been deforested, but a few patches of remnant woodland remain. Some of these woodlands can be found near the small community of Chorcha Abajo, and are one of the most convenient places to find several regional endemics and other specialties such as Bronzy Hermit, Blue-throated Goldentail, and **Black-hooded Antshrike**, among other species.

After turning off the Carretera Interamericana next to the Centro Comercial El Progreso in the village of Chiriquí (see below), take a left at the first intersection. After about 3 km the asphalt road becomes gravel. This road should be negotiable with 2WD even in the rainy season. Just past 5 km from the highway the road forks; take the right. This first part of the road passes through grassy areas that can be scanned for open-country birds. In about 1 km more you begin to enter the village of Chorcha Abajo, which consists of scattered houses strung along the road for the next 3 km. After about the first kilometer or so after reaching the village the houses become more dispersed and there are patches of woodland which can be birded from the road. The road is mostly lined by barbed wire fence, but you may be able to ask someone for permission to enter their property, which is usually freely given. At 8.4 km from the highway there is a short road to the left, opposite a Pentacostal church, that can be walked. Beyond 9 km there are few houses. The road ends at the small port of Punta Tierra where there is a small dock on a mangrove-lined inlet, 9.6 km from the highway. The forest-covered hill on the other side of the inlet is Cerro Batipa (p. 191). Scan the mangroves for **Yellow-billed Cotinga**. Although we have not seen it at this locality ourselves, the locals say it can occasionally be seen here and along the road, and it is regularly seen at Cerro Batipa.

Getting to Chorcha Abajo. The turnoff for Chorcha Abajo is at the village of Chiriquí, 29 km west of San Lorenzo and 12 km east of David. On the south side of the road, the turnoff is opposite that for Fortuna and next to the Centro Comercial El Progreso. Go around the right side of the store and follow the directions above.

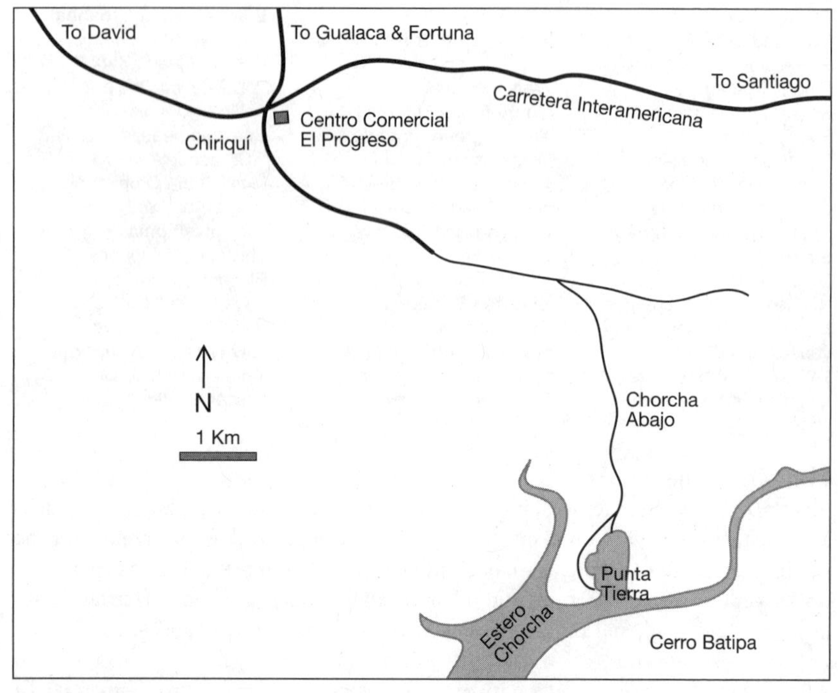

Chorcha Abajo

Accommodations and meals. The Chorcha Abajo area can be birded in an hour or two and is best done as a day trip from David (p. 201). It could also be an alternative if you find the weather bad at Fortuna or Boquete.

Bird list for Chorcha Abajo. R = rare.

Little Tinamou
Great Egret
Spotted Sandpiper
Black Vulture
Turkey Vulture
Yellow-headed Caracara
Ruddy Ground-Dove
Pale-vented Pigeon
Crimson-fronted Parakeet
Red-lored Amazon
Bronzy Hermit
Blue-throated Goldentail
Rufous-tailed Hummingbird
Blue-crowned Motmot
Red-crowned Woodpecker
Lineated Woodpecker
Cocoa Woodcreeper
Streak-headed Woodcreeper
Great Antshrike
Barred Antshrike
Black-hooded Antshrike
Southern Beardless-Tyrannulet
Yellow-crowned Tyrannulet
Mouse-colored Tyrannulet
Lesser Elaenia
Pale-eyed Pygmy-Tyrant
Panama Flycatcher
Great Kiskadee
Rusty-margined Flycatcher
Streaked Flycatcher
Piratic Flycatcher
Tropical Kingbird
Fork-tailed Flycatcher
White-winged Becard
Rose-throated Becard R
Lance-tailed Manakin
Philadelphia Vireo
Yellow-green Vireo
Black-chested Jay
Swainson's Thrush
Clay-colored Thrush
Yellow Warbler
Chestnut-sided Warbler
Crimson-backed Tanager
Golden-hooded Tanager
Blue Dacnis
Red-legged Honeycreeper
Blue-black Grassquit
Streaked Saltator
Great-tailed Grackle
Yellow-crowned Euphonia

David

David is the largest town in western Panama, and has a good range of hotels and restaurants. Although it is better to stay in the highlands if you are birding those areas, David can provide a base for birding lowland sites such as Cerro Batipa and Chorcha Abajo, or for a day trip to Fortuna if Finca La Suiza is not available for an overnight stay.

David is a major transportation hub. If you do not have a car, you can get buses here for the main localities in the highlands, including Boquete and Volcán, as well as the Oleoducto Road to Fortuna and Bocas del Toro.

David Airport

If you plan to fly to David and rent a car, there are daily flights by both Aeroperlas and AirPanama, and the airport is served by most of the major rental-car companies. The airport is south of town. To get to the Carretera Interamericana, turn left after leaving the airport. After about 3.5 km the road changes from a four-lane highway to a one-way two-lane street. At about 4 km from the airport you will reach the Parque Cervantes, the main city plaza. Continue straight for three blocks past the end of the park and turn left at the Centro Comercial 4 Esquinas. It is about 1.5 km more to the Interamericana. A right will take you east for Fortuna, and a left west for Concepción and Volcán-Cerro Punta. Continuing straight across the highway will take you to Boquete.

Accommodations and meals

The *Hotel Castilla* (ph 774-5260; fax 774-5246; hotelcastilla@cwpanama.net) in the center of David is one of the best values in town. It has 64 rooms ($27.00) with private hot-water baths and air-conditioning. There is a restaurant on premises. It is in the center of David, on the southwest corner of the Parque de Cervantes, the main city plaza. If you are coming from the east on the Interamericana, turn left just after the Super Barú supermarket as you come into town. After about 1.5 km turn right near the Centro Comercial 4 Esquinas onto Av. 3 de Noviembre. You will come to the Parque Cervantes in 3 more blocks.

The *Hotel Fiesta* (ph 775-5454, 775-5455), on the Carretera Interamericana, may be convenient if you are just passing through and don't want to go into the center of David. Rooms ($22.00) have private hot-water baths and air-conditioning. There is a restaurant on premises. If you are coming from the east, it is on the right, a bit over 1 km west of where the divided highway starts east of David.

There are several fast food places on the highway as you come into David from the east. There is a small cafeteria, on the same block as the Super Barú supermarket on the side away from the highway, that is also convenient. *Java Juice* has good hamburgers, salads, and fruit drinks. To find it turn off to the left on the block before the Super Barú supermarket (coming from the east). It is on the right after about 600 m, just before a gas station on the left.

If you plan to be staying in cabins or camping in the highlands and need to bring your own food, the *Super Barú* supermarket right on the Interamericana is an excellent place to stock up on supplies. Coming from the east, it is on the left 1.9 km from where the divided highway starts before David, opposite the turnoff to Boquete.

There are many other hotels, including several cheap *pensiones*, as well as a wide variety of restaurants in downtown David; consult one of the standard tourist guides for more choices.

Boquete

The picturesque little mountain town of Boquete, on the east side of Volcán Barú and in the heart of the coffee-growing country of the western highlands, has long been a very popular tourist destination. More recently it has also been attracting retirees from the United States and elsewhere, which has resulted in increased development. Boquete, along with the Volcán area on the west side of the volcano, is one of the best places to find many of the endemics and specialties of the western highlands. Boquete is a good deal wetter than the Volcán area, and as a result a number of species are found here that do not occur or are scarcer on other side, and vice versa.

The best birding areas around Boquete are in and above the Bajo Mono area northwest of town, including trails in the upper Río Caldera Valley and on Finca Lérida. Boquete also offers the best access to the summit of Volcán Barú, Panama's highest mountain, although the peak can also be reached by trail from near Volcán. See the section on Aguacatal (pp. 212-213) for details.

Bajo Mono. The main birding trails in the Bajo Mono area are the Pipa de Agua, Culebra, and Los Quetzales Trails, the latter in Volcán Barú National Park and connecting to the Cerro Punta area. To get to this area, take the main road through town. From the visitor center at the entrance to Boquete, it is about 2.5 km along this road to a church with two steeples, on the right, at the north end of town. About 200 m beyond the church there is a fork marked by a small statue of the Virgin Mary. Take the left fork, signposted for Bajo Lino and Palo Alto. Subsequent distances are given from this fork. At 1.5 km, take the left, signposted for Los Naranjos, at a green bus shelter. At 2.9 km, turn right at a blue bus shelter, signposted for Bajo Mono. This narrow asphalt road shortly afterward passes by Los Ladrillos ("The Bricks"), a geological formation composed of hexagonal basalt, and at 3.6 km crosses a metal cable bridge. At 4.3 km, turn left, signposted for Bajo Mono and the Sendero Los Quetzales. Immediately after this you cross a small metal bridge and enter a part of the canyon of the Río Caldera marked by spectacular formations of columnar basalt. At 7.3 km, you cross a bridge over the Río Caldera, near "The Castle," an ornate house that is a local landmark. At 8.3 km you reach a T-junction; the right turn is signposted for Volcán Barú National Park.

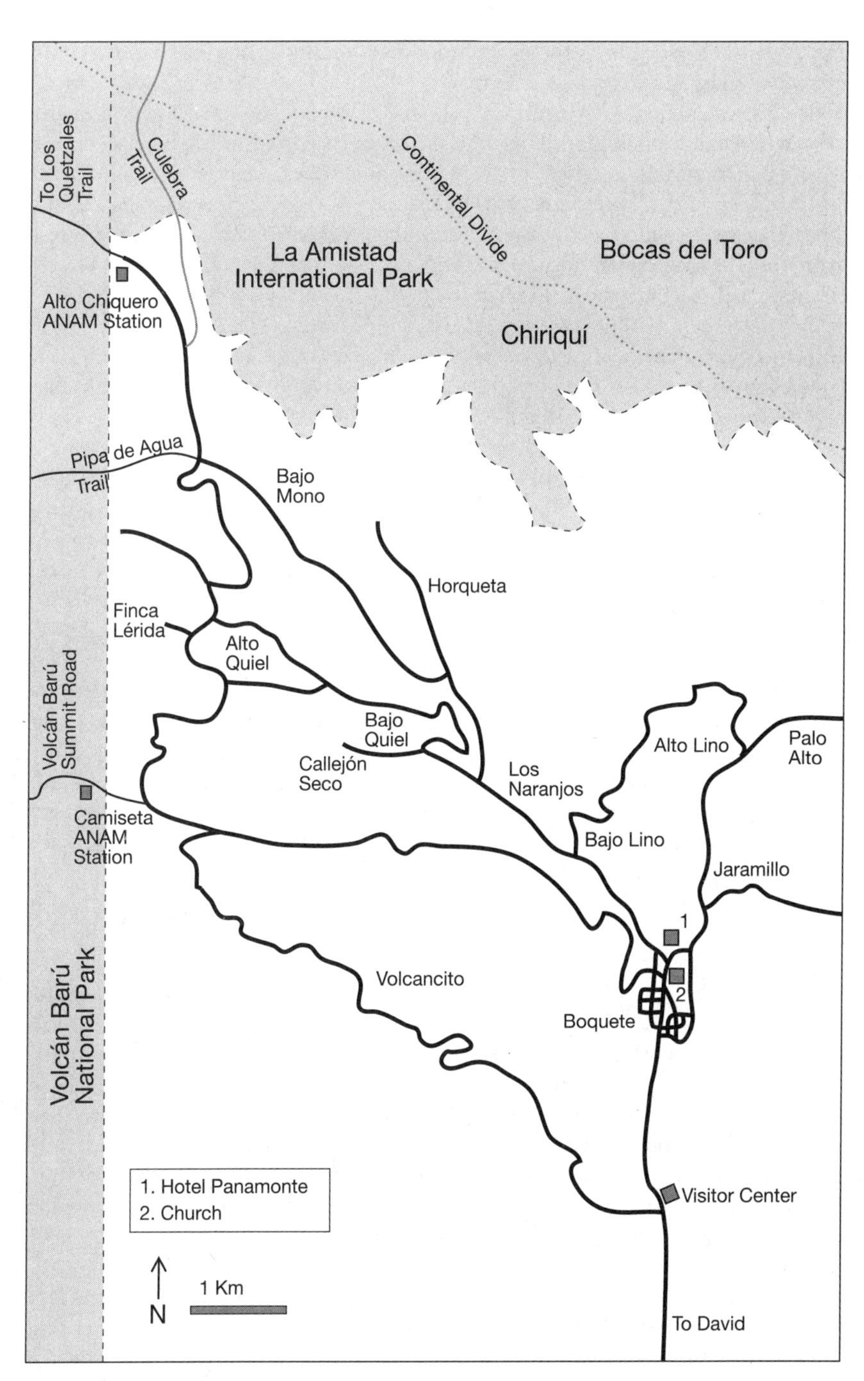

Boquete and environs

The *Pipa de Agua Trail* is the gravel road that continues straight opposite this T-junction. The road provides access to a system of pipes that provides part of Boquete's water supply. It requires 4WD, so if you only have 2WD park near the intersection and walk in. At 100 m from the entrance, there is a small house on the right; stay straight. If you drive in, there is a good place to park after about 600 m next to a section of concrete conduit. The first part of the road passes through fields, forest fragments, and second growth. After about 1.3 km from the start the forest improves. The road extends for several kilometers, starting at about 1,600 m (5.300 ft) elevation and reaching above 2,000 m (6,600 ft).

To get to the Culebra and Los Quetzales Trails, take a right at the T-junction mentioned above. At 1.3 km from this turn, there is a gravel driveway on the right that goes steeply down for about 10 m. Turn in here, where you can park next to a small shack by a swing bridge. Cross the bridge, and take the trail to the left that runs along the stream. This is the *Culebra Trail* (sometimes known as the Holcomb Trail). The first few hundred meters of this trail have recently been widened to provide access to a new resort. The trail rises slowly for about a kilometer as it follows the course of the upper Río Culebra, after which it climbs very steeply and is heavily eroded. The trail eventually crosses the Continental Divide and continues into Bocas del Toro along the upper Río Changuinola.

For the Los Quetzales Trail, continue along the same asphalt road, past the driveway that leads to the Culebra Trail. At 2.8 km from the T-junction, the asphalt ends near the ANAM ranger station at Alto Chiquero. There is a bunkroom here where you can stay; see the section on accommodations for details. Past this point 4WD is required; if you do not have it park here and continue on foot. At 1.0 km beyond the ranger station you pass the "La Chilena" trout farm, where a sign indicates that it is 8 km to Cerro Punta. Shortly after this the road descends very steeply down the Loma de los Lamentos ("Hill of Lamentations"). At about 2.0 km from the ranger station, at a bit over 1,800 m (5,900 ft) elevation, the *Los Quetzales Trail* begins on the left, just before the road reaches another trout farm. (At the time of this writing, a bridge had been washed out a few 100 m in along the trail. ANAM was planning to build another bridge higher up and relocate the start of the trail farther up along the road. Check at the ranger station for current conditions and location of the trail head.) The trail goes over a saddle on the north side of Volcán Barú, reaching over 2,400 m (7,900 ft) before arriving at the El Respingo ranger station above Cerro Punta. (See p. 215 for details on this end of the trail.) If you are interested in walking the entire trail, bear in mind that since Cerro Punta is much higher than Boquete, the hike from the Boquete side to Cerro Punta is much more strenuous than the reverse. Admission to the park is $3.50 for foreign visitors and $1.50 for Panama residents.

The T-junction mentioned above can be reached by bus from Boquete. You can also hire a taxi to take you up as far as the Alto Chiquero ranger station. On your return, you can either arrange to have the taxi meet you at a specified time, or hike down to the T-junction to catch a bus back to town.

Finca Lérida. Finca Lérida is a large coffee plantation high up on the flanks of Volcán Barú. It is mentioned frequently in Ridgely, and is well-known among ornithologists because of the extensive collections made by its owner, Tolef Mönniche, from the 1930s to the 1950s. Now operated by the Collins family, a new birding lodge was opened here in 2006. (Note that, contrary to what is mentioned in Ridgely, the management of the finca is no longer associated with that of the Panamonte Hotel in Boquete.) To reach Finca Lérida, the first part of the route is the same as that to Bajo Mono described in the previous section. At the statue of the Virgin at the north end of town, take the left, signposted for Bajo Lino and Palo Alto. At 1.5 km from this turn take the left for Los Naranjos. At 2.9 km, turn left, signposted for Callejón Seco (the right fork here goes to Bajo Mono). Finca Lérida is about 5.5 km above this turnoff.

Volcán Barú Summit Road. The higher elevations of Volcán Barú are the only place in Panama where **Timberline Wren** and **Volcano Junco** are known to occur, and **Sooty Thrush** is easier to find here than elsewhere. A host of other highland specialties can also be found. The volcano has a large array of communications towers on its summit reached via a service road above Boquete. At the time of this writing the road is in extremely poor condition, with huge gullies and exposed rocks. It requires an experienced driver with a strong 4WD vehicle with very high clearance and a winch to make it to the top. For those not so equipped, you can hire a guide in Boquete to drive you up. Transportes Volcán Barú (Gabriel Mendez, cel 6624-0577) has two high-clearance jeeps equipped to get to the summit. The road can also be hiked, although it is 13.5 km and a 1,600 m (5,300 m) change in elevation to the summit from the ranger station at the park entrance, making a round trip a rather full day.

The Camiseta ranger station is at 1,840 m (6,070 ft). The first part of the road goes through pastures and broken forest, but above, at about 2,000 m (6,600 ft) it enters better forest. **Timberline Wren** first appears with the patches of bamboo that begin at about 3,000 m (9,900 ft), and continue to near the summit. **Volcano Junco**, on the other hand, is restricted mainly to bare areas near the summit itself. **Sooty Thrush** is also common at upper elevations. Other species that are common here include **Volcano Hummingbird**, **Long-tailed** and **Black-and-yellow Silky-flycatchers**, and **Slaty Flowerpiercer**.

To get to the summit road, drive north on the main street to the church with two steeples at the north end of town (2.5 km from the visitor center at the entrance to Boquete). About 100 m past the church, take the left turn at the Folclórica Boutique store. (This road is signposted "Volcán Barú 11 km" *after* the turn; the distance is incorrect both for the ranger station or the summit.) There is a stop sign at the next intersection; go straight across it. The road immediately forks; bear right here. After 6.8 km on this pot-holed asphalt road, there is a fork. Stay right, signposted for Volcán Barú. (The left fork goes to Volcancito.) At 7.6 km, there is fork to the right that goes to Finca Lérida; keep going straight on the main road. Above this point the road becomes much rougher and 4WD is necessary. At 8.5 km is the Camiseta ranger station; the road continues another 13.5 km to the summit from here. As at Alto Chiquero, the entry fee is $3.50 for foreign visitors and $1.50 for Panama residents, but there may not be anyone here to collect it.

Getting to Boquete. The turnoff to Boquete from the Carretera Interamericana is in David, 1.9 km west of where the highway changes from two lanes to four at the east end of the city. It is well signposted, opposite the Super Barú supermarket and just before a Shell station. It is about 37 km from the turnoff to the visitor center at the entrance to Boquete.

Accommodations and meals. As tourist town, Boquete is well supplied with good hotels and excellent restaurants.

The *Coffee Estate Inn (La Montaña y Valle)* (ph/fax 720-2211; information@co feeestateinn.com; www.coffeeestateinn.com), operated by Jane Walker and Barry Robbins, two Canadian expatriates, includes three luxury bungalows ($110.00 single or double occupancy) on a working coffee estate, each with a terrace, kitchen, living room, bedroom, and hot-water bath. Gourmet dinners can be provided Monday-Saturday for an extra charge. The estate itself has a bird list of 130 species, and can arrange guided tours in the Boquete area. It is located in Alto Jaramillo, about 2.5 km from Boquete and 800 m above it; you can get directions when you book.

The *Finca Lérida Bed and Breakfast* (ph/fax 720-2285; info@fincalerida.com; www.fincalerida.com) has recently been opened by the Finca Lérida coffee estate, a well-known birding locality. The bed and breakfast has five rooms and suites ($60.00-$110.00 May 1-October 31; $65.00-$120.00 November 1-April 30). It is also planned to open an 11-room lodge on the property and to offer birding tours and other activities. These plans are still in development, so contact the hotel for current details.

The *Panamonte Inn and Spa* (ph 720-1324, 720-1327; fax: 720-2055; pana mont@cwpanama.net, hlpmonte@cwpanama.net; www.hotelpanamonte.com), is one of Boquete's most well-known institutions. It has 19 attractively furnished rooms ($60.00-$130.00), all with private hot-water bath and air-conditioning. There is an excellent restaurant attached to the hotel. The Panamonte offers various nature and birding tours. The Panamonte is at the end of the main street through town; go right at the statue of the Virgin and it is a short distance beyond.

The *La Huaca Inn* (ph 720 1343, 720-2515; www.lahuacainn.com) has rooms ($40.00, including continental breakfast) with private hot-water bath and air-conditioning. It is located diagonally across from the Panamonte.

The *Kalima Aparthotel* (ph 720-2884; cel 6673-4530; www.kalimasuites.net) has six rooms ($40.00) with private hot-water baths, air-conditioning, and kitchenette. To find it take a right off Av. Central two blocks before the plaza and continue two blocks on this street.

The *Pensión Topaz* (ph/fax 720-1005; schoeb@chiriqui.com) has eight rooms ($15.00), six with private hot-water bath, the other two with shared bath. To find it take a right off Av. Central three blocks before the plaza and continue one block on this street.

The *Hotel Rebequet* (ph 720-1365) has rooms ($15.00-$33.00 depending on size and location) with private hot-water baths. To find it take a right off Av. Central two blocks before the plaza and continue two blocks on this street.

The *Pensión Marilós* (ph 729-1380; marilos66@hotmail.com) has rooms ($15.00) with private hot-water baths. It is opposite the Rebequet; take a right off Av. Central two blocks before the plaza and continue two blocks on this street.

The *ANAM ranger station* at Alto Chiquero has two bunkrooms, each with four beds, and a kitchenette. The cost is $5.00 per person.

The *Bistro Boquete,* in the center of town on the left as you come in about a block past the plaza, serves varied international and American-style cuisine.

Java Juice, on the right as you come in on the block before the plaza, is a good place for burgers, salads, shakes, and fruit juices.

There are many other pensions, hotels, and restaurants in Boquete and the surrounding area. See one of the standard tourist guides for other options.

Bird list for Boquete. See the following section for the bird list for the Boquete-Volcán region.

Volcán and Cerro Punta

The small towns of Volcán (its official name is El Hato de Volcán) and nearby Cerro Punta are at the heart of one of the best areas for the birds of Panama's western highlands. This is one of the easiest places anywhere to see Resplendent Quetzal, by many considered to be the most beautiful bird in the world. Other sought-after species include **Black Guan**, several species of wood-quail and quail-doves, many hummingbirds, **Silvery-fronted Tapaculo**, **Wrenthrush**, and a variety of furnariids and tanagers. The area provides access to trails in both Volcán Barú National Park (14,322 ha/35,390 acres) and La Amistad International Park (207,000 ha/511,508 acres), but many other good birding areas are found lower down, including the Volcán Lakes and the Bambito and Nueva Suiza area between Volcán and Cerro Punta.

As elsewhere in the western highlands, weather can be unpredictable. During the dry season, from December to April, it is often clear in the morning but mist and drizzle start in the afternoon. During the rainy season, it may rain but it can also be beautiful all day. In either season, it is best to be prepared for rain regardless of how the

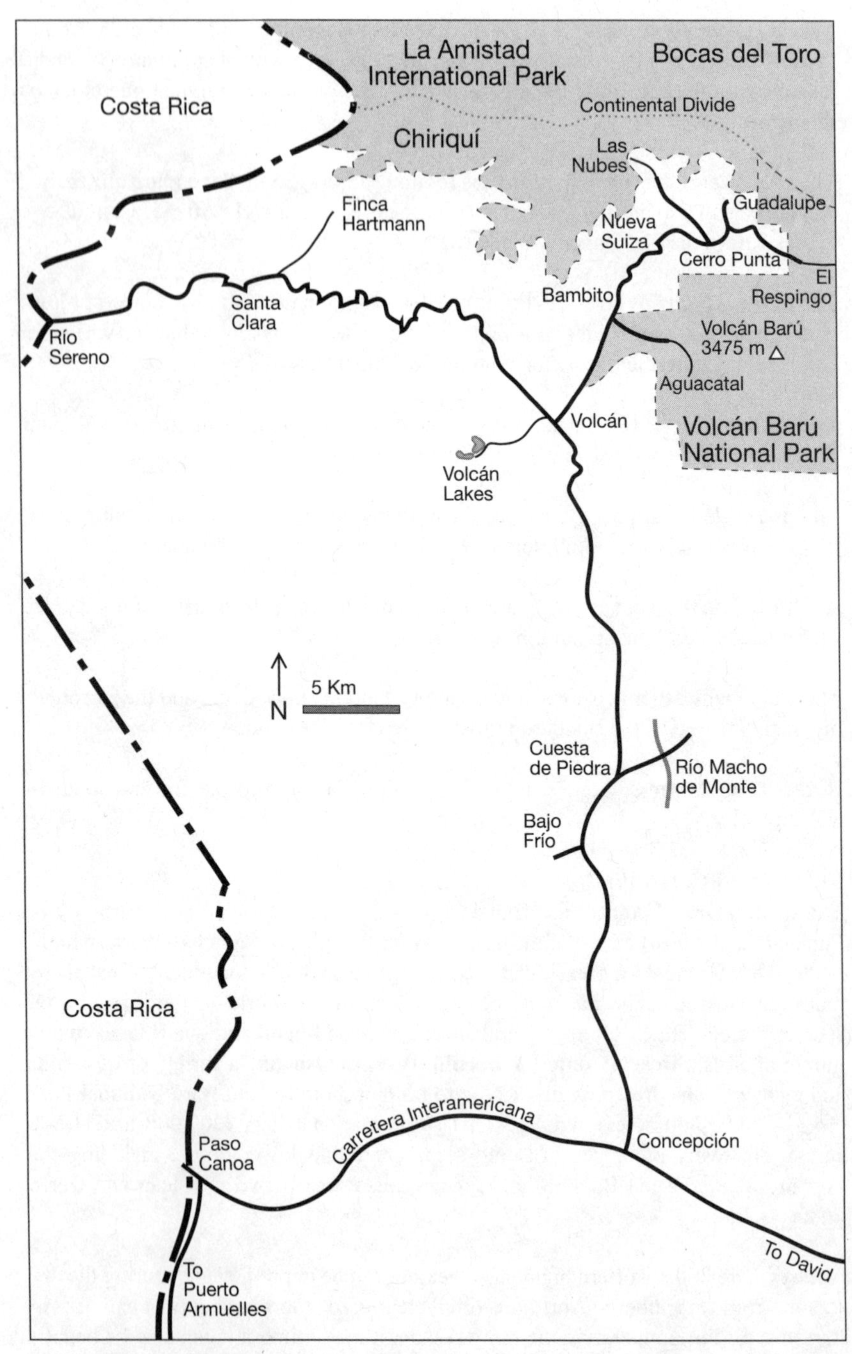

Volcán, Cerro Punta, Santa Clara, and environs

day starts out. Days are generally warm, but nights can be chilly (although it never drops to freezing), so bring a sweater or jacket. Mosquitoes are not a problem, but sometimes other insects such as blackflies can be a nuisance so it is a good idea to bring along insect repellant.

Birding areas around Volcán include sites along the road between Concepción and Volcán, Volcán Lakes near town, and various localities along the road from Volcán to Cerro Punta and the town of Guadalupe. These are discussed in turn below. The road from Concepción divides in at the police station in the center of Volcán, the left fork going to Río Sereno on the Costa Rican border and the right to Cerro Punta. Directions to most of the localities are given with reference to this police station.

Nariño Aizpurúa (ph 771-5049, cel 6704-4251) and Deibys Fonseca (ph 771-4952; cel 6532-5350) are good local bird guides for the area, and guides and tours can also be arranged by a number of the hotels.

Concepción-Volcán Road. Bajo Frío and the Río Macho de Monte are two sites on the road up to Volcán from Concepción on the Carretera Interamericana that can be good for a stopover en route, or as an alternative when the weather is bad at higher elevations. They have a number of species of lower elevations that are not found above, including some of the endemics of the western Pacific lowlands. The road to *Bajo Frío* is 15.5 km above Concepción (if coming from Volcán, it is 16.6 km from the police station), on the left next to a yellow bus shelter. The road has signs for "Bajo Frío" and "Buena Vista." The first 500 m are level and suitable for 2WD; after that it goes steeply downhill and after about 1 km becomes impassable even to 4WD. The site is mentioned mainly because **White-crested Coquette** has been seen several times on the first level stretch of the road. The road to the *Río Macho de Monte* is on the right 19.7 km above Concepción (if coming from Volcán, 12.4 km from the police station), at the town of Cuesta de Piedra, between the El Porvenir restaurant and the minisuper Meinor. Cuesta de Piedra is signposted from the direction of Concepción but not from Volcán. It is 2.5 km on a good asphalt road to a small hydro project in the canyon of the Río Macho de Monte. The canyon is lined with remnant forest at about 850 m (2,500 ft) where a number of endemic species of the western Pacific lowlands can be found, including **Riverside Wren**. Park off the road, and walk over the second small bridge, from which you can see the river running through a narrow slot canyon only 5 m wide in places. Beyond the river, the road climbs steeply and leaves the forest within a few hundred meters. Driving another kilometer brings you to a viewpoint (*mirador público*) where you can look out over the forested cliffs lining the next river valley.

Volcán Lakes (Lagunas de Volcán). These are three small shallow lakes near Volcán at about 1,300 m (4,300 ft) and surrounded by remnant woodland. Since most other forest in this zone has been cleared, it is the best place to find some species not found higher on the slopes, as well as resident and migrant wetland species. Species found here include Northern Jacana, Southern Lapwing, Laughing Falcon, Pale-billed

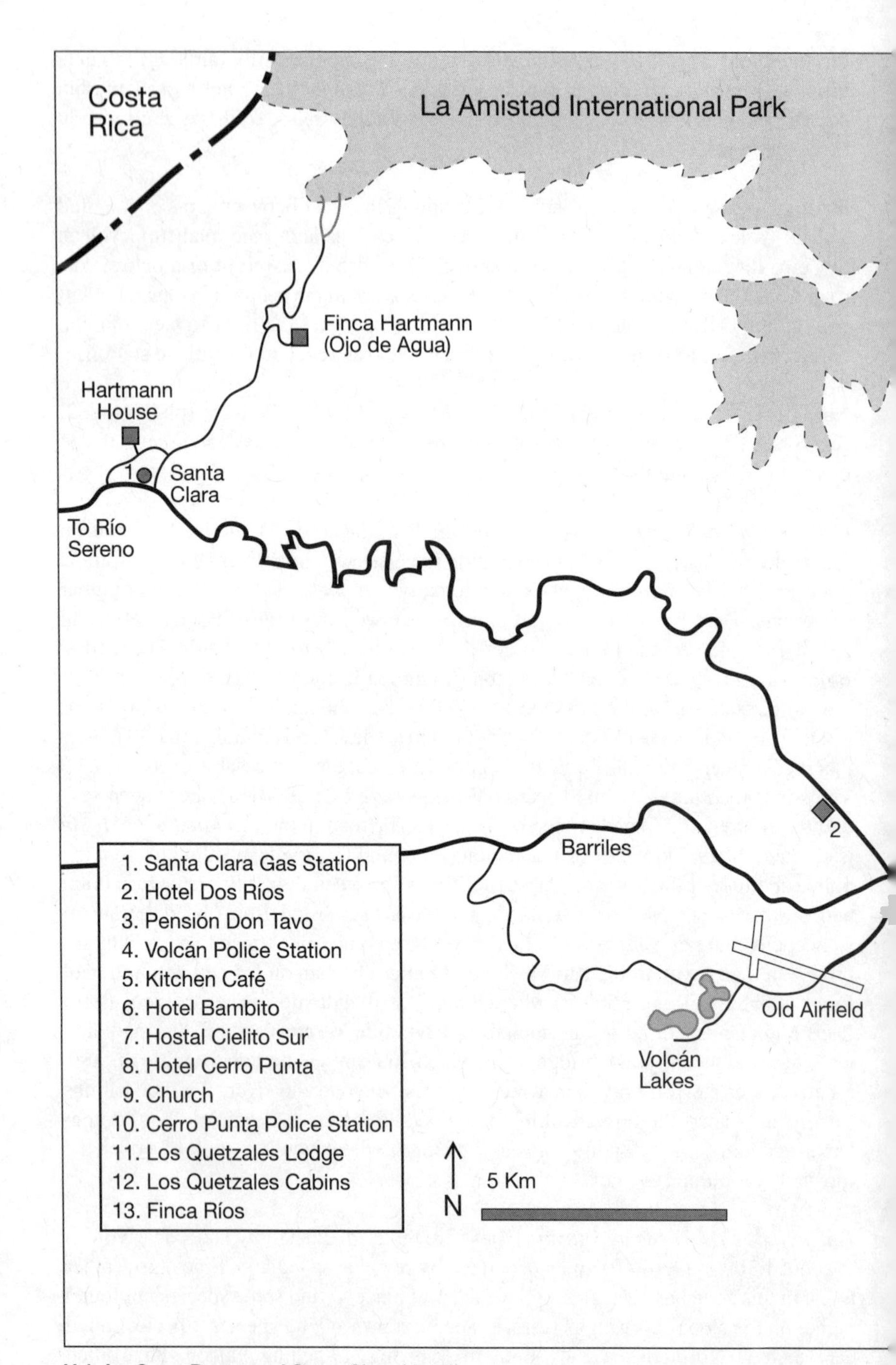

Volcán, Cerro Punta, and Santa Clara (detail)

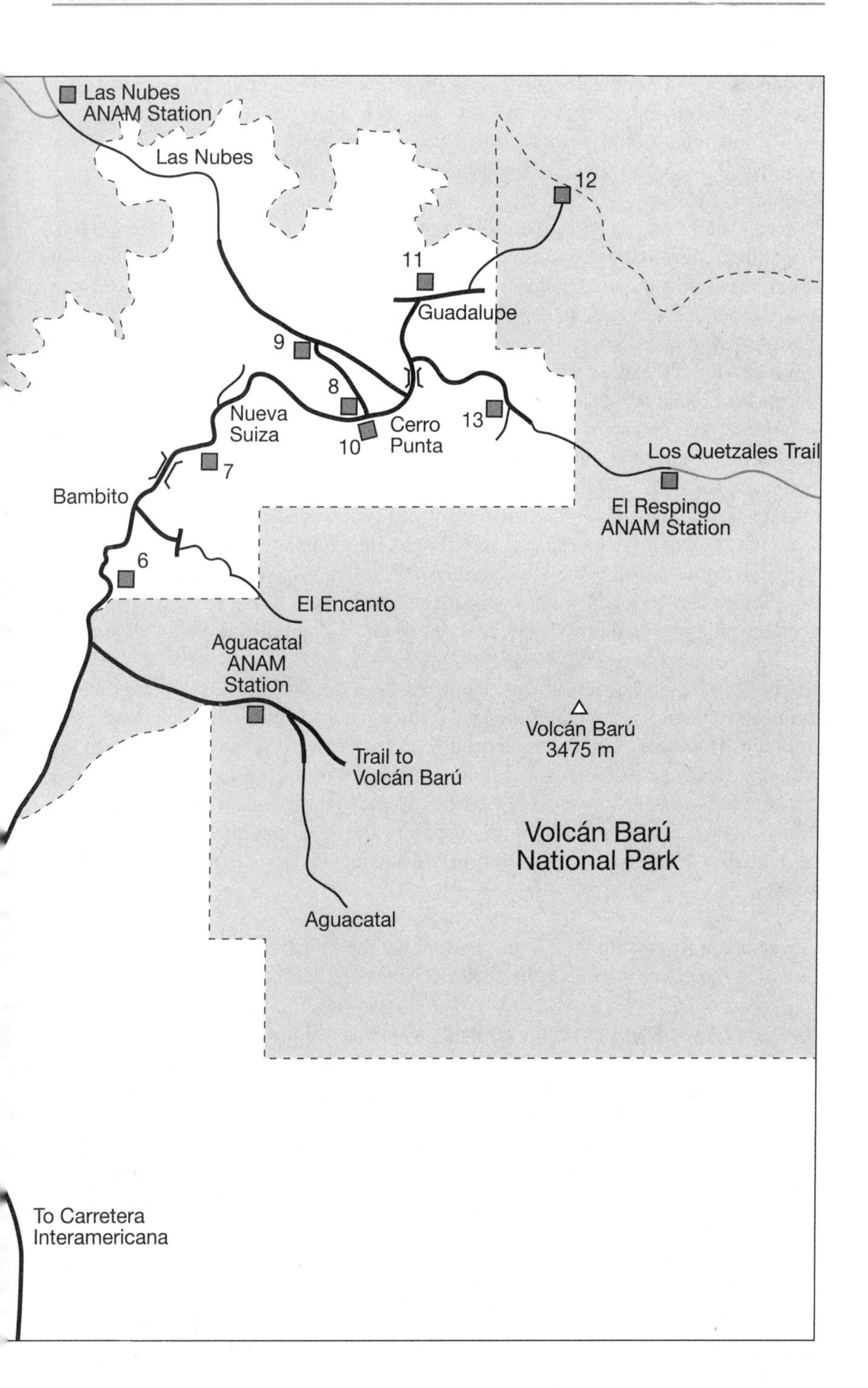
Las Nubes
ANAM Station
Las Nubes
12
11
Guadalupe
9
8
Nueva
Suiza
10
Cerro
Punta
13
Los Quetzales Trail
7
Bambito
El Respingo
ANAM Station
6
El Encanto
Aguacatal
ANAM
Station
Volcán Barú
3475 m
Trail to
Volcán Barú
Volcán Barú
National Park
Aguacatal
To Carretera
Interamericana

Woodpecker, and Slaty Antwren. This is the site in Panama at which Rose-throated Becard has been found most frequently. The form that occurs in Panama does not have a rose throat, and is thought to be a migrant from farther north in Central America.

To reach the lakes, go left at the Volcán police station on the road to Río Sereno. After 700 m, turn left at the Restaurante Yanys, where there are two tires set on top of one another as an advertising sign for a tire shop. At 800 m down this road you will come to a T-junction where you turn right. In less than 100 m the road takes a sharp left at the Restaurante Elia. At 2.2 km after the turn off the Río Sereno Road you reach the old Volcán airstrip. Drive straight across the airstrip, where a rough dirt road continues to the lakes. This road is rocky but may be negotiable with 2WD in the dry season, although it is better to have 4WD or at least high clearance. At 3.4 km from turnoff, the road enters woodland, and in another 300 comes to a sign for "Humedal Lagunas de Volcán," where the road forks. The left fork goes to the lakes. It first parallels the shore of the first lake (mostly screened by trees). After 400 m a fork goes right to private property; go left here. Immediately after this the road forks again; go right to go down to the lake shore. (If you managed to get this far in 2WD, park before this fork and walk down since the road gets worse beyond this point.) Scan the lake for Northern Jacana, Masked Duck, migrant ducks, and other species.

Going right instead of left at the "Humedal Lagunas de Volcán" sign also passes through woodland surrounding the lakes which can be birded. Beyond about 3 km from the "Humedal" sign this road becomes very narrow and runs along a steep hill with a sharp dropoff on one side. Although this part of the road is quite scenic, it is better to park and walk from here (unless you enjoy the prospect of meeting a farm truck on such a road with no room to pass). The road continues for about 9 km beyond the "Humedal" sign before to connecting to the Barriles Road west of Volcán.

The remaining sites in the area are reached via the Volcán-Cerro Punta Road. Take a right at the police station in the center of Volcán to find these sites.

Aguacatal, in Volcán Barú National Park, is an area of forest at the top of one of the largest lava flows on the side of the volcano. About two kilometers north of the Volcán police station, the Volcán-Cerro Punta Road begins to parallel rocky grassy fields that represent the lower end of this lava flow. To get to Aguacatal, turn right 5.4 km from the police station, opposite the minisuper Fu and next to the Restaurante Bambito. Remaining distances are given from this turn. At 2.6 km on this asphalt road, there is a T-junction; turn left here. In another 600 m there is a sign for Volcán Barú National Park, and shortly after that a park ranger station. At 800 m past the ranger station, there is a road to the left that goes about 1 km to a parking area where the trail to the top of Volcán Barú starts. (Note that there is easier access to the summit via the Boquete side; see p. 205 for information. If you decide to do this route, it is recommended to hire a guide because the trail can be confusing and it is easy to take a wrong turn.) For Aguacatal, continue on the main road. At 5.3 km, the asphalt ends, and the gravel road after this requires 4WD. At about this point, the road, which has been passing through open grassy habitat on the lava flow, begins to enter low scrub and patches of forest. At 6.3 km, there is a fork where you keep right; at 6.5 km there is another fork where you keep left. (The right fork here also passes through forest and can be birded if you have time.) At 7.0 km, at about 2100 m elevation, there is a sign for "Finca El Aguacatal (Ecológica)." Although the road can be driven past here, it is convenient to park and continue on foot. The road continues through forest and bamboo for about 700 m to where it terminates at a small shed. A small trail goes down on the right about 50 m to the start of a small water pipe.

After the turnoff to Aguacatal the Volcán-Cerro Punta Road enters the canyon of the Río Chiriquí Viejo. At 6.6 km, on the right, is the large Bambito Hotel (see the section on accommodations below).

El Encanto is a farm that offers rustic camping at the edge of primary forest. There is a wooden lookout tower with impressive views of Volcán Barú and the town of Volcán, and a small trail system. The turnoff from the Volcán-Cerro Punta Road is at 7.5 km north of the police station, on the right next to a vegetable shipping depot with a sign "Venta de Legumbres." This asphalt road goes up 1.2 km to a fork; take the right. About 200 m beyond this take the road that continues sharply up to the left. It is about 2.5 km more on this dirt road to El Encanto. The road is in poor condition, so it is better to contact the owner, Beltrán Martínez (ph 771-4304) who can take you up by pickup truck. Martínez eventually plans to build a cabin at the site; contact him for the current status.

A little less than 9 km from the police station the road crosses a bridge over the Río Chiriquí Viejo. You can pull off the road just after the bridge to check the river for American Dipper and Torrent Tyrannulet. At 10.1 km, on the right, is the Hostal Cielito Sur, which is well-situated for birders (see the section on accommodations below). At 10.6 km there is a side road on the left with a girder bridge over the Chiriquí Viejo. To the right of this bridge, before crossing the river, there is a gravel road that goes upstream. American Dipper and Torrent Tyrannulet can also be seen here, as well as other many other species. Blue Seedeater has been found in bamboo near the end of this road.

At 14 km from Volcán, you enter the town of Cerro Punta. The Hotel Cerro Punta (see the section on accommodations below) is on the left just after you enter town, just before the police station, which is on the right.

Las Nubes, the headquarters for La Amistad ("Friendship") International Park, has several excellent birding trails. This enormous park, shared with Costa Rica, has one of the highest concentrations of endemic species of any protected area in the world. While most of the park area is on the Atlantic slope in Bocas del Toro, it also includes the higher slopes on the Pacific side in western Chiriquí. To get there, take a left just beyond the Cerro Punta police station, signposted for the park with a yellow-and-green ANAM sign. At 1.4 km on this road, you come to a T-junction at a church. Turn right here, then take an immediate left at the next corner, which is signposted for Las Nubes and the park. Although the road beyond this is not in good condition, it can be handled in 2WD. At 2.0 km from Cerro Punta, a dirt road goes left; stay right on the main road. At 2.3 km the road crosses a bridge over the Río Chiriquí Viejo, which is very small at this point. At 3.9 km there is another dirt road on the right; stay on the main road. At 4.6 km the road crosses a girder bridge, and about 600 m beyond this changes to gravel. At 6.0 km it crosses another girder bridge; stay left on the main road after crossing. At 6.9 km you cross another girder bridge; the park entrance is just ahead. The gates are open from about 7:30 AM to 3:00 PM. Immediately after the gates is a small restaurant operated by a local community group (open 8:00 AM-4:00 PM). Just past the restaurant, keep left for park headquarters, which is 600 m from the gate, and at 2,300 m (7,600 ft) elevation. It is possible to stay at the headquarters (see the section on accommodations below). The entrance fee is $3.50 for foreign visitors and $1.50 for Panama residents.

There are three trails from park headquarters. *Panama Verde* is a short 500 m (0.3 mi) loop next to headquarters. *El Retoño,* the best birding trail, is a 2.4 km (1.4 mi) loop, mainly at one level but with some ups and downs. The trail is first along the gravel road for about 300 m, after which it branches off, eventually returning to meet the road again just below headquarters. It is good for species such as **Black Guan**, **Silvery-fronted Tapaculo**, and **Prong-billed Barbet**. The *La Cascada* trail (1.7 km/1.0 mi) is more strenuous, going up to several viewpoints, the highest at 2,500 m (8,250 ft), before reaching a waterfall. There is another trail to Cerro Pichaco, at over 3,000 m (9,900 ft), but ANAM requires that a guide be used if you wish to go there.

Continuing beyond Cerro Punta on the main road, there is a small bridge at 1.4 km past the police station. The left fork goes to the town of Guadalupe and the Los Quetzales Lodge and Cabañas, while the right goes to Bajo Grande, including Finca Ríos, the El Respingo ranger station in Volcán Barú National Park, and the Los Quetzales Trail. These are described in turn below.

Los Quetzales Lodge and Cabañas. The Los Quetzales Cabañas and the trails around them, within Volcán Barú National Park, are one of the premier birding destinations of the western highlands. To get to them, continue on the left fork after the bridge that is 1.4 km past the Cerro Punta police station. About 1 km after the bridge you start to enter the small town of Guadalupe. At 1.3 km there is a T-junction; turn right here and the Los Quetzales Lodge is on the left just beyond. While it is ideal to stay at the cabins themselves, it is possible for non-guests to arrange a visit as well. In either case, you will need to stop and check in at the lodge. The road going to the cabins is on the left about 700 m beyond the lodge. The cabins are a bit over 2 km up this road, which is extremely rocky and requires 4WD. If you only have 2WD, the hotel can take you up. See the section on accommodations for more information on the lodge and cabins.

At the cabins, hummingbird feeders attract a wealth of species, including the large and aggressive Violet Sabrewing and Magnificent Hummingbird and swarms of **White-throated Mountain-gems** and others. Seed feeders bring in **Yellow-thighed** and **Large-footed Finches**, and on rarer occasions have hosted even **Peg-billed** and **Slaty Finches**. In season Quetzals can be seen around the cabins, even nesting on the property. Buffy Tuftedcheeks, **Ruddy Treerunners**, and woodcreepers hunt through the bromeliads on epiphyte-laden trees, and **Black-billed Nightingale-Thrushes** hop about the paths.

Turning right at the small bridge beyond Cerro Punta, signposted for Parque Nacional Volcán Barú, brings you into the valley of Bajo Grande. At 2.3 km past the bridge on the right is *Finca Ríos,* a farm owned by Ade Caballero that is a good place to find Resplendent Quetzal and other specialties. Turn in by the small white house near the road, where you should pay the $2.00 entry fee to the property. Just past the house, you pass between two concrete pillars and then turn left onto a bad 4WD road. If you only have 2WD, park before this and walk up. If you have 4WD, there is a place to pull off and park about 700 m from the entrance and continue to bird on foot.

The asphalt road ends at 2.7 km from the bridge. Beyond the asphalt the gravel road is in bad condition and requires 4WD. At 4.5 km from the bridge above Cerro Punta is the *El Respingo* ranger station, at over 2,400 m (7,900 ft) elevation, which has a bunkhouse where you can stay. (See the accommodations section below). The entrance fee is $3.50 for foreign visitors and $1.50 for Panama residents. The western end of the *Los Quetzales Trail* begins here. The trail extends about 6 km to the Alto Chiquero ranger station above Boquete. See p. 204 for details on this end of the trail. (Note that this trail is often referred to as the "Boquete Trail above Cerro Punta" in Ridgely.) The first part of the trail is fairly level, but it has several steep sections later, some of them on wooden staircases, before reaching a lookout point over the upper Río Caldera Valley above Boquete. It then descends steeply into this valley. Bear in mind that if you plan to hike all the way to Boquete, the Alto Chiquero ranger station is still 11 km by road from town. However, at 2.8 km from the station you reach a road junction where you can catch a bus for Boquete.

Getting to the Volcán-Cerro Punta area. From the Carretera Interamericana, the turnoff for Volcán is in the town of Concepción, 25 km west of David. The turnoff is indicated by a sign for the Bambito Hotel. From here it is 32 km by two-lane asphalt road to Volcán. At the entrance to Volcán is a police station where the road forks. The left fork goes to Volcán Lakes and to Río Sereno; the right fork goes to Bambito, Nueva Suiza, and Cerro Punta. Volcán Barú National Park, La Amistad International Park, and the Los Quetzales Cabañas are reached from Cerro Punta. Follow the more detailed directions for each site given above.

Accommodations and meals

The *Los Quetzales Lodge and Spa* and *Cabañas Los Quetzales* (ph 771-2182, 771-2291; cel 6671-2182; fax 771-2226; stay@losquetzales.com, information@los quetzales.com; www.losquetzales.com) provide a superb base for birding the

Volcán/Cerro Punta area. They are owned by Carlos Alfaro, a prominent local conservationist. The lodge, in the town of Guadalupe, includes 10 rooms ($55.00-$65.00), 5 suites ($80.00-$90.00), and a duplex chalet near the river ($80.00 per floor), all with private hot-water baths. The chalet is particularly suitable for birding. In a separate annex there are also two hostal-type bunkrooms with shared bath ($12.00 per person), plus a family-type room with six beds ($70.00 or $16.00 per person for couples). Rates include continental breakfast. There is a restaurant and a pizzeria at the hotel. Packages are available including meals and guided birding tours. The *Cabañas* are four charming chalet-style cabins located in cloud forest in a reserve within Volcán Barú National Park, reached by a rough 4WD road. The cabins have kitchens and hot-water baths, but no electricity (lanterns are furnished at night). One cabin rents by the floor ($99.00 each); the others cost $137.50-$165.00 depending on size. There is also a more basic geodesic dome that rents for $50.00. You may either bring your own food, have meals catered from the lodge, or hire someone to cook at the cabins. Those without 4WD can get transportation from the lodge. Birding around the cabins is superb, and there are several trails in the reserve. Resplendent Quetzals nest on the grounds, and can sometimes be seen from the patios of the cabins. The lodge also rents a fifth cabin in the Bajo Grande area.

The *Hostal Cielito Sur Bed and Breakfast* (ph 771-2038; stay@cielitosur.com; www.cielitosur.com), owned by Janet and Glenn Lee, is located in Nueva Suiza, in the canyon of the Río Chiriquí Viejo about 10 kilometers above Volcán. There are four very pleasant rooms ($65.00-$75.00 May-September; $70-$80.00 November-April; closed October; breakfast included), all with private hot-water bath. The inn also has a large parlor with fireplace, dining room, and terrace. Birding is good on the attractively planted grounds and in the canyon around the inn, and birding tours can be arranged.

The *Hotel Bambito* (ph 771-4373, 771-4374; fax 771-4207; bambito@chiriqui.com; ventas5@elpanama.com; www.hotelbambito.com) is a large luxury hotel at the beginning of the canyon of the Río Chiriquí Viejo 6.6 kilometers above Volcán. The 45 rooms ($88.00) have private hot-water bath, and there is a restaurant on premises.

The *Hotel Cerro Punta* (ph 771-2020) is a moderately priced yet comfortable place to stay in Cerro Punta. It has 10 rooms ($22.00) all with private hot-water bath. There is a restaurant on premises. The hotel is located in the center of town just before the police station.

The *Hotel Dos Ríos* (ph 771-5555, 771-4271) is one of the fancier places to stay in the vicinity of Volcán. Its 14 rooms, 2 suites, and 2 bungalows (starting at $44.00 May-September, $58.00 October-April, including continental breakfast) all have private hot-water bath, and there is a restaurant on premises. It is located 2.7 km from the Volcán police station on the Río Sereno Road.

The *Pensión Don Tavo* (ph 771-5144) is the best-value moderately priced hotel in Volcán. It has 16 rooms ($25.00) with private hot-water bath. There is a Mexican restaurant on premises. The hotel is located on the right 1.0 km from the Volcán police station on the Río Sereno road.

There are bunkrooms and kitchen facilities available at both the *Las Nubes* ANAM station at La Amistad International Park and at the ANAM station at *El Respingo* in Volcán Barú National Park. There are 14 bunks in 2 rooms at Las Nubes and 7 bunks and 2 cots at El Respingo; cold-water baths. Bring a sleeping bag or your own bedding. Call the ANAM regional office in David (ph 774-6671) or in Volcán (ph 771-5383) to make reservations. You can also pitch a tent near the stations. The cost is $5.00 either for the bunkrooms or to camp, plus the park entry fee of $1.50 for Panama residents and $3.50 for foreign visitors.

Rustic camping can be arranged at *El Encanto* on the Volcán-Cerro Punta Road by calling Beltrán Martínez (ph 771-4304).

There is a wide variety of other accommodations, ranging from cheap cabins to luxury resorts, in the Volcán-Cerro Punta area. See one of the standard tourist guides for more information.

In addition to the restaurants at the Don Tavo, Dos Ríos, and Hotel Bambito, there are several others available in Volcán. The *Kitchen Café,* on the left as you come into town, serves international and American-style food. The *Acrópolis,* on a side street just off the Río Sereno Road, serves Greek food. The small restaurant next to the gas station opposite the police station in the center of Volcán serves breakfast starting at 6:00 AM (more or less).

In the Cerro Punta area there are restaurants at the Los Quetzales Lodge and the Hotel Cerro Punta. The *Restaurant Cerro Punta* just north of the police station is open for breakfast from 6:00 AM.

Just inside the *Las Nubes* entrance to La Amistad International Park there is a restaurant run by a local community group serving good local food, open from 8:00 AM-4:00 PM Monday-Saturday. Meals can be arranged at other hours for groups staying at the bunkhouse at the ANAM station.

Bird list for the Boquete-Volcán region. * = vagrant; R = rare.

Great Tinamou
Highland Tinamou R
Little Tinamou
Blue-winged Teal
Ring-necked Duck R
Lesser Scaup R
Masked Duck
Crested Guan R
Black Guan
Great Curassow R
Black-breasted Wood-Quail
Spotted Wood-Quail
Least Grebe
Pied-billed Grebe
Neotropic Cormorant
Great Blue Heron
Great Egret
Snowy Egret
Little Blue Heron
Tricolored Heron
Cattle Egret
Green Heron
Yellow-crowned Night-Heron
Black Vulture

Turkey Vulture
King Vulture
Osprey
Hook-billed Kite
Swallow-tailed Kite
White-tailed Kite
Double-toothed Kite
Plumbeous Kite
Sharp-shinned Hawk
Bicolored Hawk R
Barred Hawk
White Hawk
Gray Hawk
Great Black-Hawk
Roadside Hawk
Broad-winged Hawk
Short-tailed Hawk
Swainson's Hawk
White-tailed Hawk
Red-tailed Hawk
Black-and-white Hawk-Eagle R
Black Hawk-Eagle
Ornate Hawk-Eagle
Barred Forest-Falcon
Collared Forest-Falcon
Yellow-headed Caracara
American Kestrel
Merlin
Bat Falcon
White-throated Crake
Gray-necked Wood-Rail
Sora
Purple Gallinule
Common Moorhen
American Coot
Southern Lapwing
Killdeer
Northern Jacana
Greater Yellowlegs
Solitary Sandpiper
Spotted Sandpiper
Wilson's Snipe
Rock Pigeon
Scaled Pigeon
Band-tailed Pigeon
Ruddy Pigeon
Short-billed Pigeon
Mourning Dove
Ruddy Ground-Dove
Blue Ground-Dove
Maroon-chested Ground-Dove R
White-tipped Dove
Gray-chested Dove
Chiriqui Quail-Dove
Buff-fronted Quail-Dove
Ruddy Quail-Dove
Sulphur-winged Parakeet
Crimson-fronted Parakeet
Brown-throated Parakeet
Barred Parakeet
Orange-chinned Parakeet
Red-fronted Parrotlet R
Brown-hooded Parrot
Blue-headed Parrot
White-crowned Parrot
Red-lored Amazon
Black-billed Cuckoo
Yellow-billed Cuckoo
Squirrel Cuckoo
Striped Cuckoo
Pheasant Cuckoo
Smooth-billed Ani
Barn Owl
Tropical Screech-Owl
Vermiculated Screech-Owl
Bare-shanked Screech-Owl
Crested Owl
Spectacled Owl
Costa Rican Pygmy-Owl R
Mottled Owl
Black-and-white Owl
Unspotted Saw-whet Owl R
Short-tailed Nighthawk
Common Pauraque
Chuck-will's-widow
Whip-poor-will*
Dusky Nightjar
Chestnut-collared Swift
White-collared Swift
Vaux's Swift
Lesser Swallow-tailed Swift
Bronzy Hermit R
Green Hermit
Long-billed Hermit
Stripe-throated Hermit
Green-fronted Lancebill R
Violet Sabrewing
White-necked Jacobin
Brown Violet-ear R
Green Violet-ear
Violet-headed Hummingbird
White-crested Coquette R
Garden Emerald
Violet-crowned Woodnymph
Fiery-throated Hummingbird
Charming Hummingbird
Snowy-bellied Hummingbird
Rufous-tailed Hummingbird
Stripe-tailed Hummingbird
Black-bellied Hummingbird
White-tailed Emerald
Snowcap
White-throated Mountain-gem
Green-crowned Brilliant
Magnificent Hummingbird
Purple-crowned Fairy
Long-billed Starthroat
Magenta-throated Woodstar
Volcano Hummingbird
Scintillant Hummingbird
Violaceous Trogon
Collared Trogon
Orange-bellied Trogon
Resplendent Quetzal
Blue-crowned Motmot
Ringed Kingfisher
Green Kingfisher
White-necked Puffbird
Rufous-tailed Jacamar
Red-headed Barbet
Prong-billed Barbet
Blue-throated Toucanet
Fiery-billed Aracari
Yellow-eared Toucanet
Chestnut-mandibled Toucan
Olivaceous Piculet
Acorn Woodpecker
Red-crowned Woodpecker
Yellow-bellied Sapsucker R
Hairy Woodpecker
Smoky-brown Woodpecker
Golden-olive Woodpecker
Lineated Woodpecker
Pale-billed Woodpecker
Pale-breasted Spinetail
Slaty Spinetail
Red-faced Spinetail
Spotted Barbtail
Ruddy Treerunner
Buffy Tuftedcheek
Striped Woodhaunter
Lineated Foliage-gleaner
Scaly-throated Foliage-gleaner
Buff-fronted Foliage-gleaner
Buff-throated Foliage-gleaner
Ruddy Foliage-gleaner
Streak-breasted Treehunter
Plain Xenops
Streaked Xenops
Tawny-throated Leaftosser
Gray-throated Leaftosser R
Scaly-throated Leaftosser
Tawny-winged Woodcreeper
Ruddy Woodcreeper
Olivaceous Woodcreeper
Wedge-billed Woodcreeper
Strong-billed Woodcreeper R
Black-banded Woodcreeper R
Spotted Woodcreeper
Streak-headed Woodcreeper

Spot-crowned Woodcreeper
Brown-billed Scythebill R
Plain Antvireo
Slaty Antwren
Dusky Antbird
Immaculate Antbird
Bicolored Antbird
Black-faced Antthrush
Scaled Antpitta R
Streak-chested Antpitta
Ochre-breasted Antpitta R
Silvery-fronted Tapaculo
Yellow Tyrannulet
Greenish Elaenia
Yellow-bellied Elaenia
Lesser Elaenia
Mountain Elaenia
Torrent Tyrannulet
Olive-striped Flycatcher
Slaty-capped Flycatcher
White-fronted Tyrannulet R
Paltry Tyrannulet
Scale-crested Pygmy-Tyrant
Common Tody-Flycatcher
Eye-ringed Flatbill
Yellow-olive Flycatcher
Yellow-margined Flycatcher
White-throated Spadebill
Bran-colored Flycatcher
Common Tufted-Flycatcher
Olive-sided Flycatcher
Dark Pewee
Ochraceous Pewee R
Western Wood-Pewee
Eastern Wood-Pewee
Yellow-bellied Flycatcher
Acadian Flycatcher
White-throated Flycatcher
Hammond's Flycatcher*
Yellowish Flycatcher
Black-capped Flycatcher
Black Phoebe
Vermilion Flycatcher*
Bright-rumped Attila
Rufous Mourner
Dusky-capped Flycatcher
Panama Flycatcher
Great Kiskadee
Boat-billed Flycatcher
Social Flycatcher
Gray-capped Flycatcher
Golden-bellied Flycatcher
Streaked Flycatcher
Piratic Flycatcher
Tropical Kingbird
Fork-tailed Flycatcher
Thrush-like Schiffornis
Barred Becard
White-winged Becard
Black-and-white Becard R
Rose-throated Becard R
Masked Tityra
Bare-necked Umbrellabird R
Three-wattled Bellbird
White-ruffed Manakin
Lance-tailed Manakin
Sharpbill R
Yellow-throated Vireo
Blue-headed Vireo*
Yellow-winged Vireo
Brown-capped Vireo
Philadelphia Vireo
Red-eyed Vireo
Yellow-green Vireo
Tawny-crowned Greenlet
Lesser Greenlet
Rufous-browed Peppershrike
Black-chested Jay
Silvery-throated Jay
Gray-breasted Martin
Mangrove Swallow R
Violet-green Swallow R
Blue-and-white Swallow
Southern Rough-winged Swallow
Barn Swallow
Riverside Wren R
Rufous-breasted Wren
Rufous-and-white Wren
Plain Wren
House Wren
Ochraceous Wren
Timberline Wren
White-breasted Wood-Wren
Gray-breasted Wood-Wren
Southern Nightingale-Wren
American Dipper
Long-billed Gnatwren
Tropical Gnatcatcher
Black-faced Solitaire
Black-billed Nightingale-Thrush
Orange-billed Nightingale-Thrush
Slaty-backed Nightingale-Thrush
Ruddy-capped Nightingale-Thrush
Veery R
Gray-cheeked Thrush R
Swainson's Thrush
Wood Thrush R
Sooty Thrush
Mountain Thrush
Clay-colored Thrush
White-throated Thrush
Gray Catbird
Tropical Mockingbird*
Cedar Waxwing R
Black-and-yellow Silky-flycatcher
Long-tailed Silky-flycatcher
Blue-winged Warbler R
Golden-winged Warbler
Tennessee Warbler
Nashville Warbler*
Flame-throated Warbler
Northern Parula*
Tropical Parula
Yellow Warbler
Chestnut-sided Warbler
Magnolia Warbler
Cape May Warbler R
Black-throated Blue Warbler R
Yellow-rumped Warbler
Golden-cheeked Warbler*
Black-throated Green Warbler
Townsend's Warbler*
Hermit Warbler*
Blackburnian Warbler
Prairie Warbler*
Bay-breasted Warbler
Black-and-white Warbler
American Redstart
Prothonotary Warbler
Worm-eating Warbler
Ovenbird
Northern Waterthrush
Louisiana Waterthrush
Kentucky Warbler
Mourning Warbler
MacGillivray's Warbler
Common Yellowthroat
Masked Yellowthroat
Gray-crowned Yellowthroat
Wilson's Warbler
Canada Warbler
Slate-throated Redstart
Collared Redstart
Golden-crowned Warbler
Rufous-capped Warbler
Black-cheeked Warbler
Three-striped Warbler
Buff-rumped Warbler
Wrenthrush
Bananaquit
Common Bush-Tanager
Sooty-capped Bush-Tanager
Gray-headed Tanager
White-throated Shrike-Tanager R

White-lined Tanager
Red-crowned Ant-Tanager
Hepatic Tanager
Summer Tanager
Scarlet Tanager
Western Tanager R
Flame-colored Tanager
White-winged Tanager
Crimson-backed Tanager
Cherrie's Tanager
Blue-gray Tanager
Palm Tanager
Silver-throated Tanager
Speckled Tanager
Bay-headed Tanager
Golden-hooded Tanager
Spangle-cheeked Tanager
Scarlet-thighed Dacnis
Green Honeycreeper
Shining Honeycreeper
Red-legged Honeycreeper
Blue-black Grassquit
Variable Seedeater
Yellow-bellied Seedeater
Blue Seedeater R
Yellow-faced Grassquit
Slaty Finch R
Peg-billed Finch R
Slaty Flowerpiercer
Sooty-faced Finch R
Yellow-thighed Finch
Large-footed Finch
White-naped Brush-Finch
Chestnut-capped Brush-Finch
Stripe-headed Brush-Finch
Orange-billed Sparrow
Black-striped Sparrow
Lincoln's Sparrow*
Rufous-collared Sparrow
Volcano Junco
Streaked Saltator
Buff-throated Saltator
Black-thighed Grosbeak
Rose-breasted Grosbeak
Blue-black Grosbeak
Indigo Bunting
Red-breasted Blackbird
Eastern Meadowlark
Great-tailed Grackle
Bronzed Cowbird
Orchard Oriole
Baltimore Oriole
Yellow-billed Cacique
Chestnut-headed Oropendola
Yellow-crowned Euphonia
Thick-billed Euphonia
Yellow-throated Euphonia R
Elegant Euphonia
Spot-crowned Euphonia R
Golden-browed Chlorophonia
Yellow-bellied Siskin
Lesser Goldfinch

Santa Clara and the Río Sereno Road

The Río Sereno Road runs about 40 km westward from Volcán to the border crossing for Costa Rica at the town of Río Sereno. It mostly traverses pasture and other farmland with scattered bushes and trees and occasional woodland, from an elevation of 1,400 m (4,600 ft) at Volcán to 900 m (3,000 ft) at Río Sereno. The road can be good for foothills species not found around Volcán itself. There are areas for roadside birding at 7.8-8.2 km and at 13.8 km from Volcán. At 15.6 km there is a small waterfall sometimes frequented by hummingbirds. (If birding along the road, be careful of traffic, which includes large vegetable and cattle trucks.)

The best birding in this area is at Finca Hartmann, above the small town of Santa Clara. The Hartmann family owns two shade coffee plantations in the area, and has made part of the forest on their property a forest reserve. This is the best place in Panama to see **Turquoise Cotinga**, and **White-crested Coquette**, **Fiery-billed Aracari**, and Smoky-brown Woodpecker can also be found.

Santa Clara is marked by a gas station and grocery store at 26.5 km from Volcán. A few hundred meters beyond the gas station there is a sign for Finca Hartmann on the left side of the road; the gravel road that goes up to their property is on the right side of the road opposite the sign. The Hartmann house is about 1.6 km up this gravel road, on their lower coffee plantation, Palo Verde. The visitor's cabins are several kilometers farther up, on a higher property called Ojo de Agua. The gravel road requires 4WD, but if you do not have it the Hartmanns can take you up. There are several roads and trails through shade coffee plantations and forest that are good for birding, including one that goes up to La Amistad International Park at 2,200 m (7,260 ft).

The patriarch of the clan is Ratibor Hartmann, now in his 80s. Ratibor, a naturalist, in his youth worked with Alexander Wetmore on his many expeditions to study Panama's birds, and also with the Gorgas Memorial Laboratory. Keenly interested in the natural history of the area, he maintains a small museum with specimens of insects, reptiles and amphibians, and other animals of the area.

Accommodations and meals

Finca Hartmann (ph/fax 775-5223; cel 6676-3975; fincahartmann@hotmail.com or alice@fincahartmann.com; www.fincahartmann.com) has two cabins on the Ojo de Agua property. The larger rustic research station has five bedrooms ($20.00 per person), a kitchen with gas stove, a large living/dining room with fireplace, and a shared bathroom with hot water. The smaller cabin ($50.00 per day) is more comfortable, with a bedroom, kitchenette, and hot-water bath. There is no electricity; kerosene lamps are provided. While it is best to bring your own food, meals can be catered by the Hartmanns. If you do not have 4WD, transport from the Hartmanns' lower property in Santa Clara can be provided for $20.00 round trip. Day visits to the property are $25.00, which includes lunch. Santa Clara can be reached by bus from Volcán.

Bird list for Finca Hartmann. R = rare.

Great Tinamou R
Highland Tinamou R
Little Tinamou
Black Guan
Crested Guan R
Black-breasted Wood-Quail R
Spotted Wood-Quail
Cattle Egret
Black Vulture
Turkey Vulture
King Vulture R
Osprey R
Swallow-tailed Kite
White-tailed Kite R
Double-toothed Kite R
Sharp-shinned Hawk R
Barred Hawk
Great Black-Hawk R
Roadside Hawk
Broad-winged Hawk
Short-tailed Hawk
Swainson's Hawk
Red-tailed Hawk R
Ornate Hawk-Eagle R
Barred Forest-Falcon
Collared Forest-Falcon
Yellow-headed Caracara
Laughing Falcon
American Kestrel
Bat Falcon
Gray-necked Wood-Rail R
Scaled Pigeon R
Band-tailed Pigeon
White-tipped Dove
Chiriqui Quail-Dove
Ruddy Quail-Dove
Crimson-fronted Parakeet
Sulphur-winged Parakeet
Barred Parakeet
Orange-chinned Parakeet
Brown-hooded Parrot
Blue-headed Parrot
White-crowned Parrot
Red-lored Amazon R
Yellow-billed Cuckoo R
Black-billed Cuckoo R
Squirrel Cuckoo
Smooth-billed Ani
Crested Owl R
Mottled Owl
Black-and-white Owl R
Short-tailed Nighthawk
Common Pauraque
White-collared Swift
Vaux's Swift
Bronzy Hermit R
Green Hermit
Long-billed Hermit R
Stripe-throated Hermit
Green-fronted Lancebill R
Scaly breasted Hummingbird R
Violet Sabrewing
Brown Violet-ear R
Green Violet-ear
White-crested Coquette
Garden Emerald
Violet-crowned Woodnymph R
Fiery-throated Hummingbird R
Charming Hummingbird R
Snowy-bellied Hummingbird
Rufous-tailed Hummingbird
Stripe-tailed Hummingbird
White-tailed Emerald
White-throated Mountain-gem
Green-crowned Brilliant
Magnificent Hummingbird
Purple-crowned Fairy
Long-billed Starthroat
Magenta-throated Woodstar R
Scintillant Hummingbird
Violaceous Trogon R
Collared Trogon
Resplendent Quetzal
Blue-crowned Motmot
Green Kingfisher R
Red-headed Barbet
Prong-billed Barbet
Blue-throated Toucanet
Fiery-billed Aracari
Chestnut-mandibled Toucan R
Olivaceous Piculet
Acorn Woodpecker
Red-crowned Woodpecker
Yellow-bellied Sapsucker R
Hairy Woodpecker
Smoky-brown Woodpecker

Golden-olive Woodpecker
Lineated Woodpecker
Pale-billed Woodpecker
Pale-breasted Spinetail R
Slaty Spinetail
Red-faced Spinetail
Spotted Barbtail
Ruddy Treerunner R
Buffy Tuftedcheek R
Lineated Foliage-gleaner
Spectacled Foliage-gleaner
Buff-fronted Foliage-gleaner R
Buff-throated Foliage-gleaner
Ruddy Foliage-gleaner R
Streak-breasted Treehunter R
Streaked Xenops R
Tawny-throated Leaftosser
Gray-throated Leaftosser R
Tawny-winged Woodcreeper R
Ruddy Woodcreeper
Olivaceous Woodcreeper
Wedge-billed Woodcreeper
Spotted Woodcreeper
Streak-headed Woodcreeper
Spot-crowned Woodcreeper
Russet Antshrike
Plain Antvireo
Slaty Antwren
Bicolored Antbird R
Scaled Antpitta R
Silvery-fronted Tapaculo
White-fronted Tyrannulet R
Paltry Tyrannulet
Yellow-crowned Tyrannulet R
Greenish Elaenia
Yellow-bellied Elaenia
Lesser Elaenia
Mountain Elaenia
Olive-striped Flycatcher
Ochre-bellied Flycatcher
Slaty-capped Flycatcher
Yellow Tyrannulet R
Scale-crested Pygmy-Tyrant
Common Tody-Flycatcher
Eye-ringed Flatbill
Yellow-olive Flycatcher
White-throated Spadebill
Bran-colored Flycatcher
Common Tufted-Flycatcher R
Olive-sided Flycatcher R
Dark Pewee
Western Wood-Peewee
Eastern Wood-Peewee
Yellow-bellied Flycatcher
Willow/Alder Flycatcher R
Yellowish Flycatcher
Bright-rumped Attila
Rufous Mourner R
Dusky-capped Flycatcher
Boat-billed Flycatcher
Social Flycatcher
Gray-capped Flycatcher
Streaked Flycatcher
Sulfur-bellied Flycatcher R
Piratic Flycatcher
Tropical Kingbird
Barred Becard R
White-winged Becard
Black-and-white Becard R
Rose-throated Becard
Masked Tityra
Turquoise Cotinga
Three-wattled Bellbird
White-ruffed Manakin
Yellow-throated Vireo
Yellow-winged Vireo
Brown-capped Vireo
Philadelphia Vireo
Red-eyed Vireo
Yellow-green Vireo
Tawny-crowned Greenlet
Lesser Greenlet
Rufous-browed Peppershrike
Black-chested Jay R
Blue-and-white Swallow
Southern Rough-winged Swallow
Cliff Swallow R
Barn Swallow
Riverside Wren
Rufous-breasted Wren
Plain Wren
House Wren
Ochraceous Wren
Gray-breasted Wood-Wren
Southern Nightingale-Wren
Tropical Gnatcatcher
Black-faced Solitaire
Black-billed Nightingale-Thrush
Orange-billed Nightingale-Thrush
Ruddy-capped Nightingale-Thrush R
Gray-cheeked Thrush
Swainson's Thrush
Wood Thrush R
Mountain Thrush
Clay-colored Thrush
White-throated Thrush
Black-and-yellow Silky-flycatcher R
Long-tailed Silky-flycatcher
Blue-winged Warbler R
Golden-winged Warbler
Tennessee Warbler
Northern Parula
Tropical Parula
Flame-throated Warbler
Yellow Warbler R
Chestnut-sided Warbler
Black-throated Blue Warbler R
Black-throated Green Warbler
Blackburnian Warbler
Black-and-white Warbler
American Redstart
Prothonotary Warbler R
Worm-eating Warbler
Ovenbird
Louisiana Waterthrush R
Kentucky Warbler
Mourning Warbler
Magnolia Warbler
Masked Yellowthroat R
Gray-crowned Yellowthroat R
Wilson's Warbler
Canada Warbler R
Slate-throated Redstart
Collared Redstart R
Golden-crowned Warbler
Rufous-capped Warbler
Black-cheeked Warbler
Buff-rumped Warbler R
Bananaquit
Common Bush-Tanager
Gray-headed Tanager
White-lined Tanager R
Red-crowned Ant-Tanager R
Summer Tanager
Scarlet Tanager R
Flame-colored Tanager
White-winged Tanager
Cherrie's Tanager
Blue-gray Tanager
Palm Tanager
Silver-throated Tanager
Speckled Tanager
Bay-headed Tanager
Golden-hooded Tanager
Scarlet-thighed Dacnis
Green Honeycreeper R
Blue-black Grassquit
Variable Seedeater
Yellow-bellied Seedeater R
Yellow-faced Grassquit
Slaty Finch R
Slaty Flowerpiercer
Yellow-thighed Finch
White-naped Brush-Finch
Chestnut-capped Brush-Finch

Stripe-headed Brush-Finch R
Orange-billed Sparrow R
Black-striped Sparrow R
Rufous-collared Sparrow
Streaked Saltator
Buff-throated Saltator
Black-thighed Grosbeak
Blue-black Grosbeak R
Rose-breasted Grosbeak
Indigo Bunting
Painted Bunting R
Baltimore Oriole
Chestnut-headed Oropendola
Golden-browed Chlorophonia
Thick-billed Euphonia
Yellow-throated Euphonia R
Elegant Euphonia
Spot-crowned Euphonia
Yellow-bellied Siskin
Lesser Goldfinch

Puerto Armuelles

Several of the regional endemics and specialties of the western Pacific lowlands can be found in the remnant woodlands of the Burica Peninsula near the town of Puerto Armuelles. These include **Baird's Trogon**, **Charming Hummingbird**, **Black-hooded Antshrike**, **Orange-collared Manakin**, and **Riverside Wren**. There is also one record of **Yellow-billed Cotinga**. Other endemics have been reported from the area in the past but it is uncertain if they still persist.

Puerto Armuelles has historically been a center for the banana-growing industry in Chiriquí. The industry in the area, however, has been in decline for some time. Puerto Armuelles still features many wooden houses built by the banana company instead of the brick or adobe houses typical of most of Chiriquí.

En route from Paso Canoa to Puerto Armuelles, there are a couple of sites that may be worth checking out if you have the time. The *Progreso Marsh* is near the town of the same name 9.3 km from Paso Canoa (24.2 km north of Koco's Place in Puerto Armuelles if coming from that direction). Look for a small store named Hermanos Cáceres set back from the road, on the left heading south. Turn off the main road, double back in front of the store, and look for a small dirt road on the right. This leads in a very short distance to a small marshy area, a former arm of the Río Chiriquí Viejo, now a mostly overgrown backwater with some open water. Scrub and a few small trees line the banks. Species that can be found here include **Fiery-billed Aracari**, **Black-hooded Antshrike**, **Riverside Wren**, and **Cherrie's Tanager**.

The *Esperanza Marsh* is a little farther south, just north of the town of Esperanza. At 15.2 km there is a large ETESA (Empresa de Transmisión Eléctrica S. A.) power transmission station. Just before the power station the road crosses over the swampy course of the Río Madre Vieja. You can park on the road to the right in front the station and walk back to the marsh, being careful of traffic. Check out the trees near the small pond on the east side of the road for roosting Boat-billed Heron and other species. The marsh itself extends a considerable distance from the road.

Continuing south on the road, at 34.3 km from Paso Canoa you cross the bridge over the Río San Bartolo (check for kingfishers and waterbirds). At 35.2 km there is a column with a *"Bienvenidos"* sign from the Club de Leones by a Texaco Station. Bear right here. At 36.2 km is a large secondary school, and at 37.1 km you cross a small bridge. At 37.6 km you reach an elaborate "Bienvenidos a Puerto Armuelles" sign

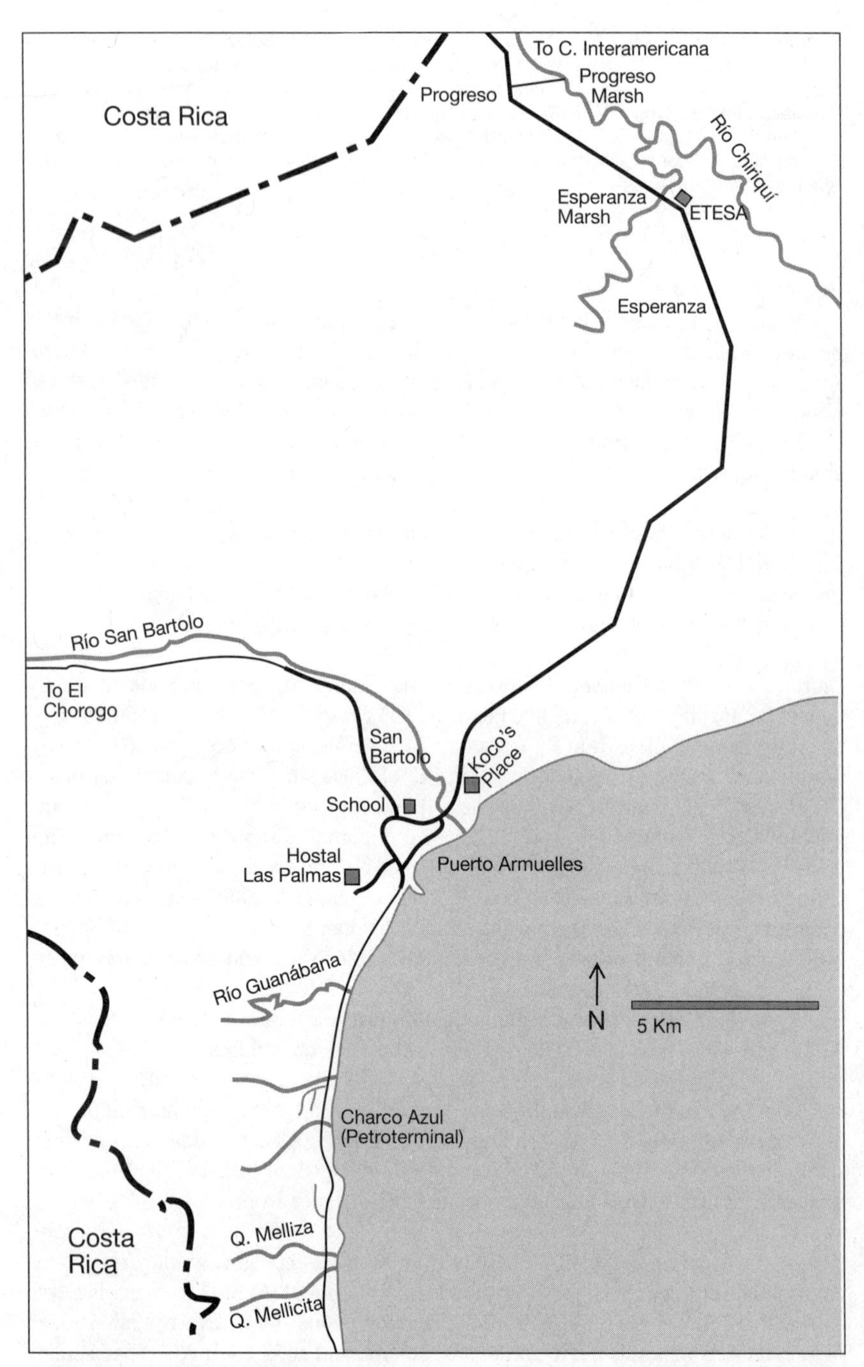

Puerto Armuelles and environs

graced by a dolphin and two pink flamingos at the entrance to the main part of town, opposite the San Antonio gas station.

The best birding areas are around *Charco Azul* south of town. It is a little complicated finding your way through Puerto Armuelles to the road south. To get to Charco Azul, take a left at the dolphin statue and then take the first right at the Banistmo Bank. Go a little over 100 m and then turn right opposite a small city park and playground. Shortly after that on this street you pass the Escuela Tomás Armuelles on the left. Continue down this straight street lined with clapboard houses. At 800 m past the school, take a right at the Templo Jehova Jireth, then take the first left at a green building, the Kiosco Estephani. After 400 m, turn right at the Templo Adventista del Séptimo Día. After another 200 m, the road goes sharply left at a sports field (campo deportivo). Remaining distances on the road south are given from this turn.

About 300 m past the campo deportivo the road becomes gravel. As of early 2006 it was in good condition for 2WD. At 0.5 km, you cross a girder bridge, and at 1.8 reach the *Río Guanábano*. The course of this river can be driven upstream during the dry season with 4WD, or with care possibly even in 2WD. You can also walk up the riverbed, especially if you have rubber boots. You can get down to the riverbed on the right of the bridge. From the bridge, on the seaward side, look for shorebirds and waterbirds. Upstream, the high banks are forested, with many fruiting trees. Understory birds, especially wrens, are common in the dense scrub. Species that have been found here include Wood Stork, Peregrine Falcon, **Black-hooded Antshrike**, Scissor-tailed Flycatcher, **Riverside Wren**, and **Cherrie's Tanager**.

At 5.2 km two concrete tracks go uphill to the right that lead to good birding trails. At about 100 from the main road is the house of Demóstenes Antoní Morale, known as "Toñín," who owns some of the adjacent woodland and is happy to have birders visit it. The road beyond requires 4WD in the dry season and may be impassable even with it in the wet. You can park near the house and walk up if the road is bad. About 400 m beyond the house a side road goes left at a metal gate. This descends about 500 m first through scrub and then forest to end at the Quebrada Seca stream. Continuing straight on the first road for another 600 m brings you to a metal gate across the road. Just past this gate on the left a small trail goes in and also descends about 500 m to the Quebrada Seca.

At 5.7 km is the front gate of Petroterminales Panamá (PTP) at Charco Azul. This is the Pacific terminus of the transisthmian oil pipeline that passes through Fortuna and extends to Chiriquí Grande on the Atlantic side. The terminal is a security zone, so you must show ID at the gate in order to pass through, and stopping is not permitted inside. However, there are several good birding trails on the property which can be used with prior permission. Permission to enter can be obtained through the Panama Audubon Society with two weeks advance notice in writing.

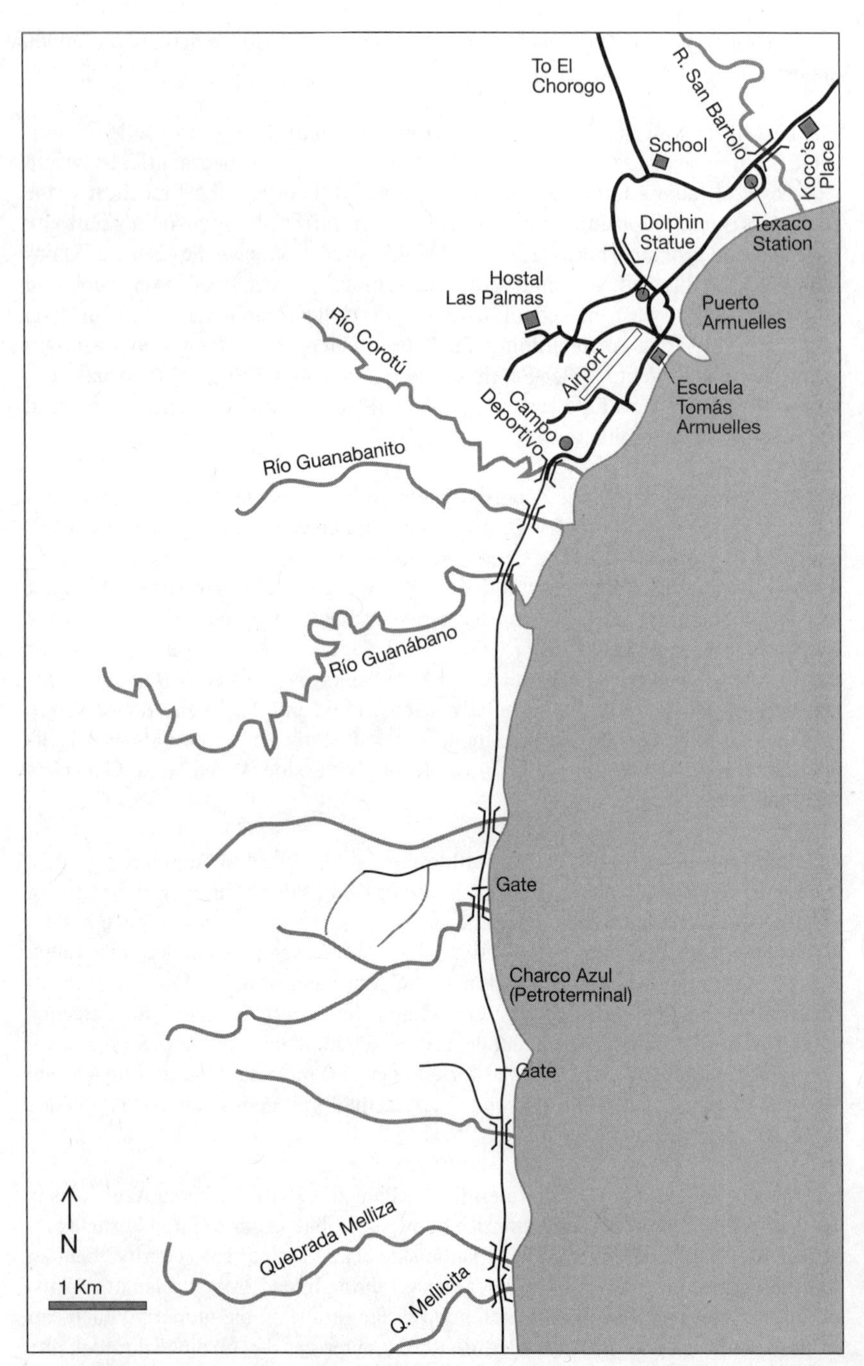

Puerto Armuelles and Charco Azul (detail)

Continuing through the PTP property, the rear gate is at 7.3 km. At 700 m beyond the rear gate there is a gravel road to the right. This road parallels the rear fence of the PTP, but there is forest along the left hand side that can be birded. After about 700 m this side road reaches pastureland.

Species that have been found at Charco Azul and adjacent woodland include Bronzy Hermit, Blue-throated Goldentail, **Charming Hummingbird**, **Veraguan Mango**, **Baird's Trogon**, **Fiery-billed Aracari**, Rufous-winged Woodpecker, Pale-billed Woodpecker, **Black-hooded Antshrike**, Bare-Crowned Antbird, **Orange-collared Manakin**, **Riverside Wren**, and **Cherrie's Tanager**.

At 8.3 km from the campo deportivo there is a small concrete bridge, and 200 m beyond it a second one. At 9.5 km there is a girder bridge over the Quebrada Melliza, and 400 m past that a concrete bridge over the Quebrada Mellicita. There are a number of scattered forest fragments in the drainages of the Melliza and Mellicita inland at a distance of about 1-2 km from the coast. Species that have been reported from this area, but not yet definitely found at Charco Azul, include Laughing Falcon, Marbled Wood-Quail, **White-crested Coquette**, Rufous-tailed Jacamar, **Golden-naped Woodpecker**, Red-rumped Woodpecker, Tawny-winged Woodcreeper, Eye-ringed Flatbill, and White-throated Shrike-Tanager. Crested Guan and Great Curassow once occurred here and may still do so. Those with more time in the Puerto Armuelles area may find these fragments worth exploring with the aid of a guide.

Those with 4WD could also try exploring the valley of the *Río San Bartolo* to the west of Puerto Armuelles. From Koco's Place go south into Puerto Armuelles. At 1.5 km take the right fork opposite the Texaco station. At 2.6 km, just past the gate to the Escuela Secundaria de Puerto Armuelles, turn sharp right (this point is 1.1 km north of the dolphin statue if coming from Puerto Armuelles). This road runs between the cemetery and the school. At about 6.3 km from the school there are small marshes on both sides of the road, and the San Bartolo comes into view on the right. At 8.0 km, just past a small concrete ball court, turn left at the crossroads and continue for 400 m, where you bear left in front of a lot bordered by white-painted tires set into the ground. All roads are dirt from this point onwards. At 600 m from this point, turn left just before a house that is sometimes used as a sawmill, and often has a tractor and lumber outside. After a further 600 m you turn left again and quickly drop down to the river. The rough pastures and scrub before getting to the river, and all points beyond the first river crossing, are usually birdy.

A track follows the river going through riverside scrub and gallery forest, pastures, and small patches of second growth. The track frequently crosses the river, and may be impassable even with 4WD in the rainy season. Species that have been found along the river include **Charming Hummingbird**, **Baird's Trogon**, **Fiery-billed Aracari**, **Black-hooded Antshrike**, **Riverside Wren**, **Cherrie's Tanager**, and various waterbirds.

If you continue straight at the sawmill instead of turning left to go to the river, after another 2.4 km you will come to an area with oil palm plantations called *Sangrillo*. There is a large and spectacular patch of *Erythrina* trees here, which when flowering during the dry season attracts a large concentration of hummingbirds including **Veraguan Mango** and Blue-throated Goldentail. As another item of interest, one of the very few pairs of working oxen left in Panama is used to haul carts in the oil palm plantations. In late afternoon, they can be seen by continuing to drive over the stream, on an iron bridge, to a point where the road turns sharp left. The oxen are housed to the right at this point.

Getting to Puerto Armuelles. From David, head west on the Interamericana Highway 53 km to Paso Canoa, the border crossing with Costa Rica. In Paso Canoa, take a sharp left just before the border checkpoint, a large shed-like building. Puerto Armuelles is 34 km south by good road. From Paso Canoa, the road follows the border for some 7-8 km. Note that just after the road leaves the border, there is a customs checkpoint. It is important to come to a complete stop here and not continue until instructed to do so.

Accommodations and meals

Koco's Place (ph 770-7049, cel 6592-3097) is a basic motel-style place ($20.00) with private cold-water bath and air-conditioning. It is on the left next to a gas station, 33.8 km after the turnoff from the Interamericana in Paso Canoa, and 3.8 km before you reach the dolphin statue in Puerto Armuelles.

The *Hostal Museo Las Palmas* (cel 6607-5967, 607-6006; anibalortiz3@hotmail.com), in a former banana-company manager's house, is a new bed and breakfast and considerably more upscale than Koco's Place. There are two rooms and a suite ($60.00-$70.00, including breakfast) with hot-water bath and air-conditioning. To find it, continue right at the dolphin statue and follow the signs. At 700 m past the statue, turn right; after another 200 m turn left, and after another 200 m turn right again. The hostal is a large white house 100 m up this road.

Enrique's restaurant, which serves Chinese and international cuisine, is just beyond the Texaco Station near the Club de Leones *"Bienvenidos"* sign at the beginning of town. *Marieth's Café*, serving breakfast (starting about 7:00 AM) and lunch, is just beyond the dolphin statue on the right near the main part of town. To find the *Don Carlo* and *Mily* restaurants, both of which serve seafood and other dishes, turn left at the dolphin statue, then take the first left after that. Take the next right in front of the police station, and go straight for 3 blocks. Turn right at the gas station and go down the street to near its end. The restaurants are near each other opposite a small park on the waterfront.

Bird list for the Puerto Armuelles area. R = rare; ? = present status uncertain

Great Tinamou
Little Tinamou
Crested Guan ?
Great Curassow ?
Crested Bobwhite
Marbled Wood-Quail
Brown Pelican
Magnificent Frigatebird
Bare-throated Tiger-Heron
Great Blue Heron
Great Egret
Snowy Egret

Little Blue Heron
Cattle Egret
Green Heron
Boat-billed Heron
King Vulture
Black Vulture
Turkey Vulture
Osprey
Swallow-tailed Kite
White-tailed Kite
Double-toothed Kite
Plumbeous Kite
White Hawk
Gray Hawk
Great Black-Hawk
Roadside Hawk
Broad-winged Hawk
Short-tailed Hawk
Zone-tailed Hawk
Black Hawk-Eagle
Crested Caracara
Yellow-headed Caracara
Laughing Falcon R
White-throated Crake
Gray-necked Wood-Rail
Semipalmated Plover
Northern Jacana
Greater Yellowlegs
Solitary Sandpiper
Willet
Spotted Sandpiper
Whimbrel
Sanderling
Pale-vented Pigeon
Short-billed Pigeon
Plain-breasted Ground-Dove
Ruddy Ground-Dove
Blue Ground-Dove
White-tipped Dove
Gray-chested Dove
Ruddy Quail-Dove
Crimson-fronted Parakeet
Brown-throated Parakeet
Orange-chinned Parakeet
Brown-hooded Parrot
Blue-headed Parrot
Red-lored Amazon
Mealy Amazon
Mangrove Cuckoo
Squirrel Cuckoo
Striped Cuckoo
Pheasant Cuckoo
Smooth-billed Ani
Groove-billed Ani
Crested Owl
Spectacled Owl
Lesser Nighthawk
Common Pauraque
Costa Rican Swift R
Lesser Swallow-tailed Swift
Bronzy Hermit
Band-tailed Barbthroat
Long-billed Hermit
Stripe-throated Hermit
Scaly-breasted Hummingbird
White-necked Jacobin
White-crested Coquette
Violet-crowned Woodnymph
Sapphire-throated Hummingbird
Blue-throated Goldentail
Charming Hummingbird
Rufous-tailed Hummingbird
Purple-crowned Fairy
Baird's Trogon
Violaceous Trogon
Black-throated Trogon
Slaty-tailed Trogon
Blue-crowned Motmot
Ringed Kingfisher
Green Kingfisher
Amazon Kingfisher
American Pygmy Kingfisher
White-necked Puffbird
Rufous-tailed Jacamar
Fiery-billed Aracari
Chestnut-mandibled Toucan
Olivaceous Piculet
Golden-naped Woodpecker R
Red-rumped Woodpecker R
Rufous-winged Woodpecker R
Lineated Woodpecker
Pale-billed Woodpecker
Pale-breasted Spinetail
Buff-throated Foliage-gleaner
Plain Xenops
Tawny-winged Woodcreeper
Wedge-billed Woodcreeper
Cocoa Woodcreeper
Black-striped Woodcreeper
Streak-headed Woodcreeper
Great Antshrike
Barred Antshrike
Black-hooded Antshrike
Russet Antshrike
Slaty Antwren
Dot-winged Antwren
Dusky Antbird
Bare-crowned Antbird
Chestnut-backed Antbird
Southern Beardless-Tyrannulet
Yellow Tyrannulet
Yellow-crowned Tyrannulet
Yellow-bellied Elaenia
Ochre-bellied Flycatcher
Paltry Tyrannulet
Northern Scrub-Flycatcher
Slate-headed Tody-Flycatcher
Common Tody-flycatcher
Eye-ringed Flatbill
Yellow-olive Flycatcher
Golden-crowned Spadebill
Ruddy-tailed Flycatcher
Sulphur-rumped Flycatcher
Yellow-bellied Flycatcher
Bright-rumped Attila
Speckled Mourner
Rufous Mourner
Dusky-capped Flycatcher
Great Crested Flycatcher
Great Kiskadee
Boat-billed Flycatcher
Social Flycatcher
Gray-capped Flycatcher
Streaked Flycatcher
Piratic Flycatcher
Tropical Kingbird
Western Kingbird R
Fork-tailed Flycatcher
Thrush-like Schiffornis
Rufous Piha
White-winged Becard
Masked Tityra
Turquoise Cotinga R
Yellow-billed Cotinga R
Orange-collared Manakin
Blue-crowned Manakin
Red-capped Manakin
Yellow-throated Vireo
Philadelphia Vireo
Yellow-green Vireo
Scrub Greenlet
Tawny-crowned Greenlet
Lesser Greenlet
Green Shrike-Vireo
Black-chested Jay
Gray-breasted Martin
Mangrove Swallow
Southern Rough-winged Swallow
Cliff Swallow
Barn Swallow
Black-bellied Wren
Riverside Wren
White-breasted Wood-Wren
Southern Nightingale-Wren
Long-billed Gnatwren
Tropical Gnatcatcher
Clay-colored Thrush
Tennessee Warbler
Yellow Warbler

Chestnut-sided Warbler
Black-and-white Warbler
Prothonotary Warbler
Worm-eating Warbler
Ovenbird
Northern Waterthrush
Mourning Warbler
MacGillivray's Warbler
Buff-rumped Warbler
Bananaquit
Bay-headed Tanager
Golden-hooded Tanager
Scarlet-thighed Dacnis
Blue Dacnis
Green Honeycreeper
Shining Honeycreeper
Red-legged Honeycreeper
Gray-headed Tanager
White-throated Shrike-Tanager R
White-shouldered Tanager R
White-lined Tanager
Summer Tanager
Cherrie's Tanager
Blue-gray Tanager
Palm Tanager
Blue Dacnis
Green Honeycreeper
Red-legged Honeycreeper
Blue-black Grassquit
White-collared Seedeater
Yellow-bellied Seedeater
Ruddy-breasted Seedeater
Yellow-faced Grassquit
Orange-billed Sparrow
Black-striped Sparrow
Streaked Saltator
Buff-throated Saltator
Painted Bunting R
Blue-black Grosbeak
Great-tailed Grackle
Bronzed Cowbird
Giant Cowbird
Baltimore Oriole
Yellow-billed Cacique
Scarlet-rumped Cacique
Chestnut-headed Oropendola
Crested Oropendola
Yellow-crowned Euphonia
Thick-billed Euphonia
Lesser Goldfinch
House Sparrow

El Chorogo

Although some of the endemics of Panama's western Pacific lowlands can be found at Chorcha Abajo or near Puerto Armuelles, to find the full complement of these species in Panama one must go to El Chorogo. El Chorogo is the largest patch of forest remaining in the lowlands of Chiriquí, and is located on the Costa Rican border to the west of Puerto Armuelles. El Chorogo is, unfortunately, difficult to visit. It is reached by very poor 4WD roads, followed by several hours on mule back or foot through cattle pasture before reaching the forest. There are no facilities, so camping is required, and you must bring all your own food and gear. This said, El Chorogo is a lovely forest with very interesting birds and other fauna, and a visit will well repay the interested birder.

Almost all of the western Chiriquí specialties can be found at El Chorogo. These include **Baird's Trogon**, **Charming Hummingbird**, **White-crested Coquette**, **Fiery-billed Aracari**, **Golden-naped** and Rufous-winged **Woodpeckers**, Tawny-winged Woodcreeper, **Black-hooded Antshrike**, Rose-throated Becard, **Riverside Wren**, and **Spot-crowned Euphonia**. **Yellow-billed Cotinga** and **Turquoise Cotinga** have each been seen once. Other birds of interest include Laughing Falcon, Great Curassow, Blue-throated Goldentail, Pale-billed Woodpecker, Northern Bentbill, and White-throated Shrike-Tanager. El Chorogo reaches almost 700 m (2,300 ft), so a few foothills and highland birds are also present, including Violet-headed Hummingbird, Slaty Antwren, White-ruffed Manakin, and even Three-wattled Bellbird in season.

Because it is the best remaining place in Panama for many species, the Panama Audubon Society has made El Chorogo a high priority for conservation. PAS, with the support of the Amos Butler Audubon Society of Indianapolis and the American Bird Conservancy, and one of its members have to date purchased 283 hectares (700 acres) for the creation of private reserves.

El Chorogo is located at the headwaters of the Río Bartolo, and is reached by traveling up the river valley from just north of Puerto Armuelles. Because of the difficulties of getting to the site, and because it is private property, arrangements for a visit must be made through the Panama Audubon Society. Contact details for PAS can be found on p. 7.

Bird list for El Chorogo. R = rare.

Great Tinamou
Little Tinamou
Gray-headed Chachalaca
Crested Guan
Great Curassow
Marbled Wood-Quail
Cattle Egret
Black Vulture
Turkey Vulture
Lesser Yellow-headed Vulture
King Vulture
Swallow-tailed Kite
Double-toothed Kite
White Hawk
Roadside Hawk
Broad-winged Hawk
Short-tailed Hawk
Black Hawk-Eagle
Laughing Falcon
Barred Forest-Falcon
Collared Forest-Falcon
White-throated Crake
Short-billed Pigeon
White-tipped Dove
Gray-chested Dove
Ruddy Quail-Dove
Crimson-fronted Parakeet
Brown-throated Parakeet
Orange-chinned Parakeet
Brown-hooded Parrot
Blue-headed Parrot
White-crowned Parrot
Red-lored Amazon
Mealy Amazon
Black-billed Cuckoo
Squirrel Cuckoo
Striped Cuckoo
Smooth-billed Ani
Tropical Screech-Owl
Spectacled Owl
Common Pauraque
Chuck-will's-widow R
Rufous Nightjar
Vaux's Swift
White-collared Swift
Bronzy Hermit R
Long-billed Hermit
Stripe-throated Hermit
Band-tailed Barbthroat
White-tipped Sicklebill R
Violet-headed Hummingbird
White-crested Coquette R
Violet-crowned Woodnymph
Violet-bellied Hummingbird
Blue-throated Goldentail
Charming Hummingbird
Snowy-bellied Hummingbird
Rufous-tailed Hummingbird
Long-billed Starthroat
Ruby-throated Hummingbird R
Baird's Trogon
Violaceous Trogon
Black-throated Trogon
Slaty-tailed Trogon
Blue-crowned Motmot
White-whiskered Puffbird
Rufous-tailed Jacamar
Fiery-billed Aracari
Chestnut-mandibled Toucan
Golden-naped Woodpecker
Red-crowned Woodpecker
Red-rumped Woodpecker
Rufous-winged Woodpecker R
Lineated Woodpecker
Pale-billed Woodpecker
Pale-breasted Spinetail
Buff-throated Foliage-gleaner
Plain Xenops
Scaly-throated Leaftosser
Tawny-winged Woodcreeper
Long-tailed Woodcreeper R
Wedge-billed Woodcreeper
Northern Barred-Woodcreeper
Cocoa Woodcreeper
Black-striped Woodcreeper
Streak-headed Woodcreeper
Great Antshrike
Black-hooded Antshrike
Russet Antshrike
Slaty Antwren
White-flanked Antwren
Dot-winged Antwren
Dusky Antbird
Chestnut-backed Antbird
Bicolored Antbird
Black-faced Antthrush
Paltry Tyrannulet
Southern Beardless-Tyrannulet
Yellow-crowned Tyrannulet
Yellow-bellied Elaenia
Ochre-bellied Flycatcher
Black-capped Pygmy-Tyrant
Scale-crested Pygmy-Tyrant
Northern Bentbill
Common Tody-Flycatcher
Olive-sided Flycatcher
Eastern Wood-Pewee
Yellow-bellied Flycatcher
Acadian Flycatcher
Bright-rumped Attila
Speckled Mourner
Rufous Mourner
Dusky-capped Flycatcher
Great Kiskadee
Boat-billed Flycatcher
Gray-capped Flycatcher
Streaked Flycatcher
Sulphur-bellied Flycatcher
Tropical Kingbird
Piratic Flycatcher
Thrush-like Schiffornis
Rufous Piha
White-winged Becard
Rose-throated Becard R
Masked Tityra
Yellow-billed Cotinga R
Three-wattled Bellbird
White-ruffed Manakin
Blue-crowned Manakin
Red-capped Manakin
Yellow-throated Vireo
Red-eyed Vireo
Yellow-green Vireo
Scrub Greenlet
Lesser Greenlet
Northern Rough-winged Swallow
Southern Rough-winged Swallow
Barn Swallow
Black-bellied Wren
Riverside Wren
Rufous-breasted Wren
Plain Wren

House Wren
White-breasted Wood-Wren
Southern Nightingale-Wren
Long-billed Gnatwren
Tropical Gnatcatcher
Swainson's Thrush
Golden-winged Warbler
Tennessee Warbler
Chestnut-sided Warbler
Bay-breasted Warbler
Black-and-white Warbler
American Redstart
Northern Waterthrush
Louisiana Waterthrush
Kentucky Warbler
Mourning Warbler
Masked Yellowthroat
Gray-crowned Yellowthroat R
Buff-rumped Warbler
Bananaquit
Gray-headed Tanager
White-throated Shrike-Tanager
White-shouldered Tanager R
Summer Tanager
Scarlet Tanager
Western Tanager
Cherrie's Tanager
Blue-gray Tanager
Bay-headed Tanager
Golden-hooded Tanager
Plain-colored Tanager
Blue Dacnis
Green Honeycreeper
Shining Honeycreeper
Red-legged Honeycreeper
Variable Seedeater
Yellow-faced Grassquit
Orange-billed Sparrow
Black-striped Sparrow
Buff-throated Saltator
Bronzed Cowbird
Baltimore Oriole
Scarlet-rumped Cacique
Chestnut-headed Oropendola
Yellow-crowned Euphonia
Thick-billed Euphonia
Spot-crowned Euphonia
White-vented Euphonia

Bocas del Toro Province

Bocas del Toro Province, the westernmost part of Panama's wet Atlantic slope, has a number of regional endemics and other specialties that can be found in Panama only here. Much of its very wet foothills and highland zones are nearly inaccessible, but the Oleoducto Road between Fortuna and Chiriquí Grande on the coast provides a good place to see otherwise hard-to-find specialties of the Atlantic slope foothills. The Bocas del Toro Archipelago is the only place in Panama where Stub-tailed Spadebill occurs, and also hosts such interesting species as Three-wattled Bellbird and Montezuma Oropendola. The Panama national endemic **Escudo Hummingbird** is found only on tiny Isla Escudo de Veraguas off the Bocas del Toro coast.

The rainfall pattern in Bocas del Toro is different from most of the rest of Panama. Instead of a marked dry season from the end of December to April followed by a wet season with increasing amounts of rain as the year progresses, there are two relatively dry and two wetter periods each year. Changuinola averages about 2,600 mm (102 in) of rain per year but has had over 4 meters (157 in, or more than 13 ft) in a single year. In February and March, and again in September and October, there is generally about half the rain as the wettest months of May to August and December. Since roads will usually be less muddy during these drier months, and migrants are present, these are the best time for a visit. In Bocas del Toro, the afternoons have much less rain than during the night and morning, so don't be discouraged if you wake up and it's raining. Just go out when you can and keep your umbrella handy.

The province and its capital city, on Isla Colón in the Bocas del Toro Archipelago, share the same name, and both are often referred to simply as "Bocas." To make things even more confusing, Isla Colón is sometimes called Bocas Island, and the Archipelago in general the Bocas Islands. Bocas del Toro Town and the Bocas Archipelago are one of Panama's most popular tourist destinations, and have a great assortment of hotels, restaurants, and other amenities. The mainland is much less developed. Places to stay are mainly limited to Changuinola, Almirante, and Chiriquí Grande, and are much more basic than those available in the Archipelago.

The best birding, on the other hand, is to be found on the mainland, where a much greater diversity of species occurs. The best areas are around Changuinola and Chiriquí Grande, along the road that connects them, and on the Oleoducto Road. However, the Archipelago also offers some good birding possibilities. Isla Escudo de Veraguas is remote and requires special arrangements to visit.

Chiriqui Grande and the Oleoducto Road

Chiriquí Grande, confusingly enough, is not in Chiriquí but in the province of Bocas del Toro. The town is the Atlantic terminus for the transisthmian oil pipeline and the Oleoducto Road[5] that parallels it. The area is one of the easiest places to find many of the endemics and specialties of the western Atlantic slope lowlands and foothills, including Rufous-tailed Jacamar, Lanceolated Monklet, Chestnut-colored Woodpecker, **Snowy Cotinga**, Black-headed Nightingale-Thrush, Ashy-throated Bush Tanager, and Black-cowled Oriole.

The area is frequently included in a birding trip to Fortuna, in Chiriquí Province, by continuing on the Oleoducto Road past the Continental Divide into Bocas del Toro. For this reason, directions and distances are given here coming from the direction of Fortuna, with the distances from Chiriquí Grande given in parentheses. See pp. 195-196 for directions to Fortuna.

As it drops from the continental divide the Oleoducto Road traverses extremely wet foothills forest, then passes through deforested and partially deforested areas on the lower slopes and in the lowlands near Chiriquí Grande. The upper part is within the Palo Seco Protection Forest (167,000 ha/412,666 acres), which forms a buffer zone for La Amistad International Park and the Fortuna Forest Reserve. The asphalt road is in good condition (except for two areas of slumping just beyond the continental divide), but steep and very winding in the foothills. If you stop to bird along this part, make sure you pull off the road as much as possible, and don't park on blind curves, since heavy trucks (some with questionable brakes) frequently use it. Avoid driving the road at night or when it is fogbound.

Verrugosa Trail. After passing the checkpoint at the Continental Divide coming from Fortuna, the Oleoducto Road begins a steep descent. The Verrugosa Trail is on the left side of the road, 900 m after the checkpoint, and directly across from the 63 km marker. The area adjacent to the road here has been scraped flat, and dirt has been pushed down the slope, partially obscuring the head of the trail. It can be found at the end of the line of undisturbed bushes along the south edge of the scraped area,

---

[5] Note that both this road and Pipeline Road in Soberanía National Park are often referred to as the *Camino del Oleoducto* in Spanish. They should not be confused.

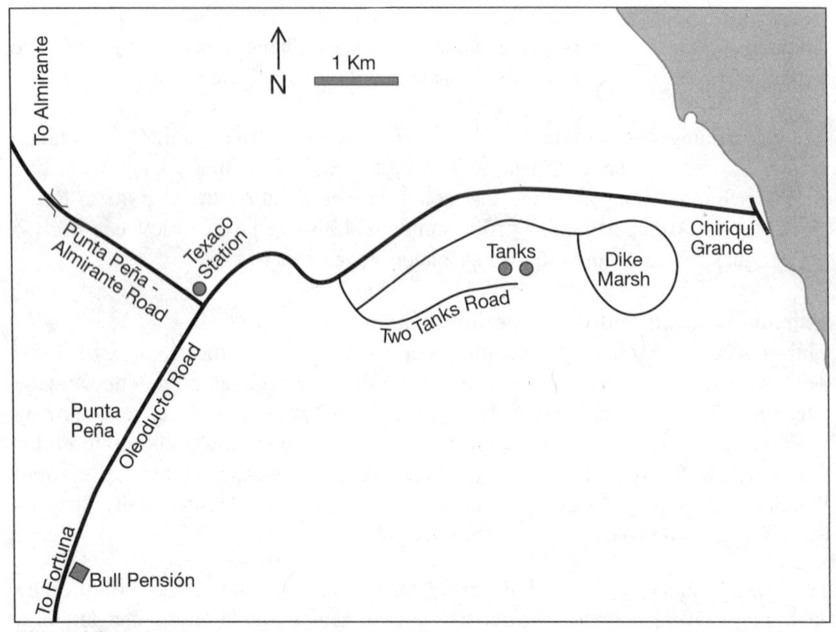

Chiriquí Grande

about 30 m from the road. The trail descends through beautiful wet foothills forest, steeply at first, but leveling out after about 20 minutes of walking. Plan to spend about one to two hours on this trail. It is good for **Lattice-tailed Trogon**, Ashy-throated Bush-Tanager, and Black-headed Nightingale-Thrush, among others.

Natural Bridge. At 5 km from the Continental Divide (30 km from Chiriquí Grande) the road crosses a culvert. The stream that passes under the road can be walked, and immediately enters a deep gorge which ends at a natural bridge and waterfall. Torrent Tyrannulets and Buff-rumped Warblers frequent the stream and White-collared Swifts nest under the falls. Sharpbill has been seen on the road just above this spot.

Rancho Ecológico Willy Mazú. This is a small rustic tourist camp operated by one of Panama's best-known bird guides, Wilberto Martínez. At 6.4 km past the Continental Divide, you will first see the ANAM headquarters, painted light yellow and green on the right. About 300 m farther along on the left is the Rancho Ecológico, which has several trails on the property. See the section on accommodations for details about staying here. There is a $5.00 fee to use the trails. A pair of Rufous-tailed Jacamars sometimes hangs about along the road near the entrance to the camp, and Lanceolated Monklet has been seen in forest at the edge of the road. The camp is near the lower limit of the Palo Seco Protection Forest and deforestation becomes increasingly prevalent below this point. Look for **Snowy Cotinga** on the slopes on the way down.

The road gradually becomes less steep, and enters flat terrain just before the bridge over the Río Guarumo at 15 km from the Continental Divide (20 km from Chiriquí Grande). Black-cowled Oriole has been seen at the bridge. At 28.2 km, the road to Almirante and Changuinola starts just beyond the small community of Punta Peña, at a Texaco gas station on the left.

Two Tanks Road. At 30.8 km from the continental divide (5 km from Chiriquí Grande) a gravel road takes off to the right, going up to two oil storage tanks on the wooded hill beyond. After 400 m a dirt road branches off to the left, going through marshy pastures and woods for another 2.2 km. In 600 m this road passes over a culvert; a path leads off to the left 20 m past it. Uniform Crake has been seen and heard at dusk along this path and in the underbrush to the right of the road here. At the end of the road a small dike, which you can walk, circles a marsh with tall trees. Check along the road and at the marsh for species such as Green Ibis, Bronzy Hermit (identify with care, since Rufous-breasted Hermit also occurs here), Chestnut-colored and Pale-billed Woodpeckers, Band-backed and **Black-throated Wrens**, Crimson-collared Tanager, and Olive-backed Euphonia. Lovely Cotinga has been reported from this site, but it has recently been found that Blue Cotinga also occurs here, so identification must be made with care. Beyond where the dirt road branches off to the left, the gravel road continues up the hill to the tanks. There is a gate prohibiting entry after 1.0 km on this road.

From Punta Peña to Chiriquí Grande the land is low and flat, mostly consisting of marsh, wet cattle pastures, clumps of woodland, and an occasional pond. The road is good for Bronzy Hermit, Montezuma Oropendola, Brown Jay, Olive-crowned Yellowthroat, and White-collared Seedeater.

At 36 km from the Continental Divide, and 7.7 km from the turnoff to Almirante, you come to Chiriquí Grande, which is rather gritty and notably unpicturesque. Chiriquí Grande used to be the departure point for launches to Bocas del Toro Town, but since the construction of the new road between Punta Peña and Almirante these all now leave from either the latter or Changuinola. There are several hotels and restaurants here and not much else of interest. There is no reason to go into town unless you intend to eat there or spend the night.

Getting to the Oleoducto Road and Chiriqui Grande. From Panama City, follow the directions for the Fortuna Forest Reserve on pp. 195-196. Once you reach the Continental Divide in Fortuna, follow the directions given in the section above. Chiriqui Grande is 35 km from the Continental Divide. From Changuinola, it is about 87 km from the bridge south of town to the intersection with the Oleoducto Road, 7.7 km south of Chiriqui Grande and 28.2 km from the Continental Divide.

Accommodations and meals. Accommodations and eating places in the Chiriquí Grande area tend to be decidedly basic. Besides the places listed below, it is also possible to bird the area from Finca La Suiza in Fortuna (p. 196), but it is difficult to arrive in the lowlands early enough for good birding.

Rancho Ecológico Willy Mazú (ph 442-1340; fax 442-8485; panabird@cwpanama.net; www.naturpanama.com, click on "More" for the Rancho) provides accommodation in four-person tents under a large thatched roof shelter ($45.00 with meals, $25.00 without). The tents have mattresses and there is a shared bathroom and hot-water showers at the site, as well as kitchen facilities. The camp is 6.7 km from the Continental Divide. Meals can be provided by prior arrangement.

There is a bunkhouse at the ANAM station ($5.00 per night) with four beds and a kitchen area. Check with the ANAM office in Changuinola (ph 758-6603) for details.

The *Bull Pensión* is located at 25.9 km past the Continental Divide (10 km north of Chiriqui Grande), on the right opposite a convenience store. It has cold-water baths; some rooms have air-conditioning ($18.00).

The *Hotel Emperador* (ph 756-9656) is in Chiriqui Grande. Rooms have air-conditioning and some have hot water ($12.50 single; $20.00 double). It is the second building to the left, just after turning left at the major intersection in town. There are several other hotels in town.

There is a small eating place about 200 m down from the ANAM station on the Oleoducto Road where you can get a simple meal and even stay for the night. There is a truck-stop type restaurant next to the Texaco station at the intersection of the Oleoducto and Almirante Roads that is open 24 hours and has decent cafeteria-style food.

Bird list for Chiriquí Grande and environs. This list includes species found in the lowlands only. For a bird list for the upper part of the Oleoducto Road in the foothills zone, see the section on Fortuna on pp. 197-199. * = vagrant; R = rare.

Little Tinamou
Black-bellied Whistling-Duck
Muscovy Duck R
Gray-headed Chachalaca
Least Grebe
Pied-billed Grebe
Neotropic Cormorant
Great Egret
Snowy Egret
Little Blue Heron
Tricolored Heron
Cattle Egret
Green Heron
Green Ibis
Wood Stork R
Black Vulture
Turkey Vulture
King Vulture R
Swallow-tailed Kite
White-tailed Kite
Double-toothed Kite
Mississippi Kite
Plumbeous Kite
Bicolored Hawk R
Common Black-Hawk
Roadside Hawk
Broad-winged Hawk
Swainson's Hawk
Collared Forest-Falcon
Laughing Falcon
Aplomado Falcon*
Bat Falcon
White-throated Crake
Gray-necked Wood-Rail
Uniform Crake R
Purple Gallinule
Sungrebe R
Sunbittern R
Northern Jacana
Willet
Spotted Sandpiper
Parasitic Jaeger
Laughing Gull
Rock Pigeon
Pale-vented Pigeon
Scaled Pigeon
Short-billed Pigeon
Ruddy Ground-Dove
Blue Ground-Dove
White-tipped Dove
Gray-chested Dove
Ruddy Quail-Dove
Crimson-fronted Parakeet
Red-fronted Parrotlet R
Blue-headed Parrot
White-crowned Parrot
Red-lored Amazon

Mealy Amazon
Squirrel Cuckoo
Striped Cuckoo
Groove-billed Ani
Common Pauraque
Common Potoo
Gray-rumped Swift
Bronzy Hermit
Rufous-breasted Hermit R
Band-tailed Barbthroat
Long-billed Hermit
Stripe-throated Hermit
White-necked Jacobin
Violet-crowned Woodnymph
Blue-chested Hummingbird
Snowy-bellied Hummingbird
Rufous-tailed Hummingbird
Purple-crowned Fairy
Long-billed Starthroat
White-tailed Trogon
Violaceous Trogon
Rufous Motmot
Broad-billed Motmot
Ringed Kingfisher
Belted Kingfisher
Amazon Kingfisher
Green Kingfisher
American Pygmy Kingfisher
White-necked Puffbird
Pied Puffbird
White-whiskered Puffbird
Rufous-tailed Jacamar R
Great Jacamar
Collared Aracari
Keel-billed Toucan
Chestnut-mandibled Toucan
Black-cheeked Woodpecker
Red-crowned Woodpecker
Cinnamon Woodpecker
Chestnut-colored Woodpecker
Lineated Woodpecker
Pale-billed Woodpecker
Slaty Spinetail
Wedge-billed Woodcreeper
Northern Barred-Woodcreeper
Cocoa Woodcreeper
Black-striped Woodcreeper
Streak-headed Woodcreeper
Fasciated Antshrike
Great Antshrike
Western Slaty-Antshrike
Pacific Antwren
Checker-throated Antwren
Bare-crowned Antbird
Chestnut-backed Antbird
Spotted Antbird
Black-faced Antthrush
Brown-capped Tyrannulet
Yellow-bellied Elaenia
Lesser Elaenia
Ochre-bellied Flycatcher
Paltry Tyrannulet
Black-capped Pygmy-Tyrant
Northern Bentbill
Common Tody-Flycatcher
Black-headed Tody-Flycatcher
Eastern Wood-Pewee
Tropical Pewee
Yellow-bellied Flycatcher
Acadian Flycatcher
Willow/Alder Flycatcher
White-throated Flycatcher*
Bright-rumped Attila
Rufous Mourner
Great Kiskadee
Boat-billed Flycatcher
Social Flycatcher
Gray-capped Flycatcher
Streaked Flycatcher
Piratic Flycatcher
Tropical Kingbird
Eastern Kingbird
Fork-tailed Flycatcher
Cinnamon Becard
White-winged Becard
Masked Tityra
Black-crowned Tityra
Lovely Cotinga
Blue Cotinga R
Snowy Cotinga R
Purple-throated Fruitcrow
Three-wattled Bellbird
Golden-collared Manakin
Red-capped Manakin
Black-chested Jay
Brown Jay
Gray-breasted Martin
Mangrove Swallow
Southern Rough-winged Swallow
Barn Swallow
Band-backed Wren
Black-throated Wren
Bay Wren
Canebrake Wren
House Wren
White-breasted Wood-Wren
Southern Nightingale-Wren
Long-billed Gnatwren
Tropical Gnatcatcher
Swainson's Thrush
Clay-colored Thrush
White-throated Thrush
Tennessee Warbler
Chestnut-sided Warbler
Bay-breasted Warbler
American Redstart
Northern Waterthrush
Mourning Warbler
Common Yellowthroat
Olive-crowned Yellowthroat
Buff-rumped Warbler
Dusky-faced Tanager
Sulphur-rumped Tanager
Tawny-crested Tanager
White-lined Tanager
Summer Tanager
Crimson-collared Tanager R
Passerini's Tanager
Flame-rumped Tanager R
Blue-gray Tanager
Palm Tanager
Golden-hooded Tanager
Blue Dacnis
Green Honeycreeper
Shining Honeycreeper
Blue-black Grassquit
Variable Seedeater
White-collared Seedeater
Lesser Seed-Finch
Yellow-faced Grassquit
Orange-billed Sparrow
Black-striped Sparrow
Buff-throated Saltator
Black-headed Saltator
Slate-colored Grosbeak
Black-faced Grosbeak
Black-thighed Grosbeak
Rose-breasted Grosbeak
Blue-black Grosbeak
Blue Grosbeak
Dickcissel
Great-tailed Grackle
Bronzed Cowbird
Giant Cowbird
Black-cowled Oriole
Orchard Oriole
Yellow-tailed Oriole
Baltimore Oriole
Yellow-billed Cacique
Scarlet-rumped Cacique
Chestnut-headed Oropendola
Montezuma Oropendola
Yellow-crowned Euphonia
Olive-backed Euphonia
Tawny-capped Euphonia
House Sparrow

Punta Peña-Almirante Road

Western Bocas del Toro has only recently been connected to the rest of Panama by road, the section between the Chiriquí Grande area and Almirante having been completed in 1999. Because of this some patches of good forest still remain along the road, especially as you get closer to Almirante coming from the east. The road starts near the community of Punta Peña, the turnoff being at a Texaco station just north of it, and 7.7. km south of Chiriquí Grande. Distances are given from this intersection, and correspond to distance markers along the road.

At 8.8 km the road crosses a bridge. There is a field to the left just past the bridge where you can pull off and park. Beyond the field there is a path, leading along the river to an Indian village, which can be birded. At the 28 km distance marker, there is a lookout on the right where **Snowy Cotinga** has been seen. Just past the 28 km marker, the right lane and part of the left lane have been lost due to slumping, a condition that has existed for several years, so be careful. Starting at about kilometer 35, there are more places along the road where the forest is closer than on the earlier part. Although some of the forest is second growth there are some patches of larger trees. At 41.7 km you cross the Río Uyama. At 600 m past the bridge, marked by a house on the right and a bus stop on the left, there is another spot where **Snowy Cotinga** has been seen. The La Escapada Eco-Lodge, which has good birding on its grounds, is on the shore of Almirante Bay at 48.5 km. See the section on accommodations for details about staying here. At 58.5 km there is a tall tree on the left that hosts a large Montezuma Oropendola colony in season. At 67.3 km you come to a T-intersection; turn left for Changuinola and right for Almirante.

Between Almirante and Changuinola, the road passes through good forest, but there are no shoulders to park on and traffic, which includes large trucks, makes it inadvisable to stop on the road. At 15.6 km from the T-intersection, there is a small thatched shelter on the right that offers one of the few places to pull off. From here you have an excellent view of the Pond Sak marshlands in the distance. At 19.6 km you reach the bridge to Changuinola.

Getting to the Punta Peña-Almirante Road. From Panama City, follow the directions for the Fortuna Forest Reserve on pp. 195-196 and Chiriqui Grande on p. 235. The intersection with the Oleoducto Road is 28.2 km from the Continental Divide in Fortuna and 7.7 km south of Chiriqui Grande, just north of the community of Punta Peña. Driving time from the Fortuna tollgate to the intersection is a little over an hour, and from the intersection to Changuinola is about 2 hours.

Accommodations and meals. The road can be birded from a base in either Chiriquí Grande (p. 236) or Changuinola (p. 245), or from the La Escapada Eco-Lodge on the road itself.

The *La Escapada Eco-Lodge* (cel 6698-9901, 6618-6106; info@laescapada.net; www.laescapada.net), located on the shores of Almirante Bay, has air-conditioned rooms ($55.00 May-November, $75.00 December-April) and a restaurant on premises. It is at 48.5 km from the turnoff from the Oleoducto Road, and about 34 km south of Changuinola.

Changuinola

The town of Changuinola is the center of the banana industry of western Bocas del Toro. Although the town itself is not very appealing and the vast banana plantations themselves are poor for birding, many endemics and specialties of the western Atlantic lowlands can be found in the surrounding area. These include White-crowned Pigeon, Olive-throated Parakeet, Chestnut-colored Woodpecker, White-collared Manakin[6], **Snowy Cotinga**, Band-backed, **Black-throated**, and **Canebrake Wrens**, Olive-crowned Yellowthroat, Black-cowled Oriole, Montezuma Oropendola, and **Nicaraguan Seed-Finch**, as well as species also found in western Chiriquí, such as White-crowned Parrot, Bronzy Hermit, Rufous-tailed Jacamar, Pale-billed Woodpecker, and Northern Bentbill. Species such as Muscovy and Masked Ducks, Least Bittern, Green Ibis, Northern Jacana, and Sungrebe can be seen in the extensive wetlands that surround the town to the north, east, and west. Swan Cay (locally called Bird Island), near Isla Colón, where Red-billed Tropicbird nests, can also be visited by boat from Changuinola. The town also provides access to Wekso, in good forest on the Río Teribe.

Most of the lowlands near town are solidly carpeted with bananas. Although the banana plantations themselves have a low diversity of birds, the roads that run south, east, and west go through swampy terrain, cattle pastures, cacao plantations, and scrubby woodlands where a variety of species can be seen. The San San-Pond Sak Wetlands (16,125 ha/39,846 acres), a protected area, surrounds Changuinola on three sides. The Río San San and its tributary the Río Negro run through forested swampland to the northwest of town, while the extensive Pond Sak marshes are to the southwest. A narrow band of coastal marshes and swamps around the mouth of the Río Changuinola are found between the town and the coast to the north. Besides having many aquatic birds, these wetlands are the best place in Panama to see manatees.

There are four good birding areas around Changuinola reachable by car: the road to El Silencio and the Río Teribe south of town; rice fields and cattle pastures to the west; the Sendero Ecológico that goes through marshlands to the northeast of town; and areas along the road that goes northwest towards Guabito on the Costa Rican border. With the exception of parts of the rice fields and pastures, all of these roads can be handled with 2WD. Boats can also be hired for trips up the Ríos Changuinola

[6] Note: A form known as "Almirante Manakin" occurs from near Changuinola to around Almirante. Males resemble Golden-collared Manakin in plumage, but with the collar paler lemon-yellow and the belly the same color (instead of olive green). While this form has traditionally been considered to be a subspecies of Golden-collared Manakin, genetic and morphological studies have shown that it is actually closer to White-collared Manakin.

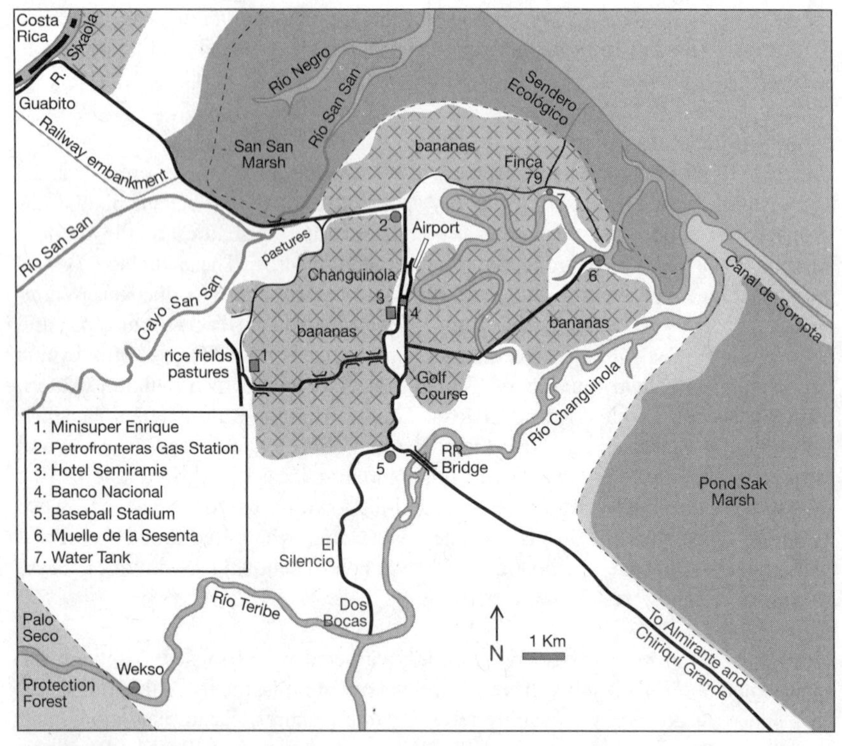

Changuinola and environs

or Teribe, to the San San-Pond Sak Wetlands, or to visit Swan Cay. Some birds may also be seen by taking the regularly-scheduled launch to Bocas del Toro Town on Isla Colón, but this usually travels too rapidly for good birding. We first describe the layout of the town of Changuinola for orientation, then cover the birding areas from south to north as you go through town. There is a final section on arranging boat trips.

Changuinola has one of the few railway systems in Panama, used to transport bananas from plantations to packing houses and ports. The road entering town from Almirante to the south crosses the Río Changuinola on a rather alarming 350 m-long railway trestle. Drivers must straddle the tracks, keeping the wheels on often loose or broken planks and trying not to think of the missing guardrails. (It is best to have an experienced driver at the wheel for this.) After exiting the bridge, the road ascends a hill. At the crest you have a view of the baseball stadium with banana plantations beyond. From here on, the route sounds complicated, but if you get confused just follow the main traffic, which will be heading toward the center of town. At 500 m past the end of the bridge, you reach the stadium and turn right, then follow the stadium wall as it curves around to the left, and turn right again in 200 m at the end of the stadium wall. In another 50 m, you reach a T-junction with an Accel gas station on the left and the Super Centro El Toro ahead. Turn right here (the road to the left goes to El Silencio), staying on the main road as it curves sharply left after 200 m. In 500 m, after some curves, continue to follow the main road as it curves left in front of a Texaco gas station. Continue another 1.3 km on this road, then turn right to cross the railroad tracks. (At the point where you turn right, going straight leads to a short, wide communications tower directly ahead). After crossing the tracks, turn sharp left, and continue north 1.8 km until you reach a small traffic circle, with a bust of military leader Omar Torrijos in the center, in front of a police station. Go around the circle to exit to the left, cross the tracks again, and drive to the next block where the Banco Nacional, on the left corner, provides a good landmark. This north-south road, the Avenida 17 de Abril (also simply known as Calle Central), is the main commercial district. Directions to sites below are given from the Banco Nacional.

Road to El Silencio. Although the road from Changuinola to El Silencio is mostly lined with houses and gardens, a variety of birds of open habitats, scrub, and edge can be found here, including Northern Bentbill, White-collared Manakin, Band-backed and **Black-throated Wrens**, Grayish Saltator, Black-cowled Oriole, Montezuma Oropendola, and Olive-backed Euphonia. To get to El Silencio from the Banco Nacional, go east across the tracks to the traffic circle in front of the police station and turn right. Drive south for 1.9 km and turn right again, cross the tracks and turn left. Go another 1.3 km to a Texaco gas station and the main road will swing to the right. In another 600 m you will come to an Accel gas station on the left and the Super Centro El Toro on the right; go straight here. (To get to the bridge over the Río Changuinola and Almirante instead, you would turn left here, drive to the baseball stadium and then follow the road to the left around the wall of the park.) In 300 m you pass a campus of the University of Panama on the right, marked by a white arch. In another 200 m, the road swings to the left as three roads come together. Take the middle road and 300 m beyond this you will see a sign for El Silencio on your right. Continue straight for 5.1 km on a good paved road to El Silencio and the port, known as Dos Bocas, near the confluence of the Ríos Changuinola and Teribe. The port is the departure point for Wekso. It may also be possible here to hire a boat to take you across the river to bird the patches of forest on the other side, which are reportedly about an hour's walk from the river. A boat might also be hired for a day of exploring on the Ríos Changuinola or Teribe for $70 to $100 or so.

Fields west of town. West of town are rice fields and pastures at the edge of the banana plantations where species such as Olive-throated Parakeet, **Canebrake Wren**, Olive-crowned Yellowthroat, **Nicaraguan Seed-Finch**, Grayish Saltator, and Black-cowled Oriole can be found. This is also the area where Paint-billed Crakes were seen in the 1980s, but none have been recorded recently. To get to this area, drive south from the Banco Nacional on the main commercial street. Follow the main paved road through a few turns as it goes around the golf course and after 1.7 km, turn right. (This side road here is broken asphalt and appears to be gravel, but is in better condition farther along.) You then drive for 3.8 km through banana plantations, crossing four bridges. Just past the fourth bridge you come to a T-junction where you turn right. Continue straight on the main asphalt road for 400 m, where the minisuper Enrique is on the right and the asphalt road turns sharply to the left. Instead of following the main road, proceed straight onto a smaller dirt road.

The first 500 m or so of this road pass through houses. In 1 km you cross a bridge, with a house with an asymmetrical roof just before it on the left. Past the bridge there are rice fields and pastures on the left and bananas on the right for the next 600 m, where another road goes left for 400 m to a dump area. Check the fields along both these roads; **Nicaraguan Seed-Finch** has been seen here. The roads to this point can be handled with 2WD with care. The road that that continues straight beyond the side road to the dump is deeply pot-holed and requires high clearance. The first 700 m have bananas on both sides; beyond that there are rice fields and pastures on the left and bananas on the right. At 1.2 km, another road forks off to the left which passes along the edge of a pasture and can also be birded. Continuing straight, at 1.7 km the road joins a larger road coming from the right. Turn left here (going right immediately brings you to a gate signposted "Propiedad Privada Finca 07," which you should not enter), where another 1.9 km will bring you to a T-junction on a curved section of road. Go right here, where after crossing a small bridge you will come to the main road to Guabito in 200 m. To get back to Changuinola, go right, where after 2.1 km you will come to the Petrofronteras de Panama Finca 6 gas station. Going right here will bring you back to the Banco Nacional in the center of town in 2.5 km.

Sendero Ecológico. The Sendero Ecológico (Ecological Trail) is a trail that goes through marshlands northeast of Changuinola and ends at the beach. It passes through tall forest with many palm trees, with coffee-colored marsh water on either side of the trail. Walking it requires crossing many log bridges, often without a handrail, so a long walking stick is very useful for support and to keep your balance. Wading is not practical due to the depth of the mud. Also, since wooden bridges may rot out quickly in this climate, use caution when crossing them in case they have not been maintained. (As a precaution, you may not want to carry expensive cameras or scopes here just in case you fall in.) Just before reaching the beach, there is a trail to the left that is easier walking than the main trail. This is a popular route to the beach on weekends, so weekdays are probably a better time for a visit.

To get to the trail, go north from the Banco Nacional for 2.5 km, where the Petrofronteras de Panama Finca 6 gas station is on the left. The main road here swings around the gas station to the left to go to Guabito. Instead of following the asphalt road, continue straight past the gas station onto a gravel road, where you will immediately pass the minisuper Popular and the Costa Rican Consulate on your left. Stay on this road, and in 1.3 km you will pass a police station on the left. At 2.6 km beyond the police station you will see a tank on a tower to the right. Just before it is a side road to the left with a sign for Finca 79. Turn onto this side road, then take a right fork immediately after turning. Keep straight for another 1.3 km, where you will see a metal gate to the left just before the road passes through a fence and bends sharply left. Park here off the side of the road to allow farm vehicles to pass by. (The last 200 m of this road may require 4WD; otherwise it can be done in a regular car.) Walk up towards where the road turns sharply left, where you will find a narrow gate (allowing people to pass through but not cattle) in the fence on the right. Go through this gate to get onto a dike, which extends for about 2 km. The Sendero Ecológico takes off to the left, about 20 m from the gate. (On your return to town, remember to go left on the asphalt road just after passing the minisuper Popular, and a right at the gas station just past the entrance to the airport to get back to the Banco Nacional.)

Road to Guabito. Guabito is a small town on the Costa Rican border about 9 km northwest of Changuinola. Although it too is surrounded by banana plantations, the middle part of the road and various side roads traverse marshy terrain and scrub where species such as Slaty Spinetail, Olive-crowned Yellowthroat, and Grayish Saltator can be found. To get to the Guabito road, go north from the Banco Nacional for 2.5 km to the Petrofronteras gas station, where the road to Guabito curves sharply around to the left. At the time of this writing the road was mostly gravel, but was being repaired and may be paved. The following distances are measured from the turn by the gas station. At 2.2 km after the turn, a gravel road goes to the left just before a bridge on the main road. This is the far end of the road to the rice fields and pastures described above. If you enter it from this direction, you can drive in for 1.9 km to Finca 07, where there is a large "Propiedad Privada" sign at the gate. (Do not enter.) If you have high clearance you can take the smaller road to the right to get back to the minisuper Enrique and eventually to Changuinola by following the directions in the section on the rice fields in reverse. (Just after the bridge on the Guabito Road another gravel road goes to the left and also leads to this same road system.)

If you stay on the main road to Guabito, at 3.6 km from the Petrofronteras gas station you will cross the Río San San; the remnants of an abandoned railroad trestle are on the right. Check the river for Sungrebe. The next few kilometers are straight with some good forest on the right, at about 5 to 6.5 km along the road, adjacent to the marshlands along the Río San San. At 5.3 km (2.6 km past the Río San San), just where the main road bends sharply to the right, a small gravel road goes off to the left. This road follows an old railroad embankment (the tracks have been removed), above the level of the fields and marshes around it. Species that have been found here include Laughing Falcon, Olive-throated Parakeet, Chestnut-colored Woodpecker, **Snowy Cotinga**, **Black-throated Wren**, and Olive-crowned Yellowthroat.

Birding by boat from Changuinola. The best areas for birding by boat are the marshes along the side channels near the mouth of the Río Changuinola, and the Río San San and its tributary the Río Negro, which are lined with heavy forest draped in Spanish Moss. There is also a canal, known as the Canal de Soropta or Changuinola Canal, that runs from the Río Changuinola and parallels the coast for about 10 km before reaching the coast near Isla Colón, that was originally used to transport bananas to Almirante. The channel is used by launches going to Isla Colón and Bocas del Toro Town, and is also the route used to reach Swan Cay from Changuinola. Although the channel has relatively little floating vegetation and is mostly lined by small farms with scattered trees rather than forest, and thus not as good as the other areas mentioned, some of the more common land and water birds can be seen here.

There are boats for hire locally, but they are not advertised and you will have to ask around. The best places to ask are at the ANAM office in town, or near the pier on the Río Changuinola, Muelle de Finca Sesenta (Plantation 60 Dock), about 20 minutes from the center of town (see directions below). One boat owner is Ernesto Quintero (ph 758-5420, cel 6629-1499) who lives directly across the street from the minisuper Cuadrante #8, house #14186, not far from the pier. His boat holds about eight or nine people.

To find the ANAM office (ph 758-6603) from the Banco Nacional in the commercial district, go north (towards Guabito) on the main street, and take the first left, just past the Banistmo bank. Go two blocks and take another left. The office is one block down this street.

To get to Muelle de Finca Sesenta from the Banco Nacional in the commercial district, go east across the tracks to the traffic circle in front of the police station and turn right. At 1.1 km south of the traffic circle, opposite the Banco de Desarrollo Agropecuario and Ministerio de Educación buildings, take the road that goes off at an angle to the left. After 200 m the road crosses a one-lane bridge. At 100 m past the bridge, stay straight where a side road branches to the right. At 1.7 km stay on the main road as it turns sharply left. At 4.5 km past the turnoff, you come to a fork where you go left; the building at the intersection is the Patsy Café. In another 200 m you pass a Centro de Salud. At 100 m beyond this is the minisuper Cuadrante #8, with a small park just before it on the right. Go through the gate for the park to get to the pier. The ticket office is in the building just past the playground equipment.

Getting to Changuinola. Changuinola can be reached from Panama City by road via the Carretera Interamericana, the Oleoducto Road, and the Punta Peña-Almirante Road. From Panama City, follow in succession the directions for "getting to Western Panama" (pp. 130-132); Fortuna (pp. 195-196); Chiriquí Grande (p. 235), and the Punta Peña-Almirante Road (p. 238). Total driving time from Panama City is about 9-10 hours. If you are coming from Costa Rica, Changuinola is about 9 km from the border crossing at Guabito.

Both Aeroperlas and AirPanama have daily flights from Panama City to Changuinola, and Aeroperlas also has flights from David and Bocas del Toro Town. The airport is near the north end of town. From the Banco Nacional, go north on the main street for about 900 m to where a road merges from the right at a Delta gas station. The entrance to the airport is 50 m ahead on the right.

Changuinola can also be reached by bus from Panama City, David, or Costa Rica. In Changuinola, the SINCOTAVECOP bus terminal, for buses to Almirante and the rest of Panama, is on Calle Central 100 m north of the Banco National, on the right. The Terminal Urracá, for buses to Costa Rica, is another 300 m north on C. Central, on the left side. Local buses also serve nearby communities such as El Silencio.

There is also regular boat service to and from Bocas del Toro Town on Isla Colón from the Muelle de Finca Sesenta, run by Bocas Marine and Tours (ph 758-9859 Changuinola, 757-9033 Bocas Town; www.bocasmarinetours.com). The fare is $5:00 per person one-way; there is a six-person minimum. There are seven trips per day, from Changuinola to Bocas Town between 7:00 AM to 4:30 PM; and from Bocas town to Changuinola between 8:00 AM to 5:30 PM. Since they do not sell round-trip tickets, you may want to buy your return ticket as soon as you arrive in Bocas Town to make sure there is space.

There are at present no regular car-rental agencies in Changuinola. We have been told there is a rather informal car-rental place near Switch 4 (a railroad switch-point) near the Banistmo bank at the north end of town (2WD cars only). Inquire at your hotel if you want to explore this option.

If you do not have a car, the best bet is to hire a taxi to take you out to a site and then come back to pick you up at a set time later; do not pay until your return. You may also be able to hire a taxi for a morning or the day. You could also no doubt get to sites along the El Silencio and Guabito Roads by local bus.

Accommodations and meals. Changuinola is not a major tourist destination, and there is not a lot of choice in hotels and restaurants. Some of the better ones are listed below.

The *Hotel Semiramis* (ph 758-6006), the best hotel in town, has rooms ($21.00) with private hot-water bath and air-conditioning, and has a restaurant on premises. It is on C. Central on the block south of the Banco Nacional, on the west side of the street.

The *Hotel Carol* (ph 758-8731) has rooms ($18.00) with private hot-water baths and air-conditioning. It is next the Hotel Semiramis on the block south of the Banco Nacional.

The *Hotel Golden Sahara* (ph 758-7478; hotelgoldensahara@yahoo.com) has rooms ($27.00) with private hot-water bath and air-conditioning. It may be somewhat quieter than the others at night because it is away from the center of town (but it is near the railroad tracks). It is on C. Central, on the east side, a few hundred meters north of the turnoff for the airport.

The *Restaurante La Herrerana*, which opens at 6:00 AM, is a good place for breakfast. It is two buildings north of the Hotel Semiramis on the same side of the street.

One of the few good restaurants in town is the *Mar del Sur*, which is on the west side of C. Central north of the Terminal Urracá. The *Patsy Café* is a good place for lunch near the Muelle de Finca Sesenta. To get there, see the directions for the Muelle in the section on birding by boat.

Bird list for the Changuinola Area. * = vagrant; R = rare; ? = present status uncertain.

Great Tinamou
Little Tinamou
Black-bellied Whistling-Duck R
Muscovy Duck R
American Wigeon R
Blue-winged Teal
Northern Shoveler R
Northern Pintail R
Lesser Scaup R
Masked Duck R
Gray-headed Chachalaca
Least Grebe
Pied-billed Grebe
Brown Booby
Neotropic Cormorant
Anhinga
Magnificent Frigatebird
Least Bittern
Rufescent Tiger-Heron
Great Blue Heron
Great Egret
Little Blue Heron
Tricolored Heron
Cattle Egret
Green Heron
Agami Heron
Yellow-crowned Night-Heron
Boat-billed Heron
Green Ibis
Roseate Spoonbill R
Jabiru*
Wood Stork R
Black Vulture
Turkey Vulture
King Vulture
Osprey
Pearl Kite R
White-tailed Kite
Double-toothed Kite
Plumbeous Kite
Northern Harrier
Bicolored Hawk R
Semiplumbeous Hawk
White Hawk
Common Black-Hawk
Great Black-Hawk
Savanna Hawk R
Harris's Hawk*
Roadside Hawk
Broad-winged Hawk
Short-tailed Hawk
Swainson's Hawk
Red-tailed Hawk
Black Hawk-Eagle
Collared Forest-Falcon
Red-throated Caracara
Laughing Falcon
American Kestrel
Bat Falcon
Peregrine Falcon
Ruddy Crake*
White-throated Crake
Rufous-necked Wood-Rail ?
Gray-necked Wood-Rail
Uniform Crake R
Sora
Yellow-breasted Crake R
Paint-billed Crake ?
Purple Gallinule
Common Moorhen
American Coot
Sungrebe R
Limpkin
Black-bellied Plover
Collared Plover
Semipalmated Plover
Killdeer
Black-necked Stilt
Northern Jacana
Greater Yellowlegs
Solitary Sandpiper
Willet
Spotted Sandpiper
Whimbrel
Ruddy Turnstone
Semipalmated Sandpiper
Western Sandpiper
Least Sandpiper
Long-billed Dowitcher R
Wilson's Snipe
Pomarine Jaeger
Parasitic Jaeger
Laughing Gull
Herring Gull R
Royal Tern
Common Tern
Rock Pigeon
Pale-vented Pigeon
Scaled Pigeon
White-crowned Pigeon
Short-billed Pigeon
Mourning Dove R
Plain-breasted Ground-Dove R
Ruddy Ground-Dove
Blue Ground-Dove
White-tipped Dove
Gray-headed Dove
Gray-chested Dove
Olive-backed Quail-Dove
Ruddy Quail-Dove
Crimson-fronted Parakeet
Olive-throated Parakeet
Brown-hooded Parrot
Blue-headed Parrot
White-crowned Parrot
Mealy Amazon
Yellow-crowned Amazon
Black-billed Cuckoo
Yellow-billed Cuckoo
Squirrel Cuckoo
Striped Cuckoo
Groove-billed Ani
Barn Owl
Vermiculated Screech-Owl
Spectacled Owl
Mottled Owl
Short-tailed Nighthawk
Lesser Nighthawk
Common Nighthawk
Common Pauraque
Chuck-will's-widow
White-collared Swift
Chimney Swift
Gray-rumped Swift
Bronzy Hermit
Band-tailed Barbthroat
Long-billed Hermit
Stripe-throated Hermit
White-tipped Sicklebill R
White-necked Jacobin
Violet-headed Hummingbird
Garden Emerald
Violet-crowned Woodnymph
Blue-chested Hummingbird

Snowy-bellied Hummingbird
Rufous-tailed Hummingbird
Bronze-tailed Plumeleteer
Purple-crowned Fairy
Violaceous Trogon
Slaty-tailed Trogon
Rufous Motmot
Ringed Kingfisher
Belted Kingfisher
Amazon Kingfisher
Green Kingfisher
Green-and-rufous Kingfisher
American Pygmy Kingfisher
White-necked Puffbird
Pied Puffbird
White-whiskered Puffbird
White-fronted Nunbird R
Rufous-tailed Jacamar
Collared Aracari
Yellow-eared Toucanet
Keel-billed Toucan
Chestnut-mandibled Toucan
Black-cheeked Woodpecker
Smoky-brown Woodpecker R
Rufous-winged Woodpecker R
Cinnamon Woodpecker
Chestnut-colored Woodpecker
Lineated Woodpecker
Pale-billed Woodpecker
Slaty Spinetail
Striped Woodhaunter
Buff-throated Foliage-gleaner
Tawny-throated Leaftosser
Scaly-throated Leaftosser
Plain-brown Woodcreeper
Long-tailed Woodcreeper
Wedge-billed Woodcreeper
Northern Barred-Woodcreeper
Cocoa Woodcreeper
Black-striped Woodcreeper
Streak-headed Woodcreeper
Fasciated Antshrike
Great Antshrike
Western Slaty-Antshrike
Spot-crowned Antvireo
Pacific Antwren
Checker-throated Antwren
White-flanked Antwren
Dot-winged Antwren
Bare-crowned Antbird
Chestnut-backed Antbird
Bicolored Antbird
Black-faced Antthrush
Black-crowned Antpitta R
Thicket Antpitta
Yellow Tyrannulet
Yellow-bellied Elaenia
Olive-striped Flycatcher
Ochre-bellied Flycatcher
Paltry Tyrannulet
Black-capped Pygmy-Tyrant
Northern Bentbill
Slate-headed Tody-Flycatcher
Common Tody-Flycatcher
Black-headed Tody-Flycatcher
Yellow-margined Flycatcher
Royal Flycatcher
Ruddy-tailed Flycatcher
Sulphur-rumped Flycatcher
Black-tailed Flycatcher
Olive-sided Flycatcher
Eastern Wood-Pewee
Tropical Pewee
Acadian Flycatcher
Willow Flycatcher
Least Flycatcher*
Long-tailed Tyrant
Bright-rumped Attila
Rufous Mourner
Dusky-capped Flycatcher
Panama Flycatcher
Great Crested Flycatcher
Great Kiskadee
Boat-billed Flycatcher
Social Flycatcher
Gray-capped Flycatcher
Streaked Flycatcher
Sulphur-bellied Flycatcher
Piratic Flycatcher
Tropical Kingbird
Eastern Kingbird
Scissor-tailed Flycatcher R
Rufous Piha
Cinnamon Becard
White-winged Becard
Masked Tityra
Black-crowned Tityra
Lovely Cotinga R
Snowy Cotinga R
Purple-throated Fruitcrow
Bare-necked Umbrellabird R
Three-wattled Bellbird
White-collared Manakin
Red-capped Manakin
White-eyed Vireo*
Philadelphia Vireo
Red-eyed Vireo
Yellow-green Vireo
Tawny-crowned Greenlet
Lesser Greenlet
Green Shrike-Vireo
Black-chested Jay
Purple Martin
Gray-breasted Martin
Brown-chested Martin R
Tree Swallow R
Mangrove Swallow
Northern Rough-winged Swallow
Sand Martin
Cliff Swallow
Barn Swallow
Band-backed Wren
Black-throated Wren
Stripe-breasted Wren
Canebrake Wren
House Wren
White-breasted Wood-Wren
Song Wren
Long-billed Gnatwren
Tropical Gnatcatcher
Veery
Gray-cheeked Thrush
Swainson's Thrush
Wood Thrush
Clay-colored Thrush
Gray Catbird
Tropical Mockingbird
Cedar Waxwing R
Blue-winged Warbler R
Golden-winged Warbler
Tennessee Warbler
Yellow Warbler
Chestnut-sided Warbler
Magnolia Warbler
Cape May Warbler R
Yellow-rumped Warbler
Blackburnian Warbler
Yellow-throated Warbler
Bay-breasted Warbler
Blackpoll Warbler R
Black-and-white Warbler
American Redstart
Prothonotary Warbler
Worm-eating Warbler
Ovenbird
Northern Waterthrush
Louisiana Waterthrush
Kentucky Warbler
Connecticut Warbler
Mourning Warbler
Common Yellowthroat
Olive-crowned Yellowthroat
Hooded Warbler R
Canada Warbler
Buff-rumped Warbler
Yellow-breasted Chat R
Bananaquit
Dusky-faced Tanager
Olive Tanager
Sulphur-rumped Tanager

White-shouldered Tanager
Tawny-crested Tanager
White-lined Tanager
Red-throated Ant-Tanager
Summer Tanager
Scarlet Tanager
Crimson-collared Tanager R
Passerini's Tanager
Blue-gray Tanager
Palm Tanager
Plain-colored Tanager
Silver-throated Tanager*
Golden-hooded Tanager
Blue Dacnis
Green Honeycreeper
Shining Honeycreeper
Red-legged Honeycreeper
Blue-black Grassquit
Slate-colored Seedeater R
Variable Seedeater
White-collared Seedeater
Nicaraguan Seed-Finch R
Lesser Seed-Finch
Yellow-faced Grassquit
Orange-billed Sparrow
Black-striped Sparrow
Grasshopper Sparrow*
Lincoln's Sparrow*
Grayish Saltator
Buff-throated Saltator
Black-headed Saltator
Black-faced Grosbeak
Rose-breasted Grosbeak
Blue-black Grosbeak
Blue Grosbeak
Indigo Bunting
Painted Bunting R
Dickcissel
Bobolink
Red-breasted Blackbird R
Great-tailed Grackle
Bronzed Cowbird
Giant Cowbird
Black-cowled Oriole
Orchard Oriole
Yellow-tailed Oriole
Baltimore Oriole
Yellow-billed Cacique
Scarlet-rumped Cacique
Chestnut-headed Oropendola
Montezuma Oropendola
Yellow-crowned Euphonia
Olive-backed Euphonia
White-vented Euphonia
House Sparrow

Wekso

The best place for forest birding in the lowlands of Bocas del Toro is *Wekso*, an ecotourism lodge on the Río Teribe operated by the Naso (also known as the Teribe), one of Panama's smallest indigenous groups and the last one in Latin America still ruled by a king. Wekso was the site of Panajungla, a well-known school for training in jungle warfare and survival operated by Panama's military regime before its overthrow by the US invasion of 1989. The remains of the concrete buildings of the school are still evident near the lodge.

The lodge is located on a hill overlooking the Río Teribe. There are two main trails at the lodge, including the short *Mirador Trail,* a 15-minute walk to a lookout over the river, and the longer *Heliconias Trail,* a loop that traverses both primary and secondary forest and takes about two hours to walk. The lodge can also arrange a tour by boat up the river to several more remote indigenous communities such as Bonyik, Cieyic, and Cieykin. Species that can be found at Wekso or on the Río Teribe include Laughing Falcon, Green-breasted Mango, Gray-rumped Swift, White-fronted Nunbird, Smoky-brown Woodpecker, Bare-crowned Antbird, Thicket Antpitta, Northern Bentbill, Black Phoebe (usually a highland bird in Panama, but found here along the river), **Snowy Cotinga**, White-collared Manakin, **Black-throated Wren**, and Montezuma Oropendola, as well as many of the more widespread forest species of the Atlantic slope.

Wekso is at the edge of the Palo Seco Protection Forest, which acts as a buffer zone for La Amistad International Park, whose boundaries are some distance upstream. It is not easy to get into the park from this direction without mounting an expedition. There is an ANAM park station, the headquarters for the Bocas del Toro side of La Amistad, near the lodge.

Getting to Wekso. The Wekso Ecolodge is operated by the Organization for the Sustainable Development of Naso Ecotourism, known by its Spanish acronym of ODESEN, a Naso community-based ecotourism project. They have an office in Changuinola (cel 6569-3869; turismonaso_odesen@hotmail.com; www.bocas.com/odesen). Boats to the lodge depart from Dos Bocas at the end of the El Silencio Road, on the Teribe south of Changuinola. ODESEN provides transportation from the airport or bus terminal in Changuinola to El Silencio ($8.00 per person for groups of 2-4, less for larger groups). If you wish, it is possible to arrange your own transportation to El Silencio from Changuinola, which could be cheaper. See p. 241 for directions to the El Silencio Road and Dos Bocas from Changuinola. The boat ride up the river, which takes about an hour, is $15.00 per person for groups of 2-4 (less per person for larger groups). Day trips or overnight stays can be arranged. While it is possible to arrange your own boat transport to Wekso from Dos Bocas, this is discouraged by ODESEN.

Accommodations and meals. Facilities at the lodge include an attractive thatched-roof cottage on the edge of the forest that contains the sleeping quarters and a dining area with a view over the river. The beds have foam mattresses and mosquito nets. The outhouse has a shower, a flush toilet and a picture window providing a good view of a White-collared Manakin lek. There is no electricity; illumination at night is by candle or flashlight and of course there is no air-conditioning or hot water. Lodging is $12.00 per night per person for groups of 2-4 (less per person for larger groups). Meals are $2.50 for breakfast, $4.00 for lunch, and $3.00 for dinner. You will be expected to hire a guide for the trails as well for $4.00 per person for 2-4 people (less per person for larger groups). You will also have to pay the national park entrance fee of $3.50 for foreign visitors and $1.50 for Panama residents.

The Bocas del Toro Archipelago

The islands of Bocas del Toro, especially Islas Colón and Bastimentos, are one of Panama's most popular tourist destinations. The town of Bocas del Toro on Isla Colón is the provincial capital and has many hotels and other tourism facilities, while much of nearby Isla Bastimentos forests and coral reefs are protected by Bastimentos Marine Park. Tourism in the islands is mainly focused on marine activities, including snorkeling, diving, surfing, and fishing. Although the Archipelago is not really a primary birding destination, since it lacks many of the birds found on the adjacent mainland, quite a few species of interest can be found here. White-crowned Pigeon can be found around the coasts, and Gray-headed Dove, Green-breasted Mango, Passerini's Tanager, Montezuma Oropendola, and Black-cowled Oriole also occur. Three-wattled Bellbird migrates here from its breeding areas in the highlands (and has recently been found breeding on Isla Colón as well). In Panama, Stub-tailed Spadebill only occurs in the Bocas Archipelago, being found on all the main islands, its next closest locality being in northwestern Costa Rica. Swan Cay, known locally as *Isla Pájaros* or "Bird Island," has the only breeding colony of Red-billed Tropicbird in the southwest Caribbean, and can easily be visited by boat from Bocas del Toro Town. During winter months, Pomarine and Parasitic Jaegers are fairly

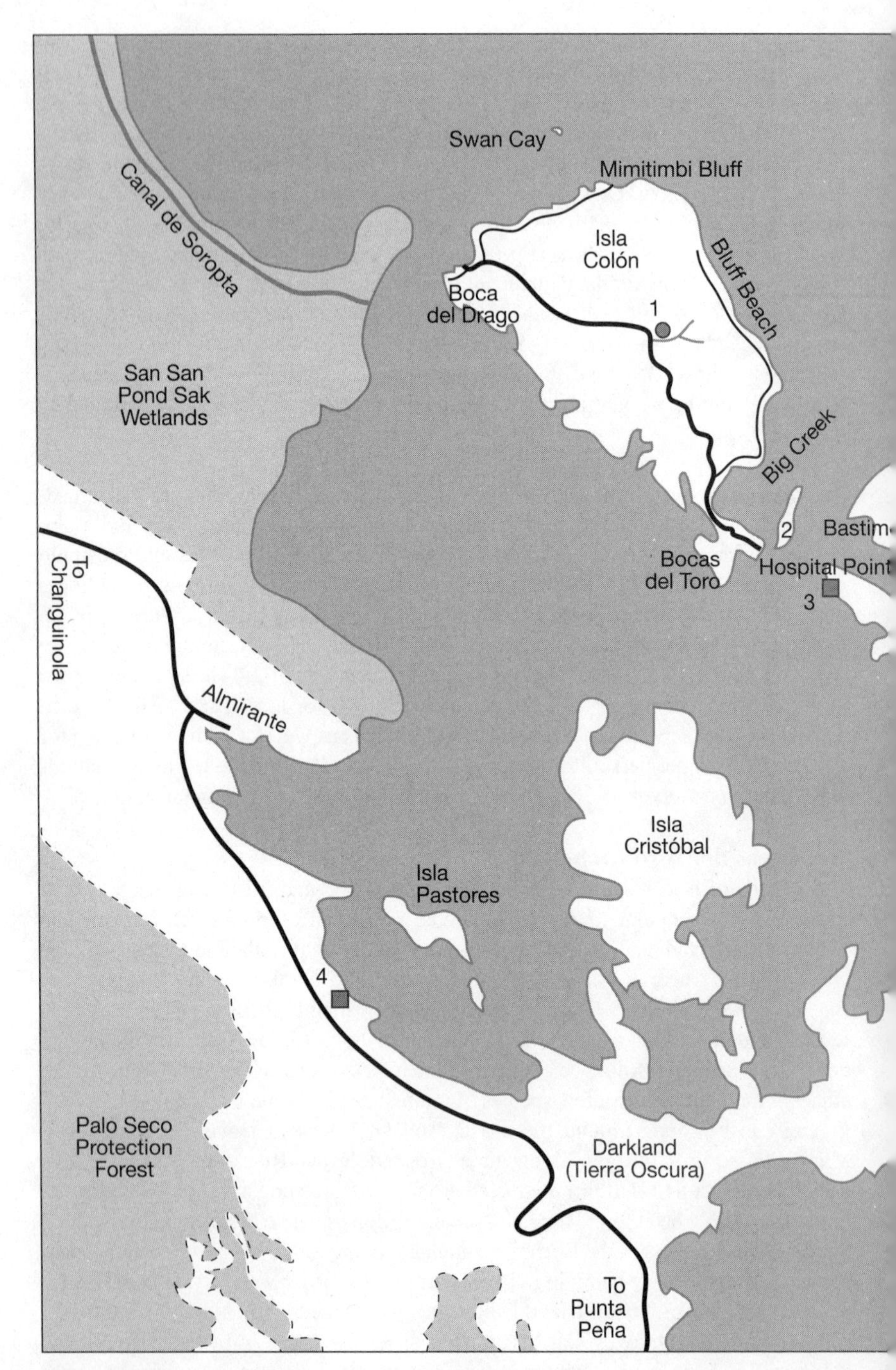

Bocas del Toro Archipelago

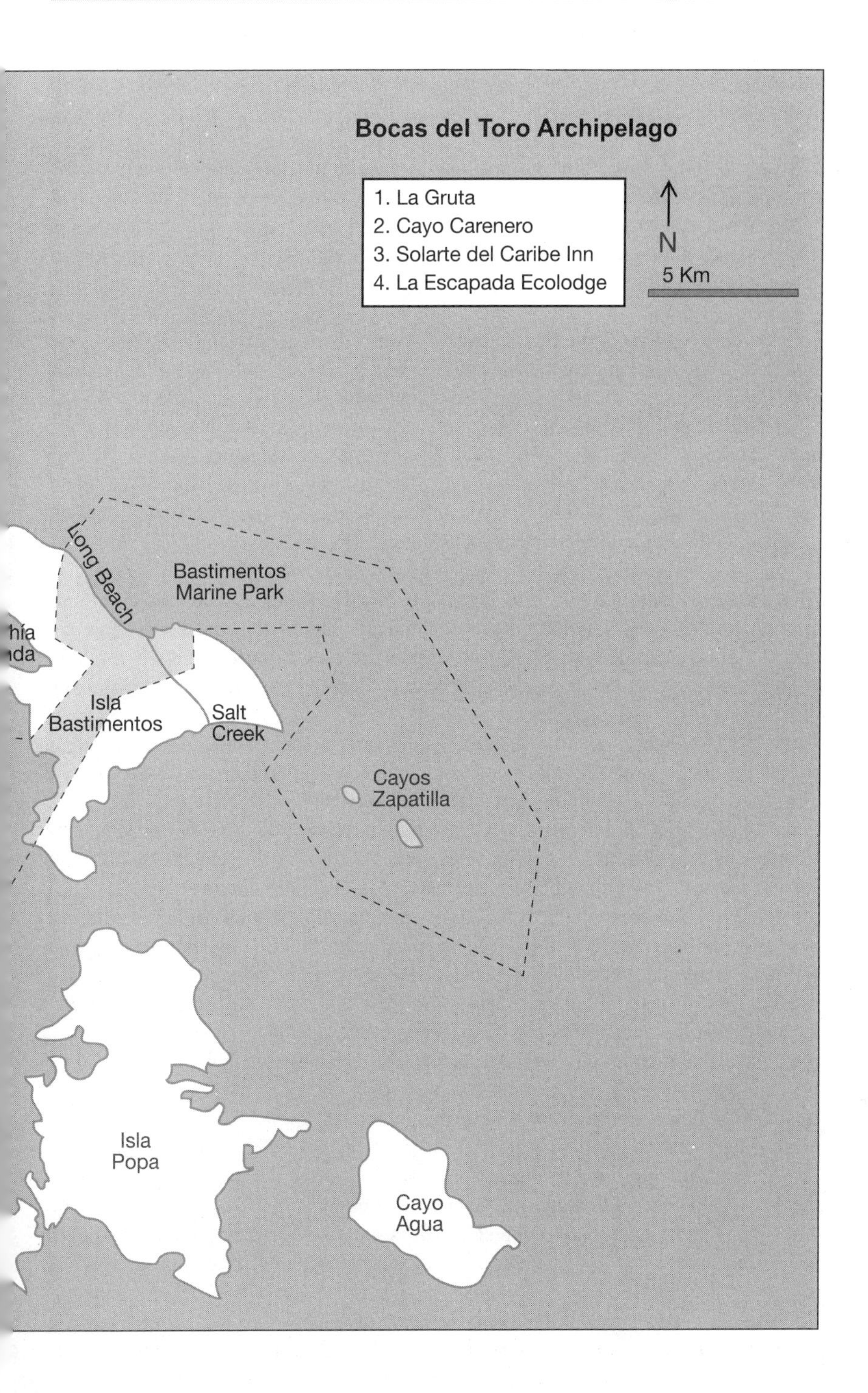
Bocas del Toro Archipelago
1. La Gruta
2. Cayo Carenero
3. Solarte del Caribe Inn
4. La Escapada Ecolodge
N
5 Km
Long Beach
Bastimentos Marine Park
Isla Bastimentos
Salt Creek
Cayos Zapatilla
Isla Popa
Cayo Agua

common on the waters of Almirante Bay, and both species can often be seen from the waterfront of Bocas Town.

Isla Colón. Isla Colón is the largest and by far the most developed island of the Archipelago. Bocas del Toro Town is located on a small peninsula at the south end of the island, connected to the rest of it by a narrow isthmus. Most of the rest of the island is undeveloped, consisting of forest, scrub, and pasture. A single asphalt road goes from Bocas town to Boca del Drago at the northwestern tip of the island.

To get to birding areas, go north on the main street (Calle 3) and then west on Av. Norte at the north end of town to reach the isthmus. Just beyond the point where the isthmus joins the main part of the island, a rough gravel road goes off to the right, signposted for Big Creek and Bluff Beach. Olive-crowned Yellowthroat can be found in the wet meadows along the first part of the road. This road continues around the east side of the island to Bluff Beach, becoming rougher the farther you go out. Most of the forest that remains along this road is fairly disturbed, and more houses are being put in all the time, so the birding is better along the Boca del Drago Road.

After passing the intersection with the road to Big Creek, the asphalt road continues through patches of forest, second growth, and pasture for the next 6 km. You can bird along the roadside here, but use care since the road is narrow and winding and often lacks shoulders, and taxi and bus drivers can come around the curves quickly.

At a little less than 6 km after the Big Creek turnoff, the road comes to the small community of Colonia Santeña, a collection of a few scattered houses. Next to a big pink house on the right there is a sign and a concrete path for *La Gruta* ("The Grotto"), a cavern used as a roost by bats that can be walked through by the adventurous. About 200 m before getting to Colonia Santeña a trail goes off to the right. It passes first through disturbed forest, reaching more intact forest after a few hundred meters. After about 700 m it forks just after climbing a small rise. The right fork descends and reaches the Río Mimitimbi after a few hundred meters; we have not followed it beyond this point. The left fork climbs slightly, passing through a fence line before also reaching the Río Mimitimbi in a few hundred meters. A short distance beyond the stream it comes to a fenced pasture. It then continues past a couple of small indigenous communities before eventually reaching the coast.

In season Bellbirds can be heard calling along the road from at least the La Gruta area north. Males can be hard to see, however, since they often call while concealed in foliage near the tops of the trees. One of best places to see them is along the road for the next 3 km past La Gruta, where there are three farm ponds in succession (more when it has been raining) on the left side. Males sometimes call from the isolated trees near these ponds where they are easier to see than in the midst of the forest.

The asphalt road reaches a T-junction on the coast about 14 km beyond the Big Creek Road intersection. The asphalt continues left to the small community of Boca del Drago. The Institute for Tropical Ecology and Conservation (ITEC) has a small biological field station here. You can get lunch in Boca del Drago at the Restaurante Yarisnori.

At the T-junction a gravel road (usually alright for 2WD) goes right for about 1 km before reaching a locked metal gate on the right. You can walk in around the side of the gate and bird along the gravel road that continues for about 4 km to Mimitimbi Bluff. This road parallels the coast through scrub and disturbed forest and can be quite birdy, especially farther out. However, although the road is public access, increasing numbers of houses are being built along here and some of the new property owners are sensitive about trespassing, so avoid straying off it.

Since vehicles are not available for rent on the island, the best option for birding is probably to arrange with a taxi driver to drop you off someplace along the road and pick you up again at a specified time. We have been quoted a round-trip price of $15.00 to La Gruta and $20.00 to Boca del Drago; don't pay until the return trip to make sure your driver comes back. There is also a bus from Bocas Town (there is a sign for it at the city park on main street) to Boca del Drago, which leaves at 8:00, 10:00, 12:00, 3:00, and 5:00, and returns from Boca del Drago at 9:00, 10:45, 1:00, 4:00, and 6:00. The price is $3.00 round-trip, $2.00 one-way to Boca del Drago, and $1.00 each way to La Gruta. Although the bus leaves a bit late to be ideal for birding, one possibility would be to hire a taxi to take you out early in the morning and then catch the bus back to town later in the day.

Isla Bastimentos. Isla Bastimentos is mostly forested, but has two small towns. Old Bank, often called simply Bastimentos Town, is on its western tip not far from Isla Colón, and is inhabited mostly by people of West Indian ancestry. Salt Creek, or Quebrada de Sal, on its eastern side, is a Ngöbe community. A major new development, however, is under construction at Red Frog Beach a few kilometers east of Old Bank. There are a few hotels and restaurants located on the southern and eastern sides of the island as well.

The central section of the island forms the land part of Bastimentos Marine Park. There are several trails on the island, but these are best visited with the aid of a guide. One area with trails is Bahía Honda, the deep embayment between the island's western peninsula and Isla Solarte. A Ngöbe community group operates a small restaurant here, and offers guided tours on the *Trail of the Sloth* (cel 6669-6269; www.bocas.com/indians/bahiahonda.htm). The route first traverses mangroves by boat, and then goes through an old cacao plantation and forest before arriving at a large cave system. We have not visited this trail ourselves, and do not have information on the price to visit. There is also a trail, about 3 km in length, that crosses the island from Salt Creek to *Long Beach (Playa Larga)* on its northern side. The first 500 meters or so are through pastures around the community. It then enters forest, and has a few fairly gentle ups and downs before reaching the coast. There is also a small trail called *Jukúble* near the village. The community currently is charging $1.00 per person to visit; $2.00 for the Jukúble Trail; $5.00 to visit the Long Beach Trail with your own guide; and $8.00 per person to be guided by someone from the community.

Tom and Ina Reichelt, who operate the Pensión Tío Tom in Bastimentos Town, are particularly good contacts for arranging a visit to Bastimentos, especially the Bahía Honda area, as well as other areas in the Archipelago. Tom is a naturalist with a particular interest in frogs, and although birding is not his area of expertise, he can certainly put you on some interesting trails. See the section on accommodations for contact information.

Boats regularly shuttle between Bocas Town and Bastimentos Town; the price is usually $2.00 each way. Getting to the more distant areas of the island where the trails are, however, entails taking a tour or hiring a boatman and guide. See below for details.

Isla Solarte. Solarte, also known as Nancy Cay, is a long thin island off the southwestern side of Bastimentos. Its eastern end, Hospital Point, is close to Bocas Town and is a popular spot for snorkeling. A little to the east of Hospital Point there is an extensive complex of small roads through forest, originally constructed as part of a planned development that is eventually supposed to include about 200 houses. However, only seven of these have been constructed so far, and at least for the time being the road complex – intended to be used by golf carts rather than cars – provides a good trail system through the forest. The system is best accessed from either of two small docks, a few hundred meters east of Hospital Point, near the Solarte del Caribe Inn. Ask your boatman to take you to the dock for the hotel; the cost from Bocas Town is usually about $5.00 each way. Be careful to keep track of forks as you walk in on the trail system, since the network is complex and it is easy to mistake your turns on the way back. The Solarte del Caribe itself is a good base for birding; see the section on accommodations for contact details.

Swan Cay. Swan Cay is a tiny but very picturesque island about 1.6 km off the north coast of Isla Colón. From September through about April about 50 to 75 pairs of Red-billed Tropicbirds nest in small alcoves and crevices on the island's cliffs. A few may be seen around the island at other times of year. These ethereal birds are an unforgettable sight as they float gently on the breeze circling the island. Brown Boobies also nest, and White-crowned Pigeons are sometimes seen. A stop at the island, most often called Bird Island locally, is a standard part of many tours around Isla Colón, which also include stops for snorkeling and swimming. Do not land on the island or approach it too closely by boat, which may disturb nesting birds.

Darkland. Darkland, or Tierra Oscura, is an area on the mainland south of Isla Cristóbal, where some species that don't occur in the Archipelago can be found. Trips here can be arranged with the aid of local guides. See the next section for details.

Other mainland areas. Days trips to mainland areas around Changuinola, including the Canal de Soropta and the San San-Pond Sak Wetlands (see the section on Changuinola for descriptions of these sites), can be arranged from Bocas Town through tour operators or local guides. See the next section for possibilities.

Tours, boat hire, and guides. Although most tourism in the Bocas Archipelago is oriented towards the sea, several tour companies offer tours to areas that may be of interest to birders. *J&J Transparente Boat Tours* (ph 757-9915; cel 6583-0351; transparentetours@hotmail.com), located on the waterfront on Calle 3 near its intersection with Calle 1, has a tour that includes either Swan Cay or the Canal de Soropta together with stops at snorkeling sites for $17.00 per person (six-person minimum). *Boteros Bocatoreños Unidos,* also known as *Boatmen United* (ph 757-9760; boterosbocas@yahoo.com), located near the south end of Calle 3 opposite the Hotel Bahía, also has a tour that includes Swan Cay in combination with snorkeling sites for $20.00 per person (five-person minimum). The *Las Brisas Hotel* (see the section on accommodations for contact information) has a tour including Swan Cay and other sites for $22.00 per person for three to four people (less per person for larger groups). They also have a tour to Salt Creek for $35.00 per person for three to four people (less per person for larger groups). Another tour that may be of interest to birders is the Manatee Tour offered by *Starfleet Eco-Adventures* (ph 757-9630; fax 757-9630; scubaStar@hotmail.com), located on Calle 1. This all-day tour goes to a manatee-viewing area on the Río San San near Changuinola via the Canal de Soropta, and would offer the possibility of seeing a variety of wetland birds in addition to manatees. The cost is $85.00 per person (four-person minimum) and includes breakfast and lunch.

If you want to set up an itinerary specifically for birding, most of the operators listed above could tailor a tour for your interests. You can also hire boatmen for specific destinations through the Boteros Bocatoreños Unidos listed above, which is a consortium of individual boatmen, or else negotiate with individual boatmen you can find along the waterfront. The Pensión Tío Tom in Bastimentos Town and Solarte del Caribe Inn on Isla Solarte (see the accommodations section below), as well as other hotels, can also arrange tours. The main tour operators, including Ancon Expeditions, Advantage Tours, EcoCircuitos, and Nattur, also either have standard Bocas del Toro packages or can arrange them; see p. 39 for contact information.

Although we are not aware or any guides currently specializing primarily in birding who are resident in the Archipelago, there are a couple of locals who have expert knowledge of trails on the islands and the adjacent mainland, and who can get you to good wildlife viewing areas. Jaime Javier Smith (ph 757-9437; cel 6674-9846; ViajesTropicales@yahoo.com) grew up on Isla Escudo de Veraguas and is very reliable. Oscar "Cuero" Gaslin (cel 6527-2254), who lives on Calle 7, on the right just before the Hotel La Veranda, in the third house on the right, has a great deal of jungle experience. We have not gone out with him ourselves, but he has been recommended to us as being very good at spotting wildlife. Both speak English.

Getting to the Bocas del Toro Archipelago. Both Aeroperlas and AirPanama have several daily flights from Panama City to Bocas del Toro Town. It is also possible to fly to Bocas Town from David on Aeroperlas and AirPanama, and from Changuinola on Aeroperlas.

You can also travel between Changuinola or Almirante and Bocas Town by motor launch. *Bocas Marine and Tours* (ph 758-9859 Changuinola, 757-9033 Bocas Town; www.bocasmarinetours.com) has seven boats daily in each direction between Bocas and Changuinola. The first boat leaves Bocas for Changuinola at 7:00 AM and the last at 4:30 PM; the respective times in the opposite direction are 8:00 AM and 5:30 PM. The price is $5.00 one-way, with a six-passenger minimum, and the trip takes about one hour. In Changuinola boats leave from the Muelle de la Sesenta; see the section on Changuinola for information on getting there. Since they do not sell round-trip tickets, you may want to buy your return ticket as soon as you arrive to make sure there is space. Although boats transit the Canal de Soropta en route, they generally go too fast to be suitable for birding. Boats go between Bocas Town and Almirante at approximately 30-40 minute intervals from 6:30 AM to 6:30 PM; the cost is $3.00.

You can drive or travel by bus to Almirante or Changuinola from Panama City (see the section on Changuinola on p. 244 for information) or from Costa Rica. If coming by car, you may prefer to take the launch from Almirante rather than Changuinola because Almirante has a guarded lot where you can leave it.

Accommodations and meals. Bocas del Toro Town has most of the hotels and restaurants in the Archipelago, as well as all of its tour companies and other services. Most of these are on Calle 3 (Main Street), which runs north and south the length of town (about eight blocks); Calle 1, which runs along the waterfront parallel to the northern part of Calle 3; and Avenida Norte (or Avenida I) which runs westward from the north ends of Calles 1 and 3 and parallels the coast. The town is so small that most places are easily found, but there are billboards with town maps at the city park near the north end of Calle 3 and at the CEFATI tourism information center on Calle 1. Other hotels and restaurants are located elsewhere on Isla Colón; Cayo Carenero a short boat ride from town; Isla Bastimentos; Isla Solarte; and Isla Cristóbal.

Accommodations in the islands range from backpacker's hostels charging under $10.00 a night to all-inclusive resorts costing hundreds of dollars a night. Restaurants are similarly diverse, from simple cafeterias to places serving excellent international cuisine, including Italian, Thai, Mexican, and others. It is beyond the scope of this guide to include the range of options available, and for a more complete list consult a more general tourist guide. A great deal of information on hotels, restaurants, tour operators, boats, as well as maps, can be found at www.bocas.com. Here we give a selection of some places that may be of particular interest to birders.

Two of the old standbys in Bocas Town, located at opposite ends of Calle 3, are the Hotel Bahía and the Hotel Las Brisas. The *Hotel Bahía* (ph 757-9626; fax 757-9692; info@hotelbahia.biz; www.hotelbahia.biz), at the south end of Calle 3, was once the headquarters building for United Fruit Company operations in the area, and was built in 1905. It has 18 rooms ($50.00 November 1 to April 30, $40.00 May 1 to October 31), all with private hot-water baths and air-conditioning. The El Encanto Bar a block away can sometimes be noisy at night, so request a room on the side of the hotel away from it. The *Hotel Las Brisas* (ph 757-9549), is at the north end of Calle 3. It has 29 rooms (starting at $15.00 with fan; $25.00 and up with air-conditioning), all with private hot-water bath. The hotel operates tours to several sites of interest for birders.

The *Solarte del Caribe Inn* (ph 757-902; cel 6488-4775; fax 757-9043; steve@solarteinn.com; www.solarteinn.com) is an attractive bed-and-breakfast on Isla Solarte, away from the bustle of Bocas Town. There are seven rooms ($71.50; two people in a room $88.00; tax and breakfast included) with fans and private hot-water baths. Dinners can be arranged for $8.00-$15.00. Daily transport is provided between the hotel and Bocas Town, which is about 10 minutes away. There are trails suitable for birding directly behind the hotel, and tours to other sites can be arranged.

The *Pensión Tío Tom* (ph 757-9831; tomima@cwp.net.pa), is a rustic *pensión* in Bastimentos Town on Isla Bastimentos. It has three rooms with shared bath and a larger room with private bath (double $13.00-$21.00; single $11.00-$20.00); cold water only. Meals can be provided. Tom and Ina Reichelt, the German owners, are excellent contacts for arranging tours specializing in frogs, insects, orchids, and photography, which would provide opportunities for birding as well.

The *Restaurante Chitré,* on Calle 3 about a block and a half south of the park, opens for breakfast at 6:00 AM. The *Restaurante Lorito* a bit north of the park has coffee available at 6:00 AM and opens for breakfast at 6:30 AM.

If you are birding on Isla Colón, about the only place available for lunch outside of Bocas Town is the *Restaurante Yarisnori* at Boca del Drago, although sodas and snacks may be available at Colonia Santeña near La Gruta. If you go to Salt Creek on the eastern end of Isla Bastimentos, you can stop for lunch at one of the three restaurants at Crawl Cay (also known as Coral Cay) at the southern tip of the island on the way back.

Bird list for the Bocas del Toro Archipelago. * = vagrant; R = rare. Some species occur on only some of the islands.

Great Tinamou
Red-billed Tropicbird
Brown Booby
Brown Pelican
Magnificent Frigatebird
Rufescent Tiger-Heron
Great Blue Heron
Great Egret
Snowy Egret
Little Blue Heron
Tricolored Heron
Cattle Egret
Green Heron
Yellow-crowned Night-Heron
Boat-billed Heron

Green Ibis
Black Vulture
Turkey Vulture
Gray-headed Kite
Hook-billed Kite R
Double-toothed Kite
Plumbeous Kite
Common Black-Hawk
Roadside Hawk
Bat Falcon
Peregrine Falcon
Gray-headed Chachalaca
Crested Guan R
White-throated Crake
Gray-necked Wood-Rail
Rufous-necked Wood-Rail
Purple Gallinule
Black-bellied Plover
Semipalmated Plover
Northern Jacana
Spotted Sandpiper
Pomarine Jaeger
Parasitic Jaeger
Laughing Gull
Pale-vented Pigeon
Scaled Pigeon
White-crowned Pigeon
Short-billed Pigeon
Blue Ground-Dove
White-tipped Dove
Gray-headed Dove
Gray-chested Dove
Crimson-fronted Parakeet
Olive-throated Parakeet
Red-fronted Parrotlet R
Brown-hooded Parrot
Blue-headed Parrot
Red-lored Amazon
Mealy Amazon
Mangrove Cuckoo R
Squirrel Cuckoo
Groove-billed Ani
Spectacled Owl
Mottled Owl
Black-and-white Owl
Lesser Nighthawk
Common Pauraque
Chuck-will's-widow
Chimney Swift
Bronzy Hermit
Band-tailed Barbthroat
Long-billed Hermit
Stripe-throated Hermit
White-necked Jacobin
Green-breasted Mango
Violet-crowned Woodnymph
Blue-chested Hummingbird
Rufous-tailed Hummingbird
Purple-crowned Fairy
Black-throated Trogon
Slaty-tailed Trogon
Rufous Motmot
Ringed Kingfisher
Belted Kingfisher
Green Kingfisher
Amazon Kingfisher
American Pygmy Kingfisher
Collared Aracari
Keel-billed Toucan
Chestnut-mandibled Toucan
Black-cheeked Woodpecker
Yellow-bellied Sapsucker R
Lineated Woodpecker
Pale-billed Woodpecker
Slaty Spinetail
Plain Xenops
Scaly-throated Leaftosser
Plain-brown Woodcreeper
Wedge-billed Woodcreeper
Cocoa Woodcreeper
Western Slaty-Antshrike
Checker-throated Antwren
White-flanked Antwren
Dot-winged Antwren
Dusky Antbird
Chestnut-backed Antbird
Brown-capped Tyrannulet
Yellow-bellied Elaenia
Ochre-bellied Flycatcher
Yellow Tyrannulet
Common Tody-Flycatcher
Stub-tailed Spadebill
Ruddy-tailed Flycatcher
Eastern Wood-Pewee
Tropical Pewee
Acadian Flycatcher
Long-tailed Tyrant
Dusky-capped Flycatcher
Panama Flycatcher
Great Crested Flycatcher
Great Kiskadee
Boat-billed Flycatcher
Social Flycatcher
Gray-capped Flycatcher
Sulphur-bellied Flycatcher
Piratic Flycatcher
Tropical Kingbird
Eastern Kingbird
Cinnamon Becard
Thrush-like Schiffornis
White-winged Becard
Masked Tityra
Black-crowned Tityra
Purple-throated Fruitcrow
Three-wattled Bellbird
Golden-collared Manakin
Red-capped Manakin
White-eyed Vireo*
Red-eyed Vireo
Lesser Greenlet
Purple Martin
Gray-breasted Martin
Mangrove Swallow
Southern Rough-winged Swallow
Barn Swallow
Bay Wren
House Wren
Long-billed Gnatwren
Tropical Gnatcatcher
Swainson's Thrush
Wood Thrush
Clay-colored Thrush
Gray Catbird
Blue-winged Warbler R
Tennessee Warbler
Yellow Warbler
Chestnut-sided Warbler
Magnolia Warbler
Yellow-rumped Warbler
Blackburnian Warbler
Bay-breasted Warbler
Cerulean Warbler
Black-and-white Warbler
American Redstart
Prothonotary Warbler
Worm-eating Warbler
Ovenbird
Northern Waterthrush
Kentucky Warbler
Mourning Warbler
Common Yellowthroat
Olive-crowned Yellowthroat
Hooded Warbler R
Canada Warbler
Bananaquit
Tawny-crested Tanager
White-lined Tanager
Red-throated Ant-Tanager
Summer Tanager
Passerini's Tanager
Blue-gray Tanager
Palm Tanager
Plain-colored Tanager
Golden-hooded Tanager
Blue Dacnis
Green Honeycreeper
Shining Honeycreeper
Blue-black Grassquit
Variable Seedeater
White-collared Seedeater

Nicaraguan Seed-Finch R
Lesser Seed-Finch
Orange-billed Sparrow
Black-striped Sparrow
Buff-throated Saltator
Rose-breasted Grosbeak
Blue-black Grosbeak
Great-tailed Grackle
Bronzed Cowbird
Giant Cowbird
Black-cowled Oriole
Orchard Oriole
Yellow-tailed Oriole
Baltimore Oriole
Yellow-billed Cacique
Montezuma Oropendola
Yellow-crowned Euphonia
Olive-backed Euphonia
House Sparrow

Escudo de Veraguas

Escudo de Veraguas is a small island, 18 km off the coast of the Valiente Peninsula and about 60 km east of Bocas del Toro Town. Only 400 hectares (988 acres) in area, it has a small but remarkably distinct fauna, including the endemic **Escudo Hummingbird** *Amazilia handleyi*. This species is very similar to the widespread Rufous-tailed Hummingbird of the mainland, but it is much larger in size, about the size of a White-vented Plumeleteer. Although originally described as a separate species by Alexander Wetmore, it is currently included as a subspecies of the Rufous-tailed Hummingbird by the AOU.

The island has only nine other species of resident land birds. Golden-collared Manakin, Bay Wren, and Blue-gray Tanager are represented by endemic subspecies, while Uniform Crake, White-crowned Pigeon, Red-lored Parrot, Tropical Kingbird, Ochre-bellied Flycatcher, and Yellow (Mangrove) Warbler are the same as those on the mainland. Uniform Crake is remarkably common, and more easily seen here than elsewhere in Panama (though still very furtive). The island has three other endemic species of vertebrates: a pygmy form of three-toed sloth, a bat, and a salamander.

There are no facilities of any kind on Escudo de Veraguas. In order to visit, it is necessary to hire a boat from Bocas del Toro Town or elsewhere in the province, and bring all your own food and gear. You can camp or else arrange to rent a vacant house from one of the island residents. During most of the year the island is inhabited by only a handful of Ngöbe, but during the fishing season in September to October hundreds may come out to dive for spiny lobster and harvest other species. There are a few dozen houses, shacks, and huts on the island, but most are empty for much of the year.

The island is extraordinarily beautiful, with steep cliffs on its southern side and a maze of blue channels eroded into its northern coast. There are many isolated rock stacks off the north side as well where Brown Boobies nest in season. The waters are clear and the fringing reef is good for snorkeling. Travel is difficult inland, however, since there are swamps just behind some of the beaches and the island's topography is very broken.

The best time to visit is from August to October when sea conditions are best. It is possible to visit at other times if the weather is good, but this may involve riding large swells much of the way. From Bocas del Toro Town the trip may take four hours or more in a good boat (depending on the size of your motor and how heavily the boat is laden), so it is best to overnight on the island. However, the hummingbird is abundant and is likely to be seen within your first few hours ashore, especially if you can stake out some of its favored *Heliconia* flowers.

Getting to Escudo de Veraguas. The best contact for arranging a trip to the island is Jaime Javier Smith (ph 757-9437; cel 6674-9846; ViajesTropicales@yahoo.com), who lives in Bocas del Toro Town and grew up on Escudo.

Accommodations and meals. The main village on the island is on its eastern end. The residents can advise you on where you can camp, and may be able to rent you a house if you prefer that to tenting. Although you should count on bringing your own food, the locals will no doubt be happy to sell you very fresh fish and other seafood.

Eastern Panama

Eastern Panama is far less developed than the Canal Area or western Panama. Roads are few, and those that exist are often in bad condition. Access to many areas is only by plane, boat, or on foot. Yet eastern Panama holds some of the best birding in the country, and will well reward the adventurous traveler. It is the best place in the country for spectacular species such as macaws and Harpy Eagle, and both the lowlands and highlands host many regional endemics restricted to eastern Panama and western Colombia. The Pearl Islands, in the Gulf of Panama, although lacking the diversity of the mainland, have several species absent or hard to find elsewhere in Panama, as well as many interesting seabird colonies.

The part of eastern Panama closest to Panama City, that is, eastern Panamá Province, the adjacent part of the Comarca de Kuna Yala, and western Darién, can easily be reached by road. Reaching the birding areas of eastern Darién requires flying, travel by boat, and/or hiking. Although a road extends into Darién as far east as Yaviza, most of the forest near it has been cut and the road is not particularly good for birding beyond the Bayano region.

Getting to eastern Panamá Province and the Comarca de Kuna Yala. From the intersection of Av. Federico Boyd and Via España, take Federico Boyd two blocks south and take a left at the traffic signal onto Av. Nicanor de Obarrio (Calle Cincuenta). Go 700 m, and turn off to the right at a sign for the Corredor Sur. Follow the signs for the airport. When you reach the end of Corredor Sur after 18 km, stay straight on the main road instead of turning off for the airport. En route to eastern Panama, you will enter the town of 24 de Diciembre (where the turnoffs for Cerro Azul and Tocumen Marsh are located) at 5.3 km after the end of the Corredor Sur. Continue straight through 24 de Diciembre to get to eastern Panamá Province and Kuna Yala. There is often a police checkpoint set up outside Chepo, 38.4 km from the end of the Corredor Sur, where you may be asked to show your license. There may be other checkpoints farther east as well, particularly at the bridge at Lake Bayano, where the driver and all passengers may be asked to show identification (such as a passport), so bring it with you.

Getting to Darién. Because of its difficulties, for the average tourist Darién is best visited by making arrangements through one of the tourism companies listed on p. 39. Ancon Expeditions has facilities at Cana and Punta Patiño. Advantage Tours has recently established a field station at Cerro Chucantí in the western part of the province and can arrange travel elsewhere in the Darién. EcoCircuitos and Nattur also have Darién options and can make arrangements.

Although eastern Darién can be visited by the independent traveler, making arrangements is not simple, since it will require hiring boats and guides (and perhaps pack horses) locally and bringing in at least part of your own food and supplies. If you plan to do so, at least one member of your party should have good Spanish language skills in order to negotiate with the locals. Few people in Darién speak English.

Aeroperlas previously had several flights per week to Darién, but at present flies to La Palma, El Real, and Sambú (all on the same flight) only on Wednesdays, and to Garachiné, Jaqué, and Piñas Bay (also the same flight) on Tuesdays and Thursdays. AirPanama goes to Garachiné, Sambú, and Jaqué (the same flight) on Mondays, Wednesdays, and Fridays. Check with the airlines for changes in service. (Note that the information on schedules on their websites may be out-of-date and incorrect.)

It is also possible to drive to Yaviza, the terminus of the Carretera Interamericana on the Río Chucunaque. The road is asphalt as far as about Metetí and then becomes gravel. The road is presently in bad condition beyond Metetí, and even with 4WD can be difficult, especially in the rainy season. There is also regular bus service to Yaviza. From there you have to hire a boat to reach El Real or other localities. See the section on El Real on pp. 286-288 for details.

Eastern Panamá Province and the Comarca de Kuna Yala

The town of Chepo historically was the "end of the road" east of Panama City, and where the Darién region traditionally began. Now part of Panamá Province, this region still retains some of its wild character, particularly once you get away from the highway. The lower Río Bayano has a large area of mangroves where many aquatic species can be found. Farther east, the Bayano region still retains some forest close to the road where many otherwise hard-to-find species can be seen. Lake Bayano, a large hydroelectric reservoir, can be explored by boat.

The Comarca de Kuna Yala is accessible by road only along the one that runs from El Llano in eastern Panamá Province to the Cartí airstrip on the Caribbean coast. Nusagandi and the Burbayar Lodge, on this road near the border between Panamá Province and Kuna Yala, are excellent places to find the foothills species of eastern Panama.

The islands off the coast of Kuna Yala are a popular tourist destination. The islands, however, are very small and have a low diversity of birds. The part of the mainland adjacent to the coast is mainly devoted to small-scale agriculture by the Kuna. Those visiting the area might be able to make arrangements through their hotels, or through one of the local tour companies, to do some birding on the mainland, though it may be difficult to get to good forest. With the exception of the areas of Mandinga on the west and Puerto Obaldía on the east, the birdlife along this coast has been scarcely investigated.

Río Bayano mangroves

The lower Río Bayano south of Chepo features the most extensive area of mangroves in eastern Panama. Species that can be found in the mangroves include Mangrove Black-Hawk, Anhinga, Cocoi Heron, Bare-throated Tiger-Heron, Yellow-crowned Night-Heron, Wood Stork, and others. There is a large nesting colony of several hundred pairs of Neotropic Cormorant and other species on the Estero Brinco off the main channel. During the period of dry-season upwelling, the sea along the coast near the mouth of the river can swarm with thousands if Brown Pelicans, Neotropic Cormorants, and other species.

The Panama Audubon Society has been working since 2000 with small communities along the coast in this area to promote ecotourism and sustainable development. A visit by small boat to the Río Bayano mangroves and to these communities can be arranged through the Society; see below.

Getting to the Río Bayano mangroves. Access to the mangroves of the Río Bayano is via Puerto Coquira, on the river south of the town of Chepo. To arrange a visit to this area call or write the Panama Audubon Society (ph 232-5977; info@panamaaudubon.org, audupan@cwpanama.net).

Nusagandi and Burbayar Lodge

The Comarca de Kuna Yala, or San Blas, is the homeland of the Kuna Indians and a semi-autonomous reserve within which they control many of their own affairs. The Kuna live mainly on small coral islands off the coast, farming on the adjacent mainland. Agriculture is confined mainly to a narrow coastal strip, and most of the mountain chain that parallels the coast retains nearly pristine rainforest used by the Kuna for hunting and gathering.

In the mid-1970s a road was constructed from El Llano on the Carretera Interamericana across the mountains to the coast near the Kuna community of Cartí. In order to prevent incursions by agricultural colonists from elsewhere in Panama, the Kuna declared the western part of their Comarca a nature reserve, and established a control point and field station at Nusagandi near the border between Panamá Province and Kuna Yala at the Continental Divide. This reserve (87,800 ha/216,958 acres) is formally known as the Área Silvestre de Narganá (Narganá Wildlands Area). Today, the part of the road in Panamá Province is largely deforested, while solid forest still prevails from Nusagandi to near the Caribbean coast.

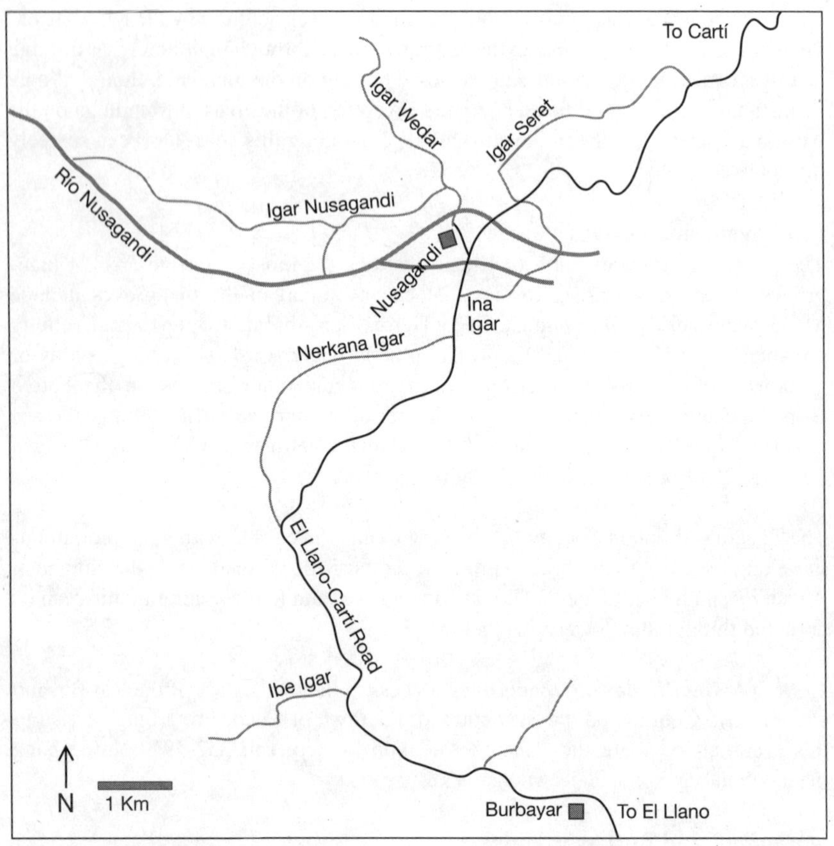

Nusagandi and Burbayar Lodge

There are two places to stay in the area: at Nusagandi itself, and at Burbayar Lodge, which is on the El Llano-Cartí Road in Panamá Province a short distance before Nusagandi. Both sites are at about 375 m (1,200 ft) in elevation. The area features a number of the endemics of the Darién lowlands and foothills as well as a several other hard-to-find species. Among the birds found here are **Russet-crowned Quail-Dove**, **Speckled (Spiny-faced) Antshrike**, Dull-mantled Antbird, **Black-crowned Antpitta**, Sapayoa, Slate-throated Gnatcatcher, and Gray-and-gold Tanager.

Burbayar Lodge. Burbayar Lodge is a small rustic ecotourism lodge located on the El Llano-Cartí Road 4 km before Nusagandi, and provides an excellent base from which to explore the area. See the section on accommodations for details on staying at Burbayar.

There is a trail system at the lodge itself, and guests accompanied by a guide also have use of trails on other nearby private property as well as those at Nusagandi. The three main trails at the lodge are *Rojo* (Red), 2.5 km (1.5 mi) long; *Blanco* (White), 400 m (0.2 mi); and *Azul* (Blue), 600 m (0.4 mi). All are fairly level, although Rojo has one long steep hill. In its final section coming back to the lodge Rojo passes through several moist ravines that are excellent for **Speckled Antshrike** and Sapayoa. There is a 1.5 km (1 mi) trail from the lodge that connects to the Ibe Igar Trail at Nusagandi and takes about 20 minutes to walk.

Nusagandi. There are six trails at Nusagandi, some of which start on the El Llano-Cartí Road and others at the headquarters. There is a $5.00 fee to use the trails, payable at the headquarters. Unfortunately, on a recent visit some of the trails were found to be somewhat overgrown or partially blocked by fallen trees. It is to be hoped that this is a temporary situation and trail maintenance will be improved in the future. In any case, you may need to check a few trails to find the ones that are most easily walkable.

Ibe Igar Trail. The Ibe Igar Trail is on the left side of the road, 1.1 km beyond the entrance to Burbayar and about 3 km before Nusagandi headquarters. **Speckled Antshrike**, Dull-mantled Antbird, and Sapayoa have been seen within the first hundred meters from the road. About 200 m in is a small stream where Sapayoa has been found nesting. It is about 3 km (1.8 mi) to a waterfall along the trail. After this the trail is overgrown but it continues at least another 3 km to a larger waterfall.

*Nergan (*or *Nerkana) Igar.* The Nergan Igar Trial starts 1.1 km beyond the entrance to Ibe Igar, up a short side road on the left, and loops back to the main road 1.3 km beyond its start. We have not birded it.

Ina Igar Trail. Ina Igar is an 800-m (0.5 mi) interpretive trail that starts on the right side of the road about 100 m before reaching the Nusagandi, then loops back to return to the road about 400 m beyond it. It is quite hilly but can have good birding. This trail and the following one run along the Continental Divide between Kuna Yala and Panamá Province.

Paseo Mariska (Igar Seret Trail). This trail begins on the left side of the road 400 m beyond Nusagandi, opposite the end of the Ina Igar Trail, and parallels the main road for 1.3 km before returning to it about 1.6 km beyond Nusagandi.

Igar Nusagandi Trail. This trail, which is 2 km (1.2 mi) long, starts at the parking lot at the headquarters and parallels the Río Nusagandi. A short side trail to the left goes to a waterfall on the Río Nusagandi. Near the junction of these trails is one of the best areas to find **Speckled Antshrike**.

Igar Wedar Trail. This trail branches off the Igar Nusagandi and extends for 1.4 km (0.8 mi). We have not birded it.

Birding is good along the road beyond Nusagandi. If the road has been maintained you can bird it by car; if not you can walk it. It is approximately 22 km from Nusagandi to the coast.

Getting to Nusagandi and Burbayar Lodge. From Panama City, follow the directions for "Getting to eastern Panama and Kuna Yala" on pp. 261-262. The turnoff for the El Llano-Cartí Road is on the left 55 km beyond the end of the Corredor Sur. The small town of El Llano is on the main road about 300 m beyond the turnoff; if you find yourself entering town, turn around and go back. The road first passes through pasture and farms. At about 7 km from the Carretera Interamericana, at about 400 m (1,300 ft) elevation, the road begins to enter fragmented forest interspersed with fields. At 13 km there is a side road to the right; stay left on the main road. The entrance to Burbayar is at 15 km, at a wooden gate on the left. At 16.2 km the Ibe Igar Trail is on the left, near a sign saying "Límite Comarca Kuna Yala – Área Silvestre Protegida." The signposted entrance to the Nusagandi headquarters is at 19 km, on the left.

The gravel El Llano-Cartí Road was repaired in early 2006, and at that time was in excellent condition past Nusagandi (and according to reports, would be improved all the way to the coast). However, in the past the road has sometimes deteriorated within a short time of being repaired. Sometimes Nusagandi has been reachable in a 2WD vehicle, while at other times a 4WD vehicle with a winch has been necessary beyond Burbayar. It would be wise to check on conditions before heading up.

Accommodations and meals

The *Burbayar Lodge* (ph 264-1679; cel 6654-0952; 6674-2964; burbayar@cwpanama.net; www.burbayar.com) has six rustic cabins that can accommodate up to 14 people; cold water only. There is also a main building with a veranda with hammocks. There is a small generator but after dark most light is provided by kerosene lamps and candles. The lodge is very ecologically conscious and designed to have as little impact on its surroundings as possible. The Spanish-born owner and host, Iñaki Ruíz, has a wealth of information about the area. Prices ($115.00 per person per night) include three meals per day, two guided hikes, and transportation from Panama City. (The price is $10.00 less if you provide your own transportation.) A variety of tours are available, including a trip by boat on Lake Bayano and a trek to the coast at Cartí. Note that it is necessary to book in advance since the lodge does not accept walk-ins.

There is a bunkhouse at *Nusagandi,* accommodating up to 30 people in screened rooms. The bunks have foam mattresses, and sheets are provided. There is also a kitchen and separate building with cold-water showers and toilets. The cost is $15.00 per person. The chief drawback of staying here during the dry season is that there is no water available then. You have to bring your own drinking water, and the showers and toilets don't work, although it is possible to bathe in a nearby stream. You also need to bring all your own food and other supplies. For reservations, call the Congreso Kuna, or Kuna Congress (ph 316-1232, 316-1234).

Bird list for Nusagandi and Burbayar Lodge. The list includes species reported from the El Llano-Cartí Road before Nusagandi as well. R = rare.

Great Tinamou
Little Tinamou
Crested Guan
Great Curassow
Marbled Wood Quail
Black-eared Wood-Quail
Least Grebe
Little Blue Heron
Black Vulture
Turkey Vulture
King Vulture
Swallow-tailed Kite
White-tailed Kite
Gray-headed Kite
Mississippi Kite
Plumbeous Kite
Tiny Hawk R
Bicolored Hawk R
Crane Hawk
Plumbeous Hawk
Barred Hawk
Semiplumbeous Hawk
White Hawk
Great Black-Hawk
Short-tailed Hawk
Swainson's Hawk
Zone-tailed Hawk
Crested Eagle R
Ornate Hawk-Eagle
Barred Forest-Falcon
Red-throated Caracara
American Kestrel
Bat Falcon
White-throated Crake
Spotted Sandpiper
Short-billed Pigeon
Ruddy Ground-Dove
Blue Ground Dove
Olive-backed Quail-Dove
Purplish-backed Quail-Dove
Russet-crowned Quail-Dove
Ruddy Quail-Dove
Orange-chinned Parakeet
Brown-hooded Parrot
Blue-headed Parrot
Red-lored Amazon
Mealy Amazon
Yellow-billed Cuckoo
Squirrel Cuckoo
Striped Cuckoo
Rufous-vented Ground-Cuckoo R
Smooth-billed Ani
Tropical Screech-Owl
Mottled Owl
Black-and-white Owl
Common Pauraque
Common Potoo
White-collared Swift
Chimney Swift
Band-rumped Swift
Rufous-breasted Hermit
Band-tailed Barbthroat
Green Hermit
Long-billed Hermit
Stripe-throated Hermit
White-tipped Sicklebill
White-necked Jacobin
Violet-headed Hummingbird
Rufous-crested Coquette
Garden Emerald
Violet-crowned Woodnymph
Violet-bellied Hummingbird
Sapphire-throated Hummingbird
Blue-chested Hummingbird
Snowy-bellied Hummingbird
Rufous-tailed Hummingbird
Bronze-tailed Plumeleteer
Purple-crowned Fairy
Long-billed Starthroat
White-tailed Trogon
Violaceous Trogon
Black-throated Trogon
Black-tailed Trogon
Slaty-tailed Trogon
Rufous Motmot
Broad-billed Motmot
Green Kingfisher
Black-breasted Puffbird
Pied Puffbird
White-fronted Nunbird
Spot-crowned Barbet
Collared Aracari
Yellow-eared Toucanet
Keel-billed Toucan
Chestnut-mandibled Toucan
Black-cheeked Woodpecker
Red-crowned Woodpecker
Stripe-cheeked Woodpecker
Cinnamon Woodpecker
Lineated Woodpecker
Crimson-bellied Woodpecker
Crimson-crested Woodpecker
Striped Woodhaunter
Slaty-winged Foliage-gleaner
Buff-throated Foliage-gleaner
Plain Xenops
Tawny-throated Leaftosser
Scaly-throated Leaftosser
Plain-brown Woodcreeper
Wedge-billed Woodcreeper
Black-striped Woodcreeper
Spotted Woodcreeper
Streak-headed Woodcreeper
Fasciated Antshrike
Barred Antshrike
Western Slaty-Antshrike
Speckled Antshrike
Russet Antshrike
Spot-crowned Antvireo
Pacific Antwren
Checker-throated Antwren
White-flanked Antwren
Dot-winged Antwren
White-bellied Antbird
Chestnut-backed Antbird
Dull-mantled Antbird
Spotted Antbird
Bicolored Antbird
Ocellated Antbird
Black-faced Antthrush
Black-headed Antthrush
Black-crowned Antpitta R
Streak-chested Antpitta
Thicket Antpitta R
Brown-capped Tyrannulet
Southern Beardless-Tyrannulet
Yellow Tyrannulet
Yellow-bellied Elaenia
Olive-striped Flycatcher
Ochre-bellied Flycatcher
Paltry Tyrannulet
Black-capped Pygmy-Tyrant
Scale-crested Pygmy-Tyrant
Yellow-margined Flycatcher
Golden-crowned Spadebill
Royal Flycatcher
Ruddy-tailed Flycatcher
Sulphur-rumped Flycatcher
Olive-sided Flycatcher
Eastern Wood-Pewee
Tropical Pewee
Acadian Flycatcher
Long-tailed Tyrant
Bright-rumped Attila
Rufous Mourner
Dusky-capped Flycatcher

Great Crested Flycatcher
Boat-billed Flycatcher
Social Flycatcher
White-ringed Flycatcher
Streaked Flycatcher
Piratic Flycatcher
Tropical Kingbird
Eastern Kingbird
Fork-tailed Flycatcher
Sapayoa
Thrush-like Schiffornis
Rufous Piha
Cinnamon Becard
White-winged Becard
One-colored Becard
Masked Tityra
Black-crowned Tityra
Blue Cotinga
Green Manakin
White-ruffed Manakin
Blue-crowned Manakin
Red-capped Manakin
Red-eyed Vireo
Yellow-green Vireo
Tawny-crowned Greenlet
Lesser Greenlet
Green Shrike-Vireo
Gray-breasted Martin
White-thighed Swallow
Southern Rough-winged Swallow
Sand Martin
Cliff Swallow
Barn Swallow
Black-bellied Wren
Bay Wren
Stripe-throated Wren
Plain Wren
House Wren
White-breasted Wood-Wren
Gray-breasted Wood-Wren
Southern Nightingale-Wren
Song Wren
Tawny-faced Gnatwren
Long-billed Gnatwren
Tropical Gnatcatcher
Slate-throated Gnatcatcher R
Swainson's Thrush
Clay-colored Thrush
Tropical Mockingbird R
Golden-winged Warbler
Tennessee Warbler
Chestnut-sided Warbler
Blackburnian Warbler
Bay-breasted Warbler
Cerulean Warbler
Black-and-white Warbler
American Redstart
Prothonotary Warbler
Northern Waterthrush
Canada Warbler
Three-striped Warbler
Bananaquit
Black-and-yellow Tanager
Dusky-faced Tanager
Olive Tanager
Sulphur-rumped Tanager
White-shouldered Tanager
Tawny-crested Tanager
White-lined Tanager
Red-throated Ant-Tanager
Hepatic Tanager
Summer Tanager
Scarlet Tanager
Crimson-backed Tanager
Flame-rumped Tanager
Blue-gray Tanager
Palm Tanager
Plain-colored Tanager
Gray-and-gold Tanager
Emerald Tanager
Bay-headed Tanager
Rufous-winged Tanager
Golden-hooded Tanager
Scarlet-thighed Dacnis
Blue Dacnis
Green Honeycreeper
Shining Honeycreeper
Red-legged Honeycreeper
Swallow Tanager R
Blue-black Grassquit
Variable Seedeater
Yellow-bellied Seedeater
Lesser Seed-Finch
Yellow-faced Grassquit
Orange-billed Sparrow
Black-striped Sparrow
Streaked Saltator
Buff-throated Saltator
Slate-colored Grosbeak
Rose-breasted Grosbeak
Blue-black Grosbeak
Eastern Meadowlark
Yellow-backed Oriole
Orange-crowned Oriole
Baltimore Oriole
Yellow-billed Cacique
Scarlet-rumped Cacique
Chestnut-headed Oropendola
Yellow-crowned Euphonia
Thick-billed Euphonia
Fulvous-vented Euphonia
White-vented Euphonia
Tawny-capped Euphonia

Bayano Region

Forty years ago, eastern Panamá Province beyond the town of Chepo was still largely forest. With the construction of the road to Darién and of the Lake Bayano hydroelectric project, much of this area has been deforested and converted to cattle pasture and agriculture. Patches of forest still remain, however, and here a remnant of the original avifauna can still be found. A number of species are close to their westernmost limit here, such as Golden-headed Manakin and White-eared Conebill. Much of the information on the birds and trails in the Bayano area was provided by José Tejada, whom we thank.

Distances to birding sites in the Bayano region are given from the Bayano Bridge, which crosses over a narrow section of the reservoir. The small village near the bridge is called Loma de Piedra, or Akwa Kuna in the Kuna language.

Just after the bridge, there are a couple of small trails where you can see some of the local specialties. Immediately after crossing the bridge, turn in on the left next to a small yellow building with signs for "Policía Nacional" and "Comarca Akua Kuna." You can park near the house with a thatched roof just beyond it, which belongs to Sr. Antonio. (If you plan to use the trail here, it would be good to make a small donation to Sr. Antonio, who takes care of the forest. There is no fixed fee but about $1.00 per person would be appropriate.) From here a short path leads down to a landing for canoes at the edge of the lake. Scan the lake shore for Striated Heron and Pied Water-Tyrant; Bare-throated Tiger-Heron and Bat Falcon have also been seen around the bridge. At the canoe landing, a small stream enters the lake. Follow the stream bed up about 10 m, where a rough path enters the forest on the left. After about 20 meters the trail crosses a small rise and divides. Either fork can be followed, but do not go more than about 200 meters down the left side to avoid entering an area restricted by the Kuna. Red-billed Scythebill, Black Antshrike, and Rufous-winged Antwren can be found in this patch of scrub. Check the area around the bridge and also farther along the road for Orange-crowned Oriole.

About 3 km beyond the bridge woodland begins to line both sides of the road, and continues for the next 16 km. All along this stretch there are many small side roads, some having been used for selective logging. Any of these may be worth exploring, although some have become overgrown and may not be passable. One particularly good spot on the road itself is near the Río Mono bridge, 6.2 km past the Bayano Bridge and just after the 134 km marker on the right side of the road. Scan the river below for Green-and-rufous Kingfisher, and the surrounding forest for Laughing Falcon and One-colored Becard; Short-tailed Nighthawk has been seen near dusk.

At about 400 m past the Río Mono is a road on the left which surprisingly is the only place in the area where Stripe-throated Wren has so far been found. One of the best side roads is at 12.3 km, on the right just past the 140 km marker. You can drive in about 300 m to where the road forks. The left fork passes through a partially cleared area before ending at a stream; the right fork goes a shorter distance before ending at the same stream. Among species that have been found here are Laughing Falcon, Pale-bellied Hermit, Black-tailed Trogon, Great Jacamar, Green-and-rufous Kingfisher, Red-rumped and Golden-green Woodpeckers, Black Antshrike, **Black-billed Flycatcher**, Barred, Pied, and Black-breasted Puffbirds, One-colored Becard, White-headed Wren, Buff-breasted Wren, White-eared Conebill, and Orange-crowned Oriole.

The main road continues through forest for the next 6 km or so beyond this side road, and then enters deforested country. For most of the rest of the way, all the way to its terminus at Yaviza in Darién, there is very little forest along the road. To the south of the road is the isolated mountain range of the Serranía de Majé, whose highest point is Cerro Chucantí at its eastern end on the border with Darién. See the section on Darién on visiting Cerro Chucantí.

About 45 km east of the Bayano Bridge is Ipetí, consisting of three separate settlements which include mestizos (Ipetí Colono), Kuna (Ipetí Kuna), and Emberá (Ipetí Chocó, or Ipetí Emberá). Ipetí Chocó is about 2 km from the main road, on a dirt road that goes right just before the bridge over the Río Ipetí. Beyond Ipetí Chocó are trails that go into the forests on the slopes of the Serranía de Majé which might well repay exploration with the aid of a guide from the community. Ipetí Chocó has recently initiated some small ecotourism projects, and it also may be possible to rent a place to sleep in a hammock at one of the houses here. You may also be able to find local guides in Ipetí Kuna on the left side of the main road.

Lake Bayano itself can make an interesting excursion. In parts of the lake there is a "ghost forest" of dead trees, still standing more than 30 years after they were drowned by creation of the reservoir. Cocoi Herons and other aquatic species are common, and large numbers of Neotropic Cormorants nest in the dead trees in some areas. The rare Black-collared Hawk can sometimes be found near the mouth of the Río Majé. The easiest way to arrange a trip on the lake is through Burbayar Lodge (see the preceding section), but it may also be possible to find a boatman at the communities near the bridge. Boat hire might be expected to cost $50.00 and up, depending on the length of the trip.

Getting to the Bayano Region. From Panama City, follow the directions for "Getting to eastern Panamá Province and Kuna Yala" (pp. 261-262). It is about 80 km from the end of the Corredor Sur to the Bayano Bridge (25 km east of the El Llano-Cartí Road). Directions for specific sites after the Bayano Bridge are given above.

Accommodations and meals. The Bayano Region is best visited as a day trip from Panama City. See pp. 74-75 for Panama City hotels. The most convenient hotel for a trip to the Bayano is the Riande Aeropuerto, just past the end of the Corredor Sur on the main road. A visit to the area can also be combined with a visit to the Burbayar Lodge or Nusagandi.

Bird list for the Bayano region, including Lake Bayano. R = rare.

Great Tinamou
Little Tinamou
Black-bellied Whistling-Duck
Muscovy Duck
Blue-winged Teal
Gray-headed Chachalaca
Crested Guan R
Least Grebe
Brown Pelican
Neotropic Cormorant
Anhinga
Magnificent Frigatebird
Rufescent Tiger-Heron
Bare-throated Tiger-Heron R
Great Blue Heron
Cocoi Heron
Great Egret
Snowy Egret
Little Blue Heron
Tricolored Heron
Cattle Egret
Green Heron
Striated Heron
Capped Heron
Black-crowned Night-Heron
Boat-billed Heron
Green Ibis
Wood Stork
Black Vulture
Turkey Vulture
King Vulture
Osprey
Gray-headed Kite
Hook-billed Kite
Swallow-tailed Kite
White-tailed Kite
Double-toothed Kite
Plumbeous Kite
Black-collared Hawk R
Bicolored Hawk R
Crane Hawk
Plumbeous Hawk R
Semiplumbeous Hawk
White Hawk
Gray Hawk
Mangrove Black-Hawk
Great Black-Hawk
Broad-winged Hawk

Short-tailed Hawk
Swainson's Hawk
Zone-tailed Hawk
Black-and-white Hawk-Eagle R
Black Hawk-Eagle
Barred Forest-Falcon
Slaty-backed Forest-Falcon R
Collared Forest-Falcon
Laughing Falcon
Merlin
Bat Falcon
Peregrine Falcon
White-throated Crake
Yellow-breasted Crake R
Purple Gallinule
Common Moorhen
Sungrebe
Sunbittern
Semipalmated Plover
Wattled Jacana
Solitary Sandpiper
Spotted Sandpiper
Least Sandpiper
Royal Tern
Pale-vented Pigeon
Scaled Pigeon
Short-billed Pigeon
Plain-breasted Ground-Dove
Ruddy Ground-Dove
Blue Ground-Dove
White-tipped Dove
Gray-chested Dove
Spectacled Parrotlet
Orange-chinned Parakeet
Brown-hooded Parrot
Blue-headed Parrot
Red-lored Amazon
Mealy Amazon
Black-billed Cuckoo
Yellow-billed Cuckoo
Squirrel Cuckoo
Little Cuckoo
Striped Cuckoo
Greater Ani
Smooth-billed Ani
Tropical Screech-Owl
Vermiculated Screech-Owl
Crested Owl
Spectacled Owl
Central American Pygmy-Owl
Mottled Owl
Black-and-white Owl
Short-tailed Nighthawk
Lesser Nighthawk
Common Nighthawk
Common Pauraque
Rufous Nightjar
Great Potoo
Common Potoo
White-collared Swift
Short-tailed Swift
Band-rumped Swift
Lesser Swallow-tailed Swift
Rufous-breasted Hermit
Band-tailed Barbthroat
Long-billed Hermit
Pale-bellied Hermit
Stripe-throated Hermit
Scaly-breasted Hummingbird
White-necked Jacobin
Black-throated Mango
Violet-headed Hummingbird
Garden Emerald
Green-crowned Woodnymph
Violet-bellied Hummingbird
Blue-chested Hummingbird
Snowy-bellied Hummingbird
Rufous-tailed Hummingbird
White-vented Plumeleteer
Purple-crowned Fairy
Long-billed Starthroat
Violaceous Trogon
Black-throated Trogon
Black-tailed Trogon
Slaty-tailed Trogon
Blue-crowned Motmot
Rufous Motmot
Broad-billed Motmot
Ringed Kingfisher
Belted Kingfisher
Amazon Kingfisher
Green Kingfisher
Green-and-rufous Kingfisher
Barred Puffbird
White-necked Puffbird
Black-breasted Puffbird
Pied Puffbird
White-whiskered Puffbird
Gray-cheeked Nunlet
White-fronted Nunbird
Great Jacamar
Collared Aracari
Yellow-eared Toucanet
Keel-billed Toucan
Chestnut-mandibled Toucan
Olivaceous Piculet
Black-cheeked Woodpecker
Red-crowned Woodpecker
Red-rumped Woodpecker
Golden-green Woodpecker
Cinnamon Woodpecker
Lineated Woodpecker
Crimson-crested Woodpecker
Buff-throated Foliage-gleaner
Plain Xenops
Scaly-throated Leaftosser
Plain-brown Woodcreeper
Ruddy Woodcreeper
Olivaceous Woodcreeper
Long-tailed Woodcreeper
Wedge-billed Woodcreeper
Northern Barred-Woodcreeper
Cocoa Woodcreeper
Black-striped Woodcreeper
Streak-headed Woodcreeper
Red-billed Scythebill
Fasciated Antshrike
Great Antshrike
Barred Antshrike
Black Antshrike
Western Slaty-Antshrike
Russet Antshrike
Spot-crowned Antvireo
Moustached Antwren
Pacific Antwren
Checker-throated Antwren
White-flanked Antwren
Rufous-winged Antwren
Dot-winged Antwren
Dusky Antbird
Jet Antbird
Bare-crowned Antbird
White-bellied Antbird
Chestnut-backed Antbird
Spotted Antbird
Bicolored Antbird
Ocellated Antbird
Black-faced Antthrush
Streak-chested Antpitta
Brown-capped Tyrannulet
Southern Beardless-Tyrannulet
Mouse-colored Tyrannulet
Yellow Tyrannulet
Yellow-crowned Tyrannulet
Forest Elaenia
Greenish Elaenia
Yellow-bellied Elaenia
Lesser Elaenia
Olive-striped Flycatcher
Ochre-bellied Flycatcher
Sooty-headed Tyrannulet
Paltry Tyrannulet
Black-capped Pygmy-Tyrant
Southern Bentbill
Common Tody-Flycatcher
Black-headed Tody-Flycatcher
Brownish Twistwing
Olivaceous Flatbill
Yellow-olive Flycatcher
Yellow-margined Flycatcher
Golden-crowned Spadebill

Royal Flycatcher
Ruddy-tailed Flycatcher
Sulphur-rumped Flycatcher
Black-tailed Flycatcher
Black-billed Flycatcher
Western Wood-Pewee
Eastern Wood-Pewee
Tropical Pewee
Yellow-bellied Flycatcher
Acadian Flycatcher
Willow/Alder Flycatcher
Black Phoebe
Pied Water-Tyrant
Long-tailed Tyrant
Bright-rumped Attila
Sirystes
Rufous Mourner
Dusky-capped Flycatcher
Panama Flycatcher
Great Crested Flycatcher
Great Kiskadee
Boat-billed Flycatcher
Rusty-margined Flycatcher
Social Flycatcher
Gray-capped Flycatcher
White-ringed Flycatcher
Streaked Flycatcher
Piratic Flycatcher
Tropical Kingbird
Eastern Kingbird
Fork-tailed Flycatcher
Thrush-like Schiffornis
Rufous Piha
Speckled Mourner
Cinereous Becard
Cinnamon Becard
White-winged Becard
One-colored Becard
Masked Tityra
Black-crowned Tityra
Blue Cotinga
Purple-throated Fruitcrow
Green Manakin
Golden-collared Manakin
Golden-headed Manakin
Red-capped Manakin
Yellow-throated Vireo
Philadelphia Vireo
Red-eyed Vireo
Yellow-green Vireo
Tawny-crowned Greenlet
Golden-fronted Greenlet
Lesser Greenlet
Green Shrike-Vireo
Black-chested Jay
Purple Martin
Gray-breasted Martin
Southern Martin
Brown-chested Martin
Mangrove Swallow
Southern Rough-winged Swallow
Sand Martin
Cliff Swallow
Barn Swallow
White-headed Wren
Black-bellied Wren
Bay Wren
Stripe-throated Wren R
Rufous-and-white Wren
Buff-breasted Wren
House Wren
White-breasted Wood-Wren
Southern Nightingale-Wren
Song Wren
Long-billed Gnatwren
Tropical Gnatcatcher
Veery
Gray-cheeked Thrush
Swainson's Thrush
Wood Thrush
Clay-colored Thrush
Tropical Mockingbird
Golden-winged Warbler
Tennessee Warbler
Tropical Parula
Yellow Warbler
Chestnut-sided Warbler
Magnolia Warbler
Cape May Warbler
Bay-breasted Warbler
Black-and-white Warbler
Prothonotary Warbler
Ovenbird
Northern Waterthrush
Louisiana Waterthrush
Kentucky Warbler
Mourning Warbler
Canada Warbler
Buff-rumped Warbler
Bananaquit
White-eared Conebill
Gray-headed Tanager
Sulphur-rumped Tanager
White-shouldered Tanager
White-lined Tanager
Red-throated Ant-Tanager
Summer Tanager
Scarlet Tanager
Crimson-backed Tanager
Flame-rumped Tanager
Blue-gray Tanager
Palm Tanager
Plain-colored Tanager
Golden-hooded Tanager
Blue Dacnis
Green Honeycreeper
Shining Honeycreeper
Red-legged Honeycreeper
Blue-black Grassquit
Slate-colored Seedeater R
Variable Seedeater
Yellow-bellied Seedeater
Lesser Seed-Finch
Yellow-faced Grassquit
Orange-billed Sparrow
Black-striped Sparrow
Streaked Saltator
Buff-throated Saltator
Black-headed Saltator
Slate-colored Grosbeak
Rose-breasted Grosbeak
Blue-black Grosbeak
Indigo Bunting
Great-tailed Grackle
Giant Cowbird
Orchard Oriole
Yellow-backed Oriole
Orange-crowned Oriole
Yellow-tailed Oriole
Baltimore Oriole
Yellow-billed Cacique
Scarlet-rumped Cacique
Yellow-rumped Cacique
Crested Oropendola
Chestnut-headed Oropendola
Thick-billed Euphonia
Fulvous-vented Euphonia
White-vented Euphonia

The Pearl Islands

The Pearl Islands, located in the Gulf of Panama, include three large islands, the Islas del Rey, San José, and Pedro González, and many smaller ones. Although the archipelago has comparatively few species relative to the mainland, the larger islands are the only place in Panama where White-fringed Antwren can be found, and Pale-bellied Hermit and Bare-throated Tiger-Heron are also easier to find here than on the mainland. There are also interesting seabird colonies in the archipelago, particularly on Pacheca, but also on other islands.

Unfortunately tourism facilities in the archipelago at present are quite limited. The island of Contadora, near the northern end of the group, has regular flights and a range of hotels. Although Contadora is a good base from which to visit the seabird colony on nearby Pacheca, and Bare-throated Tiger-Heron can be seen along its beaches, it lacks most forest birds, including the antwren and hermit.

Beyond Contadora, the rest of the archipelago has only a few scattered villages, and many islands are uninhabited. There are only two places to stay: the town of San

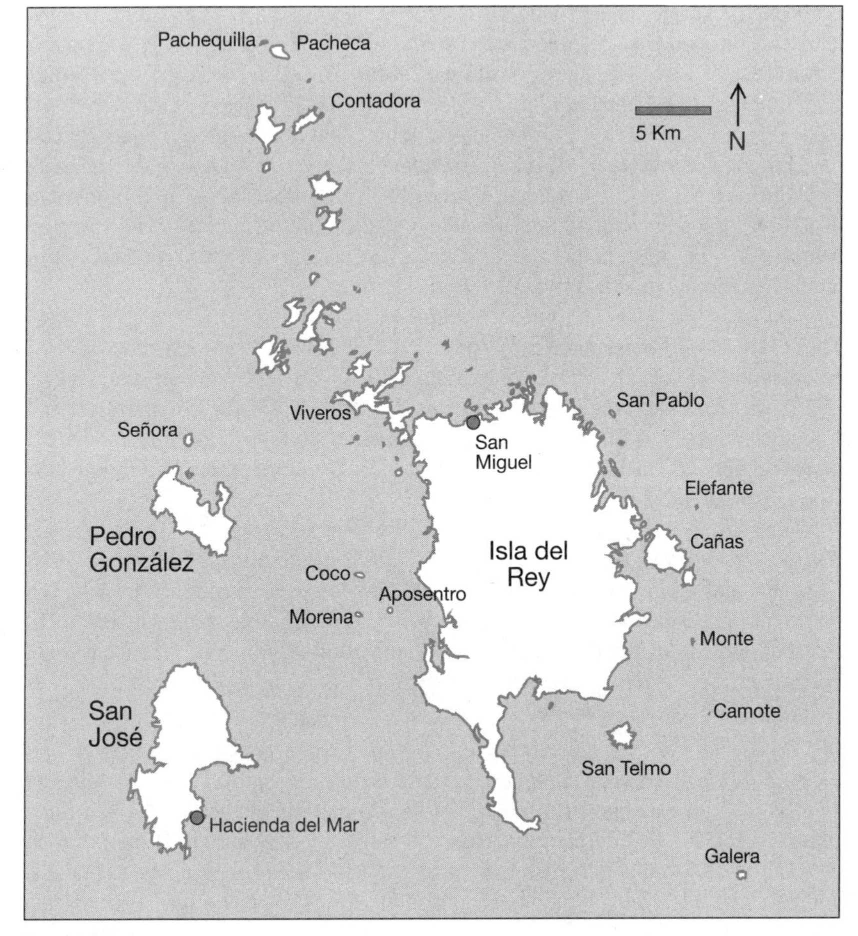

Pearl Islands

Miguel on Isla del Rey has a small *pensión*, and Isla San José has a high-end resort that caters mainly to sports fishermen, the Hacienda del Mar. See the accommodations section for more information on these hotels.

For those wanting to see the specialties of the Pearl Islands, San Miguel is the best option. White-fringed Antwren is common in second growth around the edge of town and at the airstrip, and Pale-bellied Hermit has been seen at the airstrip as well. Bare-throated Tiger-Heron has been seen on the beach and perched on buoys just offshore. Flying out on the Thursday afternoon AirPanama flight and returning on Saturday morning (see below) should give you plenty of time to bird. Hacienda del Mar is also an option for those who can afford it. We have seen Pale-bellied Hermit next to the hotel pool, and White-fringed Antwren within a few minutes walk of the hotel. Bare-throated Tiger-Heron is common on the coast of the island.

Pacheca, the northernmost island of the archipelago, has its largest seabird colony, including thousands of nesting pairs of Neotropic Cormorant as well as large numbers of Magnificent Frigatebird and Brown Pelican. Brown and Blue-footed Boobies also nest, mainly on smaller, rocky Pachequilla nearby. Pacheca is most easily visited by hiring a boat on Contadora, which is about 3 km to its south. Boatmen can be found at Playa Larga near the Hotel Contadora and at Playa Galeón near the Punta Galeón Hotel. We have been quoted a price of $20.00 per person per hour (minimum of $25.00), but a group might be able to negotiate a lower price depending on how business is. The trip should not require more than an hour or two. Do not attempt to land on Pacheca, which is privately owned.

Those who have access to a boat might find a visit to some of the other seabird colonies interesting. Thousands of Neotropic Cormorants nest on the small islands of Isla de Coco and Isla Morena between Isla del Rey and Pedro González, and sometimes the sky is filled with skeins of cormorants returning from feeding areas elsewhere in the Gulf of Panama. Boobies nest on picturesque and craggy Isla Aposentro east of Morena.

The largest colony of Brown Pelican in the Pearls is on the western end of Isla Pedro González and on nearby Isla Señora. There are also pelican colonies on Isla San Pablo off the northeast corner of Rey, and on San Telmo to the southeast. Galera, at the southern end of the archipelago, has a colony of Magnificent Frigatebird, and boobies nest on several tiny rocky islets off the east side of Rey, including Islas Elephante, Monte, and Camote.

Getting to the Pearl Islands. Aeroperlas and AirPanama both have daily flights to Contadora from Panama City. AirPanama flies to San Miguel on Isla del Rey three times a week (in the afternoon on Mondays and Thursdays and in the morning on Saturdays), and to Isla San José four times a week (Mondays, Thursdays, Fridays, and Sundays, all in the afternoon). Flights to San José can be arranged by Hacienda del Mar when you book.

Accommodations and meals

Isla Contadora

The *Hotel Punta Galeón* (ph 250-4134; fax 250-4135; reservas@puntagaleon.com; www.puntagaleon.com) has 48 rooms ($110.00), all with private hot-water bath and air-conditioning. There are two restaurants on premises.

The *Hotel Contadora* (ph 214-3719, 214-3720; fax 264-1178; reservas@hotelconta dora.com; www.hotelcontadora.com) has 300 rooms ($91.00 low season, $109.00 high season), all with private hot-water bath and air-conditioning. There is a restaurant on premises.

The *Villa Romántica* (ph 250-4067; contadora@villa-romantica.com; www.contado ra-villa-romantica.com) has a total of nine rooms and suites (starting at $62.00 low season, $72.00 high season) with air-conditioning. Some rooms have private baths and some have shared. There is a restaurant on premises.

The *Cabañas de Contadora* (ph 250 4214; cel 6674-3839; hcorrand@excite.com; http://cabanasdecontadora.tripod.com/) has four studio apartments ($30.00 double) with fans, private hot-water baths, and kitchens.

See one of the standard tourist guides for more information on Contadora.

Isla San José

The *Hacienda del Mar* (ph 269-6634, 269-6613; fax 264-1787; info@haciendadel mar.net; www.haciendadelmar.net) is basically a fishing lodge, but a very comfortable one. It has 14 cabins ($275.00-$375.00 low season, $330.00-$475.00 high season), all with private hot-water bath and air-conditioning. The lodge has an excellent restaurant.

Isla del Rey

The *Hotel San Miguel* (cel 6661-7198), operated by Camilo and Milena Zapata, has 15 basic rooms ($10.00) with fan and shared cold-water bath. The hotel is a few houses up from the beach, to the right of the Servicios Marítimos building. The *Restaurante Cacería* is a couple of streets over, and opens at 5:45 AM.

Bird list for the Pearl Islands and surrounding waters. Many of the land birds are found only on the larger islands. * = vagrant; R = rare.

Little Tinamou
Gray-headed Chachalaca
Least Grebe R
Waved Albatross*
Galapagos Petrel*
Sooty Shearwater
White-vented Storm-Petrel*
Wedge-rumped Storm-Petrel
Black Storm-Petrel
Least Storm-Petrel
Nazca Booby R
Blue-footed Booby
Peruvian Booby*
Brown Booby
Red-footed Booby*
Brown Pelican
Neotropic Cormorant
Guanay Cormorant*
Magnificent Frigatebird
Bare-throated Tiger-Heron
Great Blue Heron
Great Egret
Snowy Egret
Tricolored Heron

Little Blue Heron
Agami Heron R
Cattle Egret R
Green Heron
Black-crowned Night-Heron
Yellow-crowned Night-Heron
White Ibis
Black Vulture
Turkey Vulture
Osprey
Hook-billed Kite
Swallow-tailed Kite
Plumbeous Kite
Zone-tailed Hawk
Mangrove Black-Hawk
Roadside Hawk
Crested Caracara
Yellow-headed Caracara
American Kestrel
Merlin
Bat Falcon
Peregrine Falcon
Gray-necked Wood-Rail
Uniform Crake R
Black-bellied Plover
Wilson's Plover
Semipalmated Plover
Killdeer
American Oystercatcher
Willet
Spotted Sandpiper
Whimbrel
Ruddy Turnstone
Sanderling
Western Sandpiper
Least Sandpiper
South Polar Skua R
Pomarine Jaeger
Parasitic Jaeger
Laughing Gull
Gull-billed Tern
Royal Tern
Elegant Tern
Common Tern
Sooty Tern
Bridled Tern
Black Tern
Pale-vented Pigeon
Ruddy Ground-Dove
White-tipped Dove
Ruddy Quail-Dove
Blue-headed Parrot
Red-lored Amazon
Yellow-crowned Amazon
Black-billed Cuckoo
Smooth-billed Ani
Barn Owl
Tropical Screech-Owl
Lesser Nighthawk
Common Pauraque
Vaux's Swift
Pale-bellied Hermit
Garden Emerald
Snowy-bellied Hummingbird
Ringed Kingfisher
Belted Kingfisher
Green-and-rufous Kingfisher
Red-crowned Woodpecker
Barred Antshrike
White-fringed Antwren
Jet Antbird
Southern Beardless-Tyrannulet
Northern Scrub-Flycatcher
Greenish Elaenia
Yellow-bellied Elaenia
Lesser Elaenia
Ochre-bellied Flycatcher
Bran-colored Flycatcher
Eastern Wood-Pewee
Western Wood-Pewee
Acadian Flycatcher
Alder Flycatcher
Willow Flycatcher
Cattle Tyrant R
Panama Flycatcher
Great Crested Flycatcher
Streaked Flycatcher
Piratic Flycatcher
Tropical Kingbird
Eastern Kingbird
Yellow-throated Vireo
Yellow-green Vireo
Red-eyed Vireo
Gray-breasted Martin
Sand Martin
Cliff Swallow
Barn Swallow
Buff-breasted Wren
House Wren
Tropical Gnatcatcher
Swainson's Thrush
Cedar Waxwing R
Tennessee Warbler
Yellow Warbler (migrant forms)
Yellow (Mangrove) Warbler
Black-throated Green Warbler R
Yellow-rumped Warbler
Blackburnian Warbler
Bay-breasted Warbler
Cerulean Warbler
Black-and-white Warbler
American Redstart
Prothonotary Warbler
Northern Waterthrush
Louisiana Waterthrush
Canada Warbler
Bananaquit
Summer Tanager
Scarlet Tanager
Crimson-backed Tanager
Blue-gray Tanager
Red-legged Honeycreeper
Blue-black Grassquit
Yellow-bellied Seedeater
Lesser Seed-Finch
Streaked Saltator
Rose-breasted Grosbeak
Indigo Bunting
Dickcissel R
Orchard Oriole
Baltimore Oriole
Great-tailed Grackle

Darién Province

Darién is Panama's largest and least developed province. Although the Carretera Interamericana runs as far as Yaviza on the Río Chucunaque, other roads are few, and most travel is via its many rivers or by air. Major rivers include the Chucunaque in the western part of the province, and the Tuira, Balsas, Sambú, and Jaqué in the east. There are four main mountain ranges: the Serranía de Majé in the western part of the province; and in the eastern part, the Serranía de Darién, including Cerro Tacarcuna, paralleling the Caribbean coast; the Serranía de Pirre, in the center of the Isthmus; and the Serranía de Jungurudó, near the Pacific coast.

The lowlands and highlands of Darién both host many regional endemics that can be found only here and in eastern Colombia. Many other basically South American species are found nowhere else in the AOU checklist area. In addition, Darién is the best place in Panama for several spectacular species, including four species of macaws, Harpy and Crested Eagles, Golden-headed Quetzal, and others.

The most frequently visited sites in Darién are Cana and Punta Patiño, both with facilities operated by Ancon Expeditions. More recently a small field station has opened at Cerro Chucantí in the western part of the province. El Real and other localities can be visited with the aid of local tour operators, or by the more adventurous independent traveler.

A word about safety: some parts of Darién are not safe to visit, due to the activity of Colombian guerillas, drug traffickers, and bandits. These areas include the upper Río Tuira above Boca de Cupe (an area which includes Cerro Tacarcuna), and Jaqué and the Río Jaqué. Cerro Chucantí, Punta Patiño, Cana, and El Real are away from the zone of guerilla activity, and are regarded as safe. At Yaviza and beyond, travelers are required to check in at the local police post and register their passport information and planned itineraries. If traveling independently in Darién, it is always a good idea in any case to inquire with the local police regarding any security problems that may exist in the area.

Cerro Chucantí

Cerro Chucantí (1,439 m/4,749 ft), on the border between Panamá and Darién Provinces, is the highest peak in the Serranía de Majé. Several of the endemics of the Darién foothills and highlands can be found here, including **Russet-crowned Quail-Dove**, **Violet-capped Hummingbird**, **Beautiful Treerunner**, **Varied Solitaire**, and **Tacarcuna Bush-Tanager**. Other species of interest include Crested Guan, Great Curassow, Black-and-white Hawk-Eagle, Central American Pygmy-Owl, **Violet-throated Toucanet**, **Stripe-cheeked Woodpecker**, Ruddy Foliage-gleaner, and **Black-crowned Antpitta**.

Getting to Cerro Chucantí. Cerro Chucantí is reached from Tortí on the Carretera Interamericana, about 60 km east of the Bayano Bridge. A gravel road to the right at Tortí goes to the village of Platanilla, from which an old logging road ascends to near the peak. The Chucantí field station (see below) can be reached on foot or on horseback from there in about three to four hours, depending on conditions. Improvements to the road are planned that will make it possible to get to the station by tractor in a shorter time.

Accommodations and meals. Advantage Tours has recently opened the *Chucantí Field Station* (www.advantagepanama.com/chucantistation.html) high on Cerro Chucantí, which has four bedrooms (total capacity eight people) with private hot-water bath. Construction of a second house of equal capacity is planned. Because of the site's remote location visits must be arranged in advance. Advantage Tours offers a four-day/three-night expedition to Chucantí for $350.00 per person, which includes round-trip transportation from Panama City, bilingual guide service, meals, and lodging at the station. More than 200 hectares (500 acres) of primary forest have been purchased near the station for conservation. Contact Advantage Tours (p. 39) for more information.

Bird list for Cerro Chucantí. The list includes species found in the lowlands on the route in to Cerro Chucantí. R = rare.

Great Tinamou
Little Tinamou
Crested Guan
Great Curassow
Marbled Wood-Quail
Tawny-faced Quail R
Cattle Egret
Black Vulture
Turkey Vulture
King Vulture
Pearl Kite
White Hawk
Gray Hawk
Black-and-white Hawk-Eagle R
Black Hawk-Eagle
Ornate Hawk-Eagle
Barred Forest-Falcon
American Kestrel
Short-billed Pigeon
Ruddy Ground-Dove
Blue Ground-Dove
White-tipped Dove
Gray-chested Dove
Russet-crowned Quail-Dove
Orange-chinned Parakeet
Brown-hooded Parrot
Blue-headed Parrot
Red-lored Amazon
Mealy Amazon
Squirrel Cuckoo
Rufous-vented Ground-Cuckoo R
Smooth-billed Ani
Vermiculated Screech-Owl
Central American Pygmy-Owl R
Mottled Owl
Chimney Swift
Green Hermit
Long-billed Hermit
Stripe-throated Hermit
Scaly-breasted Hummingbird
White-necked Jacobin
Violet-headed Hummingbird
Blue-throated Goldentail
Violet-capped Hummingbird
Snowy-bellied Hummingbird
Purple-crowned Fairy
Long-billed Starthroat
Black-throated Trogon
Slaty-tailed Trogon
Tody Motmot
Green Kingfisher
Black-breasted Puffbird
Great Jacamar
Violet-throated Toucanet
Yellow-eared Toucanet
Keel-billed Toucan
Olivaceous Piculet
Black-cheeked Woodpecker
Red-crowned Woodpecker
Stripe-cheeked Woodpecker
Cinnamon Woodpecker
Lineated Woodpecker
Spotted Barbtail
Beautiful Treerunner
Buff-throated Foliage-gleaner
Ruddy Foliage-gleaner
Plain Xenops
Tawny-throated Leaftosser
Plain-brown Woodcreeper
Ruddy Woodcreeper
Long-tailed Woodcreeper
Northern Barred-Woodcreeper
Buff-throated Woodcreeper
Black-striped Woodcreeper
Spotted Woodcreeper
Brown-billed Scythebill
Fasciated Antshrike
Barred Antshrike
Western Slaty-Antshrike
Russet Antshrike
Plain Antvireo
Slaty Antwren
Dot-winged Antwren
Dusky Antbird
Spotted Antbird
Wing-banded Antbird R
Bicolored Antbird
Ocellated Antbird
Black-faced Antthrush
Black-crowned Antpitta R
Yellow-crowned Tyrannulet
Forest Elaenia
Yellow-bellied Elaenia
Olive-striped Flycatcher
Yellow-green Tyrannulet
Paltry Tyrannulet
Black-capped Pygmy-Tyrant
Scale-crested Pygmy-Tyrant
Black-headed Tody-Flycatcher
Eye-ringed Flatbill
Yellow-olive Flycatcher
Yellow-margined Flycatcher
White-throated Spadebill
Golden-crowned Spadebill
Ruddy-tailed Flycatcher
Sulphur-rumped Flycatcher
Olive-sided Flycatcher
Eastern Wood-Pewee
Black Phoebe
Bright-rumped Attila
Rufous Mourner
Dusky-capped Flycatcher
Boat-billed Flycatcher
White-ringed Flycatcher
Streaked Flycatcher
Tropical Kingbird
Cinnamon Becard
Masked Tityra
Rufous Piha
Blue Cotinga
White-ruffed Manakin
Golden-headed Manakin
Red-capped Manakin
Gray-breasted Martin
Red-eyed Vireo

Tawny-crowned Greenlet
Lesser Greenlet
Green Shrike-Vireo
Barn Swallow
Bay Wren
Stripe-throated Wren
Rufous-breasted Wren
House Wren
Ochraceous Wren
White-breasted Wood-Wren
Gray-breasted Wood-Wren
Southern Nightingale-Wren
Tropical Gnatcatcher
Varied Solitaire
Pale-vented Thrush
Tropical Mockingbird
Tropical Parula
Prothonotary Warbler
Louisiana Waterthrush
Slate-throated Redstart
Buff-rumped Warbler
Bananaquit
Tacarcuna Bush-Tanager
Black-and-yellow Tanager
Olive Tanager
Gray-headed Tanager
Hepatic Tanager
Crimson-backed Tanager
Blue-gray Tanager
Palm Tanager
Plain-colored Tanager
Silver-throated Tanager
Bay-headed Tanager
Scarlet-thighed Dacnis
Green Honeycreeper
Shining Honeycreeper
Blue-black Grassquit
Slate-colored Seedeater R
Variable Seedeater
Ruddy-breasted Seedeater
Chestnut-capped Brush-Finch
Orange-billed Sparrow
Buff-throated Saltator
Slate-colored Grosbeak
Blue-black Grosbeak
Yellow-backed Oriole
Crested Oropendola
Yellow-crowned Euphonia
Thick-billed Euphonia
Fulvous-vented Euphonia

Punta Patiño

Punta Patiño is a former cattle ranch on the southern shore of the Gulf of San Miguel. In the late 1980s it was purchased by ANCON, Panama's largest conservation organization, and converted to Panama's largest private nature reserve. Ancon Expeditions operates a lodge at the site.

Although the area lacks many of the species found in Darién National Park, such as macaws, a wide range of lowland species occur, including Humboldt's Sapphire and **Black Oropendola**. The main trail near the lodge is the *Sendero Piedra Candela* (Flintstone Trail), a loop that traverses regenerating semideciduous forest. You can also bird along roads and some shorter trails near the lodge. An old overgrown road goes from the cabins to more distant parts of the property. It is about an hour's walk on this road to good primary forest.

Ancon Expeditions offers a four-day/three-night package to Punta Patiño that includes birding at the lodge, plus an overnight trip to the nearby Emberá village of Mogue to see an active Harpy Eagle nest. Contact Ancon Expeditions for further information on this and other options.

Getting to Punta Patiño. Punta Patiño is reached by flying to La Palma, the capital of Darién province, and then traveling by small boat about 20 km along the coast to the lodge. Arrangements should be made through Ancon Expeditions.

Accommodations and meals. The main building of the lodge, which includes the dining hall, is located on a bluff overlooking the Gulf of San Miguel and has a second floor veranda. There are 10 pleasant cabins nearby, all with private cold-water bath and air-conditioning. Tours to Punta Patiño include meals and lodging. A trip to Mogue involves staying in a traditional Emberá house, sleeping either in hammocks or screened tents.

Darién National Park

Darién National Park protects 579,000 hectares (1,430,740 acres) of lowland and montane tropical rainforest, much of it wilderness, along Panama's border with Colombia. Besides a great diversity of birds, the park holds a wealth of other wildlife, including Jaguar, Puma, Baird's Tapir, White-lipped Peccary, and Brown-headed Spider Monkey.

Being in a remote and undeveloped region, the park unfortunately has very few facilities for visitors. The main areas visited by birders are Cana, a lodge on the slopes of Cerro Pirre operated by Ancon Expeditions, and El Real, a small town near the park boundaries which provides access to the Pirre ranger station. A visit to Cana is much easier (and more comfortable), since Ancon Expeditions handles all arrangements, but also much more expensive. Although the park can be visited more cheaply via El Real, it requires a substantial hike as well as carrying in your own food and equipment. It is also much easier to see the highland endemics of the Serranía de Pirre[7] from Cana, since you start out higher on the mountain. Getting to the top of the Serranía de Pirre from the El Real side requires camping and a much more strenuous hike.

Cana

Cana is the site of an ancient gold mine, Santa Cruz de Cana, which the Spanish exploited during colonial times. Raided several time by pirates, the mines were eventually abandoned in the early 1700s due to a cave-in as well as an Indian uprising. They were re-opened in the late 1800s by British interests, who constructed a tramway to the Río Tuira. Some mining activity and exploration continued until the early 1990s, when the facilities were turned over to the conservation organization ANCON. The site is now run as an ecotourism operation by Ancon Expeditions. It consists of an airstrip, bunkhouse, dining hall, and several other buildings.

Cana offers some of the most superb birding in Panama, or for that matter anywhere in the New World tropics. Four species of macaw (Great Green, Red-and-green, Blue-and-yellow, and Chestnut-fronted) can regularly be seen flying overhead or perched in trees near the bunkhouse. Crested Eagle and Black-and-white Hawk-Eagle are regularly seen, and Solitary and Harpy Eagles are possibilities. This is also probably the easiest place in Panama to see Crested Guan and Great Curassow. Regional endemics include **Choco Tinamou**, **Purplish-backed** and **Russet-crowned Quail-Doves**, **Rufous-cheeked Hummingbird**, **Dusky-backed Jacamar**, **Violet-throated Toucanet**, **Choco Toucan**, **Stripe-cheeked Woodpecker**, **Beautiful Treerunner**, **Choco Tapaculo**, **Yellowish-green Tyrannulet**, **Black-billed Flycatcher**, **Varied Solitaire**, **Pirre Warbler**, **Pirre Bush-Tanager**, **Viridian Dacnis**,

---

[7] This mountain range is officially known as the Serranía de Pirre. The peak officially named Cerro Pirre is near the northern end of the range near El Real. However, the part of the range above Cana is commonly referred to loosely as "Cerro Pirre" (including in Ridgely), and we follow that convention here. The peak closest to Cana is officially called Alturas de Nique.

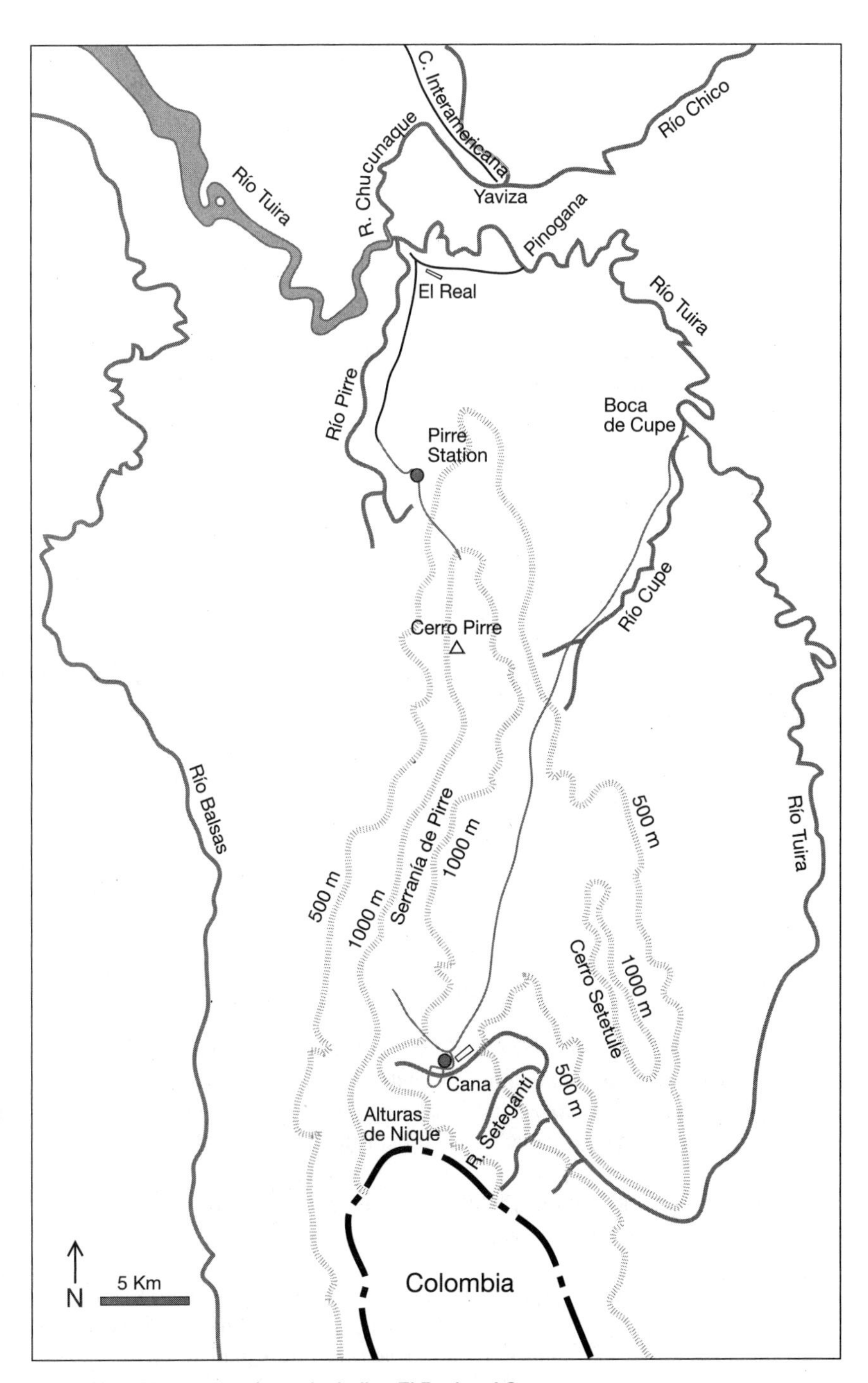

Serranía de Pirre and environs, including El Real and Cana

Green-naped Tanager, and **Yellow-collared Chlorophonia**, among others. Other species of interest found here (for many it is the best place in the AOU area to find them) include Tiny, Bicolored, and Plumbeous Hawks, Slaty-backed Forest-Falcon, Red-throated Caracara. Plumbeous Pigeon, Blue-fronted Parrotlet, Saffron-headed Parrot, Tooth-billed Hummingbird, Greenish Puffleg, Purple-throated Woodstar, Golden-headed Quetzal, Tody Motmot, Barred Puffbird, Double-banded Graytail, Rufous-winged and Rufous-rumped Antwrens, Wing-banded Antbird, Thicket, Scaled, and Ochre-breasted Antpittas, Sooty-headed Tyrannulet, Golden-crowned Flycatcher, Cinereous and One-colored Becards, Black-tipped Cotinga, Sapayoa, Sharpbill, Yellow-browed Shrike-Vireo, White-headed, Sooty-headed, and Stripe-throated Wrens, Slate-throated Gnatcatcher, White-eared Conebill, Yellow-backed, Scarlet-browed, Lemon-spectacled, and Swallow Tanagers, and Yellow-green Grosbeak.

Besides birds, Cana is also excellent for other wildlife. Tracks of Puma, Jaguar, and Baird's Tapir are common, and on rare occasions they are even glimpsed. Herds of White-lipped Peccary can sometimes be seen foraging near the bunkhouse. Mantled Howler and White-faced Monkeys are common, and Brown-headed Spider Monkeys can often be seen on the ridge near the high camp.

There are four main birding trails at Cana: Mine, Boca de Cupe, River, and Cerro Pirre. (In addition there is a short trail near the bunkhouse that goes to an abandoned locomotive.) Right around the bunkhouse and the airstrip, however, can be the best place to see macaws and raptors as they fly overhead, since they may be obscured by trees when you are within the forest. Cana is a bit over 500 m (1,650 ft) elevation. The first three trails remain at about this elevation (at least the parts that a birder is likely to cover), while the Cerro Pirre Trail goes to the top of the ridge at about 1,575 m (5,200 ft).

Mine Trail. This trail departs from the end of the clearing farthest from the airstrip. It runs at an easy grade up to the former site of the main mine pit, near which there is abandoned mine equipment shrouded in moss and ferns.

Boca de Cupe Trail. This trail leaves the clearing near the western end of the airstrip. It follows the route of the former tramway, and in some places rails can still be seen. It passes through swampy terrain and forest interrupted by clearings. The trail runs all the way to the town of Boca de Cupe on the Río Tuira, a two-day hike away.

River Trail. This trail leaves from behind the bunkhouse, first passing through open woodland and scrub before crossing a stream.

Cerro Pirre Trail. This trail departs from the western end of the airstrip before the start of the Boca de Cupe Trail. It goes steeply uphill for the first 200 meters, then levels out somewhat. It continues through beautiful forest until reaching the tent camp at about 1,300 m (4,200 ft.). The hike is long, steep in places, and can be strenuous. From there the trail continues through cloud forest to reach its high point at 1,575 m (5,195 ft). Tours to Cana usually include one or two nights at the tent camp.

How to get to Cana. Cana is operated by Ancon Expeditions, and is frequently included in the itineraries of birding tours. Access is by small plane from Panama City. Contact Ancon Expeditions (p. 39) for information on arranging a visit.

Accommodations and meals. Accommodations at Cana are rustic but comfortable. The bunkhouse has eight double-occupancy bedrooms sharing two hot-water baths. There is a separate dining hall a short distance away. Generator power is generally provided until early evening. The camp on Cerro Pirre has tents pitched under a thatched roof with cots and mattresses with blankets. There is a latrine but no bathing facilities. Meals, which are usually basic but ample, are provided as part of package tours.

Bird list for Cana. * = vagrant; R = rare. P = species recorded mainly from higher elevations on Cerro Pirre, rather than at Cana itself or just above it. Some of these may occur at Cana occasionally.

Great Tinamou
Little Tinamou
Choco Tinamou
Gray-headed Chachalaca
Crested Guan
Great Curassow
Marbled Wood-Quail
Black-eared Wood-Quail
Rufescent Tiger-Heron
Fasciated Tiger-Heron R
Cocoi Heron R
Snowy Egret R
Cattle Egret R
Agami Heron R
Black Vulture
Turkey Vulture
King Vulture
Osprey R
Gray-headed Kite
Swallow-tailed Kite
Pearl Kite R
Double-toothed Kite
Plumbeous Kite
Tiny Hawk
Sharp-shinned Hawk R
Bicolored Hawk R
Crane Hawk
Plumbeous Hawk R
Barred Hawk P
Semiplumbeous Hawk
White Hawk
Mangrove Black-Hawk
Great Black-Hawk
Solitary Eagle R
Roadside Hawk R
Broad-winged Hawk
Short-tailed Hawk
Swainson's Hawk
Crested Eagle R
Harpy Eagle R
Black-and-white Hawk-Eagle R
Black Hawk-Eagle
Ornate Hawk-Eagle
Barred Forest-Falcon
Slaty-backed Forest-Falcon R
Collared Forest-Falcon
Red-throated Caracara
Laughing Falcon
Bat Falcon
Orange-breasted Falcon R
White-throated Crake
Gray-necked Wood-Rail
Solitary Sandpiper R
Spotted Sandpiper
Pale-vented Pigeon
Scaled Pigeon
Plumbeous Pigeon
Ruddy Pigeon
Short-billed Pigeon
Dusky Pigeon R
Ruddy Ground-Dove R
Blue Ground-Dove
White-tipped Dove
Gray-chested Dove
Purplish-backed Quail-Dove
Russet-crowned Quail-Dove P
Violaceous Quail-Dove
Ruddy Quail-Dove
Chestnut-fronted Macaw
Great Green Macaw
Red-and-green Macaw
Blue-and-yellow Macaw
Spectacled Parrotlet R
Orange-chinned Parakeet
Blue-fronted Parrotlet P
Saffron-headed Parrot
Brown-hooded Parrot
Blue-headed Parrot
Red-lored Amazon
Mealy Amazon
Gray-capped Cuckoo*
Squirrel Cuckoo
Little Cuckoo
Striped Cuckoo
Rufous-vented Ground-Cuckoo R
Greater Ani
Smooth-billed Ani
Tropical Screech-Owl R
Vermiculated Screech-Owl
Bare-shanked Screech-Owl P R
Crested Owl
Spectacled Owl
Central American Pygmy-Owl R
Mottled Owl
Black-and-white Owl
Short-tailed Nighthawk
Common Pauraque
Chuck-will's-widow R
Rufous Nightjar
Great Potoo
Common Potoo

Black Swift R
White-collared Swift
Chapman's Swift R
Band-rumped Swift
Lesser Swallow-tailed Swift
Rufous-breasted Hermit
Band-tailed Barbthroat
Green Hermit
Long-billed Hermit
Stripe-throated Hermit
White-tipped Sicklebill
Tooth-billed Hummingbird P
Green-fronted Lancebill P
White-necked Jacobin
Brown Violet-ear
Black-throated Mango
Violet-headed Hummingbird
Rufous-crested Coquette
Green Thorntail R
Green-crowned Woodnymph
Violet-bellied Hummingbird
Blue-throated Goldentail
Rufous-cheeked Hummingbird P
Blue-chested Hummingbird
Snowy-bellied Hummingbird
Rufous-tailed Hummingbird
White-vented Plumeleteer
Bronze-tailed Plumeleteer R
Green-crowned Brilliant
Greenish Puffleg P
Purple-crowned Fairy
Long-billed Starthroat
Purple-throated Woodstar R
White-tailed Trogon
Violaceous Trogon
Collared Trogon P
Black-throated Trogon
Black-tailed Trogon
Slaty-tailed Trogon
Golden-headed Quetzal P
Tody Motmot
Rufous Motmot
Broad-billed Motmot
Green Kingfisher
American Pygmy Kingfisher
Barred Puffbird
White-necked Puffbird
Black-breasted Puffbird
Pied Puffbird
White-whiskered Puffbird
Lanceolated Monklet R
Gray-cheeked Nunlet
White-fronted Nunbird
Dusky-backed Jacamar
Rufous-tailed Jacamar R
Great Jacamar
Spot-crowned Barbet
Red-headed Barbet P
Violet-throated Toucanet
Collared Aracari
Yellow-eared Toucanet
Keel-billed Toucan
Chestnut-mandibled Toucan
Choco Toucan R
Olivaceous Piculet
Black-cheeked Woodpecker
Red-rumped Woodpecker
Stripe-cheeked Woodpecker
Cinnamon Woodpecker
Lineated Woodpecker
Crimson-bellied Woodpecker
Crimson-crested Woodpecker
Slaty Spinetail
Red-faced Spinetail P
Double-banded Graytail
Spotted Barbtail P
Beautiful Treerunner P
Lineated Foliage-gleaner P
Slaty-winged Foliage-gleaner
Buff-throated Foliage-gleaner
Ruddy Foliage-gleaner
Plain Xenops
Streaked Xenops R
Tawny-throated Leaftosser
Scaly-throated Leaftosser
Sharp-tailed Streamcreeper P R
Plain-brown Woodcreeper
Olivaceous Woodcreeper
Long-tailed Woodcreeper
Wedge-billed Woodcreeper
Northern Barred-Woodcreeper
Cocoa Woodcreeper
Black-striped Woodcreeper
Spotted Woodcreeper
Streak-headed Woodcreeper
Red-billed Scythebill
Brown-billed Scythebill R
Fasciated Antshrike
Great Antshrike
Western Slaty-Antshrike
Russet Antshrike
Plain Antvireo
Spot-crowned Antvireo
Moustached Antwren
Pacific Antwren
Checker-throated Antwren
White-flanked Antwren
Slaty Antwren
Rufous-winged Antwren
Dot-winged Antwren
Rufous-rumped Antwren
Dusky Antbird
Jet Antbird
Bare-crowned Antbird
Chestnut-backed Antbird
Dull-mantled Antbird R
Immaculate Antbird
Spotted Antbird
Wing-banded Antbird
Bicolored Antbird
Ocellated Antbird
Black-faced Antthrush
Rufous-breasted Antthrush
Black-crowned Antpitta R
Scaled Antpitta P R
Streak-chested Antpitta
Thicket Antpitta
Ochre-breasted Antpitta P R
Choco Tapaculo P
Brown-capped Tyrannulet
Yellow-crowned Tyrannulet R
Forest Elaenia
Gray Elaenia
Greenish Elaenia
Yellow-bellied Elaenia
Olive-striped Flycatcher
Ochre-bellied Flycatcher
Slaty-capped Flycatcher
Yellow-green Tyrannulet
Sooty-headed Tyrannulet
Paltry Tyrannulet
Bronze-olive Pygmy-Tyrant P
Black-capped Pygmy-Tyrant
Scale-crested Pygmy-Tyrant
Southern Bentbill
Common Tody-Flycatcher
Black-headed Tody-Flycatcher
Brownish Twistwing
Eye-ringed Flatbill P
Olivaceous Flatbill
Yellow-margined Flycatcher
White-throated Spadebill
Golden-crowned Spadebill
Royal Flycatcher
Ruddy-tailed Flycatcher
Sulphur-rumped Flycatcher
Black-tailed Flycatcher
Black-billed Flycatcher
Common Tufted-Flycatcher
Olive-sided Flycatcher
Western Wood-Pewee
Eastern Wood-Pewee
Yellow-bellied Flycatcher R
Acadian Flycatcher
Willow/Alder Flycatcher
Long-tailed Tyrant
Cattle Tyrant R
Bright-rumped Attila
Sirystes
Rufous Mourner

Dusky-capped Flycatcher
Great Crested Flycatcher
Great Kiskadee R
Boat-billed Flycatcher
Rusty-margined Flycatcher
Gray-capped Flycatcher
White-ringed Flycatcher
Golden-crowned Flycatcher P
Streaked Flycatcher
Piratic Flycatcher
Tropical Kingbird
Eastern Kingbird
Sapayoa R
Thrush-like Schiffornis
Rufous Piha
Speckled Mourner
Cinereous Becard
Cinnamon Becard
White-winged Becard
One-colored Becard
Masked Tityra
Black-crowned Tityra
Blue Cotinga
Black-tipped Cotinga
Purple-throated Fruitcrow
Green Manakin
Golden-collared Manakin
White-ruffed Manakin
Blue-crowned Manakin
Golden-headed Manakin
Sharpbill
Yellow-throated Vireo
Philadelphia Vireo R
Red-eyed Vireo
Yellow-green Vireo
Tawny-crowned Greenlet
Lesser Greenlet
Yellow-browed Shrike-Vireo P
Black-chested Jay
Gray-breasted Martin R
Brown-chested Martin R
White-thighed Swallow
Southern Rough-winged Swallow
Blue-and-white Swallow R
Cliff Swallow
Barn Swallow
Black-capped Donacobius R
White-headed Wren
Sooty-headed Wren
Black-bellied Wren
Bay Wren
Stripe-throated Wren
House Wren
Ochraceous Wren P
White-breasted Wood-Wren
Gray-breasted Wood-Wren P
Southern Nightingale-Wren
Song Wren
Tawny-faced Gnatwren
Long-billed Gnatwren
Slate-throated Gnatcatcher
Varied Solitaire P
Slaty-backed Nightingale-Thrush P
Gray-cheeked Thrush
Swainson's Thrush
Pale-vented Thrush
White-throated Thrush
Golden-winged Warbler
Tennessee Warbler
Tropical Parula
Yellow Warbler
Chestnut-sided Warbler
Yellow-rumped Warbler
Black-throated Green Warbler
Blackburnian Warbler
Bay-breasted Warbler
Palm Warbler R
Cerulean Warbler
Black-and-white Warbler
American Redstart
Kentucky Warbler
Prothonotary Warbler
Louisiana Waterthrush
Mourning Warbler
Canada Warbler
Slate-throated Redstart P
Pirre Warbler P
Buff-rumped Warbler
Bananaquit
White-eared Conebill
Pirre Bush-Tanager P
Yellow-backed Tanager
Black-and-yellow Tanager
Dusky-faced Tanager
Lemon-spectacled Tanager
Scarlet-browed Tanager
White-shouldered Tanager
Red-throated Ant-Tanager R
Red-crowned Ant-Tanager
Hepatic Tanager
Summer Tanager
Scarlet Tanager
Crimson-backed Tanager
Flame-rumped Tanager
Blue-gray Tanager
Palm Tanager
Plain-colored Tanager
Gray-and-gold Tanager
Emerald Tanager
Silver-throated Tanager
Speckled Tanager
Bay-headed Tanager
Golden-hooded Tanager
Green-naped Tanager R
Scarlet-thighed Dacnis
Blue Dacnis
Viridian Dacnis
Green Honeycreeper
Shining Honeycreeper
Purple Honeycreeper
Red-legged Honeycreeper
Swallow Tanager
Slate-colored Seedeater R
Variable Seedeater
Lesson's Seedeater*
Yellow-bellied Seedeater
Ruddy-breasted Seedeater
Lesser Seed-Finch
Chestnut-capped Brush-Finch P
Stripe-headed Brush-Finch
Orange-billed Sparrow
Black-striped Sparrow
Streaked Saltator R
Buff-throated Saltator
Black-headed Saltator
Slate-colored Grosbeak
Yellow-green Grosbeak P
Rose-breasted Grosbeak
Blue-black Grosbeak
Indigo Bunting
Red-breasted Blackbird R
Giant Cowbird
Yellow-backed Oriole
Yellow-tailed Oriole
Baltimore Oriole
Yellow-billed Cacique
Scarlet-rumped Cacique R
Yellow-rumped Cacique
Crested Oropendola
Chestnut-headed Oropendola
Black Oropendola R
Thick-billed Euphonia R
Fulvous-vented Euphonia
White-vented Euphonia R
Orange-bellied Euphonia
Yellow-collared Chlorophonia P

El Real

El Real de Santa María is a small and rather run-down town near the junction of the Ríos Tuira and Chucunaque. Its main attraction is that second-growth forest can be found just outside of town, and a variety of Darién endemics and specialties can be seen here within an easy walk from the town's center. Several of these, such as Spectacled Parrotlet, Yellow-breasted Flycatcher, Black-capped Donacobius, and **Black Oropendola**, do not occur at Cana or are scarce there. El Real is also a gateway to Darién National Park. For those with more time and sufficient stamina, a longer hike will bring you to Pirre Station at the edge of the Park, where many deeper forest species can be found. The Serranía de Pirre can also be climbed from this direction, but it requires a guide and carrying in all your food and equipment. Because Pirre Station is at a much lower elevation than Cana, and because the trail to the top of the Serranía de Pirre is much steeper here, this is a much more strenuous way to see the highland endemics of Darién than going to Cana.

There are two dirt roads out of El Real. One goes past the airstrip to the town of Pinogana, the other to the village of Pirre Uno, about 10 km away, from which a trail continues to Pirre Station. Most birding has been done along the road to Pirre Uno, though those with more time might also wish to explore the Pinogana Road (which also has a branch to the left that goes 1.5 km to the Río Tuira). The first few kilometers of the Pirre Uno Road pass through pastures, younger second growth, and taller secondary forest. There are occasional side trails that allow entry into the secondary forest. Birds that can be seen along this part of the road include Spectacled Parrotlet, Pale-bellied Hermit, Tody Motmot, Spot-breasted, Golden-green, and Red-rumped Woodpeckers, Double-banded Graytail, Yellow-breasted Flycatcher, White-eared Conebill, Yellow-backed Tanager, Orange-crowned Oriole, and **Black Oropendola**. Black Antshrike can be found in scrub, and Cattle Tyrant has been seen near the airstrip. Black-capped Donacobius can be found in marshy areas with tall vegetation, especially in the wet season. A Harpy Eagle was even seen once not far from the airstrip, but it can hardly be counted on!

Getting to El Real. El Real was formerly served by three Aeroperlas flights a week from Panama City via La Palma, but service has recently been cut to only one flight per week, on Wednesdays. Although it is possible more frequent service may be restored, at present if you want to go for a shorter time than a week, or on another day than the scheduled flight, you will need to go by road to Yaviza and then take a boat to El Real. A group might find it feasible to fly to El Real on Aeroperlas, and for the return, charter a small plane to come from Panama City to pick them up (or vice versa), though this would be expensive.

Yaviza is served by several buses a day from Panama City from the main bus terminal at Los Pueblos near Albrook. The trip takes about nine hours, and the fare is $11.00 one-way. You can also drive, but you will need to find a secure place to leave your car in Yaviza. It is about 260 km from Panama City to Yaviza.

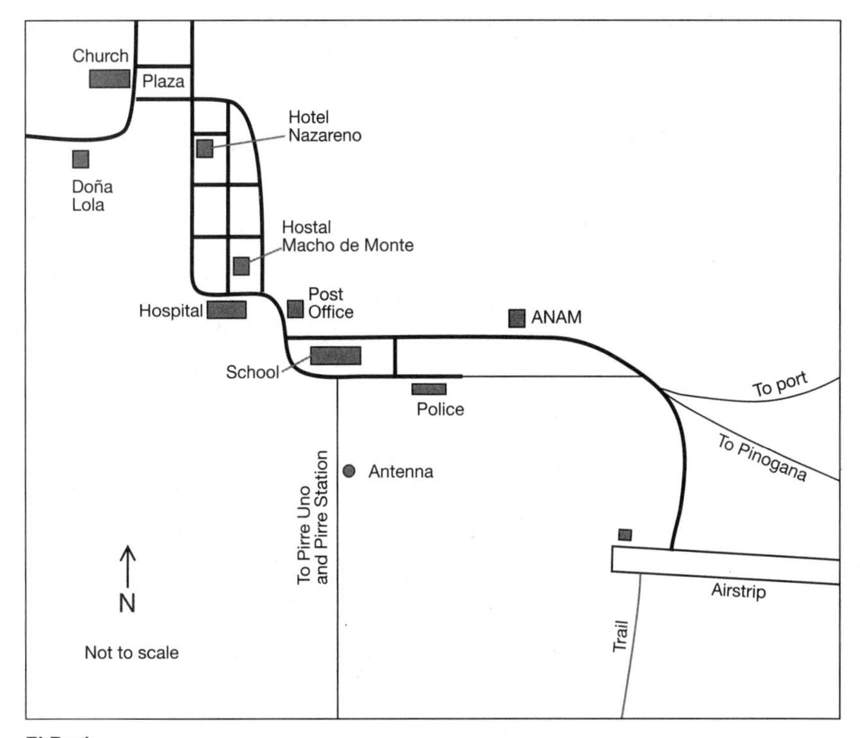

El Real

Transport between Yaviza and El Real is generally by means of *piraguas,* large motorized dugout canoes. There is no scheduled service. You will either have to hire a piragua to take you, or hitch a ride on one that is going to El Real anyway. The latter is much cheaper (probably about $5.00 per person), but you may have to wait around to find one. The cost of hiring a piragua varies with both the size of the boat and size of the motor. You will also have to pay a boatman (*motorista*) who will operate the motor plus a helper (*marinero*) who will help navigate, and will be charged for the price the gas separately, which will cost more for boats with larger motors. The following boatmen, listed in order of price, operate out of Yaviza:

Jorge Alcíbar, of the Hotel América (ph 299-4212, 293-7495; the first number is a public telephone where you can ask for Jorge), has a piragua with capacity for six people ($5.00) and a 15-HP motor ($15.00); motorista $10.00; marinero $7.00-$8.00; gas additional.

Janeth José, of the Hotel Yadarién (ph 299-4232, 299-4412), has a piragua with a capacity for 14 people ($8.00) and a 25-HP motor ($25.00); motorista $15.00; marinero $10.00; gas additional.

Ivan Lai (ph 299-4358) has a piragua with a capacity for 12 people, and uses motors of between 15-40 HP; with a 40 HP motor the price is $120.00 including his service as motorista. The marinero is an additional $20.00-$26.00; gas additional.

Accommodations and meals. Accommodations in El Real are very limited, and all places are quite basic.

Probably the best place to stay is the bunkhouse at the *ANAM station* (ph 299-6965) on the edge of town not far from the airstrip. There are two bunkrooms with a total of 10 beds with foam mattresses and a shared cold-water bath. The cost is $10.00 per night. You can also camp here for $5.00. To get to the ANAM station from the airstrip, take a left on the main road upon leaving the airstrip. The station is about 600 m along this road, on the right. It is planned to move the station to a spot closer to the airstrip in the near future.

The *Hotel Nazareno* (ph 299-6548), on the main street near the center of town, is very run-down. There are eight rooms with a total of two double ($8.00) and 15 single ($5.00) beds between them. While there are nominally cold-water baths these are sometimes (we suspect always) out of service. The hotel is also near a couple of very noisy bars and so is not the best place to stay if you want to get a good night's sleep.

The *Hostal Macho de Monte,* with eight rooms ($7.00 per person), is on the main street before you get to the Nazareno.

There is only one place to eat in town, the *Restaurante Doña Lola* (ph 299-6743), which serves basic meals for $2.00. It is near the town plaza, which is just past the Nazareno.

Yaviza. If you need to stay overnight in Yaviza en route to El Real, there are several hotels in town. The *Hotel Yadarién* (Janeth José, ph 299-4232) has eight rooms ($20.00 with air-conditioning, $15.00 with fan). There is also a house available with eight rooms with fan, toilet, and a kitchen, that rents for $5.00 per person. The *Hotel America* (Jorge Alcíbar, public phone in Yaviza, 299-4212; in Panama City 293-7495) has eight rooms ($20.00 with air-conditioning, $15.00 with fan). Restaurants include the *Rincón Dominicano* and a place run by Ivan Lai, which serve basic meals for $2.00 and under. If you will be camping around El Real or at Pirre Station, it is recommended to buy your supplies in Yaviza rather than El Real because price and selection are much better here. The best-supplied store is on the ground floor of the Hotel Yadarién.

Pirre Station

Pirre Station is at the edge of Darién National Park about 14 km south of El Real. The first 10 km or so, to the town of Pirre Uno, is along a dirt road through habitat similar to that around El Real, that is, pasture interspersed with scrub and secondary forest. Beyond Pirre Uno the route continues as a foot trail and enters better forest.

The station itself, reached in another 4 km, is inside tall primary lowland forest. The station is frequently called "Rancho Frío," but that name more properly refers to a campsite higher on the slopes of Cerro Pirre.[8]

There are several trails at Pirre Station that can be birded. One option is to walk back along the trail that goes to El Real. The *La Cascada Trail* is a loop that goes to some waterfalls along the Río Peresénico, which runs next to the station; look for Fasciated Tiger-Heron here. The *La Bruja Trail* connects with the La Cascada Trial and goes to a police outpost. The first part of this trail can be walked but closer to the outpost it is restricted. You can also bird along the first part of the trail that goes up the Serranía de Pirre, *El Estrangulador;* see the next section for a description. Species that can be found near Pirre Station include King Vulture, Crested Guan, Great Curassow, Marbled Wood-Quail, Tawny-faced Quail, Red-and-green and Great Green Macaws, Gray-cheeked Nunlet, Crimson-bellied Woodpecker, Striped Woodhaunter, Rufous-winged Antwren, White-headed and Stripe-throated Wrens, and Scarlet-browed and Lemon-spectacled Tanagers.

Getting to Pirre Station. It is 14 km from El Real to Pirre Station. On foot it will take about 3-4 hours, depending on conditions. Depending on season, you may be able to arrange car or boat transport for part of the way, or a pack horse for your gear. Call the ANAM station (ph 299-6965) in El Real for assistance in making any of the arrangements mentioned below, as well as hiring a guide. Guides that have been recommended to us include Antonio Henry and Isaac Pizarro (ph 299-6566, 299-6550). The price will have to be negotiated, but is generally about $10.00-$15.00 per day. You should also pay the park entrance fee here, which is $1.50 for Panama residents and $3.50 for foreign visitors.

If you have a lot of gear, you can hire a pack horse to bring it in, which will cost $10.00 for the horse, plus $10.00 for the handler, from El Real to Pirre Station. You can arrange to have the horse come back for your gear on the day of your return. During the dry season, you may be able to arrange through ANAM for the car from the police station to take you as far as Pirre Uno. From Pirre Uno to Pirre Station a pack horse will cost $5.00, plus the cost of the handler. During the rainy season, you can hire a boat to take you up the Río Pirre to the Emberá village of Pijibaisal, from which it is a walk of an hour or so to Pirre Station. At Pijibaisal you can hire a horse to go to Pirre Station for $5.00 (plus the handler), or else you may be able to hire a porter to carry your gear.

---

[8] When Angehr first climbed the Serranía de Pirre in 1979, before the creation of the national park and the construction of the ranger station, the name "Rancho Frío" was used by local guides for a campsite at about 640 m on the Estrangulador Trail, well beyond the present site of the station. A still higher camp on the ridge top that had been used by collectors for the Missouri Botanical Garden was called "Rancho Plástico" because of the plastic tarps they used there for shelter against the rain. In the intervening years, these names seem to have migrated downhill, and the area of the ranger station is now often called Rancho Frío, while the original Rancho Frío camp is now called Rancho Plástico.

Accommodations and meals. At Pirre Station there is a dormitory with two rooms with eight beds with foam mattresses, kitchen facilities, and a cold-water shower and flush toilets in an adjacent building. The cost is $10.00 per night. You can also camp here for $5.00. There is no electricity, so bring flashlights or headlamps and plenty of batteries. You must bring in all your own food. The rangers will cook for you for a small tip (about $5.00 per small group would be appropriate).

The Serranía de Pirre from Pirre Station

It is possible to hike to the Serranía de Pirre ridge above Pirre Station to see the highland endemics of the area, but it's not easy. In order to have enough time to bird up above, you will have to camp at least one night, and since the trails are obscure and confusing it is essential to have a local guide who knows the route. The main trail to the ridge is called *El Estrangulador* ("The Strangler") with good reason. On the first day you climb to a campsite known as Rancho Plástico (actually the original "Rancho Frío" camp; see the footnote on p. 289) at about 640 m (2,100 ft), which takes about four hours with gear. This is the last place at which water can usually be obtained. Even here it is a long steep climb down to the stream (on the left as you ascend); if possible bring lightweight collapsible water containers to haul it back to camp. Look for Sharp-tailed Streamcreeper on the stream. The next morning it is a long climb to the ridge top at 1,100 m (3,600 ft). The trail is very steep and muddy in parts, and in some places you will have to haul yourself up by your fingernails. The top of the ridge has elfin cloud forest where species such as **Russet-crowned Quail-Dove**, **Rufous-cheeked Hummingbird**, **Varied Solitaire**, **Pirre Warbler**, and **Pirre Bush-Tanager** can be found. Other species found on this route include Crested Guan, Great Curassow, Black-eared Wood-Quail, Tiny Hawk, Plumbeous Pigeon, Brown Violet-ear, Sooty-headed Wren, and Gray-and-gold Tanager.

Bird list for El Real and Pirre Station. * = vagrant; R = rare; P = species recorded from the Serranía de Pirre above Pirre Station, mainly above 500 m.

Great Tinamou
Little Tinamou
Muscovy Duck
Blue-winged Teal
Gray-headed Chachalaca
Crested Guan
Great Curassow
Marbled Wood-Quail
Tawny-faced Quail
Pied-billed Grebe
Brown Pelican
Neotropic Cormorant
Anhinga
Rufescent Tiger-Heron
Fasciated Tiger-Heron R
Cocoi Heron
Great Egret
Snowy Egret
Little Blue Heron
Cattle Egret
Green Heron
Striated Heron
Capped Heron
Yellow-crowned Night-Heron
Boat-billed Heron
White Ibis
Green Ibis R
Wood Stork
Black Vulture
Turkey Vulture
Lesser Yellow-headed Vulture R
King Vulture
Osprey
Swallow-tailed Kite
White-tailed Kite
Mississippi Kite
Plumbeous Kite
Long-winged Harrier*
Tiny Hawk R
Crane Hawk
Semiplumbeous Hawk
White Hawk
Gray Hawk
Mangrove Black-Hawk
Great Black-Hawk
Roadside Hawk
Broad-winged Hawk
Swainson's Hawk
Zone-tailed Hawk
Crested Eagle R
Harpy Eagle R
Ornate Hawk-Eagle
Slaty-backed Forest-Falcon R
Collared Forest-Falcon
Red-throated Caracara
Laughing Falcon
Bat Falcon
White-throated Crake
Gray-necked Wood-Rail

Uniform Crake R
Purple Gallinule
Sunbittern
Limpkin
Southern Lapwing
Black-bellied Plover
Black-necked Stilt
Wattled Jacana
Spotted Sandpiper
Semipalmated Sandpiper
Western Sandpiper
Wilson's Snipe
Rock Pigeon
Pale-vented Pigeon
Scaled Pigeon
Plumbeous Pigeon
Ruddy Pigeon
Short-billed Pigeon
Plain-breasted Ground-Dove
Ruddy Ground-Dove
Blue Ground-Dove
White-tipped Dove
Olive-backed Quail-Dove
Great Green Macaw
Red-and-green Macaw
Blue-and-yellow Macaw
Spectacled Parrotlet
Orange-chinned Parakeet
Blue-fronted Parrotlet P
Blue-headed Parrot
Red-lored Amazon
Mealy Amazon
Yellow-crowned Amazon R
Yellow-billed Cuckoo
Squirrel Cuckoo
Little Cuckoo
Striped Cuckoo
Rufous-vented Ground-Cuckoo R
Greater Ani
Smooth-billed Ani
Groove-billed Ani
Crested Owl
Spectacled Owl
Common Pauraque
White-collared Swift
Short-tailed Swift
Band-rumped Swift
Lesser Swallow-tailed Swift
Green Hermit P
Rufous-breasted Hermit
Band-tailed Barbthroat
Long-billed Hermit
Pale-bellied Hermit R
Stripe-throated Hermit
Scaly-breasted Hummingbird R
White-necked Jacobin
Black-throated Mango
Ruby-topaz Hummingbird*
Violet-headed Hummingbird
Green-crowned Woodnymph
Violet-bellied Hummingbird
Rufous-cheeked Hummingbird P
Blue-chested Hummingbird
Snowy-bellied Hummingbird
White-vented Plumeleteer
Bronze-tailed Plumeleteer
Greenish Puffleg P
Purple-crowned Fairy P
Long-billed Starthroat
White-tailed Trogon
Collared Trogon P
Black-throated Trogon
Black-tailed Trogon
Slaty-tailed Trogon
Tody Motmot
Blue-crowned Motmot
Rufous Motmot
Broad-billed Motmot
Ringed Kingfisher
Amazon Kingfisher
Green Kingfisher
Green-and-rufous Kingfisher
American Pygmy Kingfisher
Barred Puffbird
White-necked Puffbird
Black-breasted Puffbird
Pied Puffbird
White-whiskered Puffbird
Gray-cheeked Nunlet
White-fronted Nunbird
Dusky-backed Jacamar R
Rufous-tailed Jacamar R
Great Jacamar
Spot-crowned Barbet
Red-headed Barbet P
Collared Aracari
Yellow-eared Toucanet
Keel-billed Toucan
Chestnut-mandibled Toucan
Olivaceous Piculet
Black-cheeked Woodpecker
Red-crowned Woodpecker
Red-rumped Woodpecker
Stripe-cheeked Woodpecker P
Golden-green Woodpecker
Spot-breasted Woodpecker
Cinnamon Woodpecker
Lineated Woodpecker
Crimson-bellied Woodpecker R
Crimson-crested Woodpecker
Slaty Spinetail
Double-banded Graytail
Spotted Barbtail P
Beautiful Treerunner P
Striped Woodhaunter
Lineated Foliage-gleaner P
Slaty-winged Foliage-gleaner P
Buff-throated Foliage-gleaner
Plain Xenops
Tawny-throated Leaftosser P
Scaly-throated Leaftosser
Sharp-tailed Streamcreeper P R
Plain-brown Woodcreeper
Long-tailed Woodcreeper
Wedge-billed Woodcreeper
Northern Barred-Woodcreeper
Cocoa Woodcreeper
Black-striped Woodcreeper
Spotted Woodcreeper P
Streak-headed Woodcreeper
Red-billed Scythebill
Fasciated Antshrike
Great Antshrike
Black Antshrike
Western Slaty-Antshrike
Plain Antvireo P
Spot-crowned Antvireo
Pacific Antwren
Checker-throated Antwren
White-flanked Antwren
Slaty Antwren P
Dot-winged Antwren
Rufous-rumped Antwren
Dusky Antbird
Bare-crowned Antbird
Chestnut-backed Antbird
Dull-mantled Antbird R
Spotted Antbird
Wing-banded Antbird
Bicolored Antbird
Ocellated Antbird
Black-faced Antthrush
Black-crowned Antpitta R
Scaled Antpitta P R
Streak-chested Antpitta
Choco Tapaculo P
Brown-capped Tyrannulet
Southern Beardless-Tyrannulet
Yellow-crowned Tyrannulet
Forest Elaenia
Gray Elaenia
Greenish Elaenia
Yellow-bellied Elaenia
Olive-striped Flycatcher P
Ochre-bellied Flycatcher
Sooty-headed Tyrannulet
Paltry Tyrannulet
Scale-crested Pygmy-Tyrant P
Southern Bentbill

Black-headed Tody-Flycatcher
Eye-ringed Flatbill P
Olivaceous Flatbill
Yellow-margined Flycatcher
Yellow-breasted Flycatcher
Golden-crowned Spadebill
Ruddy-tailed Flycatcher
Sulphur-rumped Flycatcher
Common Tufted-Flycatcher P
Eastern Wood-Pewee
Tropical Pewee
Willow/Alder Flycatcher
Long-tailed Tyrant
Cattle Tyrant R
Bright-rumped Attila
Sirystes
Rufous Mourner
Dusky-capped Flycatcher
Panama Flycatcher
Great Crested Flycatcher
Lesser Kiskadee
Great Kiskadee
Boat-billed Flycatcher
Rusty-margined Flycatcher
Gray-capped Flycatcher
White-ringed Flycatcher
Streaked Flycatcher
Sulphur-bellied Flycatcher
Piratic Flycatcher
Tropical Kingbird
Eastern Kingbird
Gray Kingbird
Fork-tailed Flycatcher
Sapayoa
Thrush-like Schiffornis
Rufous Piha
Speckled Mourner
Cinnamon Becard
White-winged Becard
One-colored Becard
Masked Tityra
Black-crowned Tityra
Blue Cotinga
Purple-throated Fruitcrow
Golden-collared Manakin
White-ruffed Manakin P
Blue-crowned Manakin
Golden-headed Manakin
Sharpbill P
Red-eyed Vireo
Tawny-crowned Greenlet
Golden-fronted Greenlet
Yellow-browed Shrike-Vireo P
Rufous-browed Peppershrike
Black-chested Jay
Gray-breasted Martin
Southern Rough-winged Swallow
Sand Martin
Cliff Swallow
Barn Swallow
Black-capped Donacobius
White-headed Wren
Sooty-headed Wren
Black-bellied Wren
Bay Wren
Buff-breasted Wren
House Wren
White-breasted Wood-Wren
Gray-breasted Wood-Wren P
Southern Nightingale-Wren
Song Wren
Tawny-faced Gnatwren
Long-billed Gnatwren
Tropical Gnatcatcher
Varied Solitaire P
Slaty-backed Nightingale-Thrush P
Swainson's Thrush
Pale-vented Thrush P
Yellow Warbler
Chestnut-sided Warbler
Blackburnian Warbler
Bay-breasted Warbler
Prothonotary Warbler
Northern Waterthrush
Mourning Warbler
Common Yellowthroat
Canada Warbler
Slate-throated Redstart P
Pirre Warbler P
Buff-rumped Warbler
Bananaquit
White-eared Conebill
Pirre Bush-Tanager P
Yellow-backed Tanager
Dusky-faced Tanager
Lemon-spectacled Tanager
Scarlet-browed Tanager
White-shouldered Tanager
Tawny-crested Tanager
Summer Tanager
Scarlet Tanager
Crimson-backed Tanager
Flame-rumped Tanager
Blue-gray Tanager
Palm Tanager
Plain-colored Tanager
Gray-and-gold Tanager P
Emerald Tanager P
Silver-throated Tanager P
Speckled Tanager P
Bay-headed Tanager P
Golden-hooded Tanager
Scarlet-thighed Dacnis
Blue Dacnis
Viridian Dacnis
Shining Honeycreeper
Blue-black Grassquit
Variable Seedeater
Lesson's Seedeater*
Ruddy-breasted Seedeater
Lesser Seed-Finch
Chestnut-capped Brush-Finch P
Orange-billed Sparrow
Black-striped Sparrow
Buff-throated Saltator
Yellow-green Grosbeak P
Rose-breasted Grosbeak
Blue-black Grosbeak
Red-breasted Blackbird
Shiny Cowbird
Giant Cowbird
Orchard Oriole
Yellow-backed Oriole
Orange-crowned Oriole
Yellow-tailed Oriole
Scarlet-rumped Cacique
Yellow-rumped Cacique
Crested Oropendola
Chestnut-headed Oropendola
Black Oropendola
Thick-billed Euphonia
Fulvous-vented Euphonia
White-vented Euphonia
Orange-bellied Euphonia P

Other Localities in Darién

Other places in Darién, although even farther off the beaten track, can be visited with the assistance of local tour operators, or by the adventurous independent traveler.

Piñas Bay

Piñas Bay, on the southern coast of Darién near Jaqué, is served by two flights a week by Aeroperlas. You may be able to secure guides at the small settlement of Puerto Piñas near the airstrip to take you into forest on the upper Río Piñas. Among the species that have been found in the area are Bicolored Hawk, Great Green Macaw, Humboldt's Sapphire, Tody Motmot, Barred Puffbird, White-fronted Nunbird, and Crimson-bellied Woodpecker.

The *Tropic Star Lodge* (US ph 1-800-682-3424, (407) 423-9931; bonnie@tropicstar.com; www.tropicstar.com) is a high-end fishing resort located at Piñas Bay. Pelagic birding offshore from the bay can be excellent, including Sooty and Audubon's Shearwaters, Black, Least, and Wedge-rumped Storm-Petrels, Nazca and Brown Boobies, Common, Black, Sandwich, Royal, Bridled, and Sooty Terns, and Brown Noddy. There are also several trails near the lodge that can be birded (although the range of species here is not as good as up the Río Piñas). The lodge is primarily dedicated to fishing, but does offer non-fishing packages. (Of course, if your primary interest is pelagic birding, you will need a boat anyway.) The lodge mostly offers full-week packages; half-week packages may be available during the low season from May to September. It's not cheap; a three-day/four-night low-season non-fishing package is $1,150.00 (double-occupancy), and a similar fishing package (four passengers per boat) is $1,950.00. There is an additional transfer charge of $350.00 from Panama City to the lodge. Full-week and high-season rates, and those for fewer people per boat, are more expensive. Accommodations, in bungalows, are simple but quite comfortable, and the restaurant serves excellent food. The resort's capacity is small and it is very popular, so booking well in advance is usually necessary.

Jaqué, on the coast just south of Piñas Bay, and the Río Jaqué above it are not recommended at present due to recent kidnappings in the area, as well as the possible presence of Colombian guerillas on the upper Río Jaqué.

Sambú

Sambú, also known as Boca de Sábalo, is near the mouth of the Río Sambú, which runs between the Serranía de Pirre and the Serranía de Jungurudó. It is served by several flights per week by Aeroperlas and AirPanama. From here it is possible to hire piraguas to go upriver. You can also visit the Emberá village of La Chunga, a popular tourist destination. Most of the upper Río Sambú is in an Emberá *comarca* (indigenous reserve), and you would need to obtain permission from Emberá leaders in Sambú to make a trip to this area. Species that have been found along the river include Cocoi and Capped Herons, Green Ibis, Gray-cheeked Nunlet, White-fronted Nunbird, **Dusky-backed Jacamar**, Red-rumped Woodpecker, Thicket Antpitta, White-winged Swallow, Black-capped Donacobius, White-headed Wren, and **Black Oropendola**.

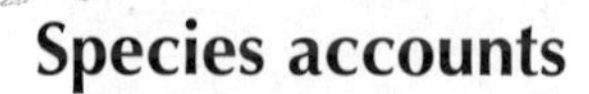

Species accounts

This section provides information on the best birding sites for finding rarities, specialties, and national and regional endemics. We have generally included here species that are rare or particularly sought after by birders, and those that have limited distributions either in Panama or in North America in general. All species that are national or regional endemics have also been included, even if common. Abbreviations for the Endemic Bird Areas these species are found in are listed below. See the Introduction (pp. 27-28) for the definitions of Endemic Bird Areas and of regional endemics used in this book. We have also taken the opportunity to publish new reports of species for which there have been few previous records (generally fewer than 10), including new species for Panama, indicated in the following list by an asterisk (*), and also include information on other species whose distribution or abundance has changed significantly since the publication of the second edition of *A Guide to the Birds of Panama,* by Robert Ridgely and John Gwynne, in 1989 (here referred to simply as Ridgely). We have generally not included information on rare species for which there have been no new records since the publication of Ridgely, with the exception of a few which may now be extirpated in Panama in order to highlight their status. See the *Annotated Checklist of the Birds of Panama* (Angehr 2006) for additional details on new species records for Panama.

The information included here is based in large part on records submitted to the Panama Audubon Society via Dodge Engleman and George Angehr as field editors of the Society's newsletter *The Toucan,* and later to Angehr as Panama correspondent for *North American Birds.* It also incorporates published records, as well as data collected as part of the Panama Audubon Society's ongoing Important Bird Areas program. We would encourage observers to submit additional records of rare species, and of other species outside their known or normal Panama ranges, to the Records Committee of the Panama Audubon Society. Records should be sent to George Angehr, Chair, Panama Records Committee (angehrg@si.edu) or to the Panama Audubon Society (audupan@cwpanama.net, or info@panamaaudubon.org).

Endemic Bird Areas of Panama

- WA (Western Atlantic) = Central American Caribbean Slope Endemic Bird Area (Atlantic slope lowlands from Honduras to Panama)
- WH (Western Highlands) = Costa Rica and Panama Highlands Endemic Bird Area (Highlands of Costa Rica and western Panama)

- WP (Western Pacific) = South Central American Pacific Slope Endemic Bird Area (Pacific slope lowlands of Costa Rica and Panama)
- EL (Eastern Lowlands) = Darién Lowlands Endemic Bird Area (Lowlands of eastern Panama and western Colombia)
- EH (Eastern Highlands) = Darién Highlands Endemic Bird Area (Highlands of eastern Panama and extreme western Colombia)
- EV (Escudo de Veraguas) = Escudo de Veraguas Secondary Area (Isla Escudo de Veraguas, Bocas del Toro)

Highland Tinamou. Rare in the western highlands, with few reports. There are records from the Los Quetzales Trail above Cerro Punta, from Finca Lérida above Boquete, Finca Hartmann above Santa Clara, and Fortuna (mainly Cerro Hornito).

Choco Tinamou (EH). The only accessible site where this species can be found is at Cana, where it is heard regularly (and, like most tinamous, seen much more rarely), on the lower part of the trail up Cerro Pirre. It is not known to occur near El Real or at Punta Patiño. It has been found in more remote areas of Darién such as the headwaters of the Ríos Tuira (Cerro Quía), Sambú, and Jaqué.

Black-bellied Whistling-Duck. Recently found to occur in Bocas del Toro in the Changuinola area (G. Angehr, D. Montañez, et al.). The species seems to be increasing on the Pacific slope despite being hunted, and now quite common in some areas, perhaps due to the spread of rice cultivation.

Fulvous Whistling-Duck. This species has recently colonized Las Macanas Marsh in Herrera, where a group of 33 appeared in 2003. It is now seen regularly there and is breeding (F. Delgado). It was previously known from Panama from two records from Panamá Province, the more recent from the Rodman Spoil Ponds near the Pacific entrance to the Panama Canal, 22 October 1998 (D. George et al.).

Comb Duck. No records since 1959. Up to five were seen at La Jagua in 1949, and one was collected on the Río Chucunaque in 1959 by A. Wetmore (Wetmore 1964). Although Wetmore was told by local residents that the species was regular there, most of them did not distinguish it from Muscovy Duck. The lack of subsequent reports suggests that these records may have pertained to vagrants rather than to a resident population.

Muscovy Duck. Recently found to occur in Bocas del Toro around Changuinola (Olson 1993; G. Angehr), where it is rare. Otherwise rare on the Pacific slope and in the Río Chagres basin.

Mallard. One recent record, a female seen at the Rodman Spoil Ponds near the Pacific entrance to the Panama Canal, 9 November 1992 (D. and L. Engleman et al.). Previous records, the most recent in 1913, have been uncertain.

Cinnamon Teal. Rare, the most recent records being several seen at Costa del Este, 23 September 2003 and for a few weeks subsequently (J. Tejada).

Northern Shoveler. Rare, the most recent records being two at Costa del Este, 23 October 1997 (D. Bradshaw, B. Paxton), and several, mostly females, at the Rodman Spoil Ponds near the Pacific entrance to the Panama Canal, between 14-25 December 1997 (various observers).

***White-cheeked Pintail.** Vagrant, one record. One was seen at Costa del Este east of Panama City, 22 September 1996 (D. and L. Engleman, J. Anguizola).

Northern Pintail. Very few recent reports, the most recent being a pair seen at the Rodman Spoil Ponds on the west bank of the Panama Canal, 14-26 December 1997 (R. Miró et al.).

***Green-winged Teal.** Vagrant, one record. A male was seen at the Rodman Spoil Ponds on the west bank of the Panama Canal, 25 December 1996 (D. and L. Engleman, D. George).

Ring-necked Duck. Rare. The most recent record is a female at Las Macanas Marsh, 18 December 2000 (D. and L. Engleman et al.), the first record from Herrera Province. Also recorded from Chiriquí, the Canal Area, and eastern Panamá Province.

Masked Duck. Now quite scarce, apparently having undergone a decline over the past 20 years. Still occurs on the Río Chagres above Gamboa, and probably elsewhere in the Chagres drainage such as on Lake Gatún. It is also regularly reported from Volcán Lakes in Chiriquí. It has recently been found near Changuinola, the first record from Bocas del Toro Province (G. Angehr).

Crested Guan. Because it is a favored game species, this species is generally scarce or absent anywhere near human habitation (or else extremely wary). In the Canal Area, a few apparently persist on Sherman in San Lorenzo National Park, and on the outer part of Pipeline Road in Soberanía National Park. They are quite common and relatively tame on Barro Colorado Island, and this is the easiest place to find them in Panama if you can arrange a visit. The species is also found at Cerro Azul-Cerro Jefe, and it is fairly common around Cana in Darién, especially at and above the high camp on Cerro Pirre.

Black Guan (WH). Not uncommon at some localities in the western highlands, such as on the *El Retoño* trail at Las Nubes in La Amistad International Park, and at Fortuna. It becomes scarcer farther east, being rare on Cerro Gaital above El Valle, and found mainly at higher elevations at El Copé. Reported from Cerro Campana in the 1980s (G. Vaucher), but is most likely gone from there now.

Great Curassow. Like the Crested Guan, this species is now very scarce anywhere near populated areas due to over-hunting. Surprisingly, even in the Canal Area a few manage to persist on Pipeline Road in the Río Limbo area in Soberanía National Park area, and perhaps also in San Lorenzo National Park. With luck it can also be found at Cerro Azul-Cerro Jefe. Probably easiest to find at Cana, especially on the trail up Cerro Pirre.

Tacarcuna Wood-Quail (EH). This species is restricted to the Tacarcuna Range near the Colombian border. Discovered only in 1963, it may not have been seen by an ornithologist since the following year, when Alexander Wetmore found it to be fairly common on Cerro Tacarcuna (Wetmore 1965). Unfortunately, the Tacarcuna area cannot now be visited safely, accounting for the lack of recent records.

Black-breasted Wood-Quail (WH). Uncommon in the western highlands. Probably most regular around Volcán and Cerro Punta, and at Fortuna. Recently found on Cerro Peña Blanca above El Copé (G. Angehr et al.). Like most wood-quails, heard much more often than seen.

Tawny-faced Quail. Rare and secretive, and most often reported from near the high camp on Cerro Pirre above Cana, where they can sometimes be found roosting at night. Also occasionally found at Pirre Station near El Real and at Cerro Azul-Cerro Jefe. Formerly reported occasionally from Pipeline Road in Soberanía National Park, but there seem to be no recent records. This species has a whistled tinamou-like call, quite unlike Panama's other forest quails.

***Christmas Shearwater.** Very rare, one record. Two were seen far offshore western Panama about 155 km south of the Burica Peninsula, 21 May 1990 (Spear and Ainley 1999).

***Manx Shearwater.** Vagrant, one record. One was found in weakened condition near shore at Isla Margarita, near Colón, 5 January 2003 (K. Aparicio et al.) and later died.

Red-billed Tropicbird. About 50-75 pairs nest on Swan Cay off the north coast of Isla Colón in Bocas del Toro, the only colony in the southwestern Caribbean. The colony is easily visited by hiring a boat from either Bocas del Toro Town or Changuinola. The species is rare on the Pacific side in the Gulf of Panama.

***White-tailed Tropicbird.** Vagrant, two records. One was reported by L. Miller off the Atlantic coast of Panama in 1936 (Miller 1937), and one was seen at sea 20 km east of Isla Escudo de Veraguas off the coast of Bocas del Toro, 29 March 1990 (S. Olson, T. Parsons; Olson 1993).

American White Pelican. One additional record, a bird seen and photographed on the Pacific coast near Pásiga in eastern Panamá Province, 28 August 2004 (J. Ortega). The only previous records, cited by Ridgely, have been from the Chitré area in 1984.

Anhinga. Recent records from Chiriquí (Playa de la Barqueta Agrícola, south of Alanje, 23 September 1996, G. Angehr et al.) and Los Santos (Pocrí, 24 April 2005, G. Angehr, J. Kushlan) are the first for these provinces.

***Great Frigatebird.** Vagrant, one record, photographed. A male and a female were seen c. 240 km SSW of Isla Coiba, within Panamanian territorial waters, 12 November 2003 (M. Force). The species may possibly occur regularly far offshore.

Least Bittern. Best known from the middle Río Chagres and around Lake Gatún, where it is probably most regularly seen at the Gamboa Ammo Dump. There are also records from the lower Chagres (at the Gatún Spillway), Tocumen Marsh, Las Macanas Marsh in Herrera, the marsh near Las Lajas Beach in Chiriquí, and on the Ríos San San and Negro near Changuinola in Bocas del Toro.

Fasciated Tiger-Heron. Few records, but this may primarily be due to its secretive nature and because its main habitat, swift-flowing streams in foothill and highland forest, is difficult to bird (although it sometimes can also be found along lowland streams). In recent years it has been reported from the Ríos Indio and Mono at Cerro Azul, the Quebrada Arena at Fortuna, at Cana, and near El Real. It has also sometimes been reported from streams along Pipeline Road in Soberanía National Park.

Bare-throated Tiger-Heron. Quite common in the Pearl Islands, where almost every small cove has a resident pair. (Individuals have even been seen by the pool at the Hotel Punta Galeón on Contadora.) Also fairly common on Isla

Coiba. Much less common on the mainland, but has been seen at Tocumen Marsh, near the Lake Bayano Bridge, along the lower Río Bayano, and on Barro Colorado Island, among other scattered localities.

Reddish Egret. This species now appears to be a casual visitor to Panama, with records of at least four additional individuals since the first record, from Isla Coiba, 12 April 1976 (R. Ridgely; Ridgely and Gwynne 1989). Records include an immature dark phase bird at the Aguadulce salt ponds, 21 September 1996 (D. George) and 2 November 1996 (D. and L. Engleman); a dark phase adult at Farfan Beach, 3 and 24 September 1997 (B. Watts) and at nearby Veracruz Beach, 23 October 1997 (S. Gauthreaux), both sites being just west of Panama City; a white-phase individual at El Agallito Beach near Chitré, 21 March 1998 (Darién and Delicia Montañez); and a juvenile at the Gatún Spillway near Colón, 26 December 2001 (Darién Montañez et al.) and 9 March 2002 (W. Adsett.).

Agami Heron. Widespread but very inconspicuous, so that seeing this species is largely a matter of chance. Most frequently reported from streams that cross Pipeline Road in Soberanía National Park, and has also been seen at the El Trogón Trail on Achiote Road. With luck it may be encountered at many other sites.

Capped Heron. This species is most frequent in Darién, especially along the Río Tuira and other streams near El Real, but may also be found on Lake Bayano and occasionally at Tocumen Marsh. It is erratic in the Canal Area proper, being most frequently reported from Summit Ponds on Old Gamboa Road.

Boat-billed Heron. Widespread but difficult to see unless you locate a day roost or spotlight along streams at night. In the Canal Area, easiest to find at Summit Ponds by arriving just at dawn, before they have gone to roost. There are also day roosts at the ponds closest to farm headquarters at Tocumen Marsh (where they also nest), and at Esperanza Marsh on the road to Puerto Armuelles in Chiriquí. The species can sometimes also be found by spotlighting the river below the Gatún Spillway near Colón at night from the roadside overlook. They can be found in the mangroves at Galeta, although these are now less easy to spotlight at night because a gate on the road is closed during hours of darkness.

Glossy Ibis. Formerly rather scarce, now spreading in Panama and recently confirmed breeding near Santa María, Herrera (G. Angehr). Can usually be found at Las Macanas Marsh, as well as in the surrounding rice fields where groups numbering in the hundreds have sometimes been seen. Also now regularly seen in eastern Panamá Province, particularly at Costa del Este, and sometimes on the Atlantic side of the Canal Area.

Green Ibis. Most frequently reported from Two Tanks Road, near Chiriquí Grande in Bocas del Toro. The species may also occasionally be seen in western Bocas del Toro, especially in the wetlands around Changuinola (including along the Canal de Soropta), on Lake Bayano, or on rivers in Darién. There has also been a recent record from the Canal Area, along the first part of Pipeline Road, 6 April 2006 (C. Batista et al.), though the species is not to be expected there.

Roseate Spoonbill. Scarce, probably being most regularly seen on the coast near Chitré. It is less common at Aguadulce than it previously was since the salt-making operations have been reduced. Individuals or small groups may also occasionally be seen at Tocumen Marsh or Costa del Este in the Canal Area, as well as at Las Lajas Beach and the adjacent marsh in Chiriquí.

Jabiru. This species is not known to occur regularly anywhere in Panama. However, apparent vagrants, perhaps from Costa Rica, occasionally turn up and seemingly wander widely, rarely staying in one place very long. Besides old reports from Cricamola in Bocas del Toro and La Jagua in eastern Panamá Province (Wetmore 1965), since 1997 there have been reports of individuals or small groups near the mouth of the Río Changuinola in Bocas del Toro (four, November 1997, J. Beytia); near Achiote, Colón Province (one, November-December 1998, M. Castro et al.); at Cenegón del Mangle in Herrera (two to four, 11-25 November 1999, F. Delgado); and on a pond about 5 km west of Divisa, Herrera (one, 12 February 2004, M. Miller, K. Winker). One was found injured on an impoundment near Los Santos in Los Santos Province on 22 October 2002, and held in captivity until it died on 20 May 2003 (F. Delgado).

Pearl Kite. While still not particularly numerous, this species is reported regularly throughout the Pacific slope, including recent records from Chiriquí (the first report being one near Puerto Armuelles, 6-7 April 2003, J. Weir). It is probably most regularly recorded at Tocumen Marsh near Panama City and around Las Macanas Marsh in Herrera.

Snail Kite. Formerly very rare in Panama, this species colonized Lake Gatún in the mid-1990s after the introduction of Apple Snails, its favored food (Angehr 1999). It is now quite common on the lake, particularly around Barro Colorado Island, and sometimes seen elsewhere in the Canal Area, particularly around the Third Locks Excavation at Mindi on the Atlantic side. It has also colonized Las Macanas Marsh, where these snails have also been introduced, and can sometimes been seen in adjacent rice fields as well.

Black-collared Hawk. Few recent records. Apparently scarce but regular on Lake Bayano, especially near the mouth of the Río Majé, and also reported from Darién on the Río Tuira and near Punta Patiño.

Tiny Hawk. Rare, being most frequently reported from Cana and from Fortuna. In the Canal Area, most likely to be found on the outer part of Pipeline Road and on Achiote Road.

***Cooper's Hawk.** Vagrant, two records. One was seen at Finca La Suiza in the Fortuna Forest Reserve, Chiriquí, 19 March 1994 (M. Letzer), and another between Boquete and David, Chiriquí, 20 March 2001 (J. Coons, J. Rowlett).

Bicolored Hawk. Rare, most recent reports coming from Cana, with a few also from Fortuna and from Nusagandi. In the Canal Area, very rare on Pipeline Road, along Escobal Road, and at Semaphore Hill near Plantation Road.

***Long-winged Harrier.** Vagrant, three records. One seen at Tocumen Marsh, 28 August 1995 (D. Rogers); one seen the El Real airstrip, Darién, 1 January 2001 (A. and N. Chartier); and one seen and photographed 24 June 2006 at Tocumen Marsh (J. A. Cubilla, R. Miró, K. Kaufmann et al.).

Plumbeous Hawk. Rare, being most commonly reported from Cana. In the Canal Area, the species occurs on Pipeline Road and Achiote Road, with also a few recent reports from the Plantation Road area. There are also records from Santa Fe, Altos del María, Cerro Azul-Cerro Jefe, Nusagandi, and the Bayano region.

Barred Hawk. Generally rare, but quite regularly reported from Fortuna, and also regularly found at higher elevations on Cerro Pirre above Cana. Also reported from above Santa Clara, the Volcán-Boquete area, above Santa Fe, El Copé, El Valle, and Cerro Azul-Cerro Jefe. Very rarely may occur in the lowlands, with a record from Sherman.

Solitary Eagle. Extremely rare, but there have been a number of recent observations from Cana, both near the airstrip and higher up on Cerro Pirre: one bird at Cana, 28 February 1992 (J. Coons); two birds near the summit of Cerro Pirre, 20 January 1996 (G. Langham et al.); one at the airstrip, 27 March 2001 (G. Angehr et al.). The species has also been observed on aerial surveys of the Serranía de Pirre by the Peregrine Fund (A. Palleroni). There have been two recent records from the Cerro Azul area: a pair between 300-400 m elevation on the Cerro Vistamares Trail, 8 April 2004 (G. Rompré et al.), and one at Birders' View in Altos de Cerro Azul, 28 April 2006 (R. Clay, O. Komar, et al.). The only recent record from western Panama has been one bird seen above Santa Fe, 25 September 1992 (D. and L. Engleman). The species has otherwise been reported from foothills in eastern Panamá Province: north of Chepo, 14 April 1949 (A. Wetmore) and north of El Llano, 9 September 1972 (N. Smith), and from Cerro Quía in Darién, 15 July 1975 (R. Ridgely). Two specimens were collected in Veraguas in the nineteenth century (labeled Calobre, but probably collected in foothills north of there).

Crested Eagle. Very rare, being most regularly reported from Cana, where this species seems to be more common (or at least less rare) than Harpy Eagle. Also recorded from near Pirre Station near El Real. There are occasional records of what may be vagrants from the Atlantic slope of the Canal Area, the most recent being immatures on Barro Colorado Island, 9 January 1998 (B. Jarvis), and on Achiote Road, 12 September 2000 (G. Rompré; Rompré 2002).

Harpy Eagle. Rare, and usually seen only by chance unless a trip can be arranged to an active nest. However, several regular nest sites are known in Darién (especially in the Mogue area near Punta Patiño) and a few in other parts of the country, and visits can often be arranged through one of the local birding tour companies. Ancon Expeditions offers a tour to Punta Patiño that includes a visit to a Harpy nest, and Adventure Tours and Nattur also can arrange such visits. (See p. 39 for contact details.) Although the species has been seen at Cana, there are few records from this site. There are a few reports from near El Real. Birds that are probably vagrants have been seen from time to time even in the Canal Area., the most recent being one at Pavón Hill on Sherman on 28 February 1998 (H. Stockwell, M. Akers et al.). However, any Harpy seen in the Canal Area at present is in all probability a captive-bred bird released as part of a re-introduction program by the Peregrine Fund in Soberanía National Park, recognizable by leg bands.

Black-and-white Hawk-Eagle. Rare, being most often reported from Cana and from Cerro Azul, where it is regularly seen at Birders' View, and can sometimes be seen soaring over the Cerro Jefe area. There are old records from Bocas del Toro, Boquete, Veraguas, and the Bayano region, but none recently from these areas.

Ornate Hawk-Eagle. Uncommon, but regularly reported from foothill areas, especially from Fortuna and from Cana and Cerro Pirre above it; also occasionally seen at Cerro Azul-Cerro Jefe. Rare in lowlands, although there are occasional records from Pipeline Road, Achiote Road, and elsewhere. Also occurs above Santa Clara, in the Volcán-Boquete area, on Isla Coiba, at Nusagandi, and in the El Real-Pirre Station area.

Slaty-backed Forest-Falcon. Scarce, most recent records being from Cana. In the Canal Area, occasionally reported from Plantation, Pipeline, and Achiote Roads. Also occurs in the Bayano region, and at El Real-Pirre Station.

Red-throated Caracara. This species remains quite common in eastern Panama, being virtually a guarantee at Cana and regularly seen around Pirre Station and at Nusagandi. It is scarce in the Bayano region, but two were seen a few kilometers beyond the Río Mamoní bridge on 26 February 2005 (W. Adsett, L. Hoquee). There have been no recent records from western Pacific Panama, including the Burica Peninsula, where it was still present in 1982 (R. Ridgely). It is possibly still present in remote parts of Bocas del Toro, where it was reported from the Río Teribe, 26 September 1993 (G. Adler). The most recent record from the Canal Area is one, certainly a vagrant, seen on Achiote Road, 20 March 1994 (G. Angehr et al.).

Crested Caracara. One near Changuinola, 16 April 2006 (D. Montañez, R. Miró) is the first record for Bocas del Toro. Otherwise common on the Pacific slope from Chiriquí to eastern Panamá Province, and scarce on the Atlantic side of the Canal Area.

Laughing Falcon. Regularly reported from Cana, and sometimes from Fortuna and the Oleoducto Road in Bocas del Toro. Also seems to be fairly regular in the lowlands of western Chiriquí, with reports from the Burica Peninsula (especially El Chorogo), Cerro Batipa, and near Las Lajas Beach. Also occasionally reported from the Changuinola area, Wekso, and Chiriquí Grande in Bocas del Toro, above Santa Clara, the Volcán Lakes, the Bayano region, and El Real-Pirre Station. Rare in the Canal Area, but has been seen at Achiote Road, Campo Chagres, and Tocumen.

American Kestrel. Previously known in Panama only as a migrant from the north, the species is now breeding in eastern Panama, apparently having colonized from Colombia. Since 1995, evidence of local breeding has been reported from Cerro Azul, Nusagandi, and Darién, and a pair was found nesting at 500 m on Cerro Chucantí on 31 March 2006 (G. Berguido et al.). Although no specimens have been collected, most individuals seen during the northern summer (when migrants are not present) have relatively unspotted breasts, similar to races found in western Colombia. Additional documentation of breeding would be desirable.

Aplomado Falcon. Not uncommon along the Carretera Nacional between Divisa and at least Chitré, and regularly reported from around Las Macanas. The species should also be looked for along the Carretera Interamericana from around Penonomé to west of Santiago, and can be found along roads through the grasslands south of Penonomé. There have been recent records in western Panamá Province from El Chirú, near the turnoff to El Valle, 6 August 2005 (M. Harvey,

D. Rodríguez) and Cerro Campana, 21 February 2003 (A. Castillo, D. Buehler, C. Tarwater), as well as an old record from the Canal Area of a vagrant on Barro Colorado Island (A. Wetmore, Wetmore 1965). There has been one recent record from Bocas del Toro, an individual seen on the Punta Peña-Almirante Road a few kilometers west of the junction with the Oleoducto Road, 2 November 1998 (W. Martínez, D. Pierpoint et al.).

Orange-breasted Falcon. A few nesting sites of this species have been found by aerial surveys by the Peregrine Fund (A. Palleroni) in remote areas of Darién (Serranías de Pirre and Junguradó) and in Omar Torrijos National Park, but these sites are nearly inaccessible on foot. In Panama the species is very unlikely to be seen except by chance, the most likely localities for a sighting probably being Cana and above El Copé.

***Ruddy Crake.** Probable vagrant, one record. One was seen in a marsh just northeast of Changuinola, Bocas del Toro, on 20 March 2006 (R. Miró).

Gray-breasted Crake. Improved knowledge of the call of this species has shown it to be more widely distributed than the four records listed in Ridgely (from Coiba Island; Sherman; Tocumen; and Puerto Obaldía, Kuna Yala) would suggest. The species seems to be regularly present at the Gatún Drop Zone on Sherman, and additional recent records have come from near Changuinola, Isla Popa (Cooper 1999), and the Palo Seco Protection Forest (along the road that goes past the Continental Divide Trail at Fortuna) in Bocas del Toro; at Volcán Lakes, Chiriquí; and the Farfan area just south of the western end of the Bridge of the Americas near Panama City.

Rufous-necked Wood-Rail. Two additional records besides those listed by Ridgely (from Almirante; near Aguadulce; and Sherman). One was seen in mangroves south of the airstrip at Sherman, 2 January 1988 (C. Quirós); and one in mangroves on Isla Colón in Bocas del Toro, near Boca del Drago, 9 March 1999 (L. Pomara). In Panama this species has been found exclusively in mangroves, although it also occurs in foothills in Costa Rica and South America. All records in Panama have come from between late December and early March, suggesting that the species may be present only seasonally.

Uniform Crake. Probably most likely to be seen in western Bocas del Toro, where it is known from the Changuinola-Almirante area and on Two Tanks Road near Chiriquí Grande, as well as on Isla Popa (Cooper 1999). The species is abundant on Isla Escudo de Veraguas. In the Canal Area, there are records only from Achiote Road and Pipeline Road.

Colombian Crake. No recent records. The species was found at Tocumen Marsh on a number of occasions in the early 1980s, where it apparently bred (and has also been recorded near Achiote, where one was collected in 1965), but there have apparently been no records either from Tocumen or anywhere else since 1984. It is possible the Tocumen records could have represented a temporary incursion.

Paint-billed Crake. No recent records. This species was reported from both the Changuinola area in Bocas del Toro and Tocumen in the early 1980s. The last record from Tocumen was apparently in 1982, and from Changuinola in 1988. Like the previous species, these records could have represented a temporary incursion.

American Oystercatcher. Recently found at Punta Patiño (B. Jiménez, K. Aparicio), the first reports from Darién. Fairly common in the Pearl Islands, and local elsewhere along the Pacific coast.

American Avocet. Two additional records besides the one cited by Ridgely (at Aguadulce, September-October 1986). A bird in breeding plumage was seen at Aguadulce, 30 March 1997 (D. Engleman et al.) and 5 April 1997 (D. George, S. Follett), and a different individual in non-breeding plumage at the same location, 6 April 1997 (D. George, S. Follett).

Northern Jacana. One seen on the Río Chagres above Gamboa, 16 May 2006 (G. Angehr) is the easternmost record. Otherwise known from western Chiriquí, Bocas del Toro, and the western Atlantic slope of Veraguas, where it is common.

Wandering Tattler. In addition to the five records listed by Ridgely (on a small island in Panama Bay; at Kobbe just west of Panama City; at Panama Viejo; at Gatún Dam; and near Isla Coiba), there have been two additional records, both from offshore rocks off Isla Coiba, 6 September 1993 (G. Angehr); 13 March 1994, (H. Loftin).

Snowy Plover. One additional record, a group of six within a large flock of sandpipers and plovers at Punta Chame, 14 December 1997 (D. Montañez et al.). Two previous reports, from Cocoplum, Bocas del Toro, in 1927, and Punta Chame in 1986-1987.

Long-billed Curlew. Since 1998 one or two individuals have been seen almost every year during the northern winter at Costa del Este or Panama Viejo. Besides those listed by Ridgely from Sherman and Coco Solo, records from elsewhere have included two birds at Aguadulce, Coclé, 22 August 1993 (G. Adler, B. Schlinger, D. Morimoto), and a flock of about 30, in Margarita, near Colón, 16 August 2004 (D. Pierpoint et al.).

Hudsonian Godwit. One additional record. Three were seen in flight from a small plane on the coast about 20 km east of Panama City, 27 September 1997 (B. Watts). One previous record from Coco Solo, near Colón, in 1983.

Dunlin. Two recent records, in addition to the five listed by Ridgely (from Panama City, on the coast of eastern Panamá Province at the mouth of the Río Pacora, Coco Solo near Colón, and Aguadulce). One was seen at Costa del Este, 7 September 1997 (G. Angehr et al.), and one at Panama Viejo, 9 March 2000 (R. Miró, K. Kaufmann.).

Ruff. One recent record, in addition to the two listed by Ridgely (from Coco Solo near Colón, and from Howard Air Force Base just west of Panama City). A female was seen at Costa del Este several times between 11-21 August 2003 (J. Tejada, D. Buehler).

Pomarine Jaeger and **Parasitic Jaeger.** Both these species are probably most easily seen in Bocas del Toro, where they are common in Almirante Bay and can sometimes be seen from the waterfront in Bocas del Toro Town. They can occasionally also be seen from shore on the Atlantic side of the Canal Area. While they occur in the Bay of Panama, on the Pacific side they are rarely seen from land.

Long-tailed Jaeger. One collected in the Gulf of Panama in November 1968 (Loftin 1991) is the first confirmed record from Panama. There has been one previous somewhat doubtful record, from Colón Harbor in 1927.

Bonaparte's Gull. In addition to three previous records from Coco Solo and the Gatún Spillway listed by Ridgely, there have been two additional records, these from the Pacific side of the isthmus. An immature was seen at Punta Paitilla, Panama City, 27 November 1988 (H. Laidlaw); and a first-winter bird was seen at the same location, 23 December 1998 (J. Carlisle et al.).

Gray-hooded Gull. Four recent records, all from Costa del Este east of Panama City. One seen on 17 October 1997 (G. Angehr, M. Allen), seen by various observers subsequently until 2 November; one seen 2 August 2000 (D. Montañez, R. Miró, C. Rhodes); two seen and photographed, 14 August 2003 (R. Miró et al.); and one seen 26 September 2004 (J. A. Cubilla et al.). The only previous record was one on the Panama City waterfront, 25 Sept 1955.

Gray Gull. A remarkable invasion occurred in 1997-1998, probably connected with the severe El Niño then taking place. The first record was a single bird at Costa del Este on 4 September 1997 (B. Paxton), with occasional birds being seen there and elsewhere in the vicinity in October 1997 and March and April 1998 (various observers). Larger groups began to be seen in May, culminating with a group of 100 at Costa del Este on 31 May (D. Montañez), then dwindling to five by 20 June (K. Kaufmann, R. Miró). In addition to those listed by Ridgely, other records include one at Paitilla Point, Panama City, 28 August 1988 (H. Laidlaw), one at Panama Viejo, 21 January 2000 (J. Tejada), and one at Costa del Este, 13 March 2005 (G. Angehr, R. Miró).

Belcher's Gull. In addition to the three previous records listed by Ridgely (from Panama Viejo, Amador, and Albrook), one additional record, an immature off Veracruz beach east of Panama City, 24 September 1997 (B. Watts). Formerly included in Band-tailed Gull.

Lesser Black-backed Gull. Four additional records besides the one listed by Ridgely, a bird that wintered at Amador from 1979-1988. Two juveniles were seen at Coco Solo, 15 November 1982 (D. and L. Engleman); a first-year bird was seen at Paitilla, Panama City, 19 December 1993 (G. Seutin, H. and E. Stockwell, K. Kaufmann); a bird in near adult plumage was seen at Costa del Este, 6 September 1997 (G. Angehr et al.) and again 20 September and 17-19 October (various observers); and an adult was seen at Punta Paitilla, 8 October 1998 (D. Montañez, R. Miró).

***Kelp Gull.** Vagrant. A group of five birds was first seen at Costa del Este east of Panamá City, 25 November, 2001 (D. Montañez et al.). Some continued to be seen at the same location by various other observers until 3 December 2001, and two were photographed.

Elegant Tern. Occasionally reported from the Panama City waterfront at Punta Paitilla and at Amador and from Costa del Este. Generally rare and irregular along the Pacific coast.

***Arctic Tern.** Vagrant, two records. An adult was seen approximately 30 km SSE of Piñas Bay, Darién, 27 May 1995 (D. Engleman). A first-winter bird was seen at Costa del Este east of Panama City, 18 October 1997 (G. Angehr, M. Allen et al.).

Forster's Tern. At least seven additional records besides the six listed by Ridgely. Most records have been from the Atlantic side of the Canal Area (Coco Solo, Isla

Margarita, Gatún, Ft. San Lorenzo), with a few from the Pacific side (Amador, Panama Viejo). Two recent records have been from El Agallito Beach, near Chitré, 8 December 1993 (F. Delgado), and 3 November 1999 (D. and C. Montañez).

Yellow-billed Tern. Two recent records. One was seen at Aguadulce, 18 July 1996 (G. Seutin), and another at El Agallito Beach near Chitré, 19 April 1998 (D. Montañez et al.). The only previous record was at Coco Solo in 1977.

***Black Noddy.** Vagrant, one sight record. One was seen on Isla Frailes del Norte, off the south coast of Los Santos, 12 September 1998 (D. Montañez et al.).

White-crowned Pigeon. Found mainly along the immediate coast in western Bocas del Toro, and infrequently reported. On rare occasions it may be seen inland, e.g. two near La Gruta, Isla Colón, on 19 August 2002 (G. Angehr, R. Miró). Sometimes seen at Swan Cay, and abundant on Isla Escudo de Veraguas. Although recorded from near El Porvenir in Kuna Yala in 1984, there have been no recent reports from there or elsewhere outside of Bocas del Toro.

Band-tailed Pigeon. Fairly common in the western highlands, above Santa Clara, in the Volcán-Boquete area, and at Fortuna. Recently found to occur on Cerro Peña Blanca above El Copé (G. Angehr et al.).

***Plumbeous Pigeon.** This species, overlooked until recently, has been found to be fairly common at Cana and on Cerro Pirre above it. (It has also been found in more remote areas in Darién, such on the Ríos Piñas, Sambú, and Jaqué.) Note that the form occurring in Panama has a distinctive white eye, by which it may be distinguished (if seen well) from the very similar Short-billed and Ruddy Pigeons, which occur in the same area but have red eyes. It also has gray underwing coverts, in contrast to these other two species, which have brown-to-rufous underwing coverts, but this character will ordinarily be difficult to see. The call is also distinctive, being three-noted (*whit-mo-go*) rather than four-noted like those of the other two species. Note that Dusky Pigeon, which has been seen once at Cana, also has a pale eye, but has two-toned underparts, with a gray head and upper breast and rufous lower breast and belly, and has rufous underwing coverts. Although Dusky also has a three-note call, it differs in quality from that of Plumbeous. See Angehr et al. (2004) and Angehr (2006) for further details on the occurrence of this species in Panama.

Dusky Pigeon (EL). One additional record besides the one cited by Ridgely (middle Tuira Valley, 1981), one seen on the trail to Cerro Pirre just above Cana, 17 April 1992 (D. and L. Engleman).

White-winged Dove. Not uncommon in its very limited Panama range in coastal Herrera and southern Coclé; most commonly reported from the El Agallito Beach area near Chitré, Chitré itself, where it can be seen perched on power lines, and from Aguadulce. It should also occur at Cenegón del Mangle, but not yet reported from there.

Common Ground-Dove. Like the previous species, most commonly reported from the El Agallito Beach area near Chitré, but also has been found at Aguadulce and is to be expected at Cenegón del Mangle. The easternmost record is one seen south of Antón, Coclé, 17 June 1997 (G. Angehr).

Plain-breasted Ground-Dove. Recently found to occur in Darién around El Real (various observers), and one recent record from Bocas del Toro, near Changuinola, 4 September 2004 (J. A. Cubilla). Generally fairly common in lowlands on the Pacific slope; scarce on the Atlantic slope in the Canal Area. Probably spreading with deforestation.

Gray-headed Dove. Recent reports have come only from Islas Colón, Bastimentos, and Solarte in the Bocas del Toro Archipelago. Although the species has been recorded on the mainland in the Changuinola-Almirante area, there have been no recent reports from there.

Brown-backed Dove (WP). Probably commonest and easiest to find on Isla Coiba, where it can be found on the Los Monos and Pozos Termales Trails. The most accessible place to find it on the mainland is the El Montuoso Forest Reserve in Herrera, though there have not been any recent records from there. The species is also found in and around Cerro Hoya National Park. Sometimes included in Gray-headed Dove or Gray-fronted Dove.

Olive-backed Quail-Dove. Probably most easily found at Pirre Station in Darién (it has not been recorded from Cana) and at Nusagandi. Possible in the Bayano region. In the Canal Area, although it has been reported from Pipeline Road, it seems to be irregular there, and there are no recent reports from Achiote Road. There are old reports from Bocas del Toro and Veraguas, but no recent ones. In Bocas del Toro it could be possible at Wekso or farther up the Río Teribe.

Chiriqui Quail-Dove (WH). Reported most frequently from Fortuna, and also found in the Cerro Punta area and above Santa Clara. Recently found to occur on Cerro Peña Blanca above El Copé (G. Angehr et al.).

Purplish-backed Quail-Dove (WA, EL). Generally scarce. Probably most readily found at Nusagandi and in the Cerro Azul-Cerro Jefe area. Also known from Fortuna, Santa Fe, El Copé, and Cerro Campana. Occurs at Cana, although apparently quite rare there.

Buff-fronted Quail-Dove (WH). Rare, but has been found in recent years at the Los Quetzales Cabins above Cerro Punta and at Fortuna and Cerro Santiago; also occurs above Boquete.

Russet-crowned Quail-Dove (EH). Regularly recorded (though, like other quail-doves, mostly by call) at upper elevations on Cerro Pirre above Cana. Although it has been recorded a few times at Cerro Jefe, it is rare there. It also occurs on Cerro Chucantí.

Violaceous Quail-Dove. Scarce, most records being from Cana, though it is also occasional at Pipeline Road, Semaphore Hill, Achiote Road, and Sherman in the Canal Area.

Azuero Parakeet (WP). This national endemic has a very small range in the southwestern Azuero Peninsula, most of which is nearly inaccessible without mounting an expedition. It can be found above El Cobachón in Cerro Hoya National Park, and, according to local residents, near Flores on the western side of the Park. It may be present only seasonally in some areas, but its local movements are poorly known. Originally described as a subspecies of Painted Parakeet.

Sulphur-winged Parakeet (WH). Fairly common at most sites in western Chiriquí, including above Santa Clara, the Cerro Punta-Volcán area, Boquete, and Fortuna and the upper part of the Oleoducto Road in Bocas del Toro, and above Santa Fe in Veraguas.

Olive-throated Parakeet. Considered rare by Ridgely, this species has increased markedly in recent years, now being quite common in the Changuinola area and on Isla Colón in Bocas del Toro. It is likely to continue to spread eastward along the Atlantic slope with deforestation, but apparently has not yet reached Chiriquí Grande.

Chestnut-fronted Macaw. Frequently seen at Cana, though perhaps the least common of the four species of macaw found there. Somewhat strangely, seems not to be present in the El Real-Pirre Station area. Occasionally escaped cage birds may be seen in Panama City and the Canal Area.

Great Green Macaw. The easiest place to see this species is at Cana, where it is practically a guarantee if you are there for any length of time, being found from near the airstrip to well up on Cerro Pirre. It is also regular around Pirre Station near El Real. The species can also be found in less easily-accessible areas, including the Río Piñas in Darién; near Cerro Hoya in the southern Azuero Peninsula (at El Cobachón and near Flores); on the upper Río Majé above Lake Bayano; in remote parts of Bocas del Toro, and perhaps in northern Veraguas and northern Coclé.

Red-and-green Macaw. Common at Cana, and regular around Pirre Station as well.

Scarlet Macaw. In Panama today confined almost entirely to Isla Coiba. It is alleged to be most common on the Playa Blanca-Playa Barco Quebrado Trail near the southern end of the island, but it has also been seen on the Pozos Termales Trail. A few pairs still persist on the mainland on the southwestern end of the Azuero Peninsula (reported by B. Cedeño and N. Julio to A. Kayser to still occur in the hills west of Guanico Abajo south of Tonosí). There is a reintroduction program for the species on the Costa Rican side of the Burica Peninsula, and some individuals apparently reach the El Chorogo area on occasion.

Blue-and-yellow Macaw. Common at Cana, apparently somewhat less so at Pirre Station. In 1993 a group of macaws that had been confiscated from wildlife smugglers, mainly consisting of Blue-and-yellow but also a few Scarlet and Great Green, were released near Portobelo on the Caribbean coast west of Colón (J. Lacs). The Blue-and-yellows at least still persist in the area, where they roost on a small offshore island near Isla Grande and can regularly be seen on the nearby mainland.

Spectacled Parrotlet. Occasionally reported from around El Real and in the Bayano region. Only one report from Cana, one seen on 3 April 2006 (F. Toldi). Rare in the Cerro Azul-Cerro Jefe area. One recent record from Barro Colorado Island, January 2000 (J. Giacacalone).

Red-fronted Parrotlet (WH). Rare, being most frequently reported from Fortuna and the upper part of the Oleoducto Road in Bocas del Toro and from El Copé. The species has also recently been recorded from Isla Popa in the Bocas del Toro Archipelago (Cooper 1999).

Blue-fronted Parrotlet. Rare, being most frequently reported from Cerro Pirre above Cana, and from the Cerro Azul-Cerro Jefe area. Also recorded from the El Llano-Cartí Road.

Saffron-headed Parrot. Reported regularly from Cana; not known from the El Real-Pirre Station area.

Yellow-crowned Amazon. The species is now surprisingly common in Panama City, being regular especially around city parks, where they can be found when they come in to roost at night. Also regularly found at Tocumen, Aguadulce, near Chitré, and in other coastal and mangrove areas along the Pacific coast.

Rufous-vented Ground-Cuckoo. Rare, most commonly found in wetter foothill areas, including El Valle, Cerro Azul-Cerro Jefe, Nusagandi, and Cana. In the Canal Area, known mostly from Pipeline and Achiote Roads.

Bare-shanked Screech-Owl (WH, EH). Rare, being most often reported from Fortuna, and from around the high camp on Cerro Pirre above Cana. Also occurs in the Volcán-Boquete area and in the Cerro Santiago, where at least three were seen or heard near Ratón on 4 March 2006 (G. Angehr, W. Adsett et al.).

Costa Rican Pygmy-Owl (WH). Rare, most commonly reported from Fortuna; also reported at the Los Quetzales Cabins, on the Los Quetzales Trail, and on the road leading up to it below El Respingo. Formerly included in Andean Pygmy-Owl.

Central American Pygmy-Owl. Rare. Most often reported from Cana, and also occurs elsewhere in Darién, in the Bayano region, and on Pipeline Road. Should be looked for at Wekso, since it has been recorded on the lower Río Changuinola. Formerly included in Least Pygmy-Owl.

Dusky Nightjar (WH). Uncommon in the highlands of Chiriquí. During the breeding season, can be found calling on the bluffs opposite the Los Quetzales Lodge in Guadalupe, and at Finca Ríos in the Bajo Grande area above Cerro Punta. The species has recently been found at Cerro Santiago in eastern Chiriquí, one being seen and heard 4-5 March 2006 (W. Adsett, G. Angehr et al.).

White-tailed Nightjar. Recently found to occur regularly at the Gatún Drop Zone at Sherman on the Atlantic side of the Canal Area. Otherwise the species is widespread but uncommon on the Pacific slope from Chiriquí to western Darién.

Oilbird. In addition to the three records listed by Ridgely (in Darién in 1959, from Pipeline Road in 1974, and from Panama City in 1987), there have been 11 more records of this species from Panama City, the southern Canal Area, and nearby localities since 1985. Localities include Panama City, Las Cumbres near Panama City, Semaphore Hill, El Charco Trail, Pipeline Road, Barro Colorado Island and elsewhere in Barro Colorado Nature Monument, and Cerro Azul. Most individuals have been found perched in trees during the day, although a few have been found in Panama City at streetlights, on buildings, or in parks. The number of records suggests there is probably an undiscovered nesting cave somewhere in Soberanía or perhaps Chagres National Park.

Black Swift. Occurs more regularly in Panama than previously supposed, and perhaps may breed. First recorded near La Esperanza, Chiriquí, in 1982 (Ridgely), more recent records have included: two at Cana, Darien, 13 February 1983 (D. Engleman. R. Johnson); a juvenile female electrocuted on a power line in Chitré in December 1995, collected (F. Delgado); a flock of about 100 at approximately 700 m on the Oleoducto Road in Bocas del Toro, 24 November 1996 (D. Engleman et al.); a flock of about 50 at Fortuna between the dam and the Continental Divide, 2 November 1998 (W. Martínez et al.); several flying past the Canopy Tower Hotel in the Canal area, 1 February 1999 (R. Ridgely et al.); one seen flying past the Canopy Tower, 28 September 2005 (D. Montañez et al.).

Chapman's Swift. Three recent records in addition to the two listed by Ridgely (from Gatún in 1911, and at Mandinga, Kuna Yala, in 1957): at least one was seen in a mixed flock of swifts and swallows above the airstrip at Cana, 30 March 1991 (D. Engleman et al.); a flock of about 40 was seen above the airstrip at Cana on 4 February 1994 (R. Ridgely); and two were seen at Jaqué, Darién, on 5 July 2005 (D. Montañez et al.). These records could pertain to a rare resident population in eastern Panama, or austral migrants (or both).

Costa Rican Swift. Apparently rather scarce in Panama, the only recent definite record being from the Río Palo Blanco, 12 km WNW of Puerto Armuelles, Chiriquí, 3-9 March 1998 (J. Wilson, A. Castillo, L. Pomara). To be looked for around Puerto Armuelles and east to Concepción. Formerly considered part of Band-rumped Swift.

Pale-bellied Hermit. Most common on the larger Pearl Islands (Rey, San José, Pedro González, Viveros, Cañas, Puercos). Uncommon in eastern Pacific Panama, including Tocumen, the Bayano region, and around El Real and Pirre Station in Darién. The species has not been reported from Cana.

***White-whiskered Hermit.** One record. One was seen at Manené, on the Río Balsas, Darién, 10 July 1996 (G. Seutin; Seutin 1998). Probably a rare resident in eastern Darién.

Green-fronted Lancebill. Rare; most often reported from Fortuna, particularly the Continental Divide Trail, and from Cerro Pirre above Cana. Also occurs above Santa Clara and in the Volcán-Boquete region. Recently found to occur on Cerro Peña Blanca above El Copé (G. Angehr et al.).

Violet Sabrewing. Fairly common in the western highlands, being found above Santa Clara, in the Volcán-Boquete area, Fortuna, and Santa Fe; also occurs at El Montuoso in Herrera. Recently found on Cerro Peña Blanca above El Copé (G. Angehr et al.) and at Altos del María (M. Harvey et al.).

Brown Violet-ear. Rare, with occasional records from above Santa Clara, above Santa Fe, above El Copé, Altos del María (M. Iliff et al.), Cerro Campana (no recent reports), Cerro Azul (Cerro Vistamares Trail), and Cana.

***Green-breasted Mango.** This species, as distinct from Veraguan Mango, has recently been found to occur in western Bocas del Toro (Olson 1993), where it has been recorded from various places on the mainland including Wekso and Darkland (Tierra Oscura), and from Islas Colón and Cristóbal. Care should be taken in identification, however, since Veraguan Mango has also been found to occur in Bocas, at least as far west as the Valiente Peninsula.

Veraguan Mango (WP). Although found throughout the Pacific slope lowlands from Chiriquí to Coclé, not particularly numerous, and best looked for near flowering trees. There are old records from the Atlantic side of the Canal Area (two collected at Gatún in 1911), and it has also been found on the Valiente Peninsula in Bocas del Toro (Olson 1993). Formerly considered part of Green-breasted Mango.

White-crested Coquette (WP). Quite scarce in westernmost Chiriquí, with recent records from near Puerto Armuelles and El Chorogo, around Santa Clara (including Finca Hartmann), and at Bajo Frío on the road from Concepción to Volcán.

Fiery-throated Hummingbird (WH). Fairly common at higher elevations in the highlands of western Chiriquí. It can be found above Santa Clara; on the La Cascada Trail at Las Nubes in La Amistad International Park; on trails above the Los Quetzales Cabins; on the Los Quetzales Trail above El Respingo; on the road to the summit of Volcán Barú above Boquete; and on the Continental Divide Trail at Fortuna, among other places.

Humboldt's Sapphire. This species frequents mangroves and coastal scrub along the southern coast of Darién. It occurs in the Jaqué area, and has been seen on the grounds of the Tropic Star Lodge, 26 and 29 May 1995 (D. and L. Engleman, G. Angehr, et al.). There is also a recent report from the village of Mogue, near Punta Patiño, in January 2005 (I. Hoyos, J. Rowlett et al.). The species probably occurs all along the coast at least this far west, and should be looked for elsewhere. Formerly considered part of Blue-headed Sapphire.

Blue-throated Goldentail. While not particularly numerous, this species can be found at most sites in the western Pacific lowlands that have at least some forest remaining, including around Puerto Armuelles, El Chorogo, Chorcha Abajo, Cerro Batipa, Cerro Hoya National Park, El Montuoso Forest Reserve, and on Isla Coiba. It also occurs at Cana in Darién. There are no recent records from the Canal Area.

Violet-capped Hummingbird (EH). Fairly common at Cerro Azul, especially around Cerro Jefe, and common on Cerro Chucantí (Angehr and Christian 2000). Not reported from Nusagandi, which may be too low for it, and does not occur at Cana.

Rufous-cheeked Hummingbird (EH). Fairly common on Cerro Pirre above Cana, and regularly seen around the high camp. It also occurs on the Serranía de Pirre above Pirre Station.

Charming Hummingbird (WP). Fairly common in the lowlands of western Chiriquí where there is some forest, including around Puerto Armuelles, El Chorogo, Chorcha Abajo, and Cerro Batipa. Scarcer at higher elevations, but sometimes reported from the Santa Clara area, the Volcán-Boquete area, and Fortuna.

Escudo Hummingbird (EV). Abundant on Isla Escudo de Veraguas, where it is the only hummingbird. Found throughout the island, from beach scrub to forest, especially at *Heliconia* patches. Usually considered part of Rufous-tailed Hummingbird.

Black-bellied Hummingbird (WH). Uncommon but regularly seen at Fortuna. Otherwise rare in the western highlands in the Volcán-Boquete area and above Santa Fe. Recently found on Cerro Peña Blanca above El Copé (G. Angehr et al.).

White-tailed Emerald (WH). Generally uncommon in the western highlands, being most often reported from Fortuna, and also found above Santa Clara, at Cerro Santiago, above Santa Fe, and above El Valle.

Snowcap. Most easily found on the aptly-named Snowcap Trail at El Copé, and also fairly regular above Santa Fe in the Alto de Piedra area. There are also recent records from El Valle and Altos del María. It has also been recorded near the Continental Divide Trail at Fortuna, in the Volcán-Boquete area, and at Cerro Santiago.

White-bellied Mountain-gem (WH). In Panama known almost exclusively from Fortuna, where it is fairly common. It also occurs at Cerro Santiago and above Santa Fe.

White-throated Mountain-gem (WH). Very common in highlands of western Chiriquí from Boquete westward, including in the Volcán-Cerro Punta area and above Santa Clara.

Purple-throated Mountain-gem. Replaces the previous species from Fortuna eastward, including Cerro Santiago, above Santa Fe, El Copé, and above El Valle, but it is rare at the latter two localities.

Green-crowned Brilliant. Most commonly found at Fortuna, El Copé, and on Cerro Pirre above Cana, but also recorded from above Santa Clara, the Volcán-Cerro Punta area, above Santa Fe, El Valle, and Cerro Azul-Cerro Jefe.

Greenish Puffleg. Quite common on Cerro Pirre above Cana.

Magenta-throated Woodstar (WH). Rather scarce in the western highlands. Most frequently reported from Fortuna, and has also been found on the Pipa de Agua Trail above Boquete. Rare in the Santa Clara area.

Purple-throated Woodstar. Rare but apparently regular at Cana, being reported from Cana itself as well as near the high camp on Cerro Pirre.

Ruby-throated Hummingbird. More common in western Panama than the records cited in Ridgely would suggest. Apparently occurs regularly in westernmost Chiriquí during the northern winter, sometimes even being fairly common near El Chorogo, and has also been recorded near Concepción, at Chorcha Abajo, and on Isla Bóquita south of David. There have been two recent records from Bocas del Toro, from Isla Popa (D. Cooper; Cooper 1999), and on the Oleoducto Road, 26 February 2006 (J. Ortega). Only one record east of Chiriquí, from Playa Coronado in western Panamá Province in 1962 (E. Eisenmann; Wetmore 1968).

Volcano Hummingbird (WH). Fairly common in the highlands of western Chiriquí, mostly at higher elevations. Can be seen along the Los Quetzales Trail above Cerro Punta, and on the summit road up Volcán Barú above Boquete.

Glow-throated Hummingbird (WH). The only place where this national endemic is seen regularly is in the Cerro Santiago area in eastern Chiriquí. Abundance here may vary seasonally, the species being common in early April 1995, when small heath-type flowers were abundant, but very scarce in early March 2006, when they were rare (G. Angehr). Note that Scintillant Hummingbird has also been recorded from this locality, so care must be taken in identification. The more extensively blackish tail of the male Glow-throated compared to Scintillant may be a more easily-seen field mark than the difference in gorget color (rose-to-purplish red in Glow-throated, reddish-orange to orange-red in Scintillant). Females and juveniles may be nearly indistinguishable in the field. According to Ridgely the species has also been seen above Santa Fe, but there are no recent records from there. Glow-throated Hummingbird might also be expected to occur at Fortuna south of the reservoir, as at Cerro Hornito above Finca La Suiza. Although female *Selasphorus* have been observed at Finca La Suiza, there have been no confirmed records so far west. A *Selasphorus* hummingbird has also been collected at Cerro Hoya (F. Delgado) but its specific identity has not yet been determined.

Scintillant Hummingbird (WH). Common in the western highlands, mainly at somewhat lower elevations than Volcano Hummingbird. It can be found above Santa Clara, in the Volcán-Boquete area, and at Fortuna.

Baird's Trogon (WP). Apparently now restricted to the Burica Peninsula, where it occurs in gallery and disturbed forest as well as better forest. It can be found at Charco Azul and along the Río San Bartolo, and is common at El Chorogo. It formerly occurred around Santa Clara, but it was last recorded from this area in 1958.

Orange-bellied Trogon (WH). Generally fairly common in the western highlands, being found in the Volcán-Boquete area, Fortuna, Santa Fe, El Copé, El Valle, Altos del María, and Cerro Campana. Perhaps commonest at Fortuna and El Copé.

Lattice-tailed Trogon (WA). This species can be difficult to find because its range is mainly confined to the mostly inaccessible western Atlantic slope. Scarce, but regularly reported, from Fortuna and the Oleoducto Road (Verrugosa Trail) and from above Santa Fe in the Alto de Piedra area and beyond it on the Atlantic slope, and also recently found at El Copé, 22 February 1992 (C. Quirós, M. Allen et al.).

Golden-headed Quetzal. Uncommon, but regularly reported from Cerro Pirre above Cana. Heard more often than seen.

Resplendent Quetzal. The western highlands of Panama are one of the easiest places in the world to see this spectacular species. During the breeding season, between February and May, usually readily found in the Volcán-Cerro Punta and Boquete areas and above Santa Clara. Found at the Los Quetzales Cabins and along the Los Quetzales Trail above Cerro Punta, and at the Pipa de Agua and Culebra Trails above Boquete, among many other localities. Although it occurs at Fortuna, here it is apparently mainly found at higher elevations than the main trails, and there are few reports.

Tody Motmot. Most readily found in Darién, at Cana and on Cerro Pirre above it, and in the El Real-Pirre Station area. It has also recently been found at El Valle, especially around the Canopy Adventure and on trails in wetter forest.

Barred Puffbird. Common at Cana, and also can be found around El Real and Pirre Station and in the Bayano region. There have also been two recent reports from farther west. One was seen at Birders' View in Altos de Cerro Azul, 27 August 2001 (J. Tejada et al.). Another was seen and photographed at the entrance to the Pipeline Road, 21 March 2004 (J. Ortega et al.) and by other observers subsequently (the first report from the Canal Area in more than 100 years). The species also occurs in northern Coclé and could be expected at lower elevations north of El Valle or El Copé, but has not been reported from there.

Lanceolated Monklet. Six recent records. One was seen at Cana, 31 December 1991 (B. Whitney et al.), the first record from eastern Panama. Two birds were seen perched together in forest at the edge of the Oleoducto Road in Bocas del Toro at an altitude of 630 m, at the present site of the Rancho Ecológico Willy Mazú, 28 May 1994 (G. Angehr, W. Martínez, et al.), and one was seen near the same locality, 4 October 1996 (W. Martínez). Two were seen perched together at 600 m on a tributary of the Río Mulabá west of Alto de Piedra above Santa Fe, 15 July 2000 (D. Montañez, R. Miró, et al.), and one was heard in the same vicinity on 16 July 2006 (R. Miró). One was seen on the Continental Divide Trail at Fortuna, 25 February 2003 (P. Coopmans). The only previous record has been a bird collected in 1926 very close to the Alto de Piedra sighting. The species is probably a rare resident in very wet foothills on the western Atlantic slope and in Darién.

Gray-cheeked Nunlet. Commonly reported from Cana and from the El Real-Pirre Station area in Darién, and also can be found in the Bayano region. The species was formerly present on Achiote Road in the Canal Area, but there have been no records since the mid-1980s and it seems to no longer occur there. There are also no recent records from the Tocumen-Chepo area where it was formerly present. The species occurs in northern Coclé, and could potentially be found in lowlands north of El Valle or El Copé.

Dusky-backed Jacamar (EL). Regularly found at Cana. Much rarer in the El Real-Pirre Station area, with only a few records from there. Also occurs on the Río Sambú.

Rufous-tailed Jacamar. Few recent records, most reports being from the foothills zone of the Oleoducto Road in Bocas del Toro, especially near the Rancho Ecológico Willy Mazú. Otherwise, the only recent reports from western Panama have been from the Burica Peninsula, including El Chorogo and the Mellicita area, but the species now seems to be quite scarce there. In Darién, has been seen occasionally near El Real, and at Cana, where it is very rare.

Prong-billed Barbet (WH). Quite common at Fortuna, and also found at Cerro Santiago. Also found at other localities in the western highlands, including above Santa Clara. Recently found on Cerro Peña Blanca above El Copé (G. Angehr et al.).

Blue-throated Toucanet (WH). Common in the western foothills and highlands, including at Finca Hartmann above Santa Clara, in the Volcán-Boquete area, and at Fortuna. Somewhat less common eastward, above Santa Fe, El Copé, El Valle; and rare at Cerro Campana. Like the following species, formerly considered part of Emerald Toucanet.

Violet-throated Toucanet (EH). Common on Cerro Pirre above Cana, and also found on Cerro Chucantí. Formerly considered part of Emerald Toucanet.

Fiery-billed Aracari (WP). While now local, this species is generally not uncommon in western Chiriquí, from the lowlands to the lower highlands wherever there are substantial patches of forest remaining, being found at Cerro Batipa and in the Burica Peninsula (Charco Azul, Mellicita area, El Chorogo) and above Santa Clara. Although there are records from the Volcán-Boquete area, it appears to be very scarce there today.

***Choco Toucan (EL).** An individual was first reported at Cana in mid-January 2000 (W. Martínez), and subsequently by many other observers over a period of several years. Although all reports from Cana may pertain to a single individual, the species may be a rare resident in eastern Darién.

Golden-naped Woodpecker (WP). Now extremely local, the only site from which this species has recently been reported regularly being El Chorogo, where it is uncommon. The only recent report from the highlands (or away from the Burica Peninsula) is one seen on a side road off the Santa Clara Road west of Volcán, 8 April 1998 (D. Rogers).

Rufous-winged Woodpecker. Rare but regular at El Chorogo in the Burica Peninsula, with several sightings since 2003, and has also been seen at Charco Azul. The only other recent reports of this very rare species have been from the Oleoducto Road in Bocas del Toro, 11 July 1986 (R. Brown), from Alto de Piedra above Santa Fe, 20-24 February 1987 (R. Brown), at Wekso, Bocas del Toro, 4 July 1996 (D. Montañez), and on the Continental Divide Trail at Fortuna, where a male and female were seen on 24 May 2003 (G. Berguido, V. Wilson).

Stripe-cheeked Woodpecker (EL). Uncommon at Cana but recorded on most trips there. Rare but regular at Cerro Azul-Cerro Jefe and Nusagandi. Mostly in foothills, but has on rare occasions occurred in lowlands in eastern Panama.

Golden-green Woodpecker. Rare, but occasionally reported from the El Real-Pirre Station area and from the Bayano region. Not recorded from Cana.

Spot-breasted Woodpecker. This species can be found in the Panama City area in the mangroves at Juan Díaz. (It also occurs in the mangroves just west of Costa del Este, but these are not easily accessible.) It can also be found around El Real and on the road to Pirre Station in Darién

Crimson-bellied Woodpecker. Rare. Most often reported from Cana, but can also be found at Cerro Azul-Cerro Jefe, Nusagandi, and around Pirre Station. In the Canal Area known from Pipeline Road, but there have been very few reports from there since the 1980s.

Coiba Spinetail (WP). Found only on Isla Coiba, where it seems to be fairly common. Recent sightings have come from the Pozos Termales Trail and the trail from Playa Blanca to Playa Barco Quebrado in the south of the island. Often considered part of Rusty-backed Spinetail.

Red-faced Spinetail. Fairly common in the highlands of western and eastern Panama, including above Santa Clara, the Volcán-Boquete area, Fortuna and the Oleoducto Road in Bocas del Toro, Santa Fe, and Cerro Pirre above Cana. Recently also found at Altos del María (J. Ortega, M. Harvey, D. Rodríguez et al.) and at El Valle (D. Rodríguez).

Double-banded Graytail. Regularly found at Cana, and has also been seen around El Real close to the airstrip, and at Pirre Station.

Spotted Barbtail. Fairly common in foothills of west and east, including above Santa Clara, the Volcán-Boquete area, Fortuna and the Oleoducto Road in Bocas del Toro, Cerro Santiago, Santa Fe, and Cerro Pirre above Cana. Recently found on Cerro Peña Blanca above El Copé (G. Angehr et al.), above El Valle (M. Harvey, D. Rodríguez et al.), at Altos del María (J. Ortega, M. Harvey), and on Cerro Chucantí (Angehr and Christian 2000).

Beautiful Treerunner (EH). Scarce, but regularly reported from upper elevations of Cerro Pirre above Cana. Also occurs on Cerro Chucantí (Angehr and Christian 2000).

Ruddy Treerunner (WH). Uncommon in the western highlands, being found above Santa Clara, in the Volcán-Boquete area, at Fortuna, and at Cerro Santiago. Recently found on Cerro Peña Blanca above El Copé (G. Angehr et al.).

Buffy Tuftedcheek. Uncommon in the western highlands, being found above Santa Clara, in the Volcán-Boquete area, at Fortuna, and at Cerro Santiago. Recently found on Cerro Peña Blanca above El Copé (G. Angehr et al.).

Striped Woodhaunter. Generally rare, being most often reported from Nusagandi and on trails in wetter forest at Cerro Azul-Cerro Jefe. It has also been reported from El Chorogo, Fortuna, El Copé, and at Pirre Station. Surprisingly, there seem to be no reports from Cana.

Lineated Foliage-gleaner. Fairly common in the western highlands, being found at Santa Clara, in the Volcán-Boquete area, at Fortuna, and at Santa Fe. Recently found on Cerro Peña Blanca above El Copé (G. Angehr et al.).

Slaty-winged Foliage-gleaner. Most frequently reported from Cana, but can also be found above Santa Fe, at El Copé, and in the Cerro Azul-Cerro Jefe area. Previously occurred regularly on Pipeline Road, but there appear to be no records from there since the late 1970s.

Buff-fronted Foliage-gleaner. Very rare, the only definite records being from the Boquete area and adjacent Bocas del Toro. The most recent report is one seen at Finca Lérida on 2 September 1990 (D. and L. Engleman, H. Laidlaw).

Streak-breasted Treehunter (WH). Uncommon in the western highlands. Most often reported from Fortuna, but can also be found above Santa Clara, in the Volcán-Boquete area, and at Cerro Santiago. Recently found on Cerro Peña Blanca above El Copé (G. Angehr et al.).

Gray-throated Leaftosser. Very rare, the only recent reports being one seen above Ojo de Agua on Finca Hartmann above Santa Clara, 7 March 1994 (G. Angehr), and one heard there 30-31 November 1995 (D. and L. Engleman). The only other definite record appears to be two specimens collected above Boquete in 1901 (Wetmore 1972).

Sharp-tailed Streamcreeper. Very rare, the most recent reports being one seen above Pirre Station at 640 m on a stream near the Rancho Plástico campsite, 4 June 1979 (G. Angehr), and one on the headwaters of the Río Sambú on the Serranía de Jungurudó, 15 August 1997 (D. Christian; Angehr et al. 2004). Although the species was collected on Cerro Pirre above Cana in 1912, there seem to be no recent reports from this area, possibly because its habitat, steep stream beds, is rarely birded. The species has also been recorded from Cerro Tacarcuna and Cerro Quía.

Tawny-winged Woodcreeper. The only locality from which this species has recently been regularly reported is El Chorogo, where it is common. Very rare at Finca Hartmann above Santa Clara, where one was seen in February 1997 (D. Christian).

Strong-billed Woodcreeper. The only localities from which this species is regularly reported are Fortuna and the Oleoducto Road just north of it in Bocas del Toro, where it is rare. There have also been single reports from above Boquete, 18 January 1996 (J. and I. Bishop), and from El Copé, 9-11 July 1984 (P. Trail). There are old records from Veraguas.

Black-banded Woodcreeper. The only recent reports are from Fortuna, where the species is rare. There are old records from Volcán Barú above Boquete, Veraguas, and western Panamá Province.

Straight-billed Woodcreeper. In the Canal Area, this species can be found in mangroves at Diablo Heights, Juan Díaz, and Tocumen on the Pacific side and at Sherman and Galeta on the Atlantic. It has also been found in western Panama at Aguadulce, and at Agallito Beach near Chitré.

Red-billed Scythebill. Most frequently reported from Cana, and also found at El Real-Pirre Station and in the Bayano region. No recent records from the Canal Area.

Brown-billed Scythebill. Most frequently reported from Cerro Pirre above Cana, and also regular at Fortuna. Can also be found at Santa Fe, Cerro Azul-Cerro Jefe, Altos del María (M. Harvey), El Copé, and Cerro Chucantí. Very rare in the Volcán-Boquete area.

Black Antshrike. Fairly common around El Real and the road to Pirre Station, and also found in the Bayano region. Apparently no definite records from Cana.

Black-hooded Antshrike (WP). Fairly common in scrubby forest and edge habitats and mangroves along the western Pacific coast throughout Chiriquí and to the western and southern Azuero Peninsula. Common at Charco Azul, El Chorogo, and other sites in the Burica Peninsula, and can also be found at Chorcha Abajo, Cerro Batipa, Las Lajas Beach, and in Cerro Hoya National Park.

Speckled Antshrike (EL). Most easily found at Nusagandi and the Burbayar Lodge, where it is uncommon but regular. It can also be found on the Xenornis Trail on Cerro Jefe, where it is also uncommon. The species is especially fond of steep wet slopes and damp ravines. Several seen at Cerro Bruja, in Portobelo National Park, Colón Province, 26-27 May 1996 (G. Angehr, D. George, M. Granados) are the westernmost record. Called Spiny-faced Antshrike by the AOU.

Rufous-winged Antwren. Common at Cana, and can also be found around El Real and Pirre Station and in the Bayano region.

White-fringed Antwren. In Panama found only in the Pearl Archipelago, where it is common on the larger islands, including Isla del Rey, San José, Pedro González, and Viveros.

Rufous-rumped Antwren. The only sites at which this species is regularly recorded are Fortuna and Cana, at both of which it is not uncommon.

Bare-crowned Antbird. Most often reported from Cana, and can also regularly be found on Achiote Road in the Canal Area. It can also be found in Bocas del Toro (on the road to El Silencio south of Changuinola, at Wekso, and around Chiriquí Grande), the Burica Peninsula (Charco Azul), El Copé, the Bayano region, and around El Real and Pirre Station.

Dull-mantled Antbird. Most often reported from Cerro Azul-Cerro Jefe and from Cana. It can also be found above Santa Fe, El Copé, El Valle, Cerro Campana, Pipeline and Achiote Roads, Nusagandi, and Pirre Station. This species can be difficult to find because it favors wet ravines. It may be necessary to wade streams in order to find it, particularly in the lowlands.

Wing-banded Antbird. Most easily found at Cana and on Cerro Pirre above it, and can also be found around Pirre Station. It is rare on Pipeline Road, being mainly recorded from the part of the road beyond the Río Limbo.

Black-headed Antthrush. Most often reported from the Cerro Azul-Cerro Jefe area. It can also occasionally be found above Santa Fe, on the Oleoducto Road in Bocas del Toro, El Valle, Cerro Campana (near the top), and at Nusagandi. Like other antthrushes, heard far more often than seen.

Rufous-breasted Antthrush. Known almost exclusively from Cerro Pirre above Cana, and from Fortuna and the Oleoducto Road in Bocas del Toro. It has also been reported from above Santa Fe. Like the previous species, heard much more often than seen.

Black-crowned Antpitta (WA, EL). Rare, most often found in wetter foothills forest, especially at Cerro Azul-Cerro Jefe, Nusagandi, and at Cana and Cerro Pirre above it. Can also be found above Santa Fe, El Copé, El Valle, Altos del María, Cerro Campana, Cerro Chucantí, and Pirre Station. In the Canal Area, formerly found on Achiote and Pipeline Roads, but there have been few if any reports from these sites since the mid-1980s. Possibly still occurs on the northern part of Pipeline.

Scaled Antpitta. Very rare. Most recent reports are from upper elevations on Cerro Pirre above Cana. It has also recently been found nesting on Cerro Gaital above El Valle (D. Rodríguez), and one was seen on the Xenornis Trail at Cerro Azul, 25 April 1999 (M. Denton, J. Wilson, J. Tejada). At Fortuna, one was seen on the Quebrada Aleman Trail, 20 July 2001 (R. Miró, Darién and Delicia Montañez), and the species has also been reported from Cerro Hornito above Finca La Suiza. There are old records from Finca Lérida above Boquete, but no recent reports from this area.

Thicket Antpitta. Common at Cana around the airstrip, and has been found in eastern Panama Province on the upper Río Majé (G. Angehr) and at Burbayar Lodge (J. Rowlett). In Bocas del Toro, it has been found at Dos Bocas south of Changuinola, Wekso, and on the Oleoducto Road. Very furtive and difficult to see without a tape recorder. Formerly considered part of Fulvous-bellied Antpitta.

Ochre-breasted Antpitta. Rare. All recent reports have been from either Fortuna or from upper elevations on Cerro Pirre above Cana.

Tacarcuna Tapaculo (EH). Restricted to the Tacarcuna range near the Colombian border. No recent records, since safety issues have precluded visits to this area in recent decades. The last reports are apparently those of Wetmore in 1964 (Wetmore 1972).

Choco Tapaculo (EH). Uncommon but regularly recorded at upper elevations on Cerro Pirre above Cana. Like other tapaculos, extremely furtive and difficult to see without a tape recorder. Formerly considered part of Nariño Tapaculo.

Silvery-fronted Tapaculo (WH). Fairly common in the western highlands, but unlikely to be found if not calling, or seen without the use of a tape recorder. Most often reported from Fortuna and the Oleoducto Road in Bocas del Toro, and also found above Santa Clara, in the Volcán-Boquete area, and Cerro Santiago. Recently found on Cerro Peña Blanca above El Copé (G. Angehr et al.).

***Yellow-bellied Tyrannulet.** Recently found in the Burica Peninsula, where individuals were seen 4-8 March 1998 at a forest fragment on the Río Palo Blanco, 12 km WNW of Puerto Armuelles (J. Wilson, L. Pomara). To be looked for elsewhere in westernmost Chiriquí. The similar Brown-capped Tyrannulet is not known from western Pacific Panama.

Gray Elaenia. Regularly reported from Cana, and can also be found around Pirre Station. Also occurs in the Canal Area on Pipeline Road, and more rarely as far south as Plantation Road and the Canopy Tower. The species can be difficult to observe since it tends to forage in the canopy. It has recently been seen from the observation tower at the Gamboa Rainforest Resort.

Sepia-capped Flycatcher. Very local in Panama, mainly being found in remnant patches of drier forest on the western Pacific slope. Known from Cerro Batipa, Santa Fe (rare), Isla Coiba, Cerro Hoya National Park, El Montuoso Forest Reserve, and El Valle (trails on the drier side of the valley). In the Canal Area, most regularly found at Metropolitan Nature Park, where it is uncommon, and more rarely as far north as Old Gamboa Road. There are also a few records from Achiote Road and Black Tank Road on Sherman on the Atlantic side.

Slaty-capped Flycatcher. Most often reported from Fortuna and the Oleoducto Road in Bocas del Toro, El Copé, and Cana. It can also be found above Santa Clara, in the Volcán-Boquete area, above Santa Fe, and at El Valle.

Yellow-green Tyrannulet (EL). This national endemic is uncommon but regularly reported from Metropolitan Nature Park and from Cana. In the Canal Area it can also be found at Campo Chagres, on Plantation Road and the first part of Pipeline Road, and there are also a few records from Achiote Road. Although the species would be expected in the Bayano region, and there is an old specimen (1927) from the lower Río Bayano, there seem to be no recent reports from here, nor any from the El Real area.

Rufous-browed Tyrannulet. Almost all recent reports have been from Fortuna and the Oleoducto Road in Bocas del Toro. Also occurs above Santa Fe, El Copé, and at Altos del María (1 July 2005, M. Harvey, D. Rodríguez, J. Pérez). Possible at Cerro Azul-Cerro Jefe, since there has been a recent record from a more remote site in Chagres National Park (Cerro Guagaral, 8 February 1994, W. Adsett, G. Seutin).

White-fronted Tyrannulet. Rare, most of the few recent reports coming from Fortuna, but has also recently been reported from Finca Hartmann above Santa Clara, at the Los Quetzales Cabins above Cerro Punta, and the road below El Respingo at the start of the Los Quetzales Trail. There are also old records from above Boquete, but apparently no recent reports.

Sooty-headed Tyrannulet. Almost all reports are from Cana, where it is moderately common, and it can also be found around El Real. Found on the lower Río Bayano by Ridgely in 1975, but there have been no reports since this west of Darién.

Bronze-olive Pygmy-Tyrant. Uncommon but regularly reported from Cerro Pirre above Cana.

Pale-eyed Pygmy-Tyrant. This species is fairly common in scrub and edge habitats throughout much of the western Pacific slope. Scarce in the immediate Canal Area, where the species is recorded regularly mainly on Old Gamboa Road and at Metropolitan Nature Park, although it is fairly common at Tocumen Marsh.

Northern Bentbill. Generally rather scarce in its limited Panama range. In Bocas del Toro, it has recently been found near Changuinola (on the road to El Silencio), at Wekso, and around Chiriquí Grande. The only recent reports from western Chiriquí are from El Chorogo, although there are older records eastward as far as San Félix.

***Yellow-breasted Flycatcher.** Evidently colonizing Darién, although so far only known from near El Real. First found 10 February 1992 (P. Coopmans), and regularly recorded by other observers since along the road to Pirre Station. To be looked for elsewhere in eastern Panama.

***Stub-tailed Spadebill.** Recently found to occur on all the major islands of the Bocas del Toro Archipelago, including Islas Colón, Bastimentos, Cristóbal, Popa, Solarte, and Cayo Agua (Olson 1993). Olson, who obtained his records through mist-netting, found the species to be most common on Islas Bastimentos and Solarte. There is one record from the mainland near Almirante, presumed to be a vagrant from the islands. The species seems to be very inconspicuous, and there have been very few records since Olson's. The next nearest population of the species is found in northwestern Costa Rica.

Tawny-breasted Flycatcher. Known in Panama only from two specimens collected high on Cerro Tacarcuna (1250 and 1460 m) in 1964 (Wetmore 1972). The species could potentially occur on Cerro Pirre as well, so that any *Myiobius* flycatcher observed at higher elevations (above 1000 m) there should be carefully documented, and photographed if possible. Both Sulphur-rumped and Black-tailed Flycatchers are known from Cana but occur mainly at lower elevations.

Black-billed Flycatcher (EL). Uncommon but regularly reported from Cana, and also found in the Bayano region.

Dark Pewee (WH). Fairly common in the western highlands, and regularly recorded above Santa Clara, in the Volcán-Boquete region, and at Fortuna.

Ochraceous Pewee (WH). Rare, but regularly reported from above Cerro Punta, on the Los Quetzales Trail above El Respingo and from Las Nubes in La Amistad International Park. One record from Fortuna in 1985; not reported from there since.

White-throated Flycatcher. Scarce in the western highlands, probably best looked for around the airstrip en route to the Volcán Lakes, from where there have been several recent reports. Also known from above Boquete, but not reported recently. One was seen near Chiriquí Grande, 15 January 1994 (D. George et al.), the second report from this area (the first being in 1926). There is also one very old specimen from the Canal Area. These latter reports from the lowlands may represent migrants or vagrants.

Black-capped Flycatcher (WH). Fairly common at higher elevations in the Volcán-Boquete region, especially above Cerro Punta, including on the Los Quetzales Trail and at Las Nubes.

Vermillion Flycatcher. Two recent records. A male was seen at Punta Patiño, Darién, on 6-11 March 1999 (M. Olmos), and a male and two females were seen at Cerro Batipa on 31 March 2002 (L. Sánchez et al.). Five previous records, listed by Ridgely, have come from Boquete, Playa Coronado in eastern Panamá Province, and the Atlantic side of the Canal Area.

Pied Water-Tyrant. Most frequently reported from Tocumen Marsh, and can also be found around the Bayano Bridge. It has also been found at Costa del Este, but since much of the wetlands there have now been drained it may no longer occur. Birds seen along the Carretera Interamericana 5 and 10 km east of the border between Panama and Darién provinces on 2 April 1994 (D. and L. Engleman, G. Angehr) are the first reports from Darién Province. Oddly, the species has not been reported from the El Real area.

Cattle Tyrant. Now apparently sparsely distributed from Darién to at least the Panama City area, but still generally rare. Since the first report from Cana in 1981, the species has been found in Darién at El Real (at the airstrip) and near Metetí on the Carretera Interamericana, on Contadora in the Pearl Islands, and in the Panama City area at Amador and Tocumen Airport. It has also been found breeding at the latter two localities. There is also one record from Coco Solo near Colón on the Atlantic side of the Canal Area. To be looked for in open areas anywhere in eastern Panama.

Golden-bellied Flycatcher (WH). Most common in the Fortuna area, and also found on the Oleoducto Road in Bocas del Toro. Can also be found in the Volcán-Boquete area and above Santa Fe. Recently found on Cerro Peña Blanca above El Copé (G. Angehr et al.).

Golden-crowned Flycatcher. Uncommon but regularly recorded at upper elevations on Cerro Pirre above Cana.

Western Kingbird. In addition to the record cited by Ridgely (from the Pacific side of the Canal Area in 1988) there have been two recent records from western Chiriquí. Two were seen near the Río San Bartolo en route from El Chorogo, 3 April 1999 (D. Montañez, W. Adsett et al.), and one was seen near Puerto Armuelles, 10 January 2006 (W. Adsett, D. Wade). The species may be of regular occurrence in the Burica Peninsula.

Sapayoa. Rare, favoring wet ravines and stream sides. Probably most easily found at Nusagandi, and also occurs at Cerro Azul-Cerro Jefe, Cana, and Pirre Station. It is also known from Pipeline Road, where it is probably necessary to walk the streams to have much chance of finding it.

Thrush-like Schiffornis. Fairly common, but heard more often than seen. It should be noted that some of the subspecies found in Panama differ in plumage and vocalizations, and research may result in their being split as different species in the future. The darker olive-brown forms that occur in lowlands and foothills of western Panama (*dumicola*), in foothills in central Panama on Cerro Campana and Cerro Azul-Cerro Jefe, and on Cerro Tacarcuna (*acrolophites*), are likely to be different from the lighter brown form (*panamensis*), which is found in the lowlands of central and eastern Panama, and at all elevations on Cerro Pirre. See Ridgely for further details.

***Gray-headed Piprites (WA).** One record, a breeding-condition female collected at 550 m on Cerro Guabo, Bocas del Toro, about 25 km SSW of Changuinola, 9 March 1994 (K. Aparicio, E. Ponce). Evidently a rare resident on the western Atlantic slope, and to be looked for at Wekso and on the Oleoducto Road.

Cinereous Becard. Regularly reported from Cana, and also occurs in the Bayano region. The species occurred in the Canal Area before 1920, but is now apparently gone, the westernmost recent record being from Campo Chagres, 17 March 2004 (E. Amengual, R. Dean), probably a vagrant.

Black-and-white Becard. Rare in the western highlands, being most frequently found at Fortuna, but also reported from above Santa Clara, in the Volcán-Boquete area, at Cerro Santiago, above Santa Fe, and at El Valle. A female seen near El Real, 5 September 1992 (L. Engleman) is the only recent record from Darién Province.

Rose-throated Becard. Apparently a rare but fairly regular migrant to western Pacific Panama. Most often reported from the Volcán Lakes, but there are also records from Finca Hartmann, El Chorogo, Chorcha Abajo, and Cerro Hoya National Park.

One-colored Becard. Regularly found at Cana, and also reported from El Real-Pirre Station. Also regularly found in the Bayano region, and reported from Nusagandi. There are no records from the Canal Area since an apparent vagrant at Summit Nature Park in 1981.

Lovely Cotinga. Since the first report of this species near Changuinola in 1978 (D. Engleman, R. Johnson), there have been a number of additional reports from Chiriquí Grande (Two Tanks Road) and from the Punta Peña-Almirante Road (various observers). However, a male Blue Cotinga was seen and photographed on Two Tanks Road, 3 July 2004 (D. and K. Wade, G. Angehr, et al.). Since some (but not all) of the previous records were based primarily on range rather than diagnostic field marks, it is now questionable how many may pertain to Blue rather than Lovely Cotinga, although it appears both species do occur at Chiriquí Grande. Any "blue" cotinga on the western Atlantic slope should be carefully studied for the field marks distinguishing these two species. Lovely lacks the black eye-ring found in Blue, has longer blue upper tail coverts usually extending almost to the end of the tail (though sometimes shorter), and has a much larger purple patch on the lower breast and belly. Differences in the color of the throat patch (slightly more purplish and less black in Lovely) are unlikely to be detectable in the field. Photographic confirmation of Lovely for Panama would be highly desirable.

Turquoise Cotinga (WP). The only site from which this species is regularly reported is Finca Hartmann in the Santa Clara area, where it is fairly common. The only recent records from the Chiriquí lowlands are one on the Río San Bartolo below El Chorogo, 3 May 1999 (D. Montañez et al.), and several individuals on the Río Palo Blanco, 12 km WNW of Puerto Armuelles, 6 and 8 March

1998 (J. Wilson et al.). Formerly found east as far as Volcán, but the last record from this area was apparently in 1960.

Black-tipped Cotinga. The only site from which this species is regularly reported is Cana and from Cerro Pirre above it, where it is uncommon but regular. It has also been seen in the Piñas Bay area near the Tropic Star Lodge. So far not reported from El Real-Pirre Station.

Yellow-billed Cotinga (WP). The only site at which this mangrove specialist has been found regularly is at Cerro Batipa, where it is often seen in trees in pastures at the edge of mangroves. It should also be looked for elsewhere in and near the mangroves south of David, such as near the end of the road at Chorcha Abajo. Aside from this area, there have been a few records from non-mangrove forest in the Burica Peninsula, including at El Chorogo (D. and L. Engleman) and in the Mellicita area (R. Ridgely). There is also one recent record from farther east, near Punuga on the eastern side of the Gulf of Montijo, 22 February 2003 (J. Weir). The species may occur elsewhere in mangroves in coastal Chiriquí and southern Veraguas.

Snowy Cotinga (WA). Most often reported from the lower part of the foothills zone of the Oleoducto Road in Bocas del Toro, but scarce even here. There have also been recent records from the Río Teribe above Wekso, and from near Changuinola on the road to Guabito.

Bare-necked Umbrellabird (WH). Rare, being most frequently reported from Fortuna (almost all records being from north of the reservoir), and from the Oleoducto Road in Bocas del Toro, mainly from February to July (though likely present later). The only recent record from the Santa Fe area has been a female or immature male near Alto de Piedra, 15 July 2000 (R. Miró et al.). There have also been three recent records from El Copé, all of females or immature males: two, 17 September 1995 (D. and L. Engleman et al.); one, 28 June 1998 (W. Adsett et al,), and one, 3 October 1999 (G. Angehr, M. Hanly). There are old records from above Boquete and on the Culebra (Holcomb) Trail in Bocas del Toro, but none recently. The species undertakes a seasonal migration to lower elevations in the non-breeding season, but these are not as wide-ranging as those of the Bellbird. There is one record from coastal Bocas del Toro, at Boca del Drago opposite Isla Colón, 25 August 1960 (Wetmore 1972).

Three-wattled Bellbird. Fairly common in the western highlands during the breeding season from March to September. Calling males, however, can be very difficult to see since their perches are often concealed by vegetation. In the highlands, Bellbirds can be found above Santa Clara, the Volcán-Boquete area, Fortuna, and Santa Fe, and more rarely at El Copé. They also probably breed at Cerro Hoya, and possibly at El Chorogo and on Isla Coiba. During the non-breeding season, Bellbirds undertake complex and poorly understood migrations (Costa Rican birds are known to migrate as far as Nicaragua and Panama), and then may be found in lowland areas, occasionally even reaching the Canal Area. They occur regularly in coastal Bocas del Toro, including on all the main islands of the Archipelago, and have even been found breeding on Isla Colón (J. Roper).

Green Manakin. Apparently rare, but probably often overlooked. Found at Cana, Nusagandi, and in the Bayano region. So far unreported from El Real-Pirre Station.

White-collared Manakin. Fairly common in the lowlands of Bocas del Toro from Changuinola west to the Costa Rican border. In the Almirante area, a form known as "Almirante Manakin" is found. It is similar to White-collared but has a pale lemon-yellow collar instead of white, thus resembling Golden-collared (but the latter has a deeper yellow collar and olive-green belly and differs in some other details), of which it has usually been considered a subspecies. Recent genetic work has shown, however, that this Almirante form is actually closer to White-collared.

Orange-collared Manakin (WP). Seems to be considerably less common and more local than Golden-collared Manakin in similar habitats. Probably most easily found in the Burica Peninsula around Puerto Armuelles, where it occurs at Charco Azul and other localities that have some forest remaining. Also can be found at Cerro Batipa, El Montuoso Forest Reserve, and Cerro Hoya National Park.

White-crowned Manakin. Rare. Most reports are from Fortuna and the Oleoducto Road in Bocas del Toro, but it has also been found above Santa Fe and at El Copé.

Sharpbill. Most reports have been from Cerro Pirre above Cana. It has also been found at Fortuna and on the Oleoducto Road in Bocas del Toro.

White-eyed Vireo. Six additional records besides the three cited by Ridgely (from Isla Bastimentos and Almirante in Bocas del Toro and Ft. Davis on the Atlantic side of the Canal Area): Ft. Davis, 2 January 1988 (D. and L. Engleman); one collected, Isla Colón, Bocas del Toro, 18 February 1988 (J. P. Angle et al.); Plantation Road, 16 December 1989 (D. Engleman); Cerro Azul, 5 January 1994 (G. and J. Clayton); Metropolitan Nature Park, 16 February 1997 (M. Allen et al.); Gatún Drop Zone, Sherman, 2 February 2006 (D. Montañez).

Blue-headed Vireo. One additional record, besides the three cited by Ridgely (from Volcán and from Fort San Lorenzo in San Lorenzo National Park): one seen at the Los Quetzales Cabins above Cerro Punta, 13 November 1999 (G. Angehr et al.). Formerly considered part of Solitary Vireo.

***Warbling Vireo.** Vagrant, two records. One was seen at approximately 450 m elevation on the Oleoducto Road, Bocas del Toro, 21 November 1990 (D. Engleman and H. Laidlaw). Another was seen at Finca Hartmann, above Santa Clara, Chiriquí, 4 February 2006 (J. Tejada).

Black-whiskered Vireo. Rare, the most recent records being from Pipeline Road, October 1997 (M. Allen), and from Semaphore Hill near the Canopy Tower, 11 April 2002 (J. Tejada et al.).

Yellow-winged Vireo (WH). Fairly common in the highlands of western Chiriquí, being found above Santa Clara and in the Volcán-Boquete area. Occurs at Fortuna mainly at higher elevations, as at Cerro Hornito.

Yellow-browed Shrike-Vireo. Reported only from Cana, and fairly rare there. Recorded mainly from the slopes of Cerro Pirre above Cana itself, but occasionally as low as the Mine Trail.

Brown Jay. Fairly common in the lowlands around Chiriquí Grande. Much less common and more local in the Changuinola area; reported from Wekso.

Azure-hooded Jay. All recent records are from Fortuna and the Oleoducto Road in Bocas del Toro, where it is uncommon but regularly reported.

Silvery-throated Jay (WH). Known mainly from the Volcán-Boquete area, especially above Cerro Punta, where it is uncommon.

***Common Raven.** Vagrant, one record; almost certainly ship-assisted. One bird was present over a period of three months near the Gatún Locks on the Panamá Canal in 1971 (D. Engleman et. al.).

***White-winged Swallow.** Vagrant or rare migrant in Darién; two records. Three were seen on the Río Tuira, downstream from Unión Chocó, 6 July 1996 (G. Seutin; Seutin 1998), and a group of seven was seen on the Río Sambú on 25 August 1997 (D. Christian, K. Aparicio; Angehr et al. 2004). To be looked for on the Tuira around El Real and Yaviza and elsewhere in eastern Panama.

Cave Swallow. One additional record besides the two previous ones cited by Ridgely (Juan Díaz and Tocumen east of Panama City): one seen at Gatún Dam, 13 October 1987 (D. and L. Engleman, C. Quirós).

Black-capped Donacobius. Most easily found in swampy areas around El Real during the rainy season. Although it has been recorded at Cana, there are very few reports from there. Can also be found around Sambú on the Río Sambú.

White-headed Wren. Most often reported from Cana, and also found at Pirre Station and in the Bayano region. In the Canal Area known mainly from Achiote Road, where it has been seen on the El Trogón Trail and elsewhere along the road.

Band-backed Wren. Most often reported from the Oleoducto Road in Bocas del Toro; around Chiriquí Grande; in the Changuinola area, especially on the road to Almirante; and above Santa Fe.

Sooty-headed Wren. Almost all reports are from Cana, where the species can be found around Cana itself and on the lower slopes of Cerro Pirre at least as far up as the high camp.

Black-throated Wren (WA). Fairly common in appropriate habitat in Bocas del Toro from the Changuinola area to Chiriquí Grande.

Riverside Wren (WP). Fairly common on the Burica Peninsula around Puerto Armuelles, at Charco Azul and in other areas. Rare above Santa Clara. Can also be found at the Río Macho de Monte below Volcán.

Stripe-throated Wren. Fairly common at Cana, and also can be found at Nusagandi. Rare in the Bayano region. The species has recently also been found at Cerro Azul-Cerro Jefe on the Xenornis and Vistamares Trails (W. Adsett et al.), which are the westernmost records.

Stripe-breasted Wren. Most frequently reported from El Copé, and can also be found on the Oleoducto Road in Bocas del Toro and above Santa Fe. In the Canal Area, the species is rare on Achiote Road, its easternmost locality.

Canebrake Wren (WA). Common in western Bocas del Toro from the Changuinola area east at least to Chiriquí Grande. Sometimes included in Plain Wren.

Ochraceous Wren (WH, EH). Fairly common in the western highlands, being found above Santa Clara, in the Volcán-Boquete area, at Fortuna and on the Oleoducto Road in Bocas del Toro, above Santa Fe, above El Valle, and at Altos del María; rare at Cerro Campana. Uncommon on Cerro Pirre above Cana. Also found on Cerro Chucantí (Angehr and Christian 2000).

Grass Wren. Formerly found in Chiriquí, but there have been no reports since 1905. Recorded from near Concepción, around Boquete, and on Volcán Barú, However, the species is very inconspicuous unless calling, and could still occur. Also called Sedge Wren.

Timberline Wren (WH). In Panama known only from near the summit of Volcán Barú, in the bamboo zone above 2,700 m (9,000 ft), where it is fairly common.

Slate-throated Gnatcatcher. Mostly recorded from Cana, where it is uncommon but regularly reported. It also occurs at Nusagandi, where it is evidently rare.

Black-faced Solitaire (WH). Common in the western highlands but difficult to see, being recorded mainly by call. Found above Santa Clara, in the Volcán-Boquete area, at Fortuna and on the Oleoducto Road in Bocas del Toro, at Cerro Santiago, and above Santa Fe. Occurs at El Copé, but mainly at upper elevations above the main trails.

Varied Solitaire (EH). Common at higher elevations on Cerro Pirre above Cana, but like the previous species difficult to see. Also common on Cerro Chucantí (Angehr and Christian 2000).

Black-billed Nightingale-Thrush (WH). Common in the highlands of Chiriquí, being found above Santa Clara, in the Volcán-Boquete area, and at Cerro Santiago. Occurs at Fortuna on Cerro Hornito.

Black-headed Nightingale-Thrush. Most often reported from the Oleoducto Road in Bocas del Toro, and also occurs above Santa Fe. The species has also been found on Cerro Peña Blanca at El Copé, two being seen at 850 m (2,800 ft) on 4 April 1992 (C. Quirós, J. Ortega).

Sooty Thrush (WH). Primarily found at higher elevations (above 2250 m/7500 ft) on Volcán Barú, where it is common. It occasionally occurs lower, on the Los Quetzales Trail above El Respingo and around Volcán.

Pale-vented Thrush. Found at Fortuna and on the Oleoducto Road in Bocas del Toro, Santa Fe, El Copé, Cerro Campana, Cerro Azul-Cerro Jefe, Cerro Chucantí, and Cana and Cerro Pirre above it. Recently also found at El Valle and Altos del María (M. Harvey, D. Rodríguez). One recent record from the Canal Area, a bird mist-netted on Howard Air Force Base on the west bank of the Canal in 1997 (D. Christian).

European Starling. One additional record, besides the one cited by Ridgely (at Albrook on the Pacific side of the Canal Area in 1979), an individual seen at Shimmy Beach at Sherman on the Atlantic side of the Canal area, 24 November 2001 (D. Montañez). Both birds probably arrived on ships traversing the Panama Canal.

Yellowish Pipit. Now apparently rather scarce and local. Best looked for around the airport at Chitré and elsewhere in Herrera such as Sarigua National Park, and in the Coclé grasslands. Few recent records from the Canal Area; probably best looked for at Tocumen.

Black-and-yellow Silky-flycatcher (WH). Uncommon in the western highlands, found above Santa Clara, the Volcán-Boquete area, Fortuna and the Oleoducto

Road in Bocas del Toro, and Cerro Santiago. Probably commonest on the Summit Road on Volcán Barú, and also regularly reported from Fortuna.

Long-tailed Silky-flycatcher (WH). Fairly common in the western Chiriquí highlands, being found above Santa Clara and in the Volcán-Boquete area. Occurs at Fortuna, but rare there.

Nashville Warbler. One additional record besides the one cited by Ridgely (west of Volcán in 1980), an individual seen near the Gatún Yacht Club near Colon, 31 March 2006 (J. Tejada).

***Virginia's Warbler.** Vagrant, one record. One was seen at Diablo Heights near Panamá City, 23 March 2001 (R. Moore).

Flame-throated Warbler (WH). Fairly common in western Chiriquí, including above Santa Clara and in the Volcán-Boquete area. One report from the Oleoducto Road in Bocas del Toro, May 1984 (P. Trail), but none from Fortuna itself.

Northern Parula. Now known to be a rare but fairly regular visitor to Panama, with at least 12 additional records besides the four listed by Ridgely. Most records have been from the Canal Area, mostly near the Atlantic coast (Sherman, Galeta, Isla Margarita, Coco Solo), but a few from the central part of the Isthmus (Old Gamboa Road, Campo Chagres) or near the Pacific coast (Farfan, Rodman). Records from elsewhere include one at the Volcán Lakes, 28 November 1995 (D. and L. Engleman); one at Boquete, 1 March 1997 (D. Montañez et al.), and two on the Cayos Zapatilla in the Bocas del Toro Archipelago, 22 October 2005 (J. A. Cubilla et al.). Earliest date 19 October, latest 26 April; reported every month between those dates, so some evidently winter.

Black-throated Blue Warbler. A rare but regular visitor to Panama, with at least 10 additional records besides the four listed by Ridgely. Most records have been from the Atlantic side of the Canal Area (Achiote Road, Sherman, Gatún Dam, Margarita, Galeta, Barro Colorado Island), but there have also been two each from Cerro Campana and Cerro Azul-Cerro Jefe. Also recorded at Volcán and Bambito in the Chiriquí highlands, and at Cana, Darién (between 12-17 December 1995, L. Augustine).

***Golden-cheeked Warbler.** Vagrant, two records. A male was seen on the Los Quetzales Trail above Cerro Punta, 3 January 2005 (J. Tejada, N. Holland, M. Harman), and a female was seen on the road below El Respingo, a little below the previous sighting, 13 February 2006 (J. Tejada).

Townsend's Warbler. Three additional records besides the four cited by Ridgely (from the Volcán-Cerro Punta area and the summit of Volcán Barú.): one at Fortuna, 24 November 1990 (D. and L. Engleman); one at Nueva Suiza, above Volcán, 13 February 1999 (J. Tejada); and one above Boquete, 10 April 2001 (P. Matthews).

Yellow-throated Warbler. Now known to be a rare but regular visitor to Panama, with at least 19 additional records besides the five listed by Ridgely. In the Canal Area, records have come from both the Atlantic and Pacific slopes, and there have been at least five records from Cerro Azul and one from Cerro Campana. In the western highlands, in addition to the record cited by Ridgely from below

Volcán, one was seen at Bambito, 12 January 1977 (A. Greensmith), and one at Jaramillo Arriba above Boquete, 1 March 1997 (D. Montañez et al.). Records from other sites include one at El Valle, 18 August 1985 (H. Stockwell, C. Meyer), the earliest date, and one at Cana, Darién, 26-28 March 1991 (D. and L. Engleman), the latest dates. Recorded every month from August to March, so some certainly winter. In Panama mostly associated with introduced pines.

Prairie Warbler. A rare but regular visitor to Panama, with at least eight additional records besides the three cited by Ridgely. Most records have been from the Canal Area, with a few from the western highlands at Volcán and Fortuna, and one from Las Lajas in the Chiriquí lowlands, 29 November 1996 (D. Engleman). Earliest record 19 November, latest 2 February; most records are from January.

Palm Warbler. A bird at Cana, 19 January 1998 (P. Coopmans), is the first record from Darién. Otherwise reported mainly from the Canal Area, especially near the Atlantic coast, and from east of Panama City, with one record from the western highlands below Boquete.

***Swainson's Warbler.** Vagrant, one record. One was mist-netted and photographed on Old Gamboa Road, 13 March 2004 (J. Brawn et al.).

Collared Redstart (WH). Common in the western highlands, and found above Santa Clara, the Volcán-Boquete area, Fortuna, Cerro Santiago, and Santa Fe. Also recently found on at higher elevations on Cerro Peña Blanca at El Copé (G. Angehr et al.).

Olive-crowned Yellowthroat. Common in appropriate habitat in Bocas del Toro, being found in the Changuinola area, at Chiriquí Grande and on the lower part of the Oleoducto Road, and on Isla Colón. One record from the Quebrada Arena in Fortuna on 2 July 1992 (D. and L. Engleman) is the only one from the Pacific slope.

Masked Yellowthroat. Reported mainly from around Volcán, and from along the road west to Río Sereno.

Gray-crowned Yellowthroat. Rather uncommon in the foothills and highlands of western Chiriquí, and recently found in the lowlands in the Puerto Armuelles area and at El Chorogo, and at Cerro Batipa.

Black-cheeked Warbler (WH). Fairly common in the western highlands, being found above Santa Clara and the Volcán-Boquete area; rare at Fortuna. Also known from Veraguas.

Pirre Warbler (EH). Rather uncommon but regularly reported from higher elevations on Cerro Pirre above Cana.

Wrenthrush (WH). Rare in the western highlands, but probably often overlooked due to its furtive habits. Reported mainly from above Cerro Punta (on the Los Quetzales Trail, at the Los Quetzales Cabins, and at Las Nubes). Rare at higher elevations at Fortuna, and recently found on Cerro Peña Blanca at El Copé (G. Angehr et al.).

White-eared Conebill. Commonest in the Bayano region; uncommon at Cana and around El Real and Pirre Station.

Tacarcuna Bush-Tanager (EH). Somewhat uncommon but regularly found in cloud and elfin forest at Cerro Azul-Cerro Jefe. Also found on Cerro Chucantí (Angehr and Christian 2000.).

Pirre Bush-Tanager (EH). Very common and usually easily found at upper elevations on Cerro Pirre above Cana. Also found on the Serranía de Junguruḋó (Angehr et al. 2004).

Sooty-capped Bush-Tanager (WH). Fairly common in the highlands of western Chiriquí, being reported from the Volcán-Boquete area, Fortuna, and Cerro Santiago.

Yellow-throated Bush-Tanager. Reported most frequently from the Oleoducto Road in Bocas del Toro, from above Santa Fe, and at El Cope. Rare at Cerro Azul (W. Adsett).

Ashy-throated Bush-Tanager. So far only known from the Oleoducto Road in Bocas del Toro and just south of the Continental Divide in Fortuna, where it is rare. First recorded in 1982, there have been several reports since along the Oleoducto Road itself, on the Verrugosa Trail, and at the Rancho Ecológico Willy Mazú, as well as a report from the Quebrada Arena in Fortuna, 7 July 1994 (W. Martínez).

Yellow-backed Tanager. Uncommon at Cana, and also found at El Real-Pirre Station.

Black-and-yellow Tanager (WA, EL). Fairly common at most of the sites at which it is found, which include Fortuna and the Oleoducto Road in Bocas del Toro, above Santa Fe, El Copé, El Valle, Cerro Campana, Cerro Azul-Cerro Jefe, Nusagandi, Cerro Chucantí, and Cana.

Lemon-spectacled Tanager. Fairly common at Cana and Pirre Station.

White-throated Shrike-Tanager. Generally rare and very local. Fairly common at El Chorogo. Otherwise most reports have been from El Copé, with a few from above Santa Fe.

Sulphur-rumped Tanager (WA, EL). Found in the Changuinola area, at Chiriquí Grande and the lower part of the Oleoducto Road, above Santa Fe (rare), on the Atlantic slope of the Canal Area, Cerro Azul-Cerro Jefe, Nusagandi, and the Bayano region.

Scarlet-browed Tanager. Uncommon at Cana, and also occurs at El Real-Pirre Station.

Crimson-collared Tanager. Generally rare and local, known from the Changuinola area, Chiriquí Grande and the lower part of the Oleoducto Road, above Santa Fe, and El Copé.

Passerini's Tanager. Very common in Bocas del Toro in the Changuinola area, Chiriquí Grande and the Oleoducto Road, and in the Bocas del Toro Archipelago.

Cherrie's Tanager (WP). Very common in the western Chiriquí lowlands, being found around Puerto Armuelles and at Cerro Batipa and many other localities; scarce in the highlands in the Volcán-Boquete area.

Blue-and-gold Tanager (WH, EH). Most frequently reported from Fortuna and the Oleoducto Road in Bocas del Toro and from El Copé, and also found above Santa Fe. Recorded at Cerro Azul-Cerro Jefe, but rare there.

Gray-and-gold Tanager. Reported mainly from Cana and the lower slopes of Cerro Pirre, where it is uncommon. Rare at Nusagandi, with only a few reports. There has also been a recent report from Cerro Azul, at Calle Maipo, 23 January 2000 (W. Adsett et al.).

Spangle-cheeked Tanager (WH). Fairly common at Fortuna, especially on the Continental Divide Trail, and also reported from the Oleoducto Road in Bocas del Toro. Generally rather rare elsewhere in the western highlands, being found in the Volcán-Boquete area, at Cerro Santiago, and above Santa Fe. Recently found on Cerro Peña Blanca above El Copé (J. Ortega, G. Angehr et al.).

Green-naped Tanager (EH). Uncommon at higher elevations on Cerro Pirre above Cana.

Viridian Dacnis (EL). Rare. Most reports are from Cana, but also found in the El Real-Pirre Station area and in the Bayano region.

Purple Honeycreeper. Uncommon at Cana. Note that the very similar Shining Honeycreeper also occurs there.

Swallow Tanager. Uncommon at Cana. There are also a few records from the El Llano-Cartí Road near Nusagandi.

Lesson's Seedeater. One additional record besides the three cited by Ridgely (from Yaviza and Cana), one seen near El Real, 27 April, 1995 (G. Rompré).

***Lined Seedeater.** Vagrant, one record. A male was seen at Sherman, 2 January 2000 (G. Seutin).

***Savannah Sparrow.** Vagrant, two records. One was seen at the Gatún Spillway near Colón, 29 December 1982 (P. and M. Akers, R. Johnson). Another at Coco Solo, also near Colón, 31 December 1976, was probably this (D. Engleman, P. Akers, A. Greensmith).

Nicaraguan Seed-Finch (WA). Generally rare in the Changuinola area, with few reports, but possibly sporadic, with many being seen west of town, 15-17 April 2006 (D. Montañez, R. Miró). Also reported from Isla Colón (V. Wilson).

Blue Seedeater. Rare, being mostly known from the western foothills and highlands. Reported from the Volcán-Boquete area, above Santa Fe, El Copé, El Valle, and Cerro Campana. The species been also recorded in the lowlands in the Canal Area, but there have been no reports from this area since 1977. Most readily found in areas where bamboo is producing seed.

Slaty Finch. Rare in the western highlands, recent reports coming from the Los Quetzales Cabins, Fortuna, and Cerro Santiago.

Peg-billed Finch (WH). The species is sporadic in the western highlands, and was briefly common in early 1979 above Cerro Punta. The most recent records have been one at the Los Quetzales Cabins, 4 July 1997 (D. Engleman et al.) and two at Las Nubes, 3 July 2001 (Darién and Delicia Montañez).

Slaty Flowerpiercer (WH). Generally common in the highlands of western Chiriquí (less common eastward), found above Santa Clara, in the Volcán-Boquete area, at Fortuna, and at Cerro Santiago.

Saffron Finch. This introduced species was previously found mainly on the Atlantic side of the Canal Area, but is now also regular at some places on the Pacific side (Albrook, perhaps Amador) and common at Panama Viejo.

Grassland Yellow-Finch. Evidently generally now quite rare and local in the grasslands of Coclé (an endemic subspecies, *eisenmanni*), with only one colony and a few scattered individuals found during surveys in June 1997 (G. Angehr). Best looked for along roads south of Penonomé. It has been found at times to be numerous on Finca Santa Mónica, 3 km west of Río Hato, Coclé, but since this is private property permission must be secured to enter. The species has also been seen at La Jagua in eastern Panamá Province and at Las Lajas in Chiriquí, but these records may represent wanderers.

Wedge-tailed Grass-Finch. Very local in grasslands in foothills, with recent reports almost exclusively from near the entrance to Altos de Campana National Park, and from just west of El Valle (M. Harvey, D. Rodríguez). Has also recently been seen in grassland to the east of Boquete (D. Wade).

Sooty-faced Finch (WH, EH). Almost all recent reports are from Fortuna. However, it has also been found, although more rarely, in the Volcán-Boquete area, Cerro Santiago, and above Santa Fe. Recently found on Cerro Peña Blanca above El Copé (G. Angehr et al.) and at El Valle (D. Rodriguez). The only record from eastern Panama is from Cerro Tacarcuna in 1964 (Wetmore et al. 1984).

Yellow-thighed Finch (WH). Very common and easily found in the highlands of western Chiriquí, being found above Santa Clara and in the Volcán-Boquete area.

Yellow-green Finch (WH). Common and usually readily found in forest in the Cerro Santiago area. It also occurs at higher elevations at Fortuna on the Quebrada Alemán Trail and on Cerro Hornito above Finca La Suiza, where it seems to be rare. (The species evidently occurs at Fortuna only south of the reservoir, not being found on the Continental Divide Trail.) Should occur at higher elevations above Santa Fe, but there seem to be no specific records from there.

Large-footed Finch (WH). Fairly common at higher elevations in the Volcán-Boquete area, particularly on the Los Quetzales Trail and at the Los Quetzales Cabins above Cerro Punta, and on the Summit Road on Volcán Barú above Boquete.

Stripe-headed Brush-Finch. Most often reported from Cana and the lower slopes of Cerro Pirre, where it is uncommon. Rare at Cerro Azul-Cerro Jefe (mainly Cerro Vistamares Trail), with few recent reports, and rare in the western highlands above Santa Clara, in the Volcán area, and at Fortuna.

Grasshopper Sparrow. Apparently there have been no confirmed records of this species in Panama since the late 1960s. Surveys in 1996 and 1997 failed to locate any in areas where it previously bred in Coclé. (The population from this area has been described as the endemic subspecies *beatriceae.*) There have also been no recent reports from other areas where it probably bred, including the Chepo-Pacora area east of Panama City (last reports in the late 1960s), and below Boquete in Chiriquí (last record in 1905). Two reports from Bocas del Toro in the 1960s evidently pertain to migrants. Should be looked for in grasslands in Chiriquí, Coclé, and eastern Panamá Province.

Volcano Junco (WH). In Panama known only from near the summit of Volcán Barú, mostly above 3,000 m (10,000 ft), rarely lower. Usually readily seen along the upper parts of the Summit Road above Boquete in grassy or bare areas and around the communications towers on the peak.

***Dark-eyed Junco.** Vagrant, one record; almost certainly ship-assisted. An individual in juvenile male plumage was captured at the Gatún Drop Zone on Sherman, 13 November 1992 (W. Martínez).

***Grayish Saltator.** This species has recently colonized western Bocas del Toro, and is fairly common from the Costa Rican border to at least the Changuinola area. The first records were near Changuinola, 30 July 2003 (D. Montañez et al.) and 13 September 2003 (G. Berguido, V. Wilson).

Black-thighed Grosbeak (WH). Uncommon in the western highlands, being known from above Santa Clara, the Volcán-Boquete area, Fortuna and the Oleoducto Road in Bocas del Toro, and above Santa Fe. Occurs at El Copé at elevations above the main trails.

Yellow-green Grosbeak. Fairly common at middle elevations on Cerro Pirre above Cana.

Black-cowled Oriole. Uncommon in western Bocas del Toro around Changuinola, at Wekso, and in the Chiriquí Grande area. Also found on Isla Colón and probably elsewhere in the Archipelago. May be spreading eastward; recently found at Palmarazo in northern Coclé (G. Angehr, A. Castillo).

Orange-crowned Oriole. Fairly common in the Bayano region, and rare at Nusagandi. Can also be found around El Real. In the Canal Area, it is rare but occasionally recorded around Gamboa, where it is believed to have been introduced by Emberá who emigrated from eastern Panama to settle on the Río Chagres some decades ago.

Montezuma Oropendola. Very common in the lowlands of Bocas del Toro in the Changuinola area and at Chiriquí Grande, and also found on the main islands of the Archipelago. In the Canal Area it can be found on Achiote Road, mainly in open areas beyond the forested part of the road, where it is uncommon.

Black Oropendola (EL). Uncommon around El Real and on the road to Pirre Station, and has also been reported from Punta Patiño. Very rare at Cana, with apparently only a single record.

Yellow-throated Euphonia. Very rare above Santa Clara and in the Volcán-Boquete area. Recent reports include one at Finca Hartmann, 9 April 1982 (D. Engleman), a male and a female at the Dos Ríos Hotel near Volcán, 26 January 1989 (D. and L. Engleman); a male and two females near Alto Jaramillo above Boquete, 26 November 1998 (L. Sánchez et al.), and one at Finca Hartmann, 15 November 1999 (M. Allen).

Spot-crowned Euphonia (WP). Uncommon but regularly reported at Finca Hartmann above Santa Clara, and also uncommon but regular at El Chorogo.

Tawny-capped Euphonia (WA, EL). Fairly common in the foothills of both west and east, being recorded from Fortuna and the Oleoducto Road in Bocas del Toro, Cerro Santiago, Santa Fe, El Copé, El Valle, Cerro Campana, Cerro Azul-Cerro Jefe, and Nusagandi. Not found on Cerro Pirre.

Olive-backed Euphonia. Common in the lowlands of Bocas del Toro around Changuinola and around Chiriquí Grande.

Orange-bellied Euphonia. Fairly common at Cana and on the lower slopes of Cerro Pirre.

Golden-browed Chlorophonia (WH). Uncommon in the western highlands, being reported from above Santa Clara, the Volcán-Boquete area, Fortuna, Cerro Santiago, and above Santa Fe. Recently recorded on Cerro Peña Blanca above El Copé (C. Quirós, J. Ortega, G. Angehr et al.).

Yellow-collared Chlorophonia (EH). Rare above 1,200 m (4,000 ft) on Cerro Pirre above Cana. First recorded in 1983, and seen on several occasions since then by various observers (P. Coopmans, W. Martínez, J. Coons), indicating that the species is a rare resident rather than a casual visitor, as suggested by Ridgely.

***American Goldfinch.** One record, probably an escaped cage bird but possibly a natural vagrant. An adult male in breeding plumage was seen near Playa Coronado in western Panamá Province, 18 March 2003 (A. Castillo, A. Krishman).

***Nutmeg Munia.** One record, probably an escaped cage bird but possibly colonizing. One was seen and photographed near Gamboa, 22 May 2003 (J. Tejada).

***Tricolored Munia.** Two records, probably escaped cage birds but possibly colonizing. One was seen just west of Summit Nature Park, Panamá Province, 6 February 2006 (M. Iliff). Another was seen and photographed in Metropolitan Nature Park in Panama City, 7 February 2006 (J. Tejada).

References

Angehr, George R. 1999. Rapid long-distance colonization of Lake Gatun, Panama, by Snail Kites. *Wilson Bulletin* 111: 265-268.

Angehr, George R. 2006. *Annotated Checklist of the Birds of Panama.* Panama Audubon Society, Panama City, Panama.

Angehr, George R., and Daniel G. Christian. 2000. An ornithological survey of the Serranía de Majé, an isolated mountain range in eastern Panama. *Bulletin of the British Ornithologists' Club* 120: 173-178.

Angehr, George R., Daniel G. Christian, and Karla M. Aparicio. 2004. A survey of the Serranía de Jungurudó, an isolated mountain range in eastern Panama. *Bulletin of the British Ornithologists' Club* 124: 51-62.

Cooper, Daniel S. 1999. Notes on the birds of Isla Popa, western Bocas del Toro, Panama. *Cotinga* 11: 22–26.

Loftin, Horace. 1991. An annual cycle of pelagic birds in the Gulf of Panama. *Ornitología Neotropical* 2: 85-94.

Miller, Loye. 1937. Winter notes on some North American birds in the tropics. *Condor* 39: 16-19.

Olson, Storrs L. 1993. Contributions to avian biogeography from the Archipelago and lowlands of Bocas del Toro, Panama. *Auk* 110:100-108.

Ridgely, Robert S., and John A. Gwynne. 1989. *A Guide to the Birds of Panama* (Second Edition). Princeton University Press, Princeton, New Jersey.

Rompré, Ghislain. 2002. Las aves del fragmento Achiote-Sur, Panamá y las implicaciones para la conservación. *Mesoamericana* 6: 14-24.

Seutin, Gilles. 1998. Two bird species new for Panama and Central America: White-whiskered Hermit *Phaethornis yaruqui* and White-winged Swallow *Tachycineta albiventer. Cotinga* 9: 22-23.

Spear, Larry B., and David G. Ainley, 1999. Seabirds of the Panama Bight. *Waterbirds* 22: 175-198.

Wetmore, Alexander. 1965. *The Birds of the Republic of Panama. Part 1. Tinamidae (Tinamous) to Rhynchopidae (Skimmers).* Smithsonian Institution, Washington D.C.

Wetmore, Alexander. 1968. *The Birds of the Republic of Panama. Part 2. Columbidae (Pigeons) to Picidae (Woodpeckers).* Smithsonian Institution, Washington D.C.

Wetmore, Alexander. 1972. *The Birds of the Republic of Panama. Part 3. Passeriformes: Dendrocolaptidae (Woodcreepers) to Oxyruncidae (Sharpbill).* Smithsonian Institution, Washington D.C.

Wetmore, Alexander, Roger F. Pasquier, and Storrs L. Olson. 1984. *The Birds of the Republic of Panama. Part 4. Passeriformes: Hirundinidae (Swallows) to Fringillidae (Finches).* Smithsonian Institution, Washington D.C.

A note on taxonomic splits

In this guide we recognize the following taxonomic splits that have taken place since the publication of Ridgely in 1989, and that involve two forms both occurring in Panama, thus increasing the country list. (We also recognize other splits between forms occurring in Panama and extralimital ones, and thus not affecting the country totals, but these are not included here.) See the *Annotated Checklist of the Birds of Panama* (Angehr 2006) for details.

Nazca Booby *S. granti,* split from Masked Booby *S. dactylatra.*

Green Heron *Butorides virescens* and Striated Heron *B. striatus,* a split of Green-backed Heron *B. virescens.*

Brown-backed Dove *L. battyi,* split from Gray-headed Dove *L. plumbeiceps.*

Costa Rican Swift *Chaetura fumosa,* split from Band-rumped Swift *C. spinicaudus.*

Veraguan Mango *Anthracothorax veraguensis,* split from Green-breasted Mango *A. prevostii*

Violet-crowned Woodnymph *Thalurania colombica* and Green-crowned Woodnymph *T. fannyi,* a split of Crowned Woodnymph *T. colombica.*

Escudo Hummingbird *Amazilia handleyi,* split from Rufous-tailed Hummingbird *Amazilia tzacatl.*

Blue-throated Toucanet *Aulacorhynchus caeruleogularis* and Violet-throated Toucanet *A. cognatus,* split from Emerald Toucanet *A. prasinus.*

Canebrake Wren *Thryothorus zeledoni,* split from Plain Wren *T. modestus.*

Cherrie's Tanager *Ramphocelus costaricensis* and Passerini's Tanager *R. passerinii,* a split of Scarlet-rumped Tanager *R. costaricensis.*

Species list for Panama

The following 972 species have been recorded in Panama and its territorial waters.

Tinamous	**Tinamidae**
Great Tinamou	*Tinamus major*
Highland Tinamou	*Nothocercus bonapartei*
Little Tinamou	*Crypturellus soui*
Choco Tinamou	*Crypturellus kerriae*
Ducks, Swans, and Geese	**Anatidae**
White-faced Whistling-Duck	*Dendrocygna viduata*
Black-bellied Whistling-Duck	*Dendrocygna autumnalis*
Fulvous Whistling-Duck	*Dendrocygna bicolor*
Comb Duck	*Sarkidiornis melanotos*
Muscovy Duck	*Cairina moschata*
American Wigeon	*Anas americana*
Mallard	*Anas platyrhynchos*
Blue-winged Teal	*Anas discors*
Cinnamon Teal	*Anas cyanoptera*
Northern Shoveler	*Anas clypeata*
White-cheeked Pintail	*Anas bahamensis*
Northern Pintail	*Anas acuta*
Green-winged Teal	*Anas crecca*
Ring-necked Duck	*Aythya collaris*
Lesser Scaup	*Aythya affinis*
Masked Duck	*Nomonyx dominicus*
Curassows, Guans, and Chachalacas	**Cracidae**
Gray-headed Chachalaca	*Ortalis cinereiceps*
Crested Guan	*Penelope purpurascens*
Black Guan	*Chamaepetes unicolor*
Great Curassow	*Crax rubra*
New World Quail	**Odontophoridae**
Crested Bobwhite	*Colinus cristatus*
Marbled Wood-Quail	*Odontophorus gujanensis*
Black-eared Wood-Quail	*Odontophorus melanotis*
Tacarcuna Wood-Quail	*Odontophorus dialeucos*
Black-breasted Wood-Quail	*Odontophorus leucolaemus*
Spotted Wood-Quail	*Odontophorus guttatus*
Tawny-faced Quail	*Rhynchortyx cinctus*
Grebes	**Podicipedidae**
Least Grebe	*Tachybaptus dominicus*
Pied-billed Grebe	*Podilymbus podiceps*

Albatrosses	**Diomedeidae**
Gray-headed Albatross	*Thalassarche chrysostoma*
Wandering Albatross	*Diomedea exulans*
Waved Albatross	*Phoebastria irrorata*
Shearwaters and Petrels	**Procellariidae**
Galapagos Petrel	*Pterodroma phaeopygia*
Parkinson's Petrel	*Procellaria parkinsoni*
Cory's Shearwater	*Calonectris diomedea*
Wedge-tailed Shearwater	*Puffinus pacificus*
Sooty Shearwater	*Puffinus griseus*
Christmas Shearwater.	*Puffinus nativitatis*
Manx Shearwater	*Puffinus puffinus*
Townsend's Shearwater	*Puffinus auricularis*
Audubon's Shearwater	*Puffinus lherminieri*
Storm-Petrels	**Hydrobatidae**
Wilson's Storm-Petrel	*Oceanites oceanicus*
White-vented Storm-Petrel	*Oceanites gracilis*
Wedge-rumped Storm-Petrel	*Oceanodroma tethys*
Black Storm-Petrel	*Oceanodroma melania*
Markham's Storm-Petrel	*Oceanodroma markhami*
Least Storm-Petrel	*Oceanodroma microsoma*
Penguins	**Spheniscidae**
Galapagos Penguin	*Spheniscus mendiculus*
Tropicbirds	**Phaethontidae**
White-tailed Tropicbird	*Phaethon lepturus*
Red-billed Tropicbird	*Phaethon aethereus*
Boobies and Gannets	**Sulidae**
Masked Booby	*Sula dactylatra*
Nazca Booby	*Sula granti*
Blue-footed Booby	*Sula nebouxii*
Peruvian Booby	*Sula variegata*
Brown Booby	*Sula leucogaster*
Red-footed Booby	*Sula sula*
Pelicans	**Pelecanidae**
American White Pelican	*Pelecanus erythrorhynchos*
Brown Pelican	*Pelecanus occidentalis*
Cormorants	**Phalacrocoracidae**
Neotropic Cormorant	*Phalacrocorax brasilianus*
Guanay Cormorant	*Phalacrocorax bougainvillii*
Darters	**Anhingidae**
Anhinga	*Anhinga anhinga*
Frigatebirds	**Fregatidae**
Magnificent Frigatebird	*Fregata magnificens*
Great Frigatebird	*Fregata minor*
Herons	**Ardeidae**
American Bittern	*Botaurus lentiginosus*
Least Bittern	*Ixobrychus exilis*
Rufescent Tiger-Heron	*Tigrisoma lineatum*
Fasciated Tiger-Heron	*Tigrisoma fasciatum*
Bare-throated Tiger-Heron	*Tigrisoma mexicanum*
Great Blue Heron	*Ardea herodias*
Cocoi Heron	*Ardea cocoi*
Great Egret	*Ardea alba*
Snowy Egret	*Egretta thula*

Little Blue Heron	*Egretta caerulea*
Tricolored Heron	*Egretta tricolor*
Reddish Egret	*Egretta rufescens*
Cattle Egret	*Bubulcus ibis*
Green Heron	*Butorides virescens*
Striated Heron	*Butorides striata*
Agami Heron	*Agamia agami*
Capped Heron	*Pilherodius pileatus*
Black-crowned Night-Heron	*Nycticorax nycticorax*
Yellow-crowned Night-Heron	*Nyctanassa violacea*
Boat-billed Heron	*Cochlearius cochlearius*
Ibises and Spoonbills	**Threskiornithidae**
White Ibis	*Eudocimus albus*
Scarlet Ibis	*Eudocimus ruber*
Glossy Ibis	*Plegadis falcinellus*
Green Ibis	*Mesembrinibis cayennensis*
Buff-necked Ibis	*Theristicus caudatus*
Roseate Spoonbill	*Platalea ajaja*
Storks	**Ciconiidae**
Jabiru	*Jabiru mycteria*
Wood Stork	*Mycteria americana*
Vultures	**Cathartidae**
Black Vulture	*Coragyps atratus*
Turkey Vulture	*Cathartes aura*
Lesser Yellow-headed Vulture	*Cathartes burrovianus*
King Vulture	*Sarcoramphus papa*
Hawks, Eagles, and Kites	**Accipitridae**
Osprey	*Pandion haliaetus*
Gray-headed Kite	*Leptodon cayanensis*
Hook-billed Kite	*Chondrohierax uncinatus*
Swallow-tailed Kite	*Elanoides forficatus*
Pearl Kite	*Gampsonyx swainsonii*
White-tailed Kite	*Elanus leucurus*
Snail Kite	*Rostrhamus sociabilis*
Slender-billed Kite	*Rostrhamus hamatus*
Double-toothed Kite	*Harpagus bidentatus*
Mississippi Kite	*Ictinia mississippiensis*
Plumbeous Kite	*Ictinia plumbea*
Black-collared Hawk	*Busarellus nigricollis*
Northern Harrier	*Circus cyaneus*
Long-winged Harrier	*Circus buffoni*
Tiny Hawk	*Accipiter superciliosus*
Sharp-shinned Hawk	*Accipiter striatus*
Cooper's Hawk	*Accipiter cooperii*
Bicolored Hawk	*Accipiter bicolor*
Crane Hawk	*Geranospiza caerulescens*
Plumbeous Hawk	*Leucopternis plumbeus*
Barred Hawk	*Leucopternis princeps*
Semiplumbeous Hawk	*Leucopternis semiplumbeus*
White Hawk	*Leucopternis albicollis*
Gray Hawk	*Asturina nitida*
Common Black-Hawk	*Buteogallus anthracinus*
Mangrove Black-Hawk	*Buteogallus subtilis*
Great Black-Hawk	*Buteogallus urubitinga*

Savanna Hawk	*Buteogallus meridionalis*
Harris's Hawk	*Parabuteo unicinctus*
Solitary Eagle	*Harpyhaliaetus solitarius*
Roadside Hawk	*Buteo magnirostris*
Broad-winged Hawk	*Buteo platypterus*
Short-tailed Hawk	*Buteo brachyurus*
Swainson's Hawk	*Buteo swainsoni*
White-tailed Hawk	*Buteo albicaudatus*
Zone-tailed Hawk	*Buteo albonotatus*
Red-tailed Hawk	*Buteo jamaicensis*
Crested Eagle	*Morphnus guianensis*
Harpy Eagle	*Harpia harpyja*
Black-and-white Hawk-Eagle	*Spizastur melanoleucus*
Black Hawk-Eagle	*Spizaetus tyrannus*
Ornate Hawk-Eagle	*Spizaetus ornatus*
Falcons and Caracaras	**Falconidae**
Barred Forest-Falcon	*Micrastur ruficollis*
Slaty-backed Forest-Falcon	*Micrastur mirandollei*
Collared Forest-Falcon	*Micrastur semitorquatus*
Red-throated Caracara	*Ibycter americanus*
Crested Caracara	*Caracara cheriway*
Yellow-headed Caracara	*Milvago chimachima*
Laughing Falcon	*Herpetotheres cachinnans*
American Kestrel	*Falco sparverius*
Merlin	*Falco columbarius*
Aplomado Falcon	*Falco femoralis*
Bat Falcon	*Falco rufigularis*
Orange-breasted Falcon	*Falco deiroleucus*
Peregrine Falcon	*Falco peregrinus*
Rails, Gallinules, and Coots	**Rallidae**
Ruddy Crake	*Laterallus ruber*
White-throated Crake	*Laterallus albigularis*
Gray-breasted Crake	*Laterallus exilis*
Black Rail	*Laterallus jamaicensis*
Clapper Rail	*Rallus longirostris*
Rufous-necked Wood-Rail	*Aramides axillaris*
Gray-necked Wood-Rail	*Aramides cajanea*
Uniform Crake	*Amaurolimnas concolor*
Sora	*Porzana carolina*
Yellow-breasted Crake	*Porzana flaviventer*
Colombian Crake	*Neocrex colombiana*
Paint-billed Crake	*Neocrex erythrops*
Spotted Rail	*Pardirallus maculatus*
Purple Gallinule	*Porphyrio martinica*
Common Moorhen	*Gallinula chloropus*
American Coot	*Fulica americana*
Finfoots	**Heliornithidae**
Sungrebe	*Heliornis fulica*
Sunbittern	**Eurypygidae**
Sunbittern	*Eurypyga helias*
Limpkin	**Aramidae**
Limpkin	*Aramus guarauna*

Plovers and Lapwings	**Charadriidae**
Southern Lapwing	*Vanellus chilensis*
Black-bellied Plover	*Pluvialis squatarola*
American Golden-Plover	*Pluvialis dominica*
Collared Plover	*Charadrius collaris*
Snowy Plover	*Charadrius alexandrinus*
Wilson's Plover	*Charadrius wilsonia*
Semipalmated Plover	*Charadrius semipalmatus*
Killdeer	*Charadrius vociferus*
Oystercatchers	**Haematopodidae**
American Oystercatcher	*Haematopus palliatus*
Stilts and Avocets	**Recurvirostridae**
Black-necked Stilt	*Himantopus mexicanus*
American Avocet	*Recurvirostra americana*
Jacanas	**Jacanidae**
Northern Jacana	*Jacana spinosa*
Wattled Jacana	*Jacana jacana*
Sandpipers and Allies	**Scolopacidae**
Greater Yellowlegs	*Tringa melanoleuca*
Lesser Yellowlegs	*Tringa flavipes*
Solitary Sandpiper	*Tringa solitaria*
Willet	*Catoptrophorus semipalmatus*
Wandering Tattler	*Heteroscelus incanus*
Spotted Sandpiper	*Actitis macularius*
Upland Sandpiper	*Bartramia longicauda*
Whimbrel	*Numenius phaeopus*
Long-billed Curlew	*Numenius americanus*
Hudsonian Godwit	*Limosa haemastica*
Marbled Godwit	*Limosa fedoa*
Ruddy Turnstone	*Arenaria interpres*
Surfbird	*Aphriza virgata*
Red Knot	*Calidris canutus*
Sanderling	*Calidris alba*
Semipalmated Sandpiper	*Calidris pusilla*
Western Sandpiper	*Calidris mauri*
Least Sandpiper	*Calidris minutilla*
White-rumped Sandpiper	*Calidris fuscicollis*
Baird's Sandpiper	*Calidris bairdii*
Pectoral Sandpiper	*Calidris melanotos*
Dunlin	*Calidris alpina*
Stilt Sandpiper	*Calidris himantopus*
Buff-breasted Sandpiper	*Tryngites subruficollis*
Ruff	*Philomachus pugnax*
Short-billed Dowitcher	*Limnodromus griseus*
Long-billed Dowitcher	*Limnodromus scolopaceus*
Wilson's Snipe	*Gallinago delicata*
Wilson's Phalarope	*Phalaropus tricolor*
Red-necked Phalarope	*Phalaropus lobatus*
Gulls, Terns, and Allies	**Laridae**
South Polar Skua	*Stercorarius maccormicki*
Pomarine Jaeger	*Stercorarius pomarinus*
Parasitic Jaeger	*Stercorarius parasiticus*
Long-tailed Jaeger	*Stercorarius longicaudus*
Laughing Gull	*Larus atricilla*

Franklin's Gull	*Larus pipixcan*
Bonaparte's Gull	*Larus philadelphia*
Gray-hooded Gull	*Larus cirrocephalus*
Gray Gull	*Larus modestus*
Belcher's Gull	*Larus belcheri*
Ring-billed Gull	*Larus delawarensis*
Herring Gull	*Larus argentatus*
Lesser Black-backed Gull	*Larus fuscus*
Kelp Gull	*Larus dominicanus*
Sabine's Gull	*Xema sabini*
Swallow-tailed Gull	*Creagrus furcatus*
Gull-billed Tern	*Sterna nilotica*
Caspian Tern	*Sterna caspia*
Royal Tern	*Sterna maxima*
Elegant Tern	*Sterna elegans*
Sandwich Tern	*Sterna sandvicensis*
Common Tern	*Sterna hirundo*
Arctic Tern	*Sterna paradisaea*
Forster's Tern	*Sterna forsteri*
Least Tern	*Sterna antillarum*
Yellow-billed Tern	*Sterna superciliaris*
Bridled Tern	*Sterna anaethetus*
Sooty Tern	*Sterna fuscata*
Large-billed Tern	*Phaetusa simplex*
Black Tern	*Chlidonias niger*
Inca Tern	*Larosterna inca*
Brown Noddy	*Anous stolidus*
Black Noddy	*Anous minutus*
White Tern	*Gygis alba*
Black Skimmer	*Rynchops niger*
Pigeons and Doves	**Columbidae**
Rock Pigeon	*Columba livia*
Pale-vented Pigeon	*Patagioenas cayennensis*
Scaled Pigeon	*Patagioenas speciosa*
White-crowned Pigeon	*Patagioenas leucocephala*
Band-tailed Pigeon	*Patagioenas fasciata*
Plumbeous Pigeon	*Patagioenas plumbea*
Ruddy Pigeon	*Patagioenas subvinacea*
Short-billed Pigeon	*Patagioenas nigrirostris*
Dusky Pigeon	*Patagioenas goodsoni*
White-winged Dove	*Zenaida asiatica*
Eared Dove	*Zenaida auriculata*
Mourning Dove	*Zenaida macroura*
Common Ground-Dove	*Columbina passerina*
Plain-breasted Ground-Dove	*Columbina minuta*
Ruddy Ground-Dove	*Columbina talpacoti*
Blue Ground-Dove	*Claravis pretiosa*
Maroon-chested Ground-Dove	*Claravis mondetoura*
White-tipped Dove	*Leptotila verreauxi*
Gray-headed Dove	*Leptotila plumbeiceps*
Brown-backed Dove	*Leptotila battyi*
Gray-chested Dove	*Leptotila cassini*
Olive-backed Quail-Dove	*Geotrygon veraguensis*
Chiriqui Quail-Dove	*Geotrygon chiriquensis*

Purplish-backed Quail-Dove	*Geotrygon lawrencii*
Buff-fronted Quail-Dove	*Geotrygon costaricensis*
Russet-crowned Quail-Dove	*Geotrygon goldmani*
Violaceous Quail-Dove	*Geotrygon violacea*
Ruddy Quail-Dove	*Geotrygon montana*
Parrots	**Psittacidae**
Azuero Parakeet	*Pyrrhura eisenmanni*
Sulphur-winged Parakeet	*Pyrrhura hoffmanni*
Crimson-fronted Parakeet	*Aratinga finschi*
Olive-throated Parakeet	*Aratinga nana*
Brown-throated Parakeet	*Aratinga pertinax*
Chestnut-fronted Macaw	*Ara severus*
Great Green Macaw	*Ara ambiguus*
Red-and-green Macaw	*Ara chloropterus*
Scarlet Macaw	*Ara macao*
Blue-and-yellow Macaw	*Ara ararauna*
Barred Parakeet	*Bolborhynchus lineola*
Spectacled Parrotlet	*Forpus conspicillatus*
Orange-chinned Parakeet	*Brotogeris jugularis*
Red-fronted Parrotlet	*Touit costaricensis*
Blue-fronted Parrotlet	*Touit dilectissimus*
Saffron-headed Parrot	*Pionopsitta pyrilia*
Brown-hooded Parrot	*Pionopsitta haematotis*
Blue-headed Parrot	*Pionus menstruus*
White-crowned Parrot	*Pionus senilis*
Red-lored Amazon	*Amazona autumnalis*
Mealy Amazon	*Amazona farinosa*
Yellow-crowned Amazon	*Amazona ochrocephala*
Cuckoos	**Cuculidae**
Dwarf Cuckoo	*Coccyzus pumilus*
Black-billed Cuckoo	*Coccyzus erythropthalmus*
Yellow-billed Cuckoo	*Coccyzus americanus*
Mangrove Cuckoo	*Coccyzus minor*
Dark-billed Cuckoo	*Coccyzus melacoryphus*
Gray-capped Cuckoo	*Coccyzus lansbergi*
Squirrel Cuckoo	*Piaya cayana*
Little Cuckoo	*Piaya minuta*
Striped Cuckoo	*Tapera naevia*
Pheasant Cuckoo	*Dromococcyx phasianellus*
Rufous-vented Ground-Cuckoo	*Neomorphus geoffroyi*
Greater Ani	*Crotophaga major*
Smooth-billed Ani	*Crotophaga ani*
Groove-billed Ani	*Crotophaga sulcirostris*
Barn Owls	**Tytonidae**
Barn Owl	*Tyto alba*
Typical Owls	**Strigidae**
Tropical Screech-Owl	*Megascops choliba*
Vermiculated Screech-Owl	*Megascops guatemalae*
Bare-shanked Screech-Owl	*Megascops clarkii*
Crested Owl	*Lophostrix cristata*
Spectacled Owl	*Pulsatrix perspicillata*
Great Horned Owl	*Bubo virginianus*
Costa Rican Pygmy-Owl	*Glaucidium costaricanum*
Central American Pygmy-Owl	*Glaucidium griseiceps*

Ferruginous Pygmy-Owl	*Glaucidium brasilianum*
Burrowing Owl	*Athene cunicularia*
Mottled Owl	*Ciccaba virgata*
Black-and-white Owl	*Ciccaba nigrolineata*
Striped Owl	*Pseudoscops clamator*
Unspotted Saw-whet Owl	*Aegolius ridgwayi*
Nightjars	**Caprimulgidae**
Short-tailed Nighthawk	*Lurocalis semitorquatus*
Lesser Nighthawk	*Chordeiles acutipennis*
Common Nighthawk	*Chordeiles minor*
Common Pauraque	*Nyctidromus albicollis*
Ocellated Poorwill	*Nyctiphrynus ocellatus*
Chuck-will's-widow	*Caprimulgus carolinensis*
Rufous Nightjar	*Caprimulgus rufus*
Whip-poor-will	*Caprimulgus vociferus*
Dusky Nightjar	*Caprimulgus saturatus*
White-tailed Nightjar	*Caprimulgus cayennensis*
Potoos	**Nyctibiidae**
Great Potoo	*Nyctibius grandis*
Common Potoo	*Nyctibius griseus*
Oilbird	**Steatornithidae**
Oilbird	*Steatornis caripensis*
Swifts	**Apodidae**
Black Swift	*Cypseloides niger*
White-chinned Swift	*Cypseloides cryptus*
Chestnut-collared Swift	*Streptoprocne rutila*
White-collared Swift	*Streptoprocne zonaris*
Chimney Swift	*Chaetura pelagica*
Vaux's Swift	*Chaetura vauxi*
Chapman's Swift	*Chaetura chapmani*
Short-tailed Swift	*Chaetura brachyura*
Ashy-tailed Swift	*Chaetura andrei*
Band-rumped Swift	*Chaetura spinicaudus*
Costa Rican Swift	*Chaetura fumosa*
Gray-rumped Swift	*Chaetura cinereiventris*
Lesser Swallow-tailed Swift	*Panyptila cayennensis*
Hummingbirds	**Trochilidae**
Bronzy Hermit	*Glaucis aeneus*
Rufous-breasted Hermit	*Glaucis hirsutus*
Band-tailed Barbthroat	*Threnetes ruckeri*
Green Hermit	*Phaethornis guy*
Long-billed Hermit	*Phaethornis longirostris*
White-whiskered Hermit	*Phaethornis yaruqui*
Pale-bellied Hermit	*Phaethornis anthophilus*
Stripe-throated Hermit	*Phaethornis striigularis*
White-tipped Sicklebill	*Eutoxeres aquila*
Tooth-billed Hummingbird	*Androdon aequatorialis*
Green-fronted Lancebill	*Doryfera ludovicae*
Scaly-breasted Hummingbird	*Phaeochroa cuvierii*
Violet Sabrewing	*Campylopterus hemileucurus*
White-necked Jacobin	*Florisuga mellivora*
Brown Violet-ear	*Colibri delphinae*
Green Violet-ear	*Colibri thalassinus*
Green-breasted Mango	*Anthracothorax prevostii*

Black-throated Mango	*Anthracothorax nigricollis*
Veraguan Mango	*Anthracothorax veraguensis*
Ruby-topaz Hummingbird	*Chrysolampis mosquitus*
Violet-headed Hummingbird	*Klais guimeti*
Rufous-crested Coquette	*Lophornis delattrei*
White-crested Coquette	*Lophornis adorabilis*
Green Thorntail	*Discosura conversii*
Garden Emerald	*Chlorostilbon assimilis*
Violet-crowned Woodnymph	*Thalurania colombica*
Green-crowned Woodnymph	*Thalurania fannyi*
Fiery-throated Hummingbird	*Panterpe insignis*
Violet-bellied Hummingbird	*Damophila julie*
Sapphire-throated Hummingbird	*Lepidopyga coeruleogularis*
Humboldt's Sapphire	*Hylocharis humboldtii*
Blue-throated Goldentail	*Hylocharis eliciae*
Violet-capped Hummingbird	*Goldmania violiceps*
Rufous-cheeked Hummingbird	*Goethalsia bella*
Blue-chested Hummingbird	*Amazilia amabilis*
Charming Hummingbird	*Amazilia decora*
Snowy-bellied Hummingbird	*Amazilia edward*
Rufous-tailed Hummingbird	*Amazilia tzacatl*
Escudo Hummingbird	*Amazilia handleyi*
Stripe-tailed Hummingbird	*Eupherusa eximia*
Black-bellied Hummingbird	*Eupherusa nigriventris*
White-tailed Emerald	*Elvira chionura*
Snowcap	*Microchera albocoronata*
White-vented Plumeleteer	*Chalybura buffonii*
Bronze-tailed Plumeleteer	*Chalybura urochrysia*
White-bellied Mountain-gem	*Lampornis hemileucus*
Purple-throated Mountain-gem	*Lampornis calolaemus*
White-throated Mountain-gem	*Lampornis castaneoventris*
Green-crowned Brilliant	*Heliodoxa jacula*
Magnificent Hummingbird	*Eugenes fulgens*
Greenish Puffleg	*Haplophaedia aureliae*
Purple-crowned Fairy	*Heliothryx barroti*
Long-billed Starthroat	*Heliomaster longirostris*
Magenta-throated Woodstar	*Calliphlox bryantae*
Purple-throated Woodstar	*Calliphlox mitchellii*
Ruby-throated Hummingbird	*Archilochus colubris*
Volcano Hummingbird	*Selasphorus flammula*
Glow-throated Hummingbird	*Selasphorus ardens*
Scintillant Hummingbird	*Selasphorus scintilla*
Trogons	**Trogonidae**
White-tailed Trogon	*Trogon viridis*
Baird's Trogon	*Trogon bairdii*
Violaceous Trogon	*Trogon violaceus*
Collared Trogon	*Trogon collaris*
Orange-bellied Trogon	*Trogon aurantiiventris*
Black-throated Trogon	*Trogon rufus*
Black-tailed Trogon	*Trogon melanurus*
Slaty-tailed Trogon	*Trogon massena*
Lattice-tailed Trogon	*Trogon clathratus*
Golden-headed Quetzal	*Pharomachrus auriceps*
Resplendent Quetzal	*Pharomachrus mocinno*

Motmots	**Momotidae**
Tody Motmot	*Hylomanes momotula*
Blue-crowned Motmot	*Momotus momota*
Rufous Motmot	*Baryphthengus martii*
Broad-billed Motmot	*Electron platyrhynchum*
Kingfishers	**Alcedinidae**
Ringed Kingfisher	*Ceryle torquatus*
Belted Kingfisher	*Ceryle alcyon*
Amazon Kingfisher	*Chloroceryle amazona*
Green Kingfisher	*Chloroceryle americana*
Green-and-rufous Kingfisher	*Chloroceryle inda*
American Pygmy Kingfisher	*Chloroceryle aenea*
Puffbirds	**Bucconidae**
Barred Puffbird	*Nystalus radiatus*
White-necked Puffbird	*Notharchus macrorhynchos*
Black-breasted Puffbird	*Notharchus pectoralis*
Pied Puffbird	*Notharchus tectus*
White-whiskered Puffbird	*Malacoptila panamensis*
Lanceolated Monklet	*Micromonacha lanceolata*
Gray-cheeked Nunlet	*Nonnula ruficapilla*
White-fronted Nunbird	*Monasa morphoeus*
Jacamars	**Galbulidae**
Dusky-backed Jacamar	*Brachygalba salmoni*
Rufous-tailed Jacamar	*Galbula ruficauda*
Great Jacamar	*Jacamerops aureus*
Barbets and Toucans	**Ramphastidae**
Spot-crowned Barbet	*Capito maculicoronatus*
Red-headed Barbet	*Eubucco bourcierii*
Prong-billed Barbet	*Semnornis frantzii*
Blue-throated Toucanet	*Aulacorhynchus caeruleogularis*
Violet-throated Toucanet	*Aulacorhynchus cognatus*
Collared Aracari	*Pteroglossus torquatus*
Fiery-billed Aracari	*Pteroglossus frantzii*
Yellow-eared Toucanet	*Selenidera spectabilis*
Keel-billed Toucan	*Ramphastos sulfuratus*
Chestnut-mandibled Toucan	*Ramphastos swainsonii*
Choco Toucan	*Ramphastos brevis*
Woodpeckers	**Picidae**
Olivaceous Piculet	*Picumnus olivaceus*
Acorn Woodpecker	*Melanerpes formicivorus*
Golden-naped Woodpecker	*Melanerpes chrysauchen*
Black-cheeked Woodpecker	*Melanerpes pucherani*
Red-crowned Woodpecker	*Melanerpes rubricapillus*
Yellow-bellied Sapsucker	*Sphyrapicus varius*
Hairy Woodpecker	*Picoides villosus*
Smoky-brown Woodpecker	*Veniliornis fumigatus*
Red-rumped Woodpecker	*Veniliornis kirkii*
Rufous-winged Woodpecker	*Piculus simplex*
Stripe-cheeked Woodpecker	*Piculus callopterus*
Golden-green Woodpecker	*Piculus chrysochloros*
Golden-olive Woodpecker	*Piculus rubiginosus*
Spot-breasted Woodpecker	*Colaptes punctigula*
Cinnamon Woodpecker	*Celeus loricatus*
Chestnut-colored Woodpecker	*Celeus castaneus*

Lineated Woodpecker	*Dryocopus lineatus*
Crimson-bellied Woodpecker	*Campephilus haematogaster*
Crimson-crested Woodpecker	*Campephilus melanoleucos*
Pale-billed Woodpecker	*Campephilus guatemalensis*
Ovenbirds and Allies	**Furnariidae**
Pale-breasted Spinetail	*Synallaxis albescens*
Slaty Spinetail	*Synallaxis brachyura*
Red-faced Spinetail	*Cranioleuca erythrops*
Coiba Spinetail	*Cranioleuca dissita*
Double-banded Graytail	*Xenerpestes minlosi*
Spotted Barbtail	*Premnoplex brunnescens*
Beautiful Treerunner	*Margarornis bellulus*
Ruddy Treerunner	*Margarornis rubiginosus*
Buffy Tuftedcheek	*Pseudocolaptes lawrencii*
Striped Woodhaunter	*Hyloctistes subulatus*
Lineated Foliage-gleaner	*Syndactyla subalaris*
Scaly-throated Foliage-gleaner	*Anabacerthia variegaticeps*
Slaty-winged Foliage-gleaner	*Philydor fuscipenne*
Buff-fronted Foliage-gleaner	*Philydor rufum*
Buff-throated Foliage-gleaner	*Automolus ochrolaemus*
Ruddy Foliage-gleaner	*Automolus rubiginosus*
Streak-breasted Treehunter	*Thripadectes rufobrunneus*
Plain Xenops	*Xenops minutus*
Streaked Xenops	*Xenops rutilans*
Tawny-throated Leaftosser	*Sclerurus mexicanus*
Gray-throated Leaftosser	*Sclerurus albigularis*
Scaly-throated Leaftosser	*Sclerurus guatemalensis*
Sharp-tailed Streamcreeper	*Lochmias nematura*
Woodcreepers	**Dendrocolaptidae**
Plain-brown Woodcreeper	*Dendrocincla fuliginosa*
Tawny-winged Woodcreeper	*Dendrocincla anabatina*
Ruddy Woodcreeper	*Dendrocincla homochroa*
Olivaceous Woodcreeper	*Sittasomus griseicapillus*
Long-tailed Woodcreeper	*Deconychura longicauda*
Wedge-billed Woodcreeper	*Glyphorynchus spirurus*
Strong-billed Woodcreeper	*Xiphocolaptes promeropirhynchus*
Northern Barred-Woodcreeper	*Dendrocolaptes sanctithomae*
Black-banded Woodcreeper	*Dendrocolaptes picumnus*
Straight-billed Woodcreeper	*Xiphorhynchus picus*
Cocoa Woodcreeper	*Xiphorhynchus susurrans*
Black-striped Woodcreeper	*Xiphorhynchus lachrymosus*
Spotted Woodcreeper	*Xiphorhynchus erythropygius*
Streak-headed Woodcreeper	*Lepidocolaptes souleyetii*
Spot-crowned Woodcreeper	*Lepidocolaptes affinis*
Red-billed Scythebill	*Campylorhamphus trochilirostris*
Brown-billed Scythebill	*Campylorhamphus pusillus*
Antbirds	**Thamnophilidae**
Fasciated Antshrike	*Cymbilaimus lineatus*
Great Antshrike	*Taraba major*
Barred Antshrike	*Thamnophilus doliatus*
Black Antshrike	*Thamnophilus nigriceps*
Black-hooded Antshrike	*Thamnophilus bridgesi*
Western Slaty-Antshrike	*Thamnophilus atrinucha*
Speckled Antshrike	*Xenornis setifrons*

Russet Antshrike	*Thamnistes anabatinus*
Plain Antvireo	*Dysithamnus mentalis*
Spot-crowned Antvireo	*Dysithamnus puncticeps*
Moustached Antwren	*Myrmotherula ignota*
Pacific Antwren	*Myrmotherula pacifica*
Checker-throated Antwren	*Myrmotherula fulviventris*
White-flanked Antwren	*Myrmotherula axillaris*
Slaty Antwren	*Myrmotherula schisticolor*
Rufous-winged Antwren	*Herpsilochmus rufimarginatus*
Dot-winged Antwren	*Microrhopias quixensis*
White-fringed Antwren	*Formicivora grisea*
Rufous-rumped Antwren	*Terenura callinota*
Dusky Antbird	*Cercomacra tyrannina*
Jet Antbird	*Cercomacra nigricans*
Bare-crowned Antbird	*Gymnocichla nudiceps*
White-bellied Antbird	*Myrmeciza longipes*
Chestnut-backed Antbird	*Myrmeciza exsul*
Dull-mantled Antbird	*Myrmeciza laemosticta*
Immaculate Antbird	*Myrmeciza immaculata*
Spotted Antbird	*Hylophylax naevioides*
Wing-banded Antbird	*Myrmornis torquata*
Bicolored Antbird	*Gymnopithys leucaspis*
Ocellated Antbird	*Phaenostictus mcleannani*
Antthrushes and Antpittas	**Formicariidae**
Black-faced Antthrush	*Formicarius analis*
Black-headed Antthrush	*Formicarius nigricapillus*
Rufous-breasted Antthrush	*Formicarius rufipectus*
Black-crowned Antpitta	*Pittasoma michleri*
Scaled Antpitta	*Grallaria guatimalensis*
Streak-chested Antpitta	*Hylopezus perspicillatus*
Thicket Antpitta	*Hylopezus dives*
Ochre-breasted Antpitta	*Grallaricula flavirostris*
Tapaculos	**Rhinocryptidae**
Tacarcuna Tapaculo	*Scytalopus panamensis*
Choco Tapaculo	*Scytalopus chocoensis*
Silvery-fronted Tapaculo	*Scytalopus argentifrons*
Tyrant Flycatchers	**Tyrannidae**
Yellow-bellied Tyrannulet	*Ornithion semiflavum*
Brown-capped Tyrannulet	*Ornithion brunneicapillus*
Southern Beardless-Tyrannulet	*Camptostoma obsoletum*
Mouse-colored Tyrannulet	*Phaeomyias murina*
Yellow Tyrannulet	*Capsiempis flaveola*
Yellow-crowned Tyrannulet	*Tyrannulus elatus*
Forest Elaenia	*Myiopagis gaimardii*
Gray Elaenia	*Myiopagis caniceps*
Greenish Elaenia	*Myiopagis viridicata*
Yellow-bellied Elaenia	*Elaenia flavogaster*
Lesser Elaenia	*Elaenia chiriquensis*
Mountain Elaenia	*Elaenia frantzii*
Torrent Tyrannulet	*Serpophaga cinerea*
Olive-striped Flycatcher	*Mionectes olivaceus*
Ochre-bellied Flycatcher	*Mionectes oleagineus*
Sepia-capped Flycatcher	*Leptopogon amaurocephalus*
Slaty-capped Flycatcher	*Leptopogon superciliaris*

Yellow-green Tyrannulet	*Phylloscartes flavovirens*
Rufous-browed Tyrannulet	*Phylloscartes superciliaris*
White-fronted Tyrannulet	*Phyllomyias zeledoni*
Sooty-headed Tyrannulet	*Phyllomyias griseiceps*
Paltry Tyrannulet	*Zimmerius vilissimus*
Northern Scrub-Flycatcher	*Sublegatus arenarum*
Bronze-olive Pygmy-Tyrant	*Pseudotriccus pelzelni*
Black-capped Pygmy-Tyrant	*Myiornis atricapillus*
Scale-crested Pygmy-Tyrant	*Lophotriccus pileatus*
Pale-eyed Pygmy-Tyrant	*Lophotriccus pilaris*
Northern Bentbill	*Oncostoma cinereigulare*
Southern Bentbill	*Oncostoma olivaceum*
Slate-headed Tody-Flycatcher	*Poecilotriccus sylvia*
Common Tody-Flycatcher	*Todirostrum cinereum*
Black-headed Tody-Flycatcher	*Todirostrum nigriceps*
Brownish Twistwing	*Cnipodectes subbrunneus*
Eye-ringed Flatbill	*Rhynchocyclus brevirostris*
Olivaceous Flatbill	*Rhynchocyclus olivaceus*
Yellow-olive Flycatcher	*Tolmomyias sulphurescens*
Yellow-margined Flycatcher	*Tolmomyias assimilis*
Yellow-breasted Flycatcher	*Tolmomyias flaviventris*
Stub-tailed Spadebill	*Platyrinchus cancrominus*
White-throated Spadebill	*Platyrinchus mystaceus*
Golden-crowned Spadebill	*Platyrinchus coronatus*
Royal Flycatcher	*Onychorhynchus coronatus*
Ruddy-tailed Flycatcher	*Terenotriccus erythrurus*
Tawny-breasted Flycatcher	*Myiobius villosus*
Sulphur-rumped Flycatcher	*Myiobius sulphureipygius*
Black-tailed Flycatcher	*Myiobius atricaudus*
Bran-colored Flycatcher	*Myiophobus fasciatus*
Black-billed Flycatcher	*Aphanotriccus audax*
Common Tufted-Flycatcher	*Mitrephanes phaeocercus*
Olive-sided Flycatcher	*Contopus cooperi*
Dark Pewee	*Contopus lugubris*
Ochraceous Pewee	*Contopus ochraceus*
Western Wood-Pewee	*Contopus sordidulus*
Eastern Wood-Pewee	*Contopus virens*
Tropical Pewee	*Contopus cinereus*
Yellow-bellied Flycatcher	*Empidonax flaviventris*
Acadian Flycatcher	*Empidonax virescens*
Alder Flycatcher	*Empidonax alnorum*
Willow Flycatcher	*Empidonax traillii*
White-throated Flycatcher	*Empidonax albigularis*
Least Flycatcher	*Empidonax minimus*
Hammond's Flycatcher	*Empidonax hammondii*
Yellowish Flycatcher	*Empidonax flavescens*
Black-capped Flycatcher	*Empidonax atriceps*
Black Phoebe	*Sayornis nigricans*
Vermilion Flycatcher	*Pyrocephalus rubinus*
Pied Water-Tyrant	*Fluvicola pica*
Long-tailed Tyrant	*Colonia colonus*
Cattle Tyrant	*Machetornis rixosa*
Bright-rumped Attila	*Attila spadiceus*
Sirystes	*Sirystes sibilator*

Rufous Mourner	*Rhytipterna holerythra*
Dusky-capped Flycatcher	*Myiarchus tuberculifer*
Panama Flycatcher	*Myiarchus panamensis*
Great Crested Flycatcher	*Myiarchus crinitus*
Lesser Kiskadee	*Pitangus lictor*
Great Kiskadee	*Pitangus sulphuratus*
Boat-billed Flycatcher	*Megarynchus pitangua*
Rusty-margined Flycatcher	*Myiozetetes cayanensis*
Social Flycatcher	*Myiozetetes similis*
Gray-capped Flycatcher	*Myiozetetes granadensis*
White-ringed Flycatcher	*Conopias albovittatus*
Golden-bellied Flycatcher	*Myiodynastes hemichrysus*
Golden-crowned Flycatcher	*Myiodynastes chrysocephalus*
Streaked Flycatcher	*Myiodynastes maculatus*
Sulphur-bellied Flycatcher	*Myiodynastes luteiventris*
Piratic Flycatcher	*Legatus leucophaius*
Tropical Kingbird	*Tyrannus melancholicus*
Western Kingbird	*Tyrannus verticalis*
Eastern Kingbird	*Tyrannus tyrannus*
Gray Kingbird	*Tyrannus dominicensis*
Scissor-tailed Flycatcher	*Tyrannus forficatus*
Fork-tailed Flycatcher	*Tyrannus savana*
Tyrant Flycatcher Allies	**Incertae sedis**
Sapayoa	*Sapayoa aenigma*
Thrush-like Schiffornis	*Schiffornis turdina*
Gray-headed Piprites	*Piprites griseiceps*
Rufous Piha	*Lipaugus unirufus*
Speckled Mourner	*Laniocera rufescens*
Barred Becard	*Pachyramphus versicolor*
Cinereous Becard	*Pachyramphus rufus*
Cinnamon Becard	*Pachyramphus cinnamomeus*
White-winged Becard	*Pachyramphus polychopterus*
Black-and-white Becard	*Pachyramphus albogriseus*
Rose-throated Becard	*Pachyramphus aglaiae*
One-colored Becard	*Pachyramphus homochrous*
Masked Tityra	*Tityra semifasciata*
Black-crowned Tityra	*Tityra inquisitor*
Cotingas	**Cotingidae**
Lovely Cotinga	*Cotinga amabilis*
Turquoise Cotinga	*Cotinga ridgwayi*
Blue Cotinga	*Cotinga nattererii*
Black-tipped Cotinga	*Carpodectes hopkei*
Yellow-billed Cotinga	*Carpodectes antoniae*
Snowy Cotinga	*Carpodectes nitidus*
Purple-throated Fruitcrow	*Querula purpurata*
Bare-necked Umbrellabird	*Cephalopterus glabricollis*
Three-wattled Bellbird	*Procnias tricarunculatus*
Manakins	**Pipridae**
Green Manakin	*Chloropipo holochlora*
White-collared Manakin	*Manacus candei*
Orange-collared Manakin	*Manacus aurantiacus*
Golden-collared Manakin	*Manacus vitellinus*
White-ruffed Manakin	*Corapipo altera*
Lance-tailed Manakin	*Chiroxiphia lanceolata*

White-crowned Manakin	*Pipra pipra*
Blue-crowned Manakin	*Pipra coronata*
Golden-headed Manakin	*Pipra erythrocephala*
Red-capped Manakin	*Pipra mentalis*
Sharpbill	**Oxyruncidae**
Sharpbill	*Oxyruncus cristatus*
Vireos	**Vireonidae**
White-eyed Vireo	*Vireo griseus*
Yellow-throated Vireo	*Vireo flavifrons*
Blue-headed Vireo	*Vireo solitarius*
Yellow-winged Vireo	*Vireo carmioli*
Warbling Vireo	*Vireo gilvus*
Brown-capped Vireo	*Vireo leucophrys*
Philadelphia Vireo	*Vireo philadelphicus*
Red-eyed Vireo	*Vireo olivaceus*
Yellow-green Vireo	*Vireo flavoviridis*
Black-whiskered Vireo	*Vireo altiloquus*
Scrub Greenlet	*Hylophilus flavipes*
Tawny-crowned Greenlet	*Hylophilus ochraceiceps*
Golden-fronted Greenlet	*Hylophilus aurantiifrons*
Lesser Greenlet	*Hylophilus decurtatus*
Green Shrike-Vireo	*Vireolanius pulchellus*
Yellow-browed Shrike-Vireo	*Vireolanius eximius*
Rufous-browed Peppershrike	*Cyclarhis gujanensis*
Jays and Crows	**Corvidae**
Black-chested Jay	*Cyanocorax affinis*
Brown Jay	*Cyanocorax morio*
Azure-hooded Jay	*Cyanolyca cucullata*
Silvery-throated Jay	*Cyanolyca argentigula*
Common Raven	*Corvus corax*
Swallows	**Hirundinidae**
Purple Martin	*Progne subis*
Gray-breasted Martin	*Progne chalybea*
Southern Martin	*Progne elegans*
Brown-chested Martin	*Progne tapera*
Tree Swallow	*Tachycineta bicolor*
Mangrove Swallow	*Tachycineta albilinea*
White-winged Swallow	*Tachycineta albiventer*
Violet-green Swallow	*Tachycineta thalassina*
Blue-and-white Swallow	*Pygochelidon cyanoleuca*
White-thighed Swallow	*Neochelidon tibialis*
Northern Rough-winged Swallow	*Stelgidopteryx serripennis*
Southern Rough-winged Swallow	*Stelgidopteryx ruficollis*
Sand Martin	*Riparia riparia*
Cliff Swallow	*Petrochelidon pyrrhonota*
Cave Swallow	*Petrochelidon fulva*
Barn Swallow	*Hirundo rustica*
Donacobius	**Incertae sedis**
Black-capped Donacobius	*Donacobius atricapilla*
Wrens	**Troglodytidae**
White-headed Wren	*Campylorhynchus albobrunneus*
Band-backed Wren	*Campylorhynchus zonatus*
Sooty-headed Wren	*Thryothorus spadix*
Black-throated Wren	*Thryothorus atrogularis*

Black-bellied Wren	*Thryothorus fasciatoventris*
Bay Wren	*Thryothorus nigricapillus*
Riverside Wren	*Thryothorus semibadius*
Stripe-throated Wren	*Thryothorus leucopogon*
Stripe-breasted Wren	*Thryothorus thoracicus*
Rufous-breasted Wren	*Thryothorus rutilus*
Rufous-and-white Wren	*Thryothorus rufalbus*
Buff-breasted Wren	*Thryothorus leucotis*
Plain Wren	*Thryothorus modestus*
Canebrake Wren	*Thryothorus zeledoni*
House Wren	*Troglodytes aedon*
Ochraceous Wren	*Troglodytes ochraceus*
Grass Wren	*Cistothorus platensis*
Timberline Wren	*Thryorchilus browni*
White-breasted Wood-Wren	*Henicorhina leucosticta*
Gray-breasted Wood-Wren	*Henicorhina leucophrys*
Southern Nightingale-Wren	*Microcerculus marginatus*
Song Wren	*Cyphorhinus phaeocephalus*
Dippers	**Cinclidae**
American Dipper	*Cinclus mexicanus*
Old World Warblers and Allies	**Sylviidae**
Tawny-faced Gnatwren	*Microbates cinereiventris*
Long-billed Gnatwren	*Ramphocaenus melanurus*
Tropical Gnatcatcher	*Polioptila plumbea*
Slate-throated Gnatcatcher	*Polioptila schistaceigula*
Thrushes	**Turdidae**
Black-faced Solitaire	*Myadestes melanops*
Varied Solitaire	*Myadestes coloratus*
Black-billed Nightingale-Thrush	*Catharus gracilirostris*
Orange-billed Nightingale-Thrush	*Catharus aurantiirostris*
Slaty-backed Nightingale-Thrush	*Catharus fuscater*
Ruddy-capped Nightingale-Thrush	*Catharus frantzii*
Black-headed Nightingale-Thrush	*Catharus mexicanus*
Veery	*Catharus fuscescens*
Gray-cheeked Thrush	*Catharus minimus*
Swainson's Thrush	*Catharus ustulatus*
Wood Thrush	*Hylocichla mustelina*
Sooty Thrush	*Turdus nigrescens*
Mountain Thrush	*Turdus plebejus*
Pale-vented Thrush	*Turdus obsoletus*
Clay-colored Thrush	*Turdus grayi*
White-throated Thrush	*Turdus assimilis*
Mockingbirds and Allies	**Mimidae**
Gray Catbird	*Dumetella carolinensis*
Tropical Mockingbird	*Mimus gilvus*
Starlings	**Sturnidae**
European Starling	*Sturnus vulgaris*
Pipits and Wagtails	**Motacillidae**
Yellowish Pipit	*Anthus lutescens*
Waxwings	**Bombycillidae**
Cedar Waxwing	*Bombycilla cedrorum*
Silky-flycatchers	**Ptilogonatidae**
Black-and-yellow Silky-flycatcher	*Phainoptila melanoxantha*
Long-tailed Silky-flycatcher	*Ptilogonys caudatus*

Wood-Warblers	**Parulidae**
Blue-winged Warbler	*Vermivora pinus*
Golden-winged Warbler	*Vermivora chrysoptera*
Tennessee Warbler	*Vermivora peregrina*
Nashville Warbler	*Vermivora ruficapilla*
Virginia's Warbler	*Vermivora virginiae*
Flame-throated Warbler	*Parula gutturalis*
Northern Parula	*Parula americana*
Tropical Parula	*Parula pitiayumi*
Yellow Warbler	*Dendroica petechia*
Chestnut-sided Warbler	*Dendroica pensylvanica*
Magnolia Warbler	*Dendroica magnolia*
Cape May Warbler	*Dendroica tigrina*
Black-throated Blue Warbler	*Dendroica caerulescens*
Yellow-rumped Warbler	*Dendroica coronata*
Golden-cheeked Warbler	*Dendroica chrysoparia*
Black-throated Green Warbler	*Dendroica virens*
Townsend's Warbler	*Dendroica townsendi*
Hermit Warbler	*Dendroica occidentalis*
Blackburnian Warbler	*Dendroica fusca*
Yellow-throated Warbler	*Dendroica dominica*
Prairie Warbler	*Dendroica discolor*
Palm Warbler	*Dendroica palmarum*
Bay-breasted Warbler	*Dendroica castanea*
Blackpoll Warbler	*Dendroica striata*
Cerulean Warbler	*Dendroica cerulea*
Black-and-white Warbler	*Mniotilta varia*
American Redstart	*Setophaga ruticilla*
Prothonotary Warbler	*Protonotaria citrea*
Worm-eating Warbler	*Helmitheros vermivorum*
Swainson's Warbler	*Limnothlypis swainsonii*
Ovenbird	*Seiurus aurocapilla*
Northern Waterthrush	*Seiurus noveboracensis*
Louisiana Waterthrush	*Seiurus motacilla*
Kentucky Warbler	*Oporornis formosus*
Connecticut Warbler	*Oporornis agilis*
Mourning Warbler	*Oporornis philadelphia*
MacGillivray's Warbler	*Oporornis tolmiei*
Common Yellowthroat	*Geothlypis trichas*
Olive-crowned Yellowthroat	*Geothlypis semiflava*
Masked Yellowthroat	*Geothlypis aequinoctialis*
Gray-crowned Yellowthroat	*Geothlypis poliocephala*
Hooded Warbler	*Wilsonia citrina*
Wilson's Warbler	*Wilsonia pusilla*
Canada Warbler	*Wilsonia canadensis*
Slate-throated Redstart	*Myioborus miniatus*
Collared Redstart	*Myioborus torquatus*
Golden-crowned Warbler	*Basileuterus culicivorus*
Rufous-capped Warbler	*Basileuterus rufifrons*
Black-cheeked Warbler	*Basileuterus melanogenys*
Pirre Warbler	*Basileuterus ignotus*
Three-striped Warbler	*Basileuterus tristriatus*
Buff-rumped Warbler	*Basileuterus fulvicauda*
Wrenthrush	*Zeledonia coronata*
Yellow-breasted Chat	*Icteria virens*

Bananaquit	**Incertae sedis**
Bananaquit	*Coereba flaveola*
Tanagers	**Thraupidae**
White-eared Conebill	*Conirostrum leucogenys*
Common Bush-Tanager	*Chlorospingus ophthalmicus*
Tacarcuna Bush-Tanager	*Chlorospingus tacarcunae*
Pirre Bush-Tanager	*Chlorospingus inornatus*
Sooty-capped Bush-Tanager	*Chlorospingus pileatus*
Yellow-throated Bush-Tanager	*Chlorospingus flavigularis*
Ashy-throated Bush-Tanager	*Chlorospingus canigularis*
Yellow-backed Tanager	*Hemithraupis flavicollis*
Black-and-yellow Tanager	*Chrysothlypis chrysomelas*
Rosy Thrush-Tanager	*Rhodinocichla rosea*
Dusky-faced Tanager	*Mitrospingus cassinii*
Olive Tanager	*Chlorothraupis carmioli*
Lemon-spectacled Tanager	*Chlorothraupis olivacea*
Gray-headed Tanager	*Eucometis penicillata*
White-throated Shrike-Tanager	*Lanio leucothorax*
Sulphur-rumped Tanager	*Heterospingus rubrifrons*
Scarlet-browed Tanager	*Heterospingus xanthopygius*
White-shouldered Tanager	*Tachyphonus luctuosus*
Tawny-crested Tanager	*Tachyphonus delatrii*
White-lined Tanager	*Tachyphonus rufus*
Red-crowned Ant-Tanager	*Habia rubica*
Red-throated Ant-Tanager	*Habia fuscicauda*
Hepatic Tanager	*Piranga flava*
Summer Tanager	*Piranga rubra*
Scarlet Tanager	*Piranga olivacea*
Western Tanager	*Piranga ludoviciana*
Flame-colored Tanager	*Piranga bidentata*
White-winged Tanager	*Piranga leucoptera*
Crimson-collared Tanager	*Phlogothraupis sanguinolenta*
Crimson-backed Tanager	*Ramphocelus dimidiatus*
Passerini's Tanager	*Ramphocelus passerinii*
Cherrie's Tanager	*Ramphocelus costaricensis*
Flame-rumped Tanager	*Ramphocelus flammigerus*
Blue-gray Tanager	*Thraupis episcopus*
Palm Tanager	*Thraupis palmarum*
Blue-and-gold Tanager	*Bangsia arcaei*
Plain-colored Tanager	*Tangara inornata*
Gray-and-gold Tanager	*Tangara palmeri*
Emerald Tanager	*Tangara florida*
Silver-throated Tanager	*Tangara icterocephala*
Speckled Tanager	*Tangara guttata*
Bay-headed Tanager	*Tangara gyrola*
Rufous-winged Tanager	*Tangara lavinia*
Golden-hooded Tanager	*Tangara larvata*
Spangle-cheeked Tanager	*Tangara dowii*
Green-naped Tanager	*Tangara fucosa*
Scarlet-thighed Dacnis	*Dacnis venusta*
Blue Dacnis	*Dacnis cayana*
Viridian Dacnis	*Dacnis viguieri*
Green Honeycreeper	*Chlorophanes spiza*
Shining Honeycreeper	*Cyanerpes lucidus*

Purple Honeycreeper	*Cyanerpes caeruleus*
Red-legged Honeycreeper	*Cyanerpes cyaneus*
Swallow Tanager	*Tersina viridis*
Buntings and Allies	**Emberizidae**
Blue-black Grassquit	*Volatinia jacarina*
Slate-colored Seedeater	*Sporophila schistacea*
Variable Seedeater	*Sporophila americana*
White-collared Seedeater	*Sporophila torqueola*
Lesson's Seedeater	*Sporophila bouvronides*
Lined Seedeater	*Sporophila lineola*
Yellow-bellied Seedeater	*Sporophila nigricollis*
Ruddy-breasted Seedeater	*Sporophila minuta*
Nicaraguan Seed-Finch	*Oryzoborus nuttingi*
Lesser Seed-Finch	*Oryzoborus angolensis*
Blue Seedeater	*Amaurospiza concolor*
Yellow-faced Grassquit	*Tiaris olivaceus*
Slaty Finch	*Haplospiza rustica*
Peg-billed Finch	*Acanthidops bairdii*
Slaty Flowerpiercer	*Diglossa plumbea*
Saffron Finch	*Sicalis flaveola*
Grassland Yellow-Finch	*Sicalis luteola*
Wedge-tailed Grass-Finch	*Emberizoides herbicola*
Sooty-faced Finch	*Lysurus crassirostris*
Yellow-thighed Finch	*Pselliophorus tibialis*
Yellow-green Finch	*Pselliophorus luteoviridis*
Large-footed Finch	*Pezopetes capitalis*
White-naped Brush-Finch	*Atlapetes albinucha*
Chestnut-capped Brush-Finch	*Buarremon brunneinucha*
Stripe-headed Brush-Finch	*Buarremon torquatus*
Orange-billed Sparrow	*Arremon aurantiirostris*
Black-striped Sparrow	*Arremonops conirostris*
Lark Sparrow	*Chondestes grammacus*
Savannah Sparrow	*Passerculus sandwichensis*
Grasshopper Sparrow	*Ammodramus savannarum*
Lincoln's Sparrow	*Melospiza lincolnii*
Rufous-collared Sparrow	*Zonotrichia capensis*
White-crowned Sparrow	*Zonotrichia leucophrys*
Volcano Junco	*Junco vulcani*
Dark-eyed Junco	*Junco hyemalis*
Cardinals and Allies	**Cardinalidae**
Streaked Saltator	*Saltator striatipectus*
Grayish Saltator	*Saltator coerulescens*
Buff-throated Saltator	*Saltator maximus*
Black-headed Saltator	*Saltator atriceps*
Slate-colored Grosbeak	*Saltator grossus*
Black-faced Grosbeak	*Caryothraustes poliogaster*
Yellow-green Grosbeak	*Caryothraustes canadensis*
Black-thighed Grosbeak	*Pheucticus tibialis*
Rose-breasted Grosbeak	*Pheucticus ludovicianus*
Blue-black Grosbeak	*Cyanocompsa cyanoides*
Blue Grosbeak	*Passerina caerulea*
Indigo Bunting	*Passerina cyanea*
Painted Bunting	*Passerina ciris*
Dickcissel	Spiza americana

American Orioles and Allies	**Icteridae**
Bobolink	*Dolichonyx oryzivorus*
Red-breasted Blackbird	*Sturnella militaris*
Eastern Meadowlark	*Sturnella magna*
Yellow-headed Blackbird	*Xanthocephalus xanthocephalus*
Great-tailed Grackle	*Quiscalus mexicanus*
Shiny Cowbird	*Molothrus bonariensis*
Bronzed Cowbird	*Molothrus aeneus*
Giant Cowbird	*Molothrus oryzivorus*
Black-cowled Oriole	*Icterus prosthemelas*
Orchard Oriole	*Icterus spurius*
Yellow-backed Oriole	*Icterus chrysater*
Orange-crowned Oriole	*Icterus auricapillus*
Yellow-tailed Oriole	*Icterus mesomelas*
Baltimore Oriole	*Icterus galbula*
Yellow-billed Cacique	*Amblycercus holosericeus*
Scarlet-rumped Cacique	*Cacicus uropygialis*
Yellow-rumped Cacique	*Cacicus cela*
Crested Oropendola	*Psarocolius decumanus*
Chestnut-headed Oropendola	*Psarocolius wagleri*
Montezuma Oropendola	*Psarocolius montezuma*
Black Oropendola	*Psarocolius guatimozinus*
Goldfinches and Allies	**Fringillidae**
Yellow-crowned Euphonia	*Euphonia luteicapilla*
Thick-billed Euphonia	*Euphonia laniirostris*
Yellow-throated Euphonia	*Euphonia hirundinacea*
Elegant Euphonia	*Euphonia elegantissima*
Fulvous-vented Euphonia	*Euphonia fulvicrissa*
Spot-crowned Euphonia	*Euphonia imitans*
Olive-backed Euphonia	*Euphonia gouldi*
White-vented Euphonia	*Euphonia minuta*
Tawny-capped Euphonia	*Euphonia anneae*
Orange-bellied Euphonia	*Euphonia xanthogaster*
Yellow-collared Chlorophonia	*Chlorophonia flavirostris*
Golden-browed Chlorophonia	*Chlorophonia callophrys*
Yellow-bellied Siskin	*Carduelis xanthogastra*
Lesser Goldfinch	*Carduelis psaltria*
American Goldfinch	*Carduelis tristis*
Old World Sparrows	**Passeridae**
House Sparrow	*Passer domesticus*
Waxbills and Allies	**Estrildidae**
Nutmeg Munia	*Lonchura punctulata*
Chestnut Munia	*Lonchura malacca*

General index

The main accounts (including accommodations, restaurants, and bird lists) for birding sites, other major localities of interest, and tour companies are indicated in **bold**, and maps where the site is shown are indicated in *italics*.

Index of bird names